D0593838

COMPUTER SIMULATION IN GEOLOGY

COMPUTER SIMULATION IN GEOLOGY

by
JOHN W. HARBAUGH
PROFESSOR OF GEOLOGY
STANFORD UNIVERSITY
and
GRAEME BONHAM-CARTER
ASSISTANT PROFESSOR OF GEOLOGY
UNIVERSITY OF ROCHESTER

WILEY-INTERSCIENCE, a Division of John Wiley & Sons, Inc.
New York / London / Sydney / Toronto

Library of Congress Catalogue Card Number: 70–112848
SBN 471 35136 9
Printed in the United States of America
10 9 8 7 6 5 4 3 2 1

Preface

This book is an outgrowth of work begun at Stanford in 1964, when dynamic computer-simulation models of shallow-water marine-sedimentation processes began to be developed. In August 1966 Graeme Bonham-Carter came to Stanford as a postdoctoral scholar to join in the work, which at that time began to receive financial support from the Geography Branch of the Office of Naval Research and encouragement from the head of the Geography Branch, Miss Evelyn Pruitt. As the work progressed, we became increasingly aware that many of the ideas that we were employing in our sedimentation-simulation work could also be employed in many other fields in geology. Furthermore we could see that the simulation work taking place in fields outside geology—for example, in business, physiology, and forest ecology—could have strong relevance to geology.

It seemed to us that geologists probably could make effective use of simulation methods if they were provided with a general introduction to the subject written from their point of view. We argued that, since geologists deal with complex dynamic systems in the earth, they need to learn something about the methods for constructing mathematical models of dynamic systems. This volume, then, is our response to this need, and we hope that it will advance the "systems viewpoint" in geology. We hasten to point out, however, that geologists are no strangers to the systems viewpoint. They have always made use of conceptual models of dynamic systems. Our present goal is to provide them with some of the mathematical and computational tools that may be used to extend these thought processes.

When we planned this volume in early 1968, we realized that we were faced with a major problem in communication. The book was to be addressed to geologists and yet much of it must necessarily deal with topics that are in themselves neither geological nor familar to geologists. Our task at the outset was to show that geologists have a genuine need for simulation methods and then to try to explain and illustrate the tools of simulation in language that they could understand and appreciate.

Our plan in this volume is simple. First, we have tried to make a philosophical case for simulation in view of the fact that it involves some important changes in mode of thinking, particularly since it alters the traditional "cause and effect" type of scientific thinking that pervades geology. Second, we have tried to illustrate the objectives and elementary ideas of simulation with examples that are fairly simple and convincing. Third, we have set out to explain, in some depth, the major classes of simulation tools and to show how these tools may be incorporated as components into a wide variety of simulation models. Finally, we have illustrated the use of these tools in a variety of geological applications.

This volume has been planned as a combination textbook and reference work. It is designed for users who have some knowledge of computing. Brief FORTRAN IV computer programs are used as examples in many parts of the book. Adoption of the systems viewpoint, however, is independent of whether computers are used. On the other hand, we should make clear that most of the techniques are feasible only with fast, large computers. Furthermore the book contains some mathematics. Simulation-model building is inherently mathematical; it cannot be treated adequately without reference to topics in statistics, linear algebra, calculus, and numerical analysis. Although the level of mathematical sophistication varies, readers will not make maximum use of the book if they fail to immerse themselves in the mathematical details. Furthermore full use of the book will require a working knowledge of computer programming. We have, however, attempted to develop the mathematical details so that they can be understood by readers with modest mathematical backgrounds, and many of the applications can be appreciated without a background in computing.

Most of the computer programs included in this volume were distributed in *FORTRAN IV Programs for Computer Simulation in*

Geology, by John W. Harbaugh and Graeme Bonham-Carter, Office of Naval Research Technical Report, June 30, 1970, Office of Naval Research, Geography Branch, Contract N 00014–67–A–0112–0004, Task No. NR 388–081. In addition, Chapters 3, 4, 6, 7, and 9 reflect, in substantial part, work supported by the Office of Naval Research.

The book has a threefold division. Chapters 1 and 2 deal with "systems philosophy" and provide an introduction to the classes, uses, and constructions of simulation models. Chapters 3 to 8 deal with the fundamentals of the mathematical and computational "tools" of simulation. Chapters 9 and 10 deal with applications of the tools to geological problems. Five sedimentation models are described in detail in Chapter 9. Other applications are surveyed in Chapter 10, including examples from ecology and paleoecology, evolution and morphology, structural geology, fluvial geomorphology, hydrology, and geochemistry and petrology.

The "tool chapters" (Chapters 3 through 8) treat a variety of mathematical techniques that we think have broad application in simulation modeling. The tools include (Chapter 3) generation and application of random variables, (Chapter 4) Markov chains, (Chapter 5) an introduction to some methods for numerically solving equations, (Chapter 6) methods of treating fluid flow and diffusion, (Chapter 7) application of elementary control-systems ideas, and (Chapter 8) optimization methods. We do not claim that these form all of the mathematical tools that are relevant to simulation in geology, but we suggest that many geological simulation models will make strong use of these tools. Finally, we are frank to admit that not all of the tools that we present have been applied in geological simulation to date, but we forecast that they will be applied in the future.

We are indebted to many persons for assistance. Foremost we would like to thank Mrs. Marriam Ring of Stanford University for assistance in preparing much of the book, ranging from organizing the bibliography to typing and retyping several drafts of the manuscript. We also thank Mrs. Chloe Trimble, Mrs. Madeleine Roumbanis, and Mrs. Pauline Jeffers, also of Stanford University, for typing and bibliographic assistance. Some of the drawings were prepared by Perfecto Mary, of Stanford University. Jay Woods of Stanford University helped with some of the computer programming, and Gary Brahms of John Wiley and Sons, Inc., helped with many of the publishing details. A number of persons critically read parts of the manuscript. They include Dr. John C. Davis of the State Geological Survey of Kansas, Dr. L. J. Drew of Cities Service Oil Company, Professor William T. Fox of Williams College, Professor Robert M.

Garrels of the Scripps Institute of Oceanography, Professor John C. Griffiths of Pennsylvania State University, Professor William C. Krumbein of Northwestern University, Dr. M. K. Horn of Cities Service Oil Company, Professor George Hornberger of the University of Virginia, Professor Ray Linsley of Stanford University, Professor Duane Marble of Northwestern University, Professor Donald B. McIntyre of Pomona College, Dr. Daniel F. Merriam of the State Geological Survey of Kansas, Fred Molz of Stanford University, Professor Henry N. Pollack of the University of Michigan, Professor Paul E. Potter of Indiana University, Professor David M. Raup of the University of Rochester, Professor Irwin Remson of Stanford University, Dr. Walther Schwarzacher of the Queen's University, Belfast, and Professor Paul Switzer of Stanford University.

JOHN W. HARBAUGH
GRAEME BONHAM-CARTER

November 1969
Stanford, California
Rochester, New York

Contents

CHAPTER 1

Introduction

Until recent years to "simulate" meant to feign or imitate, or to assume the appearance of something without representing it in reality. In the 1940s, however, simulation took on a new sense, beginning with the work of John von Neumann, who applied "Monte Carlo analysis" in dealing with problems related to the shielding of nuclear reactors. These problems could not be solved by physical experimentation because of the costs and physical hazards involved, and their solution by conventional mathematical methods was too complicated. An alternative approach was the use of Monte Carlo methods, in which random processes with known statistical probability distributions were simulated mathematically.

When high-speed digital computers appeared in the early 1950s, the meaning of "simulation" began to change, because "simulation" then came to signify the use of mathematical models that could be manipulated by computers to perform kinds of experiments. Almost for the first time social scientists and business theorists found that they could perform controlled experiments by using computers to carry out the arithmetic and logic operations embodied in simulation models. The programs used to control the computers were in effect mathematical representations of the simulation models.

Simulation, viewed in this manner, is thus a class of techniques that involve setting up a model of a real situation and then performing experiments on the model. In this sense simulation is extremely broad, ranging from such seemingly unrelated applications as business-management games used with digital computers to representation of groundwater discharge and recharge with electrical analog devices. Other examples of simulation include wind-tunnel tests with physical scale models of aircraft, Link trainers, digital models of fluid flow, and digital models of sociological phenomena and forest ecology. In this book we are concerned entirely with simulation techniques that involve digital computers. Although we touch on static simulation models to some extent, our prime concern is with dynamic simulation models.

With the explosive rise of simulation applications in fields outside geology, there is need to sit back and take stock. Are the tools being developed in these fields applicable in geology? Much of the simulation literature is concerned with such topics as queuing, inventory, and routing of vehicular traffic. Is the philosophy behind these methods similar to that employed, for example, in a computer simulation model of marine sedimentation? It seems to us that a common thread linking these diverse applications is the systems philosophy. Even if not explicitly stated, computer simulation of any process or collection of processes entails three basic steps: the definition of a system, the construction of a model, and the use of the model to imitate the behavior of the system (Figure 1-1). Let us examine these steps a little more closely.

We may define a system as a set of dynamically interrelated components. If any part of a system is changed, repercussions are felt throughout the system. Interdependence of components is thus generally necessary in the definition of a system. Establishing the boundaries of a system is also part of its definition. For example, it could be argued that the entire universe is a system; but within such a universal system it is expedient to define smaller systems—for example, the solar system; the earth's atmosphere, hydrosphere, and lithosphere; a river-drainage system; and the biological components of the soil. Likewise the gross economy of the world is a system, but a factory and a supermarket are also systems. Thus each system must be arbitrarily defined by specifying its boundaries and its components.

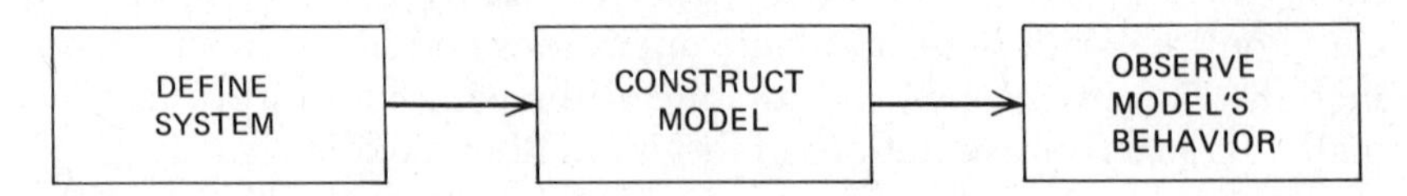

Figure 1-1 Three basic steps in simulation.

At this point it is obvious that we are invariably faced with a hierarchy of systems, lesser systems being nested within larger ones. A hierarchy of systems defined for a sea coast provides an example (Figure 1-2). The gross system might be considered to include the sea, beaches, streams, deltas, and cliffs. Within this large system, however, one might wish to isolate a delta as a system in itself; on a still smaller scale a single distributary within the delta could be considered as a system—and so on, *ad infinitum.*

In defining a system the manner of defining the boundaries of the system is critical. We can distinguish physical boundaries and abstract boundaries. For example, the physical boundaries of a sedimentary basin might be defined by the areal dimensions of the body of water enclosing the basin. An abstract boundary, on the other hand, might be defined by the inclusion of certain geological variables and omission of others. Furthermore boundaries may be arbitrarily defined so that the system is either "open" or "closed." Closed systems, as their name implies, are isolated in the sense that they neither receive input from, nor provide output to, the "outer world" or to the larger systems that surround them. Although models of closed systems may be simpler to create, it is obvious that most closed systems are unrealistic. Real-world systems are almost invariably open, being continually affected by external factors and in a state of dynamic

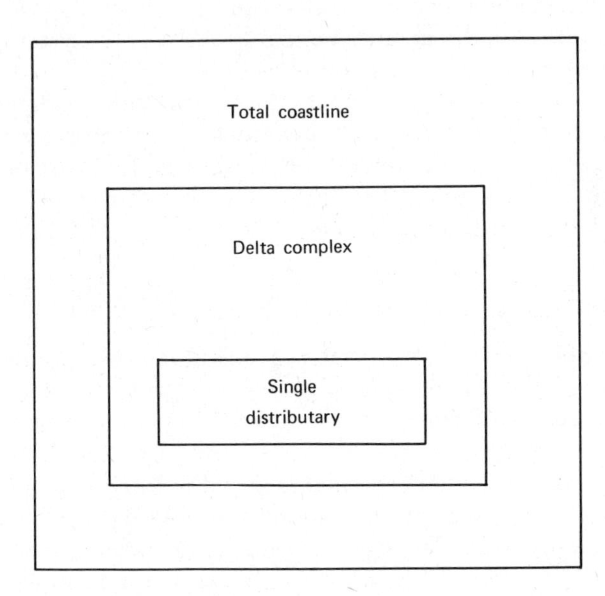

Figure 1-2 Nesting of systems.

equilibrium. Factories have inputs (raw materials) and outputs (finished products); their performance adjusts to the market need for products and the rate of inflow of materials. Similarly a coastal system has inputs (streams, longshore currents) and outputs (turbidity currents); its performance—embodied as beaches, deltas, lagoons, and so on—is continually being adjusted to "external" changes, such as rates of sediment input, eustatic fluctuations, and storms.

Because a system consists of an assemblage of parts that are interrelated in a complex manner, it may be exceedingly hard to predict the effect of altering the state of a particular variable or of changing the system structure. Man has first to simplify the system conceptually and then to represent it with a model, which is an artificial system that attempts to portray the characteristics of a real system. There are many types of models—such as physical, conceptual, and graphic—but in many cases the most powerful and flexible way of representing a system is with mathematical models. Mathematical models may either seek complete and precise solutions to certain aspects of systems or they may be used for computer simulation, in which case "solutions" are obtained by observing the model's performance on a computer. Dynamic simulation is the operation of the model of a system in such a way that the behavior of the real system is reproduced to some degree as the model moves through time.

The literature on the general concepts of dynamic systems is large. Articles and books—such as those by Beer (1959), Von Bertalanffy (1956), and Chorafas (1965), among others—provide readable general introductions to the subject. Applications to business problems are highly advanced [e.g., the techniques described by Forrester (1961)]. Descriptions of ecological applications include those of Watt (1964, 1966) and Odum and Odum (1959).

REPRESENTATIVE APPLICATION OF A SIMULATION MODEL

At this point it is appropriate to examine the application of a simulation model to an actual geological problem. The problem concerns the flow of seawater into the Michigan basin when the salt deposits of the Cayugan series were formed in the Late Silurian. Paleogeographic reconstructions interpret the Michigan basin to have been surrounded by a series of shallow shoals and reefs in the Late Silurian (Figure 1-3). The reefs and shoals served as partial barriers, admitting seawater and confining the denser brine produced through evaporation to the deeper, central part of the basin. The problem is to

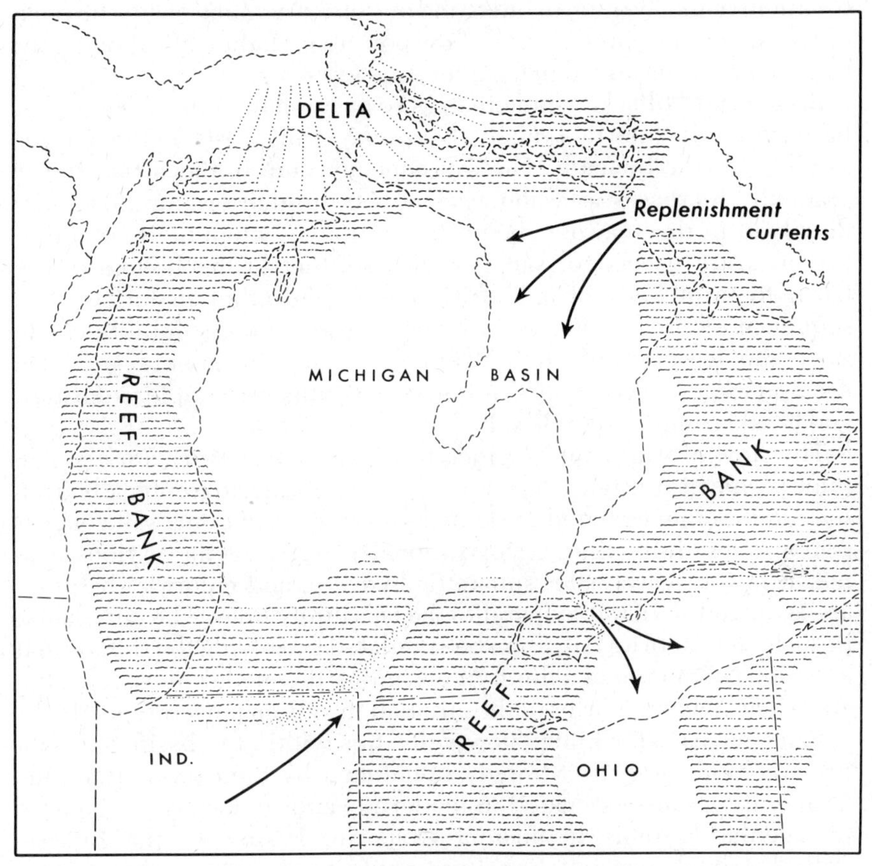

Figure 1-3 Map showing inferred geography in, and adjacent to, the Michigan basin during Late Silurian time. From Briggs and Pollack (1967). Copyright 1967 by the American Association for the Advancement of Science.

determine the location and size of inlets to the basin so that the flow pattern of brine and the accompanying precipitation of salt accord with the observed distribution of salt.

Briggs and Pollack (1967) developed a dynamic, quantitative model of evaporite sedimentation with which they performed a series of experiments in an attempt to interpret the salt distribution in the Michigan basin. Their model incorporates a semirestricted marine basin from which water is continually removed by evaporation. Brine with the salinity of ordinary seawater is drawn through the inlets and is progressively concentrated until salt precipitates. The model permits experimentation with (a) shape of the depositional basin,

(b) number and location of inlets and outlets, (c) volume of flow through inlets and outlets, and (d) rate of evaporation. Mathematical details of the model are discussed in Chapter 9.

Briggs and Pollack assume that, as seawater flows into a brine-filled basin, it tends to form a relatively thin layer that rests on the denser, saturated brine beneath. The salt concentration in the surface layer gradually increases as water is removed through evaporation. Salts dissolved in the seawater begin to precipitate when their solubility products are exceeded, salt particles settling to the bottom of the basin to form beds of anhydrite, halite, and other materials. The sequence of precipitates is in inverse order to the solubility of the salts, the least soluble being removed first. The general order of precipitation is carbonates, followed by sulfates, simple chlorides, and finally complex double salts.

The flow of water into an idealized evaporite basin with a single inlet is shown in Figure 1-4. Flow lines diverge geographically from the inlet, being arranged at right angles to lines of equal salt concentration (isosalinity lines). Depositional patterns of the various evaporite facies—carbonate, sulfate (anhydrite), and chloride (halite)—are arranged so that they parallel the isosalinity lines. The circulation in such an evaporite basin can be simulated by solving the differential equations of motion of a fluid. In designing the model and solving the equations the boundaries of the circulation system must be specified.

Rates of seawater influx, flow velocities within the basin, and salt-precipitation rates are strongly influenced by the evaporation rate. Evaporation rates estimated in modern evaporite basins can be used in calculating rates in ancient evaporite basins. In the Briggs–Pollack model, precipitation is assumed to occur whenever a prescribed saturation concentration is reached. In making calculations an appropriate mass of dissolved material is converted to the thickness of "deposited" precipitate.

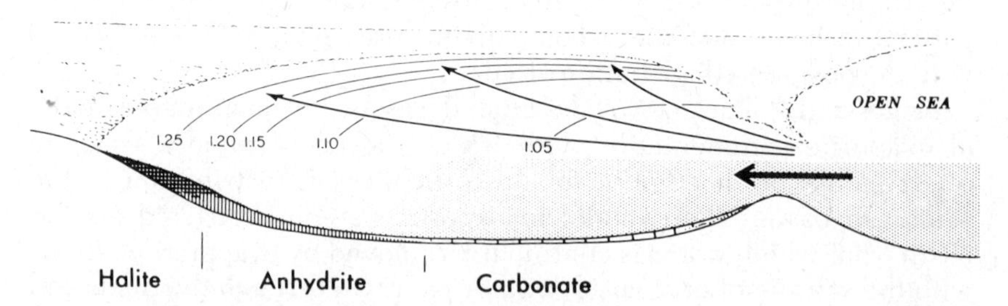

Figure 1-4 Diagram showing flow lines in idealized evaporite basin connected with the open sea by a single inlet. From Briggs and Pollack (1967). Copyright 1967 by the American Association for the Advancement of Science.

The observed thickness of salt in the Michigan basin is shown in Figure 1-5. In an attempt to simulate the observed salt distribution, Briggs and Pollack initially arranged their model so as to make the simulated flow pattern (Figure 1-6) accord with the generalized paleogeographic interpretation of Figure 1-3. It is assumed that two principal inlets admitted seawater to the evaporite basin, and a single outlet existed toward the southeast, feeding into a local basin underlying what is now part of Ohio and Lake Erie. Their model makes use of a meshwork of square, two-dimensional cells. The direction and

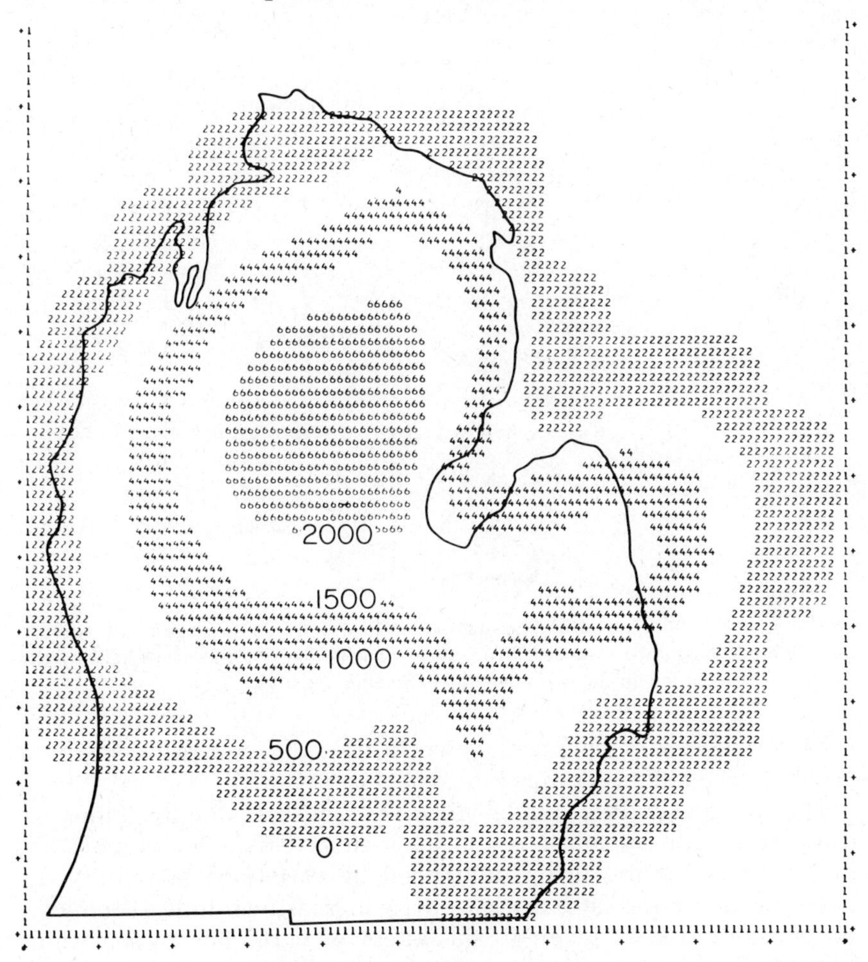

Figure 1-5 Thickness countour map based on borehole data showing distribution of halite in Cayugan series (Upper Silurian) in the Michigan basin. Bands of symbols denote countor lines on computer-printed maps. Values are in feet. From Briggs and Pollack (1967). Copyright by the American Association for the Advancement of Science.

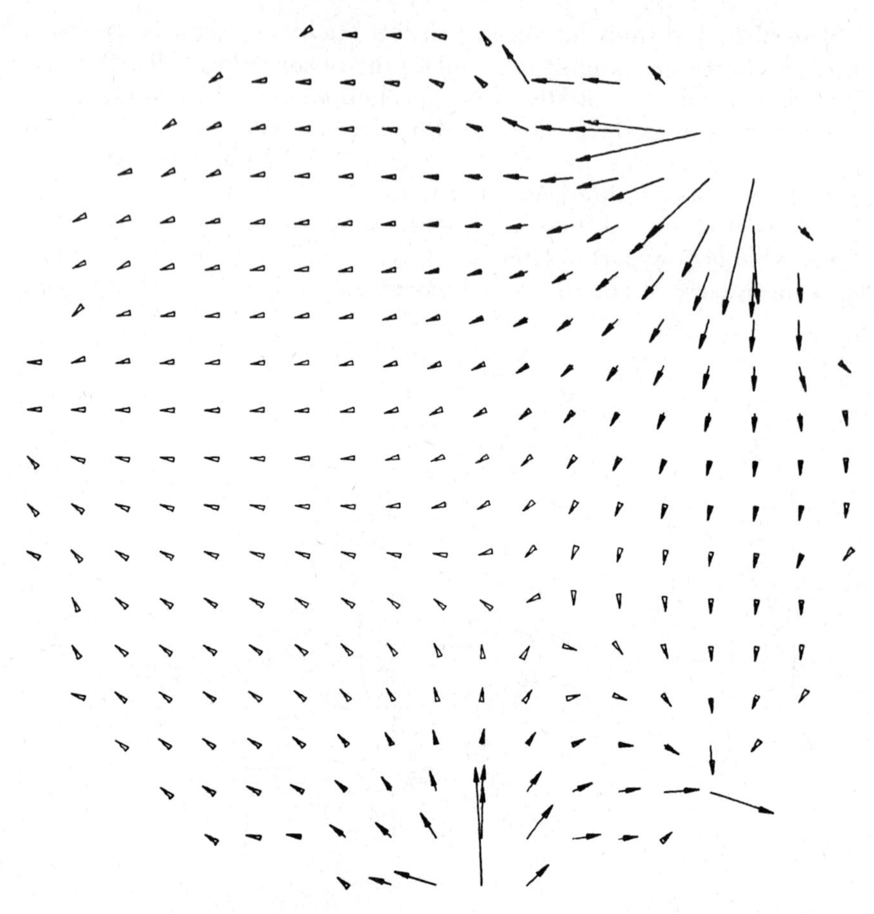

Figure 1-6 Flow diagram from simulation run in which two inlets and one outlet to the Michigan basin are assumed. From Briggs and Pollack (1967). Copyright by the American Association for the Advancement of Science.

relative velocity of flow are shown in the center of each cell by a vector whose length is proportional to velocity.

The Briggs–Pollack model involves several simplifications as compared with an actual evaporite system. First, the evaporation rate is assumed not to be influenced by variations in salinity. In reality, evaporation rates decrease with increasing salinity. Secondly, no consideration is given to the effect of variations in sea water density on the flow patterns. The higher density of water of high salinity would of course, affect the flow pattern in an actual evaporite basin. In the model, however, the flow is treated essentially as a two-dimensional system in which the uppermost layer of water, from which evaporation occurs, is assumed to be of negligible thickness. Because these assumptions affect the behavior of the model, the results of the model must be interpreted in accord with the assumptions that are incorporated in it.

The pattern of salinity variations that accompany the flow system of this initial model is such that saturation with respect to sodium chloride (311 grams per liter) is attained only in the western part of the basin (Figure 1-7). Precipitation of halite under these circumstances

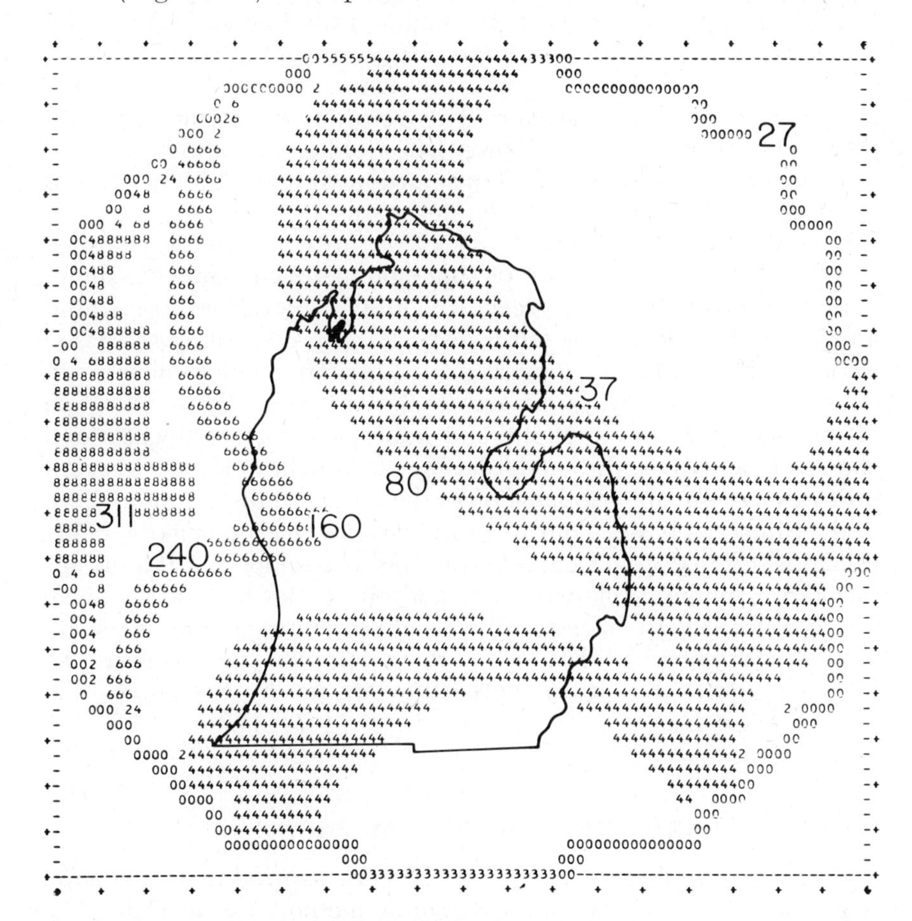

Figure 1-7 Isosalinity contours produced after a steady state has been attained in simulation run assuming two inlets and one outlet to the Michigan basin (Figure 1-6). Values are in grams of sodium chloride per liter. Saturation values of 311 occur in western part of area. From Briggs and Pollack (1967). Copyright by the American Association for the Advancement of Science.

would be confined to the extreme western, and not to the central, part of the Michigan basin. Clearly this model is not in accord with the observed distribution of salt, and the assumed volumes and positions of inflow and outflow are not acceptable. Furthermore Briggs and Pollack found that no modifications of the relative flow volumes through the two inlets and one outlet in the model were sufficient to move the site of salt deposition to the central part of the basin. Accordingly it was necessary to modify the basic assumptions in the model.

In a subsequent series of simulation experiments, Briggs and Pollack assumed that, in addition to the two main inlets and one outlet, there were a chain of "leaks" in the surrounding reefs that permitted radial inflow of seawater. The resulting isosalinity patterns obtained with the model (Figure 1-8) yielded saturation sites that provide for salt accumulation in proportions (Figure 1-9) that accord relatively well with observed salt thicknesses.

In retrospect the Briggs–Pollack evaporite model provides an excellent example of the use of simulation. It incorporates assumptions that have varying degrees of validity. The equations for fluid flow and relationships for salt transport and precipitation are based on sound physical and chemical principles. The geological assumptions are, of necessity, less rigorous. Nevertheless, by modification of the geological assumptions and adherence to the physical and chemical principles, the model has been brought to a state where it provides a reasonable explanation of the observed distribution of salt in the Michigan basin. Thus its principal value has been in deducing relationships that stem from different sets of assumptions. It indicates that the assumption of two inlets and one outlet for the Michigan basin, coupled with an assumed constant evaporation rate, is incompatible with observed data. When the assumptions are modified by incorporation of the "leaky reefs," they are much more compatible with observations.

SIMULATION AND THE SCIENTIFIC METHOD

Before proceeding further, we need to be assured that simulation is in complete accord with the scientific method. Let us examine a hypothetical geological problem and consider the parallel between a traditional and a simulation approach. Assume that the problem is concerned with a sandstone deposit that varies in average grain size from place to place. The problem is to explain the origin of the varia-

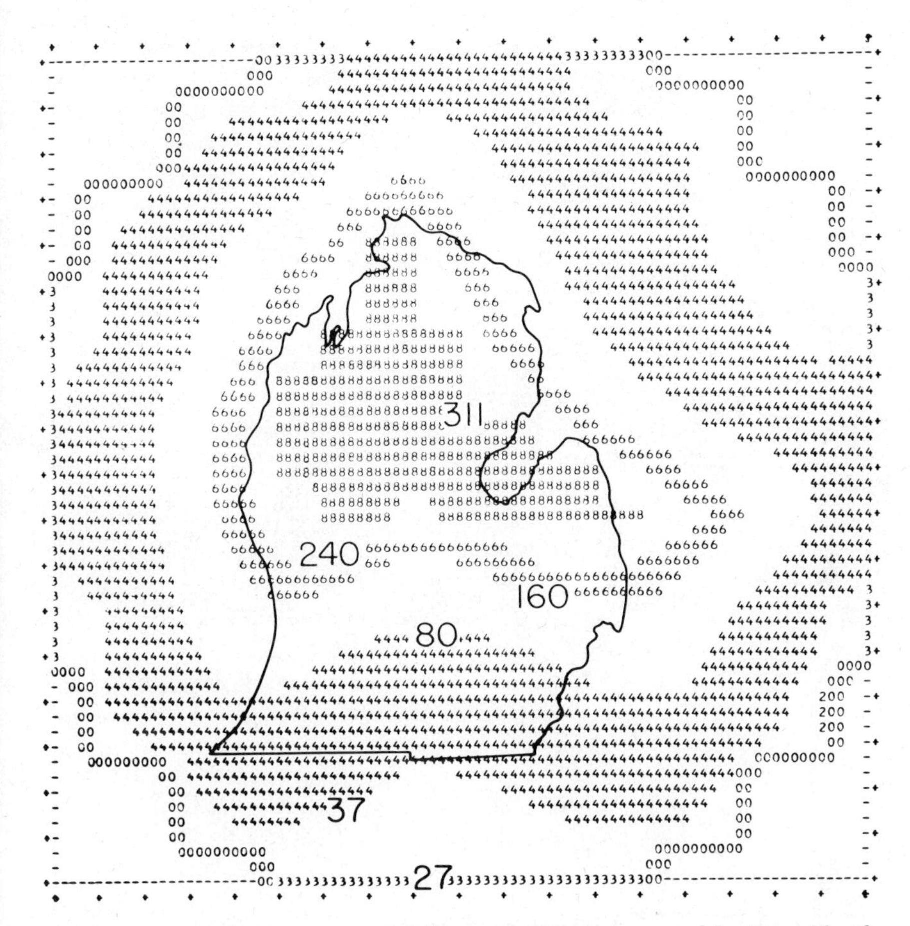

Figure 1-8 Isosalinity map showing salt concentrations, in grams of sodium chloride per liter, after 4000 scale years of operation of modified model. Assumptions include presence of two major seawater inlets and one outlet, as well as inflow of seawater through "leaky reefs" that surround basin. From Briggs and Pollack (1967). Copyright by the American Association for the Advancement of Science.

tions in grain size. A geologist studying this sandstone might proceed as follows:

1. Initially he would observe the sandstone, collecting samples that are subsequently analyzed to determine grain size. Employing a descriptive classification method, he then prepares a description of the grain-size variations.

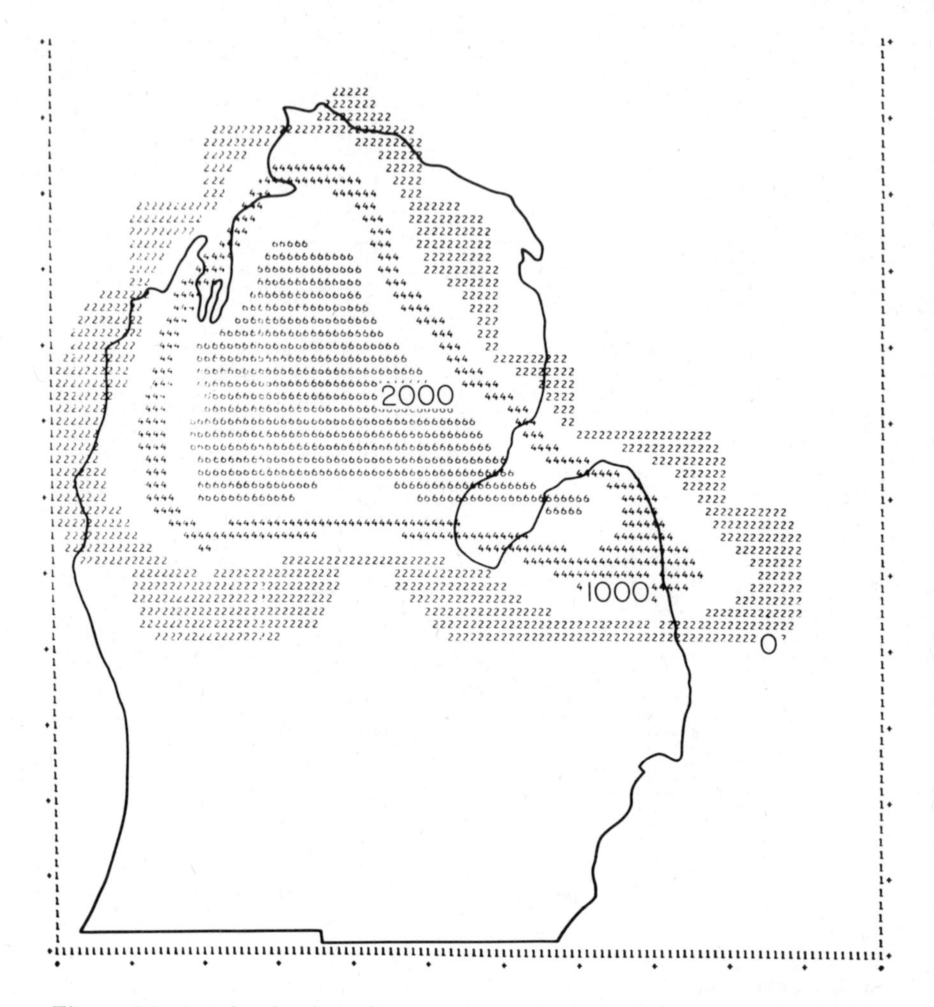

Figure 1-9 Simulated salt-thickness patterns employing "leaky reef" hypothesis. Thickness values are in feet. Compare this isopach map with isopach map constructed with observed data (Figure 1-5). From Briggs and Pollack (1967). Copyright by the American Association for the Advancement of Science.

2. Next he develops several hypotheses or conceptual models that may explain the origin of the size variations. This is a process of induction and involves generalization as well as an element of uncertainty. The geologist may be strongly influenced by previous experience and will undoubtedly be influenced by the views of other geologists who have dealt with similar problems. Perhaps several

models or hypotheses emerge at this stage that seem equally feasible. These correspond to Chamberlin's "multiple working hypotheses."

3. By deductive reasoning, the geologist then "thinks through" the consequences of each of his conceptual models. If the models are complex and contain components that are interdependent, the deductive process may be difficult. Nevertheless the objective is to arrive at the logical consequences that follow from the assumptions incorporated into each conceptual model.

4. The geologist then compares the results obtained by this deductive process with the observational data, or he may obtain new data in response to critical questions raised in the deductive process. In other words, he tests the hypotheses.

5. He draws conclusions concerning the validity of the hypotheses or models. If the hypotheses are not wholly satisfactory, which is usually the case, the geologist (or his successors who are attracted to the problem) proceeds through the cycle again. This involves making additional observations, producing new or revised hypotheses by induction, deducing the consequences of the hypotheses, and testing the results by observation or experimentation. The cycle may be repeated many times, possibly to the point at which a single explanation is sufficient.

The philosophy behind simulation parallels these steps of the scientific method. The person who constructs a simulation model initially deals with observations of a real-world system (Figure 1-10). By inductive reasoning he then creates a model that he thinks will imitate the system. He performs simulation experiments that give him "solutions" for given sets of starting conditions and then compares the artificially produced data with real-world data, modifying the model accordingly. If he decides that the model is satisfactory, he can answer specific questions and draw conclusions.

The point we wish to make is that geologists are fully accustomed to "mental simulation" consisting of the process of deductive reasoning that follows the inductive development of a theory or conceptual model. Any theory that attempts to explain the workings of a real-world system, however, tends to be complex, and logical manipulation of it by sheer thought alone soon becomes difficult. Computer simulation overcomes some of this difficulty by substituting machine memory and an exceedingly fast computational ability to make logical and mathematical deductions. Although computer simulation is a new technique, the equivalent steps in the scientific method are well established.

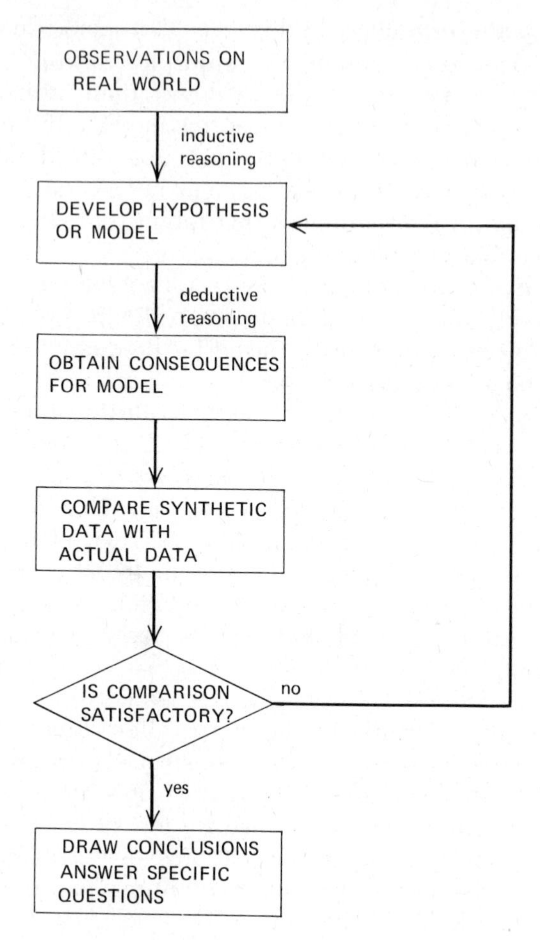

Figure 1-10 Flow chart illustrating steps in simulation modeling as an aspect of the scientific method.

MAIN FEATURES OF MODELS OF DYNAMIC SYSTEMS

Now that we have dwelled on the rationale for simulation, let us consider the basic features of many dynamic system models. Five main kinds of features can be distinguished:

1. System elements
2. System boundaries
3. Structure of interrelationships, including feedback
4. Transport and accounting of materials and energy
5. Time flow

A Dynamic Geology-Student Model

The main features of a dynamic system can best be described by illustrating their role in an idealized simulation model. The model is hypothetical and deals with the flow of geology students through universities into the geological job market. Such a model is simple to construct and understand, and yet is realistic in some of its aspects. In fact the model could probably be expanded and improved to the point where it would be useful in forecasting the supply of students who enter the geological profession.

The model incorporates a number of assumptions, which are summarized in Figure 1-11. Students who enter universities are presumed to make a decision either to study geology or not to study geology. This represents a vast oversimplification of the decision-

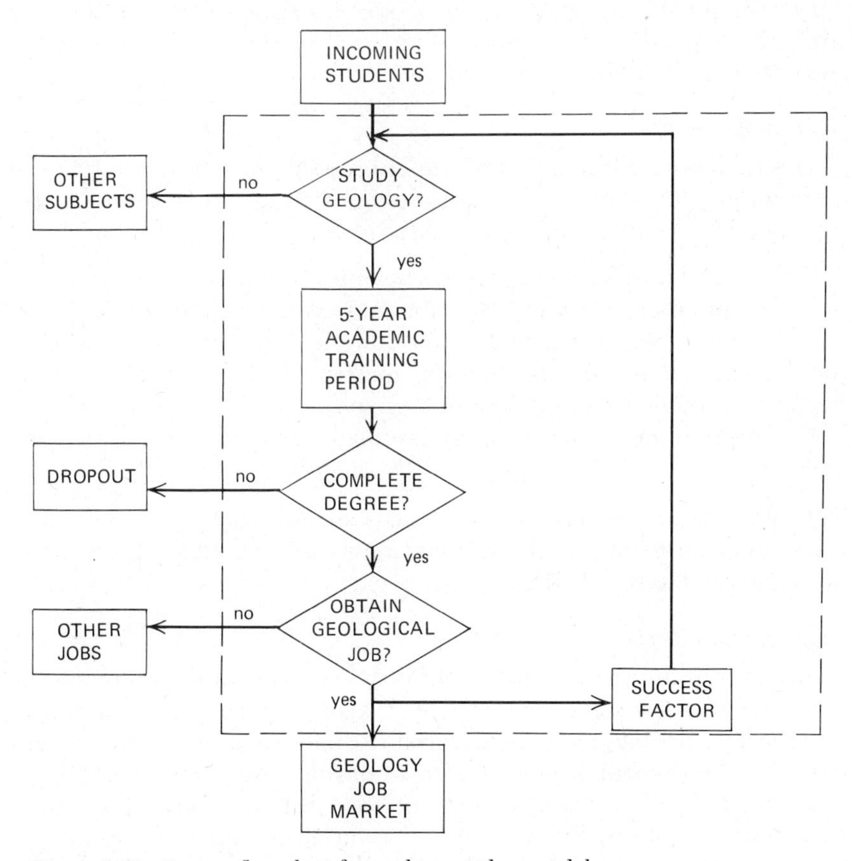

Figure 1-11 System flow chart for geology-student model.

making processes in which incoming college students engage, but it is useful for constructing the model. The students who elect to major in geology are then assumed to pass through an academic training program that averages 5 years. Some do not complete the program and are regarded as "dropouts" as far as geology is concerned, passing from the system. Others obtain degrees and are faced with the challenge of finding a job. Some are absorbed by the geology job market, whereas others go into nongeological work. A final, major assumption is that the relative degree of success of the new graduates in finding jobs strongly influences the decisions of incoming students to major in geology or in other fields. For example, the "success" of newly graduated geologists in finding jobs (measured in terms of numbers of jobs, starting salaries, etc.) in some universities probably has a strong influence on students choosing to major in geology. Furthermore the behavior of the model, in spite of its simplicity, is difficult to predict in advance. The components of the model are described in detail below.

System Elements

The elements of the geology-student model consist of two types of components. One type is "material," consisting of persons who are aggregated into different categories, as follows:

1. The total number of students who enter universities per year.
2. The number of students who begin to major in geology per year.
3. The number of students who complete the requirements for their terminal degree (bachelor's, master's, or doctor's) and thus become available for employment, per year.
4. The number of new geology graduates who take geological jobs per year.

The other type of component consists of attributes that are "nonmaterial"—for example, the ratio of newly graduated students to the number who take geological jobs per year.

System Boundaries

Any system must be defined not only in terms of the components and variables being considered but also in terms of its limiting boundaries, or "edges." Some variables may lie outside the system boundary, shown in Figure 1-11 as a dashed line, yet are still considered to be important aspects of the system. These are called *exogenous* variables. For example, the variable "incoming students" is independent of what goes on inside the system in this representation. Although it influences the system, it is not changed within it. The variables inside the system are termed *endogenous*; not only do they influence other parts of the system but they are also affected by changes within the system.

Of course the system boundary (also the model boundary) is artificial in that it merely sets limits to a small part of a much larger system. It would be possible to consider a larger model in which the flow of incoming students is a function of, say, national birthrate and tuition fees. In other words, the boundaries of the system model could be stretched to include a number of factors not considered in the present model.

Structure of Interrelationships, Including Feedback

The central aspect of any system, or the model of one, is the structure of interrelationships among elements in the system. How does the behavior of one element, for example, affect another element, and so on? It is common practice to show the structure of models with flow diagrams (Figure 1-11) that are similar to the flow charts used to describe computer programs. This is appropriate, of course, because the models of systems in this book are computer programs. Although there are no strict conventions for preparing flow diagrams, system elements are usually contained in rectangular boxes, and interrelationships among elements are represented by arrows. An arrow represents a direction or sequence and implies a direct, causal linkage, which may consist of the flow of material, money, people, or information in the system. In the geology-student model diagram (Figure 1-11) some of the arrows indicate the flow of people, and one of them represents the flow of information that stems from the success-factor element.

At this point we should point out that interrelationships also can be shown by diagrams that represent statistical relationships. Lines in a correlation diagram (Figure 1-12) may represent significant correlations in a statistical sense, but they need not represent a direct causal linkage. Although there may be a strong superficial resemblance between a correlation diagram and a system flow chart, the two represent quite different outlooks. The correlation diagram illustrates interrelationships present in observed data that have been empirically deduced by statistical analysis. The construction of a correlation diagram is a postmortem, after-the-event approach. A flow diagram, on the other hand, is a representation of the way the system works. To emphasize these differences consider the correlation between the elements "geology dropout" and "all incoming students" in Figure

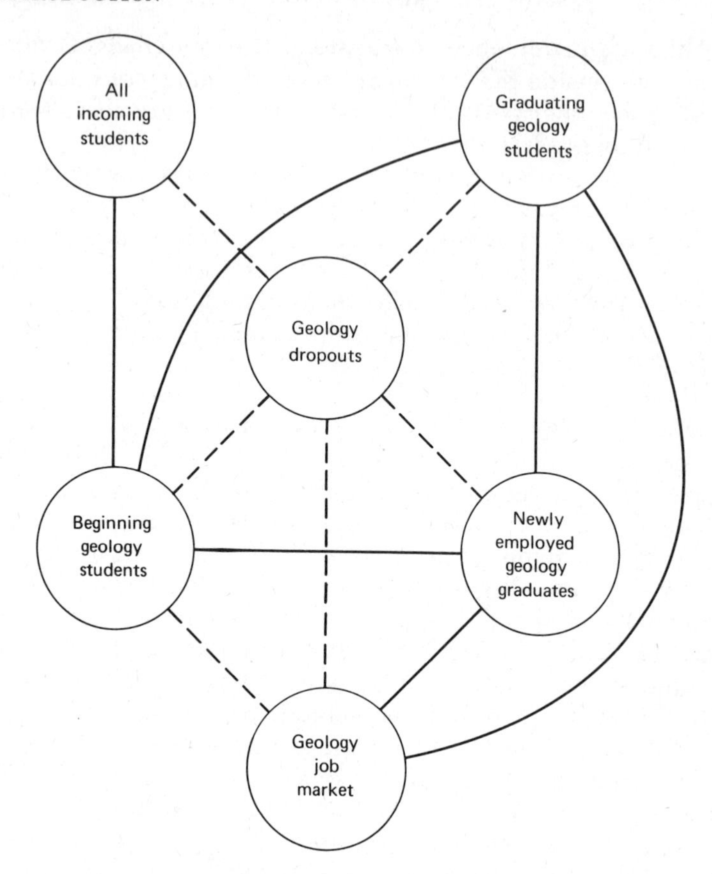

Figure 1-12 Correlation diagram showing general statistical associations in hypothetical system. Lines connecting elements indicate significant statistical correlation. Continuous lines indicate positive correlations; dashed lines indicate negative or inverse correlations. Diagram is schematic; distances between elements are not related to strength of correlations.

1-12, which we assume shows a significant statistical association between the two. The system diagram (Figure 1-11), on the other hand, does not show a direct link between these elements. Instead the linkage is indirect, involving elements that lie between. The structure of the system diagram is a product of inductive reasoning, whereas the structure of the correlation diagram is deductive.

We should emphasize that the system diagram must explain the correlation network if it is to have any validity. A number of system models might be tried and discarded, but the intercorrelations in the data will remain. The important point is that the correlation network does not attempt to explain how the system works, whereas the system diagram provides an explicit interpretation of the functional relationships among the components of the system.

An exceedingly important type of dependency relationship is one that involves feedback. In Figure 1-11 the "success factor" is affected by the number of new geology graduates obtaining geological jobs, and it directly influences incoming students in their choice to study geology or not. This is a negative-feedback loop that *controls* the system. Negative feedback, as its name implies, negates or reverses a previous condition. The behavior of a system consisting of a household furnace and thermostat is a familiar example. If too many new geological graduates are produced, the success factor will reflect a glut on the job market and will have the effect of discouraging newcomers from majoring in geology. Conversely, if too few geology students enter the system, the success factor will stimulate newcomers to enter geology. With this control properly adjusted, the system cannot get very far out of hand.

Feedback loops occur in every system. In addition to negative feedback, which exerts control, there is also positive feedback, which has the reverse effect. Instead of exerting a damping effect on large fluctuations, it tends to amplify. If, for instance, the success factor in Figure 1-11 had the effect of increasing the inflow of students when Ph.D. graduates were glutting the market, the situation would escalate rapidly and finally "blow up" with general unemployment of geologists. A classical example of positive feedback is economic inflation. Rising wages beget rising prices, which beget rising wages, and so on, until eventually an economic crash occurs and negative feedback again begins to exert control.

Transport and Accounting of Materials and Energy

Transportation processes are involved in most dynamic systems. In natural systems both materials and energy flow between different parts of the system. In the student model incoming students undergo a transformation and filtering process that results in an output of qualified geology graduates. The students are analogous to "materials." The transformation process is crudely represented by a series of decision boxes, which permit transport along a number of different paths. Although the analogy to the flow of energy is more difficult to discern, we might consider that "energy" is transformed during student work into usable knowledge, the tangible result being the progressive transformation of incoming students to students with degrees.

In this example the distribution of material components in space is not considered. In many geological systems, however, the transport of materials from one location to another is extremely important. For example, the transport of sediment in a river-delta system or the flow of water and dissolved salts in a marine evaporite basin involves the transport of materials as well as the flow and transformation of energy.

Implicit in the structure of any system is an accounting mechanism. This is poorly represented in the geology-student example, but it nevertheless exists. The variables that record the status of material components are part of a budget or accounting system. All movement in the system is reflected in the "accounts." The accounting aspects of geological dynamic systems are important and are treated later in this volume.

Flow of Time

Implicit in any dynamic system is the flow of time. All the processes are time dependent; the transport of materials and the transformation of energy are meaningless without the flow of time. In representing a dynamic system it is convenient to freeze time and look initially at a static "snapshot" of the system. The condition of the system during a "snapshot" is given by the "state" of each variable. In dealing with digital models, time can then be moved forward in a series of discrete steps, the system state being altered by an increment at each step. In this way a dynamic system can be built up from a series of static snapshots. In the geology-student example the state of the system is recorded annually. With each new year new students arrive, graduating geologists leave, and the variables are adjusted to new values.

ROCK CYCLE AS A DYNAMIC SYSTEM

All geologic processes are part of a dynamic system that embraces the entire earth. The most familiar aspect is known as the rock cycle, which can be represented in abstract form (Figure 1-13) by linking its major components with arrows that indicate transformations within the system. The arrows can be regarded as feedback loops that link the components together to make them totally interdependent. For example, processes of deposition are linked, on the one hand, with uplift, erosion, and weathering—and, on the other hand, with compaction, deep burial, metamorphism, and granitization. None of these processes can operate without affecting all others, because all are linked in a gross "loop" that is constantly undergoing change.

The rock cycle overall is hopelessly large to be dealt with in any realistic level of detail. To create an effective model of part of the

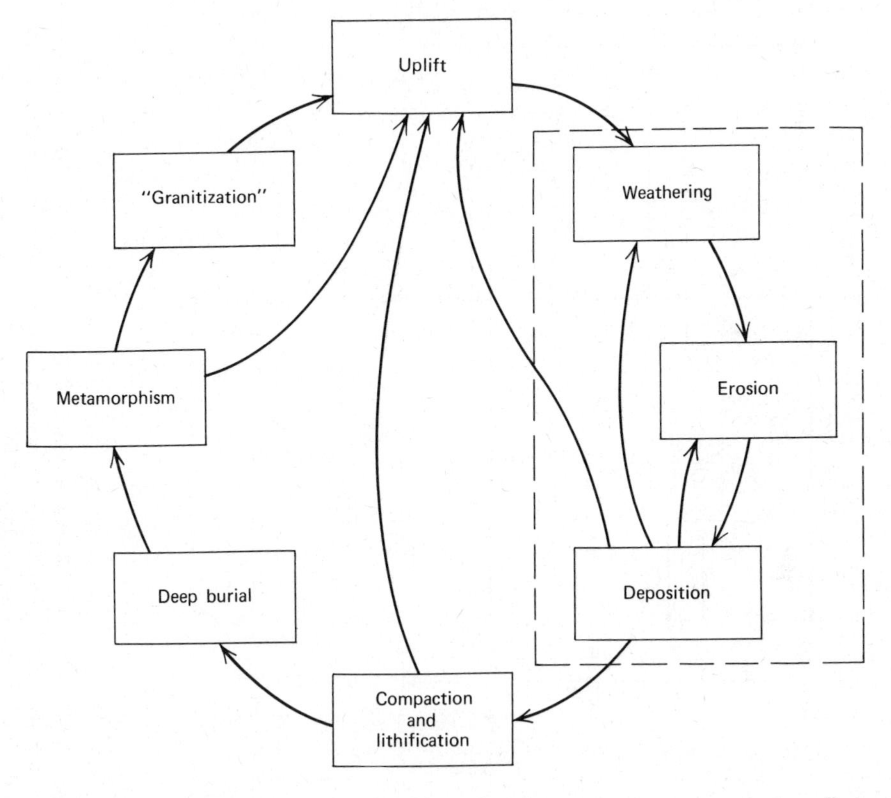

Figure 1-13 Simplified diagram of rock cycle, showing main processes that affect transformation of rocks in the earth's crust. Dashed line defines smaller system, embracing weathering, erosion, and deposition, separated from remainder of system.

rock cycle, that part must be excised from the overall system and treated in isolation, or semi-isolation. This demands of course that the isolated segment be surrounded by a boundary. For example, if we wish to create a model dealing only with weathering, erosion, and deposition, we can isolate these processes from the gross system by enclosing them within a boundary, shown by the dashed line in Figure 1-13. Although this is a convenience, it is obviously very artificial, because in the real system weathering and erosion cannot be separated from uplift nor from the other major processes. The motivation to separate parts of the gross system is to be able to develop a model of them in greater detail, treating them as separate systems.

An expanded version of part of the weathering–erosion–deposition segment of the rock cycle is illustrated in Figure 1-14, which relates

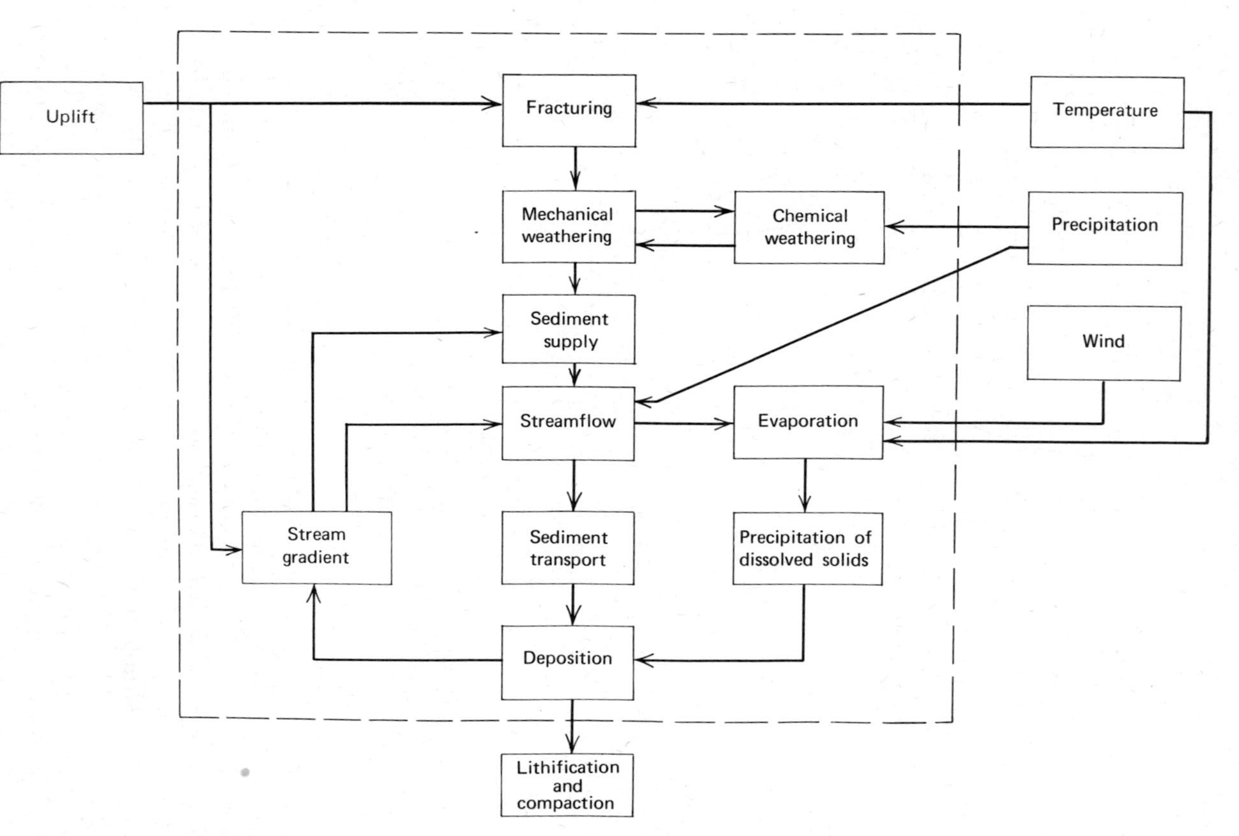

Figure 1-14 Diagram of fluvial erosion system. Uplift and climatic factors of temperature, precipitation, and wind are outside the system, feeding into the system, but not receiving feedback from the system. Lithification and compaction are also outside the system, but receive output from the system instead of feeding into the system.

some of the interrelationships linking weathering and deposition in a fluvial erosion system. The boundaries of the local system are represented by dashed lines, whereas the linkage among components is shown by arrows. Components inside the boundary are interlocked so that none can undergo change without influencing, directly or indirectly, all other endogenous components. Exogenous components are not interlocked. Regardless of its construction, our representation is a simplified abstraction, lacking many of the components of the real system.

Annotated Bibliography

Beer, S., 1959, *Cybernetics and Management,* John Wiley and Sons, New York, 214 pp.

Very readable account of the nature of systems and cybernetics.

Chorafas, D. N., 1965, *Systems and Simulation,* Academic Press, New York, 503 pp. Early chapters contain a discussion of simulation philosophy.

Briggs, L. I., and Pollack, H. N., 1967, "Digital Model of Evaporite Sedimentation," *Science,* v. 155, pp. 453–456.

A classic paper describing a dynamic-deterministic simulation model that has been convincingly used in interpreting ancient geologic conditions in the Michigan basin.

Forrester, J. W., 1961, *Industrial Dynamics,* John Wiley and Sons, New York, 464 pp.

Presents details for constructing dynamic simulation models of business organizations.

Harbaugh, J. W., and Merriam, D. F., 1968, *Computer Applications in Stratigraphic Analysis,* John Wiley and Sons, New York, 282 pp.

Chapter 8 provides an introduction to simulation in geology.

Odum, E. P., and Odum, H. T., 1959, *Fundamentals of Ecology,* W. B. Saunders, Philadelphia, 546 pp.

An important reference that deals, in part, with systems concepts in ecology.

Pitts, F., 1967, *Simulation Bibliography,* unpublished report, Department of Geography, University of Hawaii.

Lists about 1050 references to simulation in many fields of application, including psychology, political science, sociology, business, geography, biology, and engineering.

Von Bertalanffy, L., 1956, "General Systems Theory," *General Systems,* v. 1, pp. 1–10.

Watt, K. E. F., 1964, "The Use of Mathematics and Computers To Determine Optimal Strategy and Tactics for a Given Insect Pest Control Problem," *The Canadian Entomologist*, v. 96, pp. 202–220.

A classic paper describing the systems approach, employing computer-simulation models in various ecological applications.

Watt, K. E. F., ed., 1966, *Systems Analysis in Ecology*, Academic Press, New York, 276 pp.

A collection of papers by individual authors.

CHAPTER 2

Models and Simulation

CLASSIFICATION OF SIMULATION MODELS

The various types of models that can be used to simulate geological dynamic systems are listed in Table 2-1. There are two main classes, physical models and symbolic models. Most physical models are material scale models, which may be either larger or smaller than the

TABLE 2-1 Classification and Examples of Types of Geological Models That May Be Used for Simulation

Type of Model	*Example*
Physical models:	
Material scale	Sedimentation tank models
Electrical analog	Groundwater withdrawal and recharge
Symbolic models:	
Graphic	Beach process-response models
Mathematical:	
Deterministic – static	Simulation of gravity response of buried body
Deterministic – dynamic	Simulation of deltaic sedimentation
Probabilistic – static	Simulation of oil search by grid drilling
Probabilistic – dynamic	Reef-growth simulation

actual prototype; for example, material scale models of crystal lattices are larger than their actual prototypes. Most material scale models in geology, however, are smaller than the actual prototype—for example, the soft clay models used by Cloos (1968) for studying Gulf Coast fracture patterns. Electrical analog models (not including analog computers) form another special type of physical model, in which electric currents represent the flow of fluids or heat.

Symbolic models consist of two general forms—namely, graphic and mathematical. Graphic models—such as maps, sections, and block diagrams—are commonly small-scale pictorial representations of larger features. Graphic models are not ordinarily used for simulation in themselves, but they may be used to describe simulation models (e.g., the fluvial erosion system of Figure 1-14). Other examples are provided by the process-response diagrams (Figure 2-1) of Krumbein (1964) and Whitten (1964). It can be argued that these system diagrams, as well as more formal system flow charts (Figure 1-11), facilitate "mental simulation" by depicting the interrelationships among the components of a system. A map or a section, on the other hand, generally depicts only the product formed by a sequence of events and does not provide for direct representation of the processes themselves.

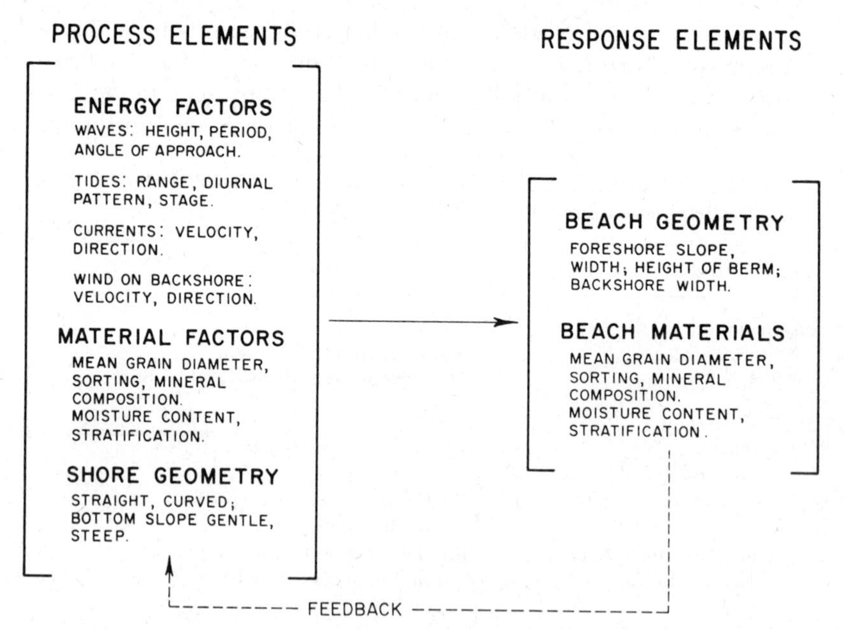

Figure 2-1 Process-response model of beach. From Krumbein (1964).

In this volume we are principally concerned with mathematical models applied in a simulation context. Mathematical models, of course, may be applied in a variety of ways, such as the use of statistical models to summarize observational data. Most simulation models, however, actually generate a great deal of new data, and therefore their roles are quite different from those of models used for data reduction.

It should be stressed that many mathematical models involve little or no computation, regardless of whether they are used in a simulation or in a data-reduction context. Other mathematical models, however, involve so much computation, either because of their complexity or because of the need for a large number of calculations or logic operations, that they can be conveniently manipulated only with computers. Computer-simulation models are in effect mathematical models in the form of computer programs. General treatment of the role of models in science is provided by Chorafas (1965), Chorley and Haggett (1967), Cole and King (1968), Kendall (1968), and Rosenblueth and Wiener (1945). The various roles of models in geology have been discussed elsewhere by Bonham-Carter and Harbaugh (1968), Griffiths (1967), Krumbein and Graybill (1965), Whitten (1964), and several other authors.

Although this volume is concerned entirely with applications of digital computers, it should be pointed out that analog computers have important and, as yet, only slightly exploited applications in geological simulation. Most of the geological analog-computer applications to date have been concerned with the flow of underground fluids, although analog computers have been used with notable success by Raup (1965) in modeling the geometry of shell coiling.

Simulation models can be classified according to whether they are static or dynamic, deterministic or probabilistic (or hybrid in that they combine both deterministic and probabilistic components). A deterministic model is one in which there is no element of chance. In a dynamic-deterministic model the state of the model at any point in time is completely predetermined. In a probabilistic model, however, there is a degree of uncertainty. The state of a dynamic-probabilistic model at any subsequent moment in time cannot be precisely predicted because of one or more components of uncertainty (stochastic components).

Many deterministic models may be regarded as special cases of stochastic models. For example, the process by which heat is conducted from a warm object to a cooler one is in essence a stochastic process that depends on the random motion of molecules. On the other hand, the number of molecules in any real situation is so large that, with virtually absolute certainty, heat will flow from warm to cold. Thus for practical purposes we can assume that the process of heat conduction is deterministic and can effectively represent it with a deterministic model.

All the types of mathematical simulation models listed in Table 2-1 are useful in geology. Static-deterministic simulation models have been widely used in geophysics—for example, in the computation of a series of gravity responses of hypothetical buried bodies whose shapes, depths, and densities may be assumed so that they yield various alternative geological "solutions" to an observed gravity anomaly (Isaacs, 1967). Although they may be of virtually any degree of complexity, most static-deterministic models in geology have been relatively simple. More complex dynamic-deterministic geological models have been developed, such as the Briggs–Pollack model for evaporite sedimentation described in Chapter 1. We describe several other dynamic-deterministic sedimentation models in Chapter 9.

Static-probabilistic models are represented by Monte Carlo methods, which involve sampling from frequency distributions by using random numbers. An example is provided by the work of Griffiths and Drew (1966), which deals with the simulation of grid drilling in mineral exploration.

Dynamic-probabilistic models are of particular importance in geology because most geologic processes can be regarded as possessing random components. Most dynamic-probabilistic models also contain deterministic components and are therefore "hybrid" in that respect. An example is provided by the carbonate-ecology model developed by Harbaugh (1966) and described in Chapter 9, which contains a number of interdependent deterministic and probabilistic components.

GOALS OF SIMULATION

In common with scientific endeavor as a whole, simulation is concerned with three main objectives: understanding, prediction, and control. In geology we are concerned principally with understanding and prediction. Control of most geologic processes is not feasible, although such processes as river flow can be modified by man. On the other hand, man-made systems can be wholly or partly controlled. For example, the traffic-flow system in a city can be controlled by regulating the direction of movement on certain streets and by adopting policies and practices that have a major influence on the system. To provide another example, man can influence weather and climate to a slight

TABLE 2-2 Classification of System Types with Respect to the Degree to Which They Can Be Influenced by Man and the Relative Importance of Simulation in Achieving Objectives

	Natural	*Hybrid*	*Artificial*
Examples	Solar system Rock cycle	Forest ecology Mineral exploration	Traffic flow Mine operation
Degree of man's influence	None	Some	Major
Goals			
Understanding	Important	Important	Some importance
Prediction and forecasting	Important	Important	Important
Decision-making to control system	Not relevant	Important	Important

degree, but he cannot be said to "control" the system to the extent that he can control city traffic. Obviously we are dealing with a spectrum of systems in which the individual systems can be arranged according to the degree to which they can be influenced by man (Table 2-2). A purely man-made system, such as a factory at one end of the spectrum, may be defined as a system created and controlled by man. A strictly natural system at the other end of the spectrum, such as the solar system, is a system in which man plays no part. Between these end members of "natural" and "man-made" systems there are a variety of hybrid systems, which provide examples of some of the most effective simulation applications. Good examples occur in the study of man's influence in ecology. Natural ecological systems are delicately balanced by exceedingly complex interlocking relationships. Simulation can help to answer questions about the possible effects of man's intrusion on this balance (Watt, 1964).

If the objective of a simulation study is to provide specific answers, it is important for the computer program that embodies the simulation model to be capable of yielding clear-cut answers. Framing the questions and devising a simulation model that will yield suitable answers are of course closely related. An example of a question that might be "asked" of a particular simulation model is, "could oil entrapment occur within a particular geologic structure under a given set of hydrodynamic and oil-density conditions?" It would be reasonable to "ask" this type of question of some simulation models, such as the hydrodynamic oil-trap model described by Harbaugh and Merriam (1968, pp. 257–259). On the other hand, it is very difficult to devise simulation models that can deal with clear-cut questions, the reason being that there are relatively few important geological problems for which "yes" or "no" answers are feasible. Instead of seeking simple answers, the objectives are, more commonly, to explore the consequences of different sets of assumptions.

DEVELOPING COMPUTER MODELS

It is difficult to lay down rigorous guidelines for the model builder to follow. The model builder must create models by intuition, including components that he regards as important and slighting those he considers as minor. Furthermore he must specify the functional interrelationships among the components. Ultimately these tasks involve devising ways of representing geologic processes with algebraic equations and logic operations. Devising an algebraic representation for a collection of geologic processes, such as glaciation or hydrothermal ore deposition, is extremely challenging in itself and may lead to new insight into the processes themselves. The mechanisms of the processes must be well defined before they can be reduced to sets of equations and logic statements. This is not to imply that their representation is necessarily "correct" because it is accomplished mathematically. It does imply, however, that the representation is explicit and unambiguous. The task of observing the effects of actual geologic processes, formulating hypotheses about how these processes work, and then reducing them to a level of abstraction that permits formulation of mathematical models is in many respects an intuitive trial and error process. We point out, however, that construction of an explicit and unambiguous, but totally wrong, mathematical model may be far worse than a vague and imprecise representation that is more correct. The danger is that mathematical models tend to have a formal, "scientific" appearance that may lend an unwarranted credibility.

Types of Relationships

Before proceeding further we need to consider the types of relationships that we deal with in geological models. These relationships can be thought of as forming a spectrum (Table 2-3) ranging from purely theoretical, abstract relationships at one end to heuristic relationships at the other end, with empirical relationships occurring between. Theoretical relationships tend to be simpler, more versatile in application, and better understood than other types of relationships. Empirical relationships are based on observations and often can be viewed statistically. An example is the relationship between

TABLE 2-3 Types of Relationships Arranged in a Continuous Spectrum Ranging from Theoretical to Heuristic

	Type of Relationship	*General Example*	*Well-Known Sedimentological Example*	*More Complex Sedimentological Example*
Increasing understanding ↑; Increasing versatility of application ↑; Increasing complexity of actual relationship ↓	Theoretical	Continuity equation in fluid flow	Stoke's law relating velocity of fall of spherical particles to diameter	Theoretical longshore-current model of Putnam et al. (1949)
	Empirical	Maximum optical extinction angles and proportions of sodium and calcium in plagioclase feldspars	Hjulstrom's (1935) sediment transport-deposition relationship between flow velocity and grain size	Empirical longshore-current model of Inman and Quinn (1951)
	Heuristic	Warping of earth's crust in response to loading or unloading of sediment	Relationship of sediment-size sorting to wave energy	Markov models of marine-organism-community interactions of Harbaugh (1966)

the maximum optical extinction angle and the proportions of sodium and calcium in plagioclase feldspars. Underlying theory can be brought to bear on this relationship, but it has stood for many years as an empirical relationship, without a theoretical framework. Finally, heuristic relationships can be expressed without an understanding of why they exist. Obviously there is no sharp boundary between heuristic and empirical relationships, the difference being one of degree.

Clearly there are pitfalls inherent in the indiscriminate use of simulation models, and these pitfalls may stem from the careless choice of functional relationships. Although theoretical relationships are the most versatile in application (e.g., the application of thermodynamics to the crystallization of igneous rocks), there are many aspects of geology where little physical theory exists. Lacking theoretical relationships, the model builder must use either empirical or heuristic ones. Empirical relationships tend to lack versatility, perhaps requiring the model builder to devise a completely heuristic relationship. If the heuristic relationship "works" in an empirical sense, this may give a greater insight into the process being modeled, possibly leading to development of theory. Although heuristic and empirical relationships may work for a given set of conditions, they may be inflexible and unable to respond appropriately under other sets of conditions. For example, let us suppose that we have developed a model for glacial outwash fans that incorporates a number of heuristic and empirical relationships. After adjusting the input parameters we can produce a very good analogy to a particular outwash fan. If, however, many experimental adjustments on the model fail to produce other reasonable types of outwash fans, we might suspect that the model is a poor analogy. Any general conclusions drawn from such a model might be quite erroneous. Obviously the simulation-model builder needs to be aware of the dangers of inappropriate assumptions.

Programming the Model

Once the structure of the model is established, the next step is to develop a computer program that represents the various components and processes to be simulated. We do not propose to dwell on the details of computer programming in this book. We should point out, however, that there are a wide variety of computer languages from which to choose, some of them especially designed for simulation. Most of the geological simulation models that have been published to date have been written in FORTRAN, which is a general-purpose language, principally designed for scientific (as opposed to commercial) computing. We have found that FORTRAN is sufficiently versatile for the needs of most geological simulation models, and we have exclusively adopted it for this book. Readers unfamiliar with FORTRAN are advised to consult some of the numerous books on the subject, such as those by McCracken (1965) or Hull (1966).

The special-purpose simulation computing languages have been developed for a wide variety of applications. Several have been designed principally for operations research and are concerned with the movement through the system of discrete elements (e.g., cars or people), incorporating such topics as queuing and inventory control. SIMSCRIPT, DYNAMO, and GPSS are examples of such languages. Besides languages designed for modeling discrete elements, there are those for simulating systems with continuous variables, such as IBM's Continuous System Modeling Program (CSMP), designed for the 360 series computers (IBM, 1968). Languages of this type are particularly suitable for dealing with problems that involve the solution of sets of differential equations. Teichroew and Lubin (1966) discuss the relative merits of different types of simulation languages.

Another decision that the model builder faces is how to display the simulation results. Most simulation studies produce large volumes of data, often representing a number of variables in time and space. Accordingly graphic-display techniques form an important aspect of computer simulation. Both the line printer and computer-driven digital plotters (and ultimately cathode-ray-tube devices) can be used for the effective display of graphs, maps, and cross sections. Furthermore program packages for the automatic plotting of contour maps and three-dimensional perspective or isometric diagrams are increasingly available at most computation centers (e.g., Peikert, 1969).

Assessing the Results

In many cases the results of a simulation study are very difficult to assess. Computer models normally produce large volumes of output consisting of theoretically derived data, which must then be compared with actual data. By matching the output with real-world observations, the model can be evaluated, and its ability to provide a useful analog to the real system can be judged. Although this sounds straightforward, in practice there are many difficulties. To begin, real-world data may be difficult to obtain for many geological dynamic systems, and, even when available, the data may be sparse or unsuitable in form. For example, suppose that we are modeling a sedimentary basin and that the program output yields a number of contour maps and cross sections pertaining to the basins. Objective, quantitative comparison of this output with maps and cross sections derived from observations of a real sedimentary basin may be difficult. One must simply rely on a qualitative comparison, to see if the results "look right."

For some models, however, quantitative comparison between theoretical and observed data is possible with numerical optimization techniques. Suppose that a particular model responds in a highly sensitive way to changes in the input variables. By "tuning" the model, adjusting the input variables in a sequence of experiments, the behavior of the model may be optimized, making the "fit" of the theoretical and observed data values as close as possible.

PROGRAM FOR GEOLOGY STUDENT MODEL

It is now appropriate to go through the steps in translating a conceptual model to a computer program. The geology-student model outlined in Chapter 1 is useful for this purpose. We can begin by expanding the simple flow chart (Figure 1-11) of the geology-student system into a more detailed program flow chart (Figure 2-2) that outlines the computational steps. Both diagrams can be regarded as abstract models of the real system. Before getting involved in programming details, however, let us first work through the algorithm, or sequence of instructions by hand. The steps in the algorithm are outlined in the same order as that shown in Figure 2-2. The sets of capital letters in parentheses are names given to the different variables, or identifiers, in the FORTRAN program that we will develop subsequently. It is convenient to use these identifiers in both our manual calculations and the FORTRAN program. In the description that follows the manner in which the identifiers are linked together is shown with the isolated FORTRAN assignment statements that are interspersed in the text.

Steps in the Algorithm

The steps in the algorithm are as follows:

1. The first step consists of entering the data that control the operation of the program and provide information needed at the outset, as follows:

   a. The number of students admitted as freshmen to the geology program in years 1 through 5, NGEOL(1), NGEOL(2), · · · , NGEOL(5) (the first year of actual calculations begins in year 6).

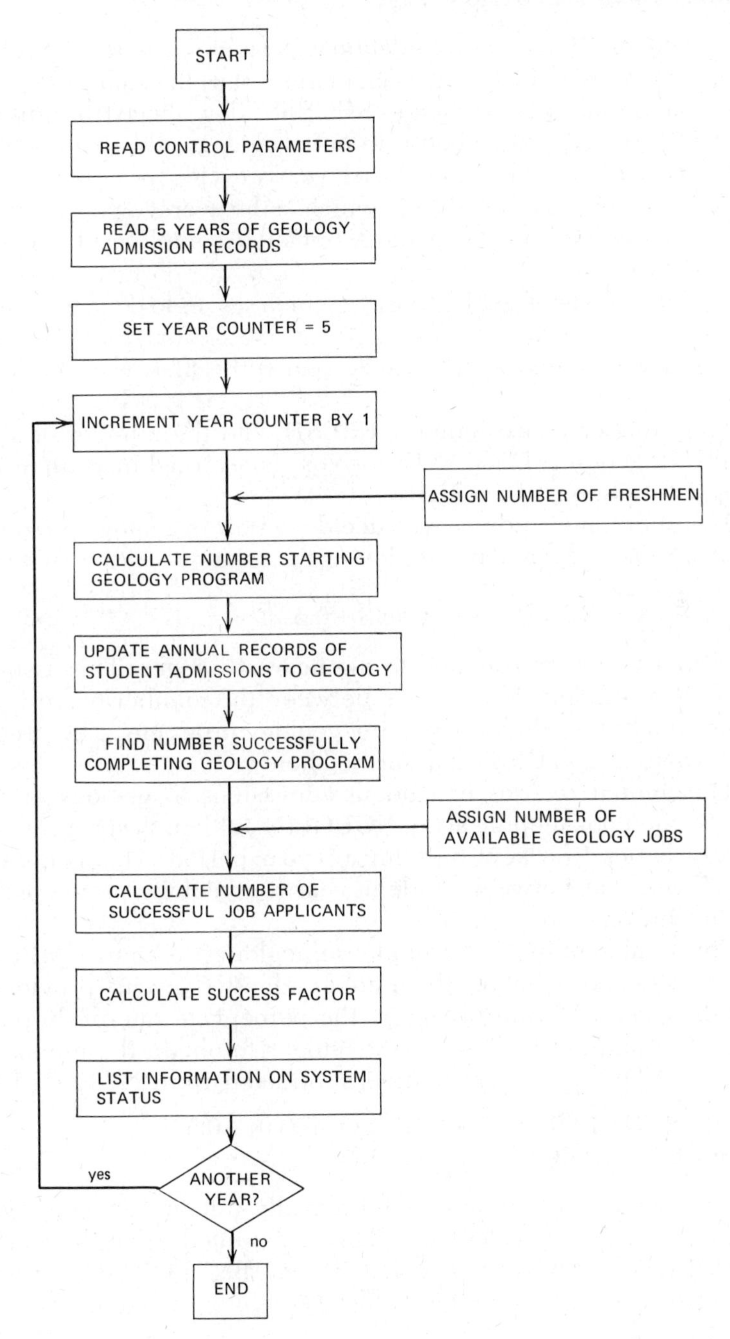

Figure 2-2 Flow chart of computer-simulation model for geology-student system.

b. The duration of the simulation experiment in years, NYEARS, allowing for the fact that calculations start in year 6.

c. An initial success factor, SUCESS, that affects the number of freshmen who choose to enter geology; this parameter is recalculated at the end of each yearly cycle.

d. A dropout rate, DRPRAT, which is the average percentage of students who start in geology but fail to complete all 5 years of the course.

2. In the next step the identifier N, which serves to count the years, is set to 5.

3. N is incremented by 1, making year 6 the first year of calculations.

4. The number of freshmen, NFRESH, and the number of available jobs in geology, DEMAND, for year N are read in as input (see appropriate columns of Table 2-4).

5. The number of students who decide to start in geology programs, NGEOL(N), is calculated by employing the relation

```
NGEOL(N)=NFRESH*SUCESS+0.5
```

where NGEOL(N) is rounded to the nearest integer. This assumes that there is a linear dependence between the number of freshmen and the number of students starting geology programs. The proportionality constant SUCESS is described below.

6. The annual records of student admissions to geology are retained in the bookkeeping array NGEOL(N), although strictly speaking records need be kept only for a 5-year period. This is because of the 5-year lag between students starting geology programs and those leaving on graduation.

7. The number of students completing geology programs, CMPLET, is then calculated by taking the number admitted 5 years previously, NGEOL(N–5), and multiplying by the percentage rate of dropouts, DRPRAT, rounding to the nearest integer to obtain the number of dropouts, NDROP, which is then subtracted from NGEOL(N–5):

```
NDROP=NGEOL(N-5)*(DRPRAT/100.0)+0.5
CMPLET=NGEOL(N-5)-NDROP
```

8. The number of students who actually obtain geological jobs, NJOBS, for year N is calculated. This is assumed to be the smaller of CMPLET or DEMAND. MIN0 is an IBM FORTRAN library function for finding the smallest value of two integer arguments.

```
NJOBS=MIN0(CMPLET, DEMAND)
```

9. The number students who fail to obtain geological jobs is calculated:

JBLESS=CMPLET−NJOBS

10. A new value of the proportionality factor SUCESS is calculated. From the past experience of some hypothetical admissions and employment records it is decided to try the heuristic relationship

SUCESS=(DEMAND/CMPLET)/10

This assumes that on the average 10 percent of all freshmen start geology programs. When DEMAND=CMPLET and JBLESS=0, SUCESS=0.1, which ensures that exactly one-tenth of the freshmen class will enter geology. If DEMAND>CMPLET and JBLESS=0, SUCESS>0.1, which causes more than 10 percent of the freshmen to enter geology. On the other hand, if DEMAND<CMPLET and JBLESS>0, SUCESS<0.1, which reduces the percentage to a figure of less than 10. This factor thus acts as a controller, employing negative feedback to prevent either a glut or a dearth of geology graduates.

11. The status variables N, NFRESH, NGEOL(N), NDROP, CMPLET, DEMAND, NJOBS, JBLESS, and SUCESS are printed out.

12. Finally, if the value of N is equal to NYEARS, the calculations are stopped. Otherwise there is a return to step 3, and the calculations are repeated for another year.

A set of trial calculations for the first 2 years of a simulation run are outlined below, and the results of more extended calculations are listed in Table 2-4, in which the italicized figures are those assigned as input, whereas those shown in conventional type are calculated in the simulation run. Each row corresponds to calculations for a single year. The calculations start at year N = 6, but the previous 5 years are included in the table because admissions records, NGEOL(N), for 5 years in advance of the simulation run are required. Figures in Table 2-4 could pertain to a large university or to an aggregation of several smaller universities.

Trial Calculations for Year 6

NFRESH=1000
DEMAND=60
NGEOL(6)=NFRESH * SUCESS+0.5=1000 * 0.05+0.5=50
NDROP=NGEOL(N-5) * (DRPRAT/100.0)+0.5=51 * (4/100)+0.5=2
CMPLET=NGEOL(N-5)−NDROP=51−2=49

TABLE 2-4 Data Pertaining to a 17-Year Run of the Geology-Student Simulation Model[a]

Year N	*Number of Incoming Students (Input)* NFRESH	*Number of Starting Geology Students* NGEOL(N)	*Number of Students Who Drop out Who Started in Year* N − 5 NDROP[b]	*Number of Students Who Graduate in Geology* CMPLET	*Number of Geology Jobs Available for New Graduates (Input)* DEMAND	*Number of New Graduates Who Obtain Geology Jobs* NJOBS	*Success Factor* SUCESS
1	–	*51*	–	–	–	–	–
2	–	*47*	–	–	–	–	–
3	–	*50*	–	–	–	–	–
4	–	*51*	–	–	–	–	–
5	–	*49*	–	–	–	–	*0.050*
6	*1000*	50	2	49	*60*	49	0.122
7	*1002*	122	2	45	*60*	45	0.133
8	*941*	125	2	48	*60*	48	0.125
9	*1127*	141	2	49	*60*	49	0.122
10	*1050*	129	2	47	*60*	47	0.128
11	*987*	126	2	48	*60*	48	0.125
12	*991*	124	5	118	*60*	60	0.051
13	*1032*	52	5	120	*60*	60	0.050
14	*1241*	62	6	135	*60*	60	0.044
15	*849*	38	5	124	*60*	60	0.048
16	*987*	48	5	121	*60*	60	0.050
17	*1001*	50	5	119	*60*	60	0.050

[a]Figures in italics represent input; other figures were obtained by hand calculation during manual simulation run.
[b]Dropout rate = 4 percent.

```
NJOBS=MIN0(CMPLET,DEMAND)=MIN0(49,60)=49
JBLESS=CMPLET–NJOBS=49–49=0
SUCESS=(DEMAND/CMPLET)/10=(60/49)/10=0.122
```

Trial Calculations for Year 7

```
NFRESH=1002
DEMAND=60
NGEOL(7)=1002*0.122+0.5=122
NDROP=47*(4/100)+0.5=2
CMPLET=47–2=45
NJOBS=MIN0(45,60)=45
JBLESS=45–45=0
SUCESS=(60/45)/10=0.133
```

Notice that NGEOL(7)=122 in our calculations, but in a subsequent computer run a value of 123 was obtained. This discrepancy is due to a rounding error caused by retaining insufficient digits in the hand calculation.

Computer Program

The transition from hand calculation to computer calculation is a simple one, particularly since the names assigned to the variables can be used as FORTRAN variable identifiers, and the steps in the computer program can be arranged essentially to parallel the manual calculations. The FORTRAN program listed in Table 2-5 is almost identical to the collection of FORTRAN statements illustrated above. Declaration and format statements have been added, however, and a few operations have been necessary to convert integer variables to real variables.

Results of a run with input data listed in Table 2-6 are shown graphically in Figure 2-3. Note that with a supply of freshmen fluctuating about a mean of 1000 per year and with a constant job demand at 60, the present form of the success ratio, coupled with the 5-year lag, causes more or less regular oscillation in the number of jobless. Figure 2-4 portrays another run in which the supply of freshmen was held constant at 1000, but the job demand fluctuated on a 7-year period.

What has been gained by building this model? The model is exceedingly simple, of course, and it is relatively easy to predict its behavior under various conditions. We could, however, make things much more complex by including multiple options for students pursuing different terminal degrees (bachelor's, master's, Ph.D.) by making the dropout rate vary year-to-year, by causing the annual supply of

TABLE 2-5 FORTRAN Program Representing Dynamic Simulation Model of Geology-Student System

```
C.....GEOLOGY STUDENT MODEL
      DIMENSION NGEOL(100)
      INTEGER  DEMAND, CMPLET
    1 FORMAT(2F5.2, I5)
    2 FORMAT(5I5)
    3 FORMAT(1H1, 4X, 'GEOLOGY STUDENT MODEL'/ 5X, 'DROPOUT RATE = ',
     1 F5.2,'  PERCENT PER YEAR'/5X,'INITIAL SUCCESS FACTOR =',F4.2//
     2 4X,' N    NFRESH    NGEOL    NDROP    CMPLET  DEMAND  NJOBS   JBLESS
     3 SUCESS')
    4 FORMAT(2I5)
    5 FORMAT(1H , I5,  9X, I8)
    6 FORMAT(1H , I5, 1X, 7I8, F9.3)
C.....READ CONTROL PARAMETERS
      READ(5,1) SUCESS, DRPRAT, NYEARS
C.....READ GEOLOGY ADMISSION RECORDS FOR FIRST 5 YEARS
      READ(5,2) (NGEOL(N), N=1,5)
C.....WRITE OUT HEADINGS AND FIRST 5 YEARS RECORDS
      WRITE(6,3) DRPRAT, SUCESS
      DO 10 N=1,5
   10 WRITE(6,5) N, NGEOL(N)
C.....BEGIN ANNUAL CYCLE
      DO 40 N=6, NYEARS
C.....READ FRESHMAN INPUT AND JOB DEMAND
      READ(5,4) NFRESH, DEMAND
C.....CALCULATE NEW ADMISSIONS TO GEOLOGY
      NGEOL(N)=FLOAT(NFRESH)*SUCESS+0.5
C.....CALC. NO. OF DROPOUTS AND STUDENTS DUE TO COMPLETE THAT STARTED
C.....FIVE YEARS AGO
      NDROP=FLOAT(NGEOL(N-5))*DRPRAT/100.0+0.5
      CMPLET=NGEOL(N-5)-NDROP
C.....CALCULATE NUMBER OF JOBS FILLED
      NJOBS=MIN0(CMPLET,DEMAND)
      JBLESS=CMPLET-NJOBS
C.....DETERMINE SUCESS FACTOR
      DEM=DEMAND
      CMP=CMPLET
      SUCESS=DEM/CMP/10.0
C.....WRITE OUT VALUES FOR YEAR
   40 WRITE(6,6) N, NFRESH, NGEOL(N), NDROP, CMPLET, DEMAND, NJOBS,
     1 JBLESS, SUCESS
      RETURN
      END
```

freshmen to have a normal frequency distribution with a particular mean and variance, and so on. The behavior of the model would not be easy to predict under these more complex circumstances.

The behavior of the present model is strongly influenced by the success factor, which is defined as demand divided by the number of students completing geology programs. This relationship may be reasonable for universities in major oil-producing regions, where the oil industry employs many of the new geology graduates. On the other hand, this assumption might be very unrealistic for universities in other regions, such as the northeastern United States.

Despite its shortcomings, creating such a model has forced us to think more clearly about the processes involved in the supply and

TABLE 2-6 Example of Input Data for Geology-Student-Model Program, Assuming Steady Demand and Fluctuating Supply of Freshmen[a]

```
0.05 4.00   50
  51   47   50    51    49
1000   60
1002   60
 941   60
1127   60
1050   60
 987   60
 991   60
1032   60
1241   60
 849   60
 987   60
1001   60
1000   60
1010   60
 951   60
1031   60
 941   60
1087   60
 979   60
1036   60
1260   60
 844   60
 977   60
1112   60
 999   60
 888   60
1227   60
 942   60
1127   60
1050   60
 987   60
 991   60
1032   60
1241   60
 849   60
 987   60
1001   60
1000   60
1002   60
 941   60
1127   60
1050   60
 987   60
 991   60
1032   60
```

[a]Numbers in upper two rows are control parameters. In remaining rows numbers in left column represent incoming freshmen each year, and those in right column represent employers' demand. Response of model based on this input is shown in Figure 2-3.

demand of geology graduates. We now have a framework for future research on the problem. The next steps might be to examine historical records to test the model, to explore the decision-making processes that freshmen undergo in deciding which curriculum to pursue, and to subdivide the model to allow several optional paths, such as

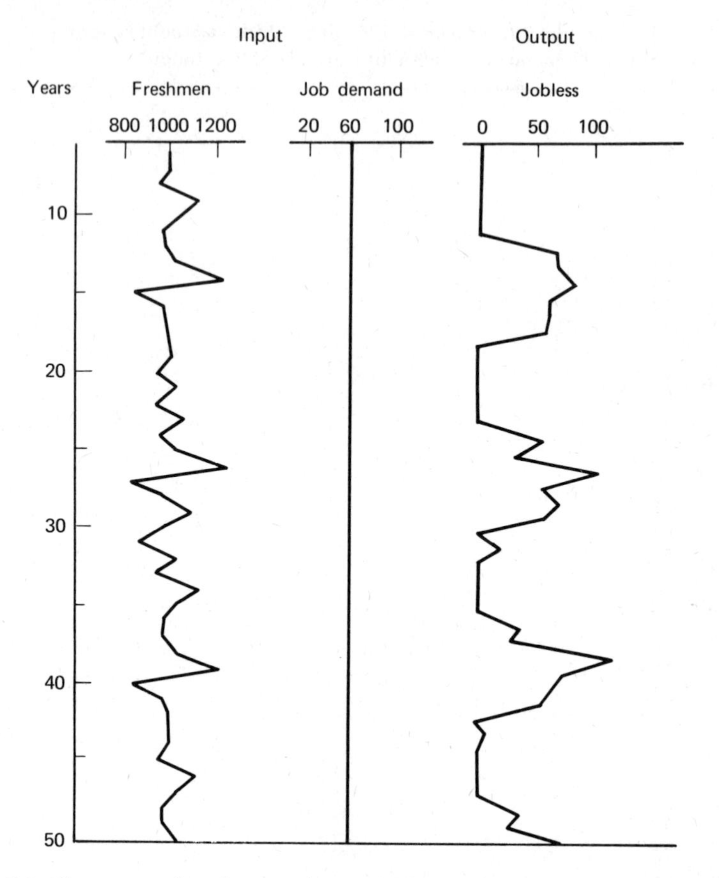

Figure 2-3 Response of geology-student model based on input data of Table 2-6. Output consists of the number of jobless. Input consists of the number of entering freshmen, which fluctuates irregularly from year to year, and job demand, which remains constant.

bachelor's, master's, and Ph.D. programs. A more sophisticated model might be valuable both for predicting and controlling the flow of graduates. Understanding, prediction, and control are the principal goals of most simulation models.

This particular model is sufficiently simple to be presented in terms of a few mathematical equations instead of a computer program. If, however, the model were to be made more complex, strictly mathematical representation might become exceedingly complex. A computer program is easier to formulate and often gives insight into the way in which a problem may be structured mathematically.

In retrospect it is appropriate to view the geology-student model as

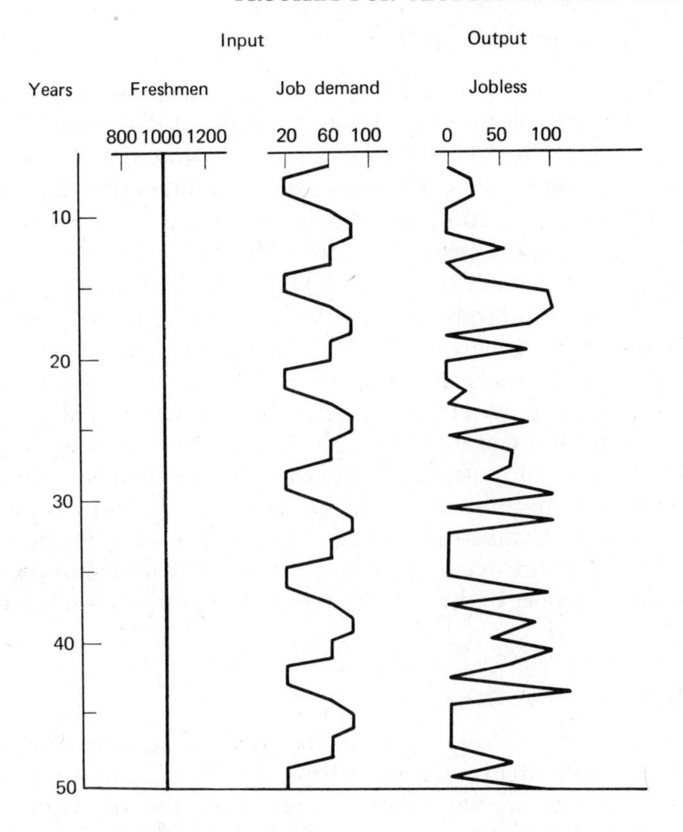

Figure 2-4 Simulation run of geology-student model in which the input of freshmen is held constant at 1000 and job demand fluctuates regularly around 60. Model responds with extreme fluctuations in the number of jobless.

an example of a simulation model that contains the following features that are analogous to those in many more advanced computer-simulation models:

1. Components of the system are represented in the model by FORTRAN variables.
2. People and information moving through the system are represented in the sequence of FORTRAN statements.
3. The flow of time is represented by discrete yearly steps, a year being represented by each iteration of an outer DO-loop.
4. The status of the system is given for any instant in time by the state of the variables, which perform a bookkeeping task.
5. The system is controlled by negative feedback.

REPRESENTATION OF TIME

The geology-student model is an example of a simulation model that exists in the time dimension. Each iteration of the program corresponds to 1 year. Time is not represented continuously in the model; the transition from one year to the next is instantaneous. The model is like a motion-picture film, with each successive frame in the film 1 year younger than the preceding one. The choice of a year per iteration is arbitrary. A more detailed and more realistic model might employ smaller time increments, such as a month or a week. Regardless of their duration, however, the time increments still represent discrete steps. Continuous time can be conveniently approximated in a digital simulation model by a sequence of increments or steps.

Establishing the duration of each time increment is an important consideration. If computer time (and therefore cost) is not a consideration, each time increment should be of relatively short duration, so as to approximate most effectively the passage of continuous time. However, computer machine-time costs are usually major considerations and are almost in direct proportion to the number of time increments.

REPRESENTATION OF SPACE

The computer representation of the geology-student model is not concerned with spatial relationships. Many geological problems, however, involve spatial relationships. Thus the representation of space is of prime importance in a geological simulation model.

The builder of geological models has a number of choices concerning representation of space. He must consider the number of dimensions to be represented (one, two, or three) and whether space is to be represented in continuous or discrete form. Figure 2-5 provides an example of the representation of spatial relationships with a continuous curving line. Although the line represents a topographic profile that occupies two dimensions, the line itself can be thought of as a function of the single independent variable y. If we can write a function that represents the line, we can represent the line in continuous form, graphing it by supplying different values of the independent variable y, and in turn obtaining values of the dependent variable z.

An alternative method is to approximate a line in discrete form. For example, the same topographic profile can be divided into a series of columns of equal width along the y-axis, with the height of each column approximating the line at that location (Figure 2-6). The height of columns can be represented by a sequence of numbers. If the shape of

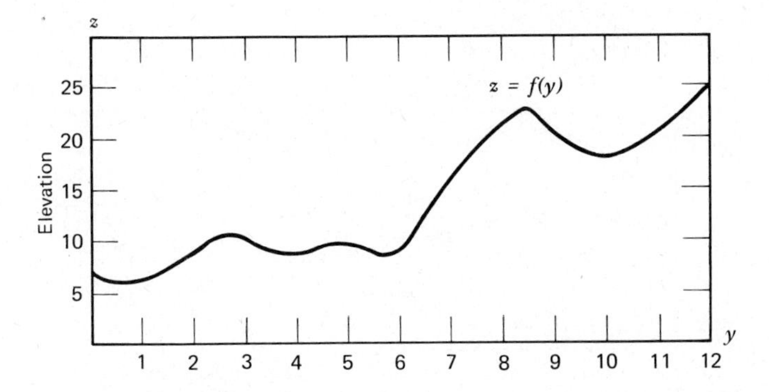

Figure 2-5 Topographic profile represented by line in which z is a function of y.

the profile is complex, it might be difficult to find a function that describes it, but it is simple to represent the profile with a set of numbers, no matter how complicated its shape. Furthermore computing languages, such as FORTRAN, readily provide for representation of sequences of numbers in array form.

Representation of surfaces, in contrast to lines, involves an additional dimension. The choices for representation are similar, however. Surfaces can be represented by continuous functions, such as polynomials (Figure 2-7a) involving two independent variables or by harmonic functions (Figure 2-7b); they may also be approximated in discrete form by a series of rectangular cross-sectioned prisms of different height (Figure 2-8). The height of the prisms can be represented as a series of numbers stored in a two-dimensional array, the indices of the elements in the array specifying the geographic location of each prism.

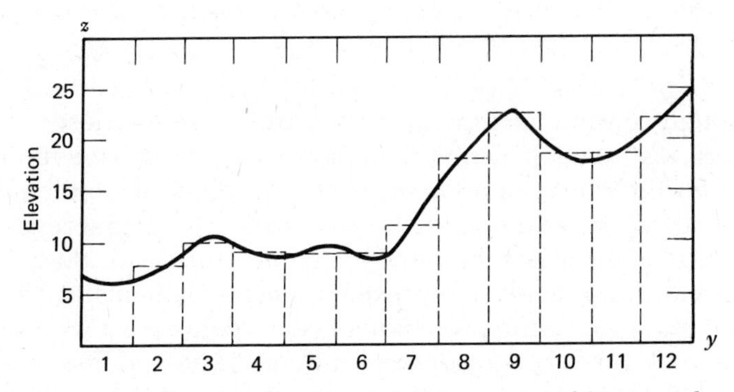

Figure 2-6 Representation of profile (Figure 2-5) by a series of discrete columns. The height of each column can be specified by a numerical value in an array.

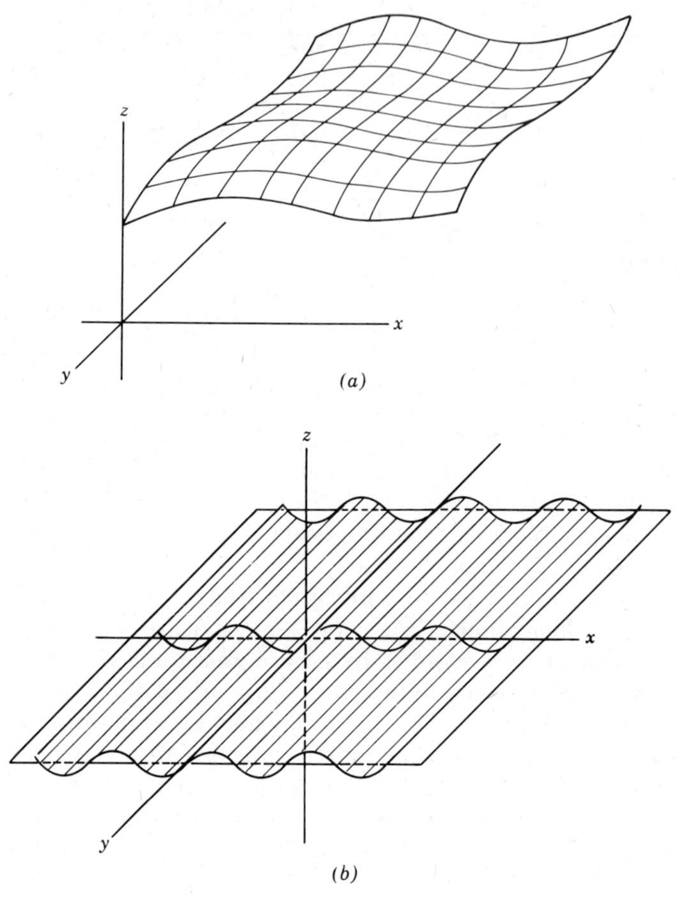

Figure 2-7 Examples of surfaces that are represented by continuous functions: (*a*) surface described by third-degree polynomial; (*b*) corrugated surface described by harmonic function.

Although continuous functions are widely used to describe spatial relationships in many computing applications, they have limited usefulness for the representation of space in geological simulation applications. There are two reasons for this. First, the spatial relationships of many geological features are so complex that it would be difficult or practically impossible to represent them by mathematical functions. Second, many geological relationships are concerned with qualities, such as rock types or presence of different kinds of fluids. Qualities are readily represented in a discrete system, but they are difficult to represent with conventional functional relationships. Finally, most

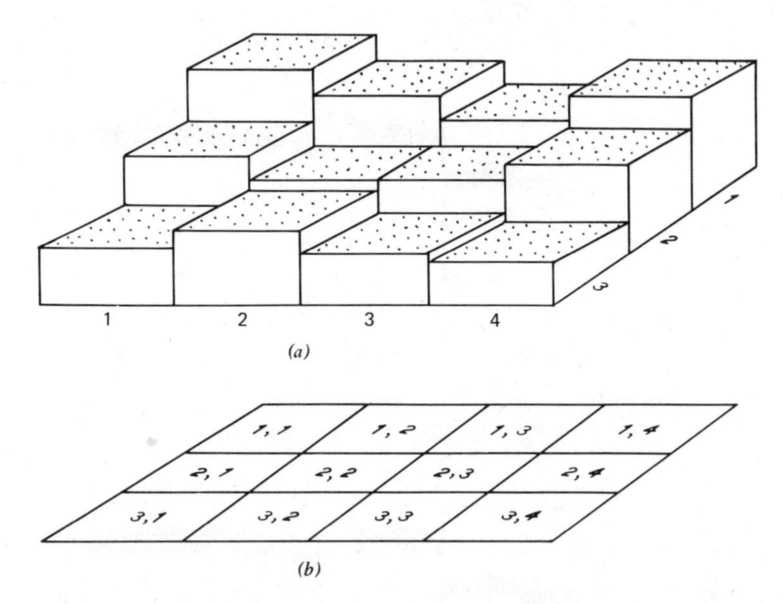

Figure 2-8 Meshwork of square cross-sectioned prisms (*a*) whose heights provide approximation to irregular surface, and two-dimensional array (*b*) whose elements represent heights of prisms. Column and row indices of array also serve as geographic coordinates.

dynamic geological simulation models need to incorporate materials-accounting systems. The nature of accounting, which involves crediting and debiting of discrete quantities of materials, inherently requires discrete representation of space. Figure 2-9, which deals with the representation of three-dimensional spatial relationships, also portrays the manner in which qualities can be represented. When qualities are represented by numbers, they can be stored and manipulated in computer arrays. Most of the simulation examples in this book employ discrete representation of space. We should point out, however, that there are many inherent advantages in continuous functions if they can be employed and that simulation-model builders should consider their use if feasible.

The choice of the number of dimensions and the fineness of the meshwork are critical factors in the construction of digital simulation models. If space is to be represented, the number of dimensions is either 2 or 3. Compared with two-dimensional representation, there is a major increase in the number of cells required for full-scale representation of three-dimensional space in discrete form. Not only does the number of cells increase sharply in going from two to three

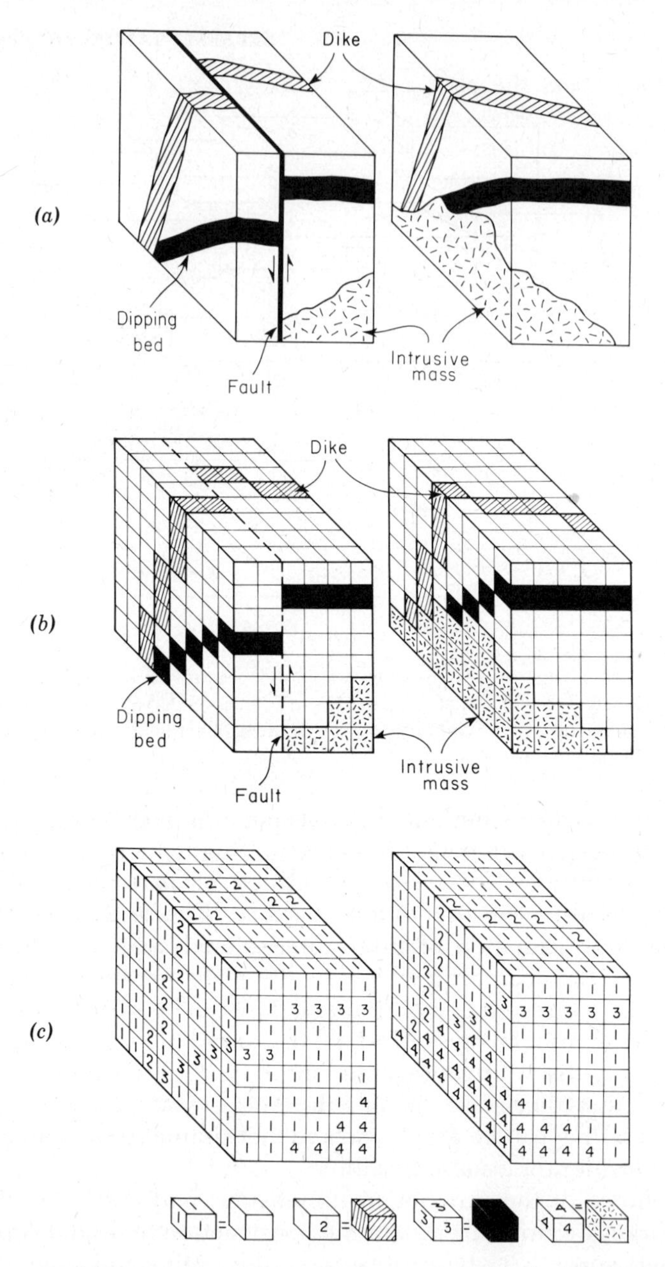

Figure 2-9 Method of representing qualitative relationships in three-dimensional space. (*a*) Block diagram (split apart to reveal details) employing conventional graphic symbols. (*b*) Block diagram in which space has been compartmented into rectangular cells. Conventional graphic symbols are still employed. (*c*) Substitution of numbers, which form three-dimensional array, for conventional graphic symbols. From Harbaugh (1966).

dimensions, but the complexity of interrelationships also increases. In a two-dimensional meshwork each interior cell touches eight cells (four in contact at edges, four in contact at corners). In a three-dimensional meshwork, however, each interior cell is in contact with 26 other cells (6 are in contact along surfaces, 12 along edges, and 8 at corners). If the transport of material from cell to cell is incorporated into the model, the number of arithmetic and logic operations per cell may be significantly higher in a three-dimensional model than in a two-dimensional one. Thus the combination of more cells and increased complexity per cell tends to cause three-dimensional models to be several orders of magnitude more complex than two-dimensional ones.

We should note, however, that a two-dimensional array can store information about bodies that exist in three-dimensional space. The two-dimensional array of Figure 2-8b stores information that can be regarded as representing either a surface (the tops of the prisms) or a solid body (the aggregation of prisms themselves), both of which occupy three-dimensional space.

As we have pointed out, arrays provide effective means of storing information about the properties of materials. A simulation program representing a marine sedimentation model, for example, might contain quantitative information on porosity, permeability, sorting, and the proportions of sand, silt, and mud. In addition, it might contain qualitative information on sediment color and the presence or absence of Foraminifera. This information could be stored in various ways. A simple and direct one would be to create an array for each property (Table 2-7). The various arrays could then be thought of as coinciding in space, each occupying the same area or volume as the others.

Types of Grid Frameworks

The grid systems that we have described thus far represent frameworks that are fixed in space. Materials, energy, or information may move through such a framework, but the framework itself is rigid. An alternative is to employ a distortable framework (Figure 2-10). Displacement of materials can then be represented by movement of the grid framework itself, rather than by transporting material from cell to cell.

There are important advantages to each of the two grid systems. The advantages of the rigid-frame system lie principally in the ease with which it can be represented. The indices of arrays that describe the space also serve as its Cartesian coordinates. If the coordinate

TABLE 2-7 Example of Representation of Sediment Properties in Three-Dimensional Space in FORTRAN Programs

Property Represented	*FORTRAN Array*[a]
Porosity in percent	PORE(I,J,K)
Permeability in millidarcys	PERM(I,J,K)
Sorting	SORT(I,J,K)
Sand in percent	SAND(I,J,K)
Silt in percent	SILT(I,J,K)
Mud in percent	MUD(I,J,K)
Color (red, gray, green, black)	COLOR(I,J,K)
Foraminifera (presence or absence)	FORAM(I,J,K)

[a] I, J, and K are identifiers representing array indices in the three dimensions.

axes are assumed to be rectilinear, the array-element indices can be regarded as automatically defining a rigid framework. The only remaining assumption is that of a scale for each dimension. Of course the same scale can be used for all dimensions, but it may be convenient to employ a vertical scale that differs from the horizontal scale.

The advantage of a flexible framework lies principally in increased versatility in representing spatial relationships. In spite of its flexibility, the coordinate system is retained, permitting ready representation of various properties by the use of multiple arrays. The coordinate system, however, can be deformed, additional information being

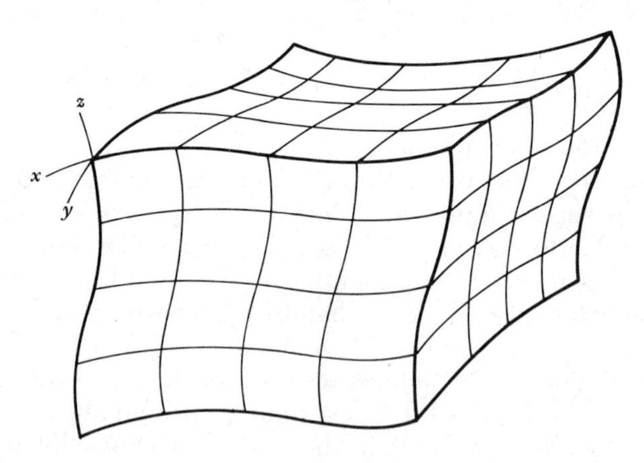

Figure 2-10 Three-dimensional flexible grid system.

needed to represent the deformation. A major disadvantage is that the framework can undergo only a limited amount of deformation before the shapes of the individual cells in the framework are so distorted as to be difficult to portray graphically. Furthermore special steps may be needed if "space is conserved" – that is, if the overall area or volume of the grid undergoes neither expansion nor contraction as it is deformed.

A third alternative is to employ a hybrid grid system in which two of the dimensions of the grid framework are regarded as rigid and the third is capable of undergoing deformation. This type is especially useful in many geological simulation models – for example, in the carbonate-ecology model described in Chapter 9, where the two geographic dimensions are represented by a rigid framework and the third dimension can undergo distortion.

MATERIALS ACCOUNTING

The conservation of matter is one of the fundamental principles of science; matter can neither be created nor destroyed, except in nuclear processes. As a consequence dynamic simulation models that purport to be realistic should adhere to the provision that they neither symbolically "create" nor "destroy" matter. This requirement can be met by employing accounting methods that contain information on all materials that enter, leave, or remain within the boundaries of the system represented by the model. The net difference between materials entering and those leaving a system during a simulation run should be equal to the overall gain or loss of materials. In other words,

$$\text{accumulation} = \text{input} - \text{output}$$

This concept is extremely simple and is widely applied, commonly being termed *mass balance* or *materials balance.* Furthermore the requirements are such that not only must the gross amounts of materials moving into and through the system be accounted for but each different type of material must be subject to the same bookkeeping rigor. Bowen and Inman (1966), Horn and Adams (1966), and Neumann and Land (1968) provide examples of geological applications of materials-accounting methods. Applications in other fields include input–output analysis (Leontief, 1966), as developed by economists, which provides an approach that might be adapted for accounting of energy relationships in geology.

Although the techniques used in materials accounting are simple in principle, they tend to make large demands on computer-storage requirements. Consider, for example, the problem of storing information on sedimentary materials in a modest sedimentation model that employs a geographic grid containing 50×50 cells. If there are 5 different types of sediment to be accounted for over an interval of 10 time increments, the total number of items to be stored in the accounting system after the 10 time increments have elapsed is $50 \times 50 \times 5 \times 10 = 125{,}000$. This is a staggering volume of information for a small simulation model. Furthermore this figure is merely the amount of information to be stored at the end of the run. It does not include the large number of arithmetic and logic operations involved in the accounting operations themselves.

The delta-deposition model of Bonham-Carter and Sutherland (1968) provides a good example of an accounting system and is more fully discussed in Chapter 9. It employs a geographic "accounting grid" to receive sediment deposited at the mouth of the river and provides for sediment of various specified grain sizes. As the river pours out into the sea, the settling trajectories of sediment particles are calculated (Figure 2-11). The particles are assigned to particular cells in the accounting grid, depending on where they come to rest.

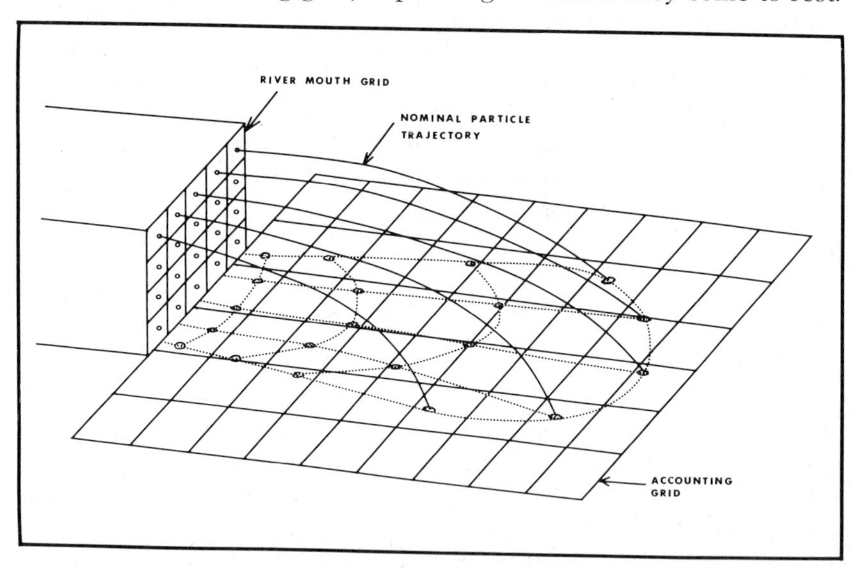

Figure 2-11 Use of accounting grid in delta-sedimentation model. River flowing in rectangular channel at left pours into sea (upper edge of river-mouth grid coincides with sea level). Distribution of sediment-size fractions in river channel at its mouth is accounted for by river-mouth accounting grid. During each time increment the amount of each sediment-size fraction leaving the river mouth is equal to the amount deposited on the sea floor. From Bonham-Carter and Sutherland (1968).

During each time increment the amount of sediment of each grain size that enters each cell is tabulated.

The delta model employs a four-dimensional FORTRAN array SED(I,J,L,N) for accounting purposes. The indices I and J pertain to the rows and columns in the horizontal plane of the geographic grid. Index L denotes the particular sediment-size fraction, and N is an index of the time increment. Array SED is thus capable of containing all essential information about the sediment that has been deposited. The numerical values of elements in the array describe the amount of sediment of each size deposited in thickness units per cell. The array is indexed according to location, type of sediment, and time increment. The time increment also serves as a stratigraphic index, because the sediment can be regarded as deposited in consecutive layers, each layer corresponding to a time increment.

It will be instructive to observe the role of array SED. Initially let us specify a particular time increment, say N = 1, so that part of the array, SED(I,J,L,1), can be regarded as a three-dimensional array. Furthermore, let us specify a certain sediment-size fraction, say sand, for which L = 1. Then the part of the array that is represented by SED(I,J,1,1) can be regarded as a two-dimensional array, with I and J as variables. We can readily represent a two-dimensional array graphically. Figure 2-12 portrays a pair of two-dimensional arrays,

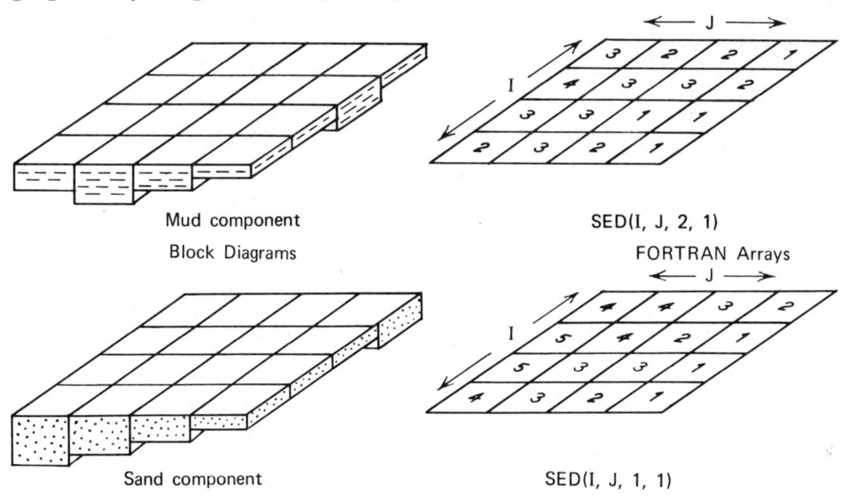

Figure 2-12 Diagrams illustrating method of storing sediment-size fractions within four-dimensional array SED for given time increment. I and J represent geographic-coordinate indices. The two arrays shown can be regarded as two-dimensional slices of a four-dimensional array. Thicknesses of prisms in block diagrams correspond to numerical values assigned to elements in arrays.

one to contain the content of sand, SED(I,J,1,1) and the other to contain the content of mud (L = 2), SED(I,J,2,1). If sand and mud are the only sediment-size fractions treated in the system, the total amount of sediment deposited (or eroded) from each cell during each time increment is the algebraic sum of the amount in the corresponding cells in the two arrays. A negative sum would imply that erosion, rather than deposition, had occurred.

Information contained in specific elements in an array can be combined to yield still other kinds of information. For example, the information pertaining to the amounts of sand and mud deposited per time increment per cell in Figure 2-12 can be combined to yield the total thickness of sediment deposited per time increment per cell (Figure 2-13). In turn the aggregate thickness for any specified number of time increments can be obtained by summation. Finally, thickness information can be coupled with other information to compute changes in the depth of water (Figure 2-14).

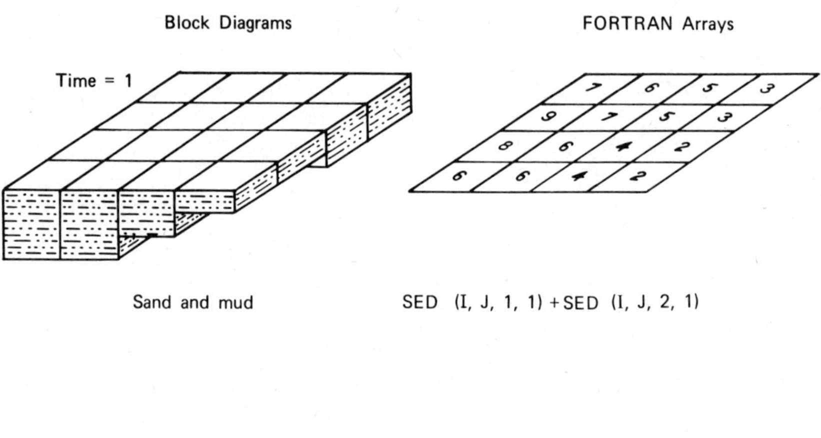

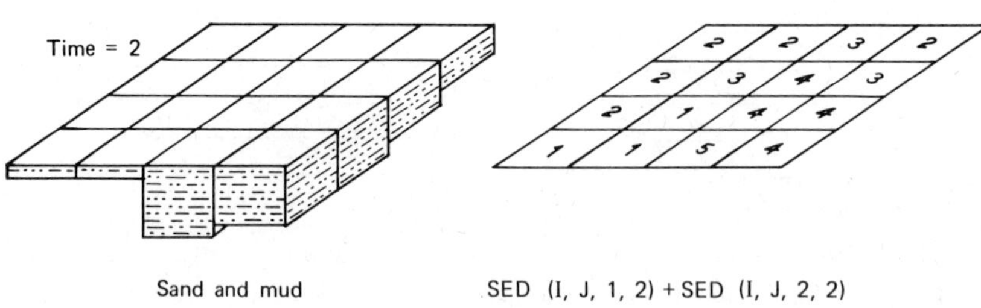

Figure 2-13 Block diagrams and arrays to illustrate the effect of combining thicknesses of mud and of sand (Figure 2-12) to obtain total thickness for time step. In turn thicknesses for two or more time steps can be combined to yield aggregate thickness.

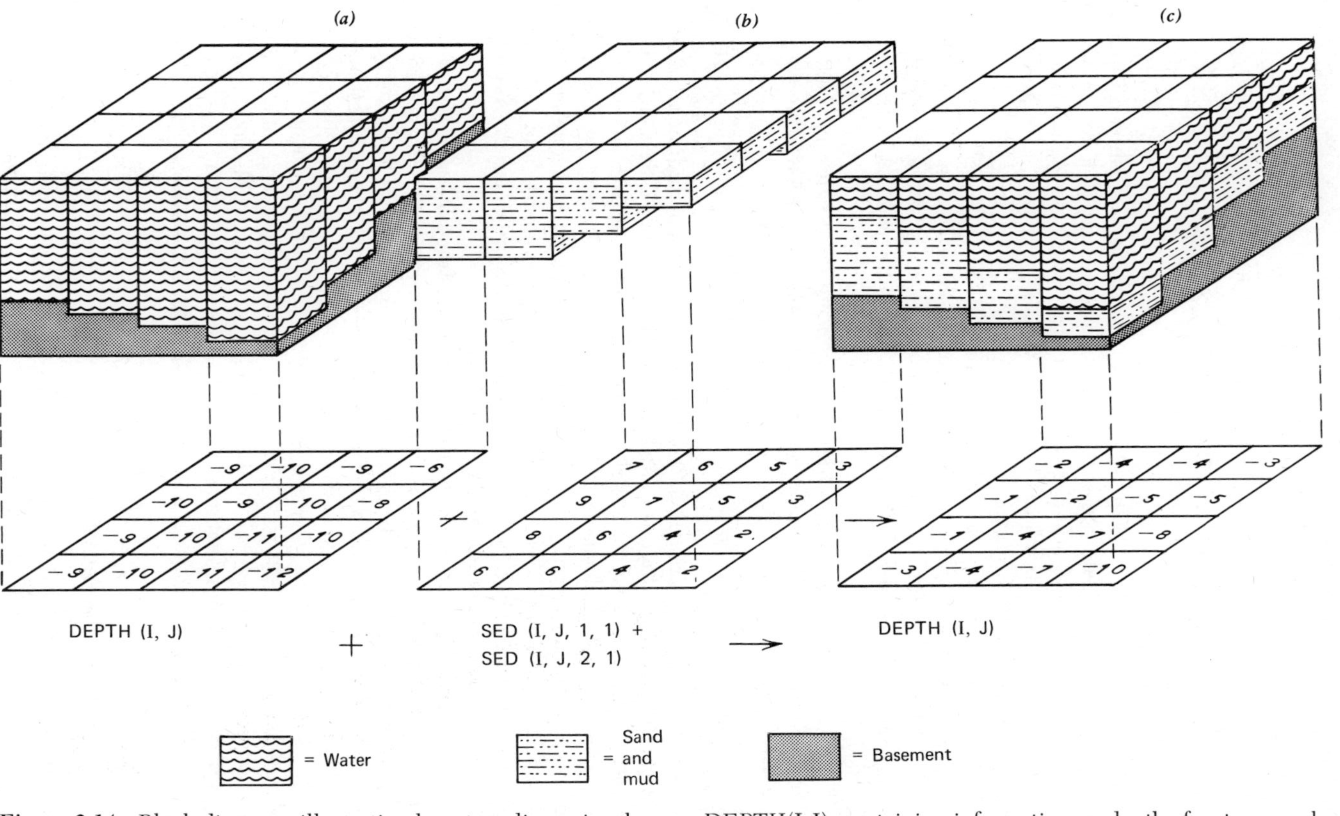

Figure 2-14 Block diagrams illustrating how two-dimensional array, DEPTH(I,J), containing information on depth of water, may be coupled with four-dimensional array SED(I,J,L,N) to compute new depth. Previous depth (*a*) is added algebraically to sediment deposited (*b*) to yield depth in next time increment (*c*).

Problems

Problem 2-1

Employing the FORTRAN program for the geology-student model, perform the following simulation experiments:

a. Use the input data of Table 2-4 with the exception of freshmen input, NFRESH, which can be altered so that it is characterized by a normal distribution with mean of 100 and a standard deviation of 20 (see Chapter 3 for generation of normally distributed random variables). Draw a frequency-distribution curve of JBLESS.

b. What constant values of NFRESH and DEMAND will minimize JBLESS? This is an optimization problem that can be solved in a number of ways (see Chapter 8). For this problem explore the consequences of different values of NFRESH and DEMAND, approaching the problem by trial and error.

c. If NFRESH is increased suddenly from a constant value of 1000 to 1500 and assuming that DEMAND remains constant at 60, what is the effect on CMPLET and JBLESS?

Problem 2-2

Assume that the depositional history and distribution of sediment in an area is stored in a four-dimensional FORTRAN array called SED(I,J,K,L). Each entry in SED represents a thickness of sediment, measured in feet. The indices I, J refer to the location of a square cell (50×50 feet) by row and column number in a map grid. The index K refers to the time-increment number in a time scale involving discrete intervals of fixed length, and L refers to the sediment size, where size is represented by discrete classes. Assume that $L = 1$ = sand, $L = 2$ = silt, and $L = 3$ = mud.

For a grid with nine square cells, three grain-size fractions, and with two time increments write down the 54 numbers that represent the distribution of sediment in a hypothetical region. Then write and run a FORTRAN program to (a) calculate the average sediment thickness in each cell (I,J); (b) calculate the ratio of sand to mud in each cell (I,J); (c) print out a map displaying the contents of all the cells where $K = 1$, $L = 2$; (d) print out a stratigraphic slice section, fixing $J = 1$, $L = 3$, and denote the vertical exaggeration; and (f) assume a bulk density of the sediment (density of grains plus pore space) and calculate the total sediment mass for each value of K.

Annotated Bibliography

Bonham-Carter, G. F., and Harbaugh, J. W., 1968, "Simulation of Geologic Systems: An Overview," Computer Contribution 22, Kansas Geological Survey, pp. 3–10.

Discusses goals of simulation in relation to the type of system under study.

Bonham-Carter, G. F., and Sutherland, A. J., 1968, "Mathematical Model and FORTRAN IV Program for Computer Simulation of Deltaic Sedimentation," Computer Contribution 24, Kansas Geological Survey, 56 pp.

Uses a fixed grid for accounting purposes.

Bowen, A. J., and Inman, D. L., 1966, *Budget of Littoral Sands in the Vicinity of Point Arguello, California*, U.S. Army Coastal Engineering Research Center, Technical Memorandum No. 19, 41 pp.

Systems principles applied to a field study of a section of California coast. A sand budget is developed on the basis of the transport rates of all significant littoral processes. Each process is examined to assess the sedimentary contributions (credits) and losses (debits). To balance sediment transports region is subdivided into five cells, and a quantitative transport rate is determined for each process in each cell.

Chorafas, D. N., 1965, *Systems and Simulation*, Academic Press, New York, 503 pp.

The first three chapters in this book provide a good discussion of the philosophy of simulation and classification of models.

Chorley, R. J., and Haggett, P., eds., 1967, *Models in Geography*, Methuen, London (Barnes and Noble, New York), 816 pp.

An extensive collection of articles that are very useful as a source of references, particularly on models in geomorphology and on the general use of models in science.

Cloos, E., 1968, "Experimental Analysis of Gulf Coast Fracture Patterns," *Bulletin of the American Association of Petroleum Geologists*, v. 52, pp. 420–444.

Describes use of material scale models constructed of soft clay for physical simulation of major geological structures.

Cole, J. P., and King, C. A. M., 1968, *Quantitative Geography—Techniques and Theories in Geography*, John Wiley and Sons, New York, 692 pp.

Chapter 2 contains a useful discussion of models in geography, well documented with diagrams and examples.

Griffiths, W. C., 1967, *Scientific Method in the Analysis of Sediments*, McGraw-Hill, New York, 508 pp.

Chapter 1 contains a clear exposition of the scientific method. Chapters 13 through 21 pertain to the use of statistical models in sedimentology.

Griffiths, J. C., and Drew, L. J., 1966, "Grid Spacing and Success Ratios in Exploration for Natural Resources," in *Proceedings of Symposium and Short Course on Computers and Operations Research in Mineral Industries*, April 1966, Pennsylvania State University, Mineral Industries

Experimental Station Special Publication 2-65, v. 1, pp. Q1–Q24.

A good example of a static-probabilistic model, employing Monte Carlo sampling techniques. This model pertains to a study of a largely man-made sampling system for mineral exploration.

Harbaugh, J. W., 1966, "Mathematical Simulation of Marine Sedimentation with IBM 7090/7094 Computers," Computer Contribution 1, Kansas Geological Survey, 52 pp.

A complex dynamic-probabilistic model containing interdependent probabilistic and deterministic components.

Harbaugh, J. W., and Merriam, D. F., 1968, *Computer Applications in Stratigraphic Analysis*, John Wiley and Sons, New York, 282 pp.

Chapter 8 includes a discussion of the objectives of simulation.

Hjulstrom, F., 1935, "Studies in the Morphological Activity of Rivers as Illustrated by the River Fyris," *Bulletin of the Geological Institute of the University of Uppsala*, v. 25, pp. 221–528.

Horn, M. K., and Adams, J. A. S., 1966, "Computer-Derived Geochemical Balances and Element Abundances," *Geochimica et Cosmochimica Acta*, v. 30, pp. 279–297.

An instructive application of mass-balance principles.

Hull, T. E., 1966, *Introduction to Computing*, Prentice-Hall, Englewood Cliffs, N. J., 212 pp.

A good introductory book on programming and computers.

IBM, 1968, *System 360 Continuous System Modeling (360-CX-16X), Users Manual*, Report No. H20-0367-2, 61 pp.

Program package for modeling systems described by differential equations.

Inman, D. L., and Quinn, W. H., 1951, "Currents in the Surf Zone," in *Proceedings of the Second Conference on Coastal Engineering*, Council on Wave Research, Richmond, Calif., pp. 24–36.

Formula for predicting longshore-current velocity is developed on the basis of theoretical analysis, thus providing an example of a theoretical relationship.

Isaacs, K. N., 1967, "The Simulation of Magnetic and Gravity Profiles by Digital Computer," *Geophysics*, v. 31, 773–778.

Example of a static-deterministic simulation model, used for exploring possible shapes, depths, and densities of buried rock and mineral bodies given magnetic and gravity profiles.

Kendall, M. G., 1968, "An Introduction to Model Building and Its Problems," in *Mathematical Model Building in Economics and Industry*, ed., M. G. Kendall, Hafner, New York, pp. 1–14.

A short but interesting discussion of model building, covering the classification of models, problems of model building, and relations among variables.

Krumbein, W. C., 1964, *A Geological Process-Response Model for Analysis of Beach Phenomena,* Northwestern University Geology Department, Technical Report No. 8, pp. 1–15.

A conceptual model, embodying the philosophy of systems and simulation in written and graphic form.

Krumbein, W. C., and Graybill, F. A., 1965, *An Introduction to Statistical Models in Geology,* McGraw-Hill, New York, 475 pp.

Chapter 2 contains an important discussion of models in geology, with a particularly relevant section on process-response models.

Leontief, W., 1966, *Input–Output Economics,* Oxford University Press, New York, 257 pp.

Input–output analysis is a technique developed by economists for describing the movement of goods and services between different sectors of the economy.

McCracken, D. D., 1965, *A Guide to FORTRAN IV Programming,* John Wiley and Sons, New York, 151 pp.

A good introduction to FORTRAN IV.

Neumann, C. A., and Land, L. S., 1968, *Algal Production and Lime Mud Deposition in the Bight of Abaco: A Budget,* paper presented at the Geological Society of America annual meeting, Mexico City, November 1968 (abstract).

Budget principles employed to show that algal productivity can account for the entire mass of aragonite mud present in the Bight of Abaco. Interesting application of materials accounting to problems in carbonate sedimentation.

Peikert, E. W., 1969, "Interactive Graphics—a New Dimension," *Computing Report,* v. 5, pp. 4–7.

Describes a time-sharing computing system which uses a cathode-ray tube for input and output of geologic maps and cross sections.

Putnam, J. A., Munk, W. H., and Traylor, M. A., 1949, "The Prediction of Longshore Currents," *Transactions of the American Geophysical Union,* v. 30, pp. 337–345.

Empirical relationship for predicting longshore-current velocities is given, in contrast to the theoretical relationship derived by Inman and Quinn (1951).

Raup, D. M., 1965, "Theoretical Morphology of the Coiled Shell," *Science,* v. 147, pp. 1294–1295.

Example of use of analog computer for geological simulation.

Rosenblueth, A., and Wiener, N., 1945, "The Role of Models in Science," *Philosophy of Science*, v. 12, pp. 316–321.

A classic paper on this subject.

Teichroew, D. and Lubin, J. F., 1966, "Computer Simulation—Discussion of the Technique and Comparison of Languages," *Communications of the Association for Computing Machinery*, v.9, No. 10, pp. 723–741.

Watt, K. E. F., 1964, "Computers and the Evaluation of Resource Management Strategies," *American Scientist*, v. 52, pp. 408–418.

Readable paper dealing with the general concepts of representing ecological systems with dynamic computer-simulation models.

Whitten, E. H. T., 1964, "Process-Response Models in Geology," *Bulletin of the Geological Society of America*, v. 75, pp. 455–464.

Describes graphic and written models that embody systems and simulation philosophy.

CHAPTER 3

Generating Random Variables

Many dynamic simulation models require variables that are supplied from an external source to form the exogenous variables. These inputs may be either deterministic or stochastic (Table 3-1), or they may combine both properties. A deterministic variable, as its name implies, is one whose behavior is completely predetermined. Deterministic variables often may be represented by simple mathematical functions; for example, an oscillatory waveform may be represented by a simple sine function. There is no element of chance in a purely deterministic function. The state of the variable at any subsequent point in time or space is completely predictable.

This chapter is concerned with methods of producing stochastic variables. Many natural processes can be regarded as stochastic in that they exhibit random behavior. The word "stochastic" is derived from Greek, meaning literally "to guess" or "to project with uncertain aim." A stochastic variable, therefore, is one that varies in an uncertain manner.

The behavior of a stochastic variable, however, can be described statistically. For example, if a random variable is described by a Gaussian distribution, its statistical behavior is characterized by its mean and standard deviation. Besides the Gaussian, or normal, distribution, there are other types of theoretical frequency distributions, each characterized by a set of parameters. Generally speaking,

TABLE 3-1 General Classification of Types of Exogenous Inputs to Dynamic Simulation Models, with Hypothetical Variables That Could Be Used as Examples

General Category	*Form*	*Example*
Stochastic	Discrete empirical distribution	Number of eruptions per century of active volcano
	Discrete theoretical distribution	Poisson frequency distribution of sizes of marine invertebrate skeletons in a "death" population
	Continuous theoretical distribution	Lognormal frequency distribution of bed thicknesses in a stratigraphic section
	Markov transition probability matrix	Succession of fixed bottom-dwelling marine organisms through time in a local area on the sea floor
Deterministic	Constant	Average salinity of the ocean
	Deterministic function of time	Variations in length of day throughout the year

if the statistical properties of a real-world variable are known, it is possible to produce an artificial variable whose statistical behavior parallels that of the actual variable. Thus, although the behavior of a random variable for a single or a small number of experiments or observations can be predicted with only a small measure of confidence, the overall behavior in a large number of experiments can be predicted with a high degree of certainty.

In many simulation applications it is desirable to create variables that combine both deterministic and random components. An example (Figure 3-1) might be a variable whose gross behavior is that of a sine wave, on which a lesser proportion of random "noise" is superimposed. Such a variable could be readily synthesized with a digital computer by periodically evaluating a sine function whose argument could be time or distance and algebraically adding to it a sequence of random numbers of prescribed statistical properties. Methods of generating and combining random numbers are vital for simulating stochastic components in dynamic simulation models.

Generating values of a deterministic function generally is a simple matter. Generating sequences of numbers that satisfy tests for

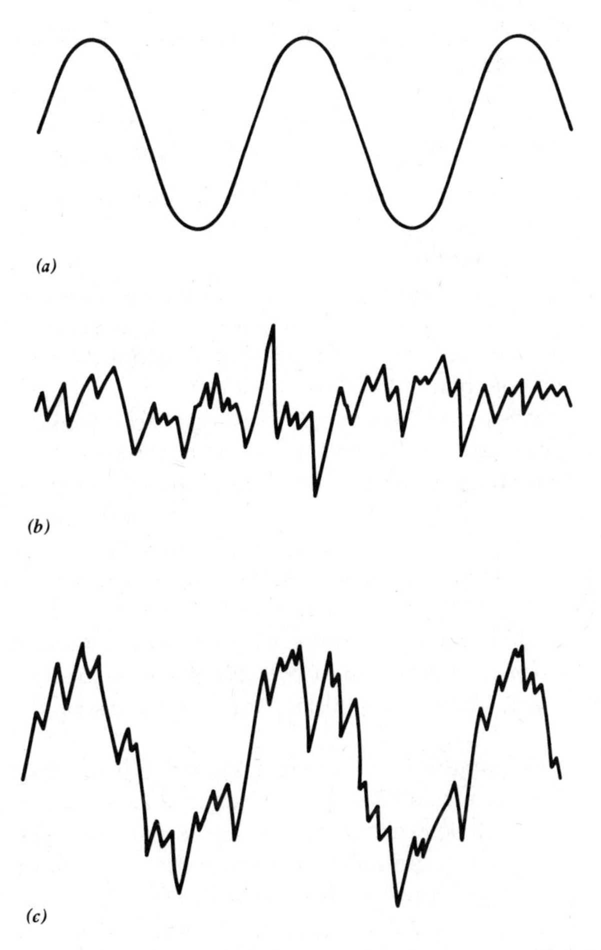

Figure 3-1 Diagram to illustrate combination of variables to create "hybrid" variable: (*a*) purely deterministic variable (sine wave); (*b*) random variable ("noise"); (*c*) algebraic addition, combining sine wave and random noise.

randomness and possess desirable statistical properties, however, is considerably more complex. There are many problems connected with the generation of random numbers. In fact the subject is so large that random-number generation has grown to be a respectable branch of computer science.

Material covered in this chapter is concerned with (a) methods of generating random numbers that form uniform (rectangular) frequency distributions, (b) transformation of uniform distributions to produce a variety of prescribed frequency distributions, and

(c) illustrating the use of numbers thus generated with a simple simulation model.

METHODS OF GENERATING RANDOM NUMBERS

Three principal classes of methods for generating random numbers have been used: (a) manual or mechanical methods, (b) electrical-analog-computer methods, and (c) digital-computer methods. Of these three, only the first two classes yield numbers that are truly random. The latter class yields sequences of numbers that do not completely conform with the definition of a random sequence.

Manual and mechanical methods include dice rolling, card shuffling, and roulette-wheel spinning. Unfortunately these methods are much too slow and cumbersome to be used directly with digital computers. Electrical-analog computers also may be used to generate random numbers. By converting electronic noise (random components in the behavior of an electric current) to numerical values, a sequence of random numbers is readily obtained. Furthermore electrical-analog methods avoid the slowness and cumbersome nature of mechanical methods. However, since most digital computers make no provision for direct electrical-analog input, analog methods are not generally suitable for producing random numbers for digital simulation models.

Both electrical-analog and mechanical methods produce sequences of numbers that are truly random. Thus any two long sequences of random numbers will never be the same. Surprisingly, this is a disadvantage for many simulation experiments. It is often desirable, particularly in developing or testing simulation models, to use the same sequence of "random" numbers repeatedly, so that variations due to different sequences are avoided. Digital methods of generating random numbers are superior in this respect, and it is to these methods that we now turn.

DIGITAL GENERATION OF PSEUDORANDOM NUMBERS

Digital methods can produce only sequences of "random" numbers that are repeatable or reproducible. In other words, given the same starting value, any two sequences generated by the same method will be identical. "Random" numbers produced in this way are called pseudorandom, because the mere aspect of repeatability implies the possibility of predicting the sequence and therefore proof of nonrandomness. The term "pseudorandom number" is difficult to define. Lehmer (1951) has defined it as "a vague notion embodying the idea of a sequence in which each term is unpredictable to the uninitiated and whose digits pass a certain number of tests traditional with statisticians and depending somewhat on the use to which the sequence is to be put."

Mid-Square Technique

Historically, the first method of generating pseudorandom numbers with a digital computer was the mid-square method, proposed by Von Neumann in 1946. In the mid-square method each number in the sequence is obtained by selecting the mid-digits of the square of the previous number in the sequence. For example, assume that the starting number is x_0. The square of this number is x_0^2, and x_1 is taken as the mid-digits of x_0^2. For example, if $x_0 = 12$ and two digits are selected for the generation of each subsequent number, the sequence that results is as follows:

Sequence of Numbers	*Squared Values*
$x_0 = 12$	$x_0^2 = 144$
$x_1 = 14$	$x_1^2 = 196$
$x_2 = 19$	$x_2^2 = 361$
$x_3 = 36$	$x_3^2 = 1296$
$x_4 = 29$	$x_4^2 = 841$
$x_5 = 84$	$x_5^2 = 7056$
$x_6 = 05$	$x_6^2 = 25$
$x_7 = 25$	

Any such sequence must ultimately repeat some previous value, and from that point on the sequence is repeated, or cyclic. Under some circumstances a zero value may be produced, at which point the sequence must be halted, because only zero values will ensue subsequently. Therefore the method has long been abandoned.

Congruential Methods

In light of the shortcomings of the mid-square technique, it is expedient to draw up criteria for acceptable methods of generating sequences of pseudorandom numbers. The numbers should be (a) uniformly distributed (i.e., form a rectangular frequency distribution in which each number has an equal probability of occurrence) within the specified range, (b) statistically independent, each number being completely independent of previous numbers in the sequence, (c) nonrepeating within a desired span, and yet (d) reproducible within this span, in addition (e) the method of generating the numbers should be fast, because some simulation methods require millions of numbers, and (f) ideally it should require a minimum amount of computer memory.

Congruential methods for generating pseudorandom numbers were developed to satisfy these criteria. These methods are completely deterministic in the sense that each number in a sequence of numbers is uniquely determined by the arithmetic processes that are involved. In spite of the fact that the processes are not random, we may regard the generated sequence of numbers as being random from a pragmatic point of view. If the sequences consistently pass statistical tests of randomness, and the numbers can be shown to be uniformly distributed, the numbers can be used as effectively as if they were truly random.

Congruential methods are based on the general relationship

$$x_i = ax_{i-1} + c(\text{modulo } m) \tag{3-1}$$

where x_i = a pseudorandom integer
i = subscript of successive pseudorandom integers produced, $i-1$ being the immediately preceding integer
m = a large integer used as the modulus
a and c = integer constants used to govern the relationship, in company with m

Repeated evaluation of this expression results in the production of a sequence of integers whose possible range of values is from zero to $m-1$. The modulus notation indicates that x_i is the remainder when $(ax_{i-1}+c)$ has been divided a whole number of times by m. If the constants are suitably chosen, the sequence of integers generated by this method has a uniform distribution within the range 0 to $m-1$.

Such a sequence will repeat itself at most after m steps and in some cases much sooner, depending on the values of m, a, and c. For example, if $m=16$, $a=3$, and $c=1$, and if the initial value of x is 2, the sequence of x's generated is 2, 7, 6, 3, 10, 15, 14, 11, 2, 7, 6, 3, 10, $\cdots$. The calculations for this case are listed in Table 3-2. As can be observed, the sequence is repeated every eight numbers.

A period length of eight would be far too short for any realistic application. In fact the longest possible period length is generally preferable, other factors remaining unchanged. As we have pointed out, the maximum possible period length is equal to m. If the constant $c=0$, then a special case of the congruential relationship, known as

TABLE 3-2 Calculation of Pseudorandom-Numbers when $x_0 = 2$, $a = 3$, $m = 16$, and $c = 1$[a]

i	$ax_{i-1}+c$	x_i	x_i/m
0	–	2	0.125
1	7	7	0.437
2	22	6	0.375
3	19	3	0.187
4	10	10	0.625
5	31	15	0.937
6	46	14	0.875
7	43	11	0.687
8	34	2	0.125
9	7	7	0.437
10	22	6	0.375
11	19	3	0.187
.	.	.	.
.	.	.	.

[a]Numbers in column labeled x_i are pseudorandom integers; numbers in column labeled x_i/m are pseudorandom numbers in the range 0.0 to 1.0.

the *multiplicative-congruence method,* is defined by the following relationship:

$$x_i = ax_{i-1} \text{ (modulo } m) \tag{3-2}$$

Most computer methods use this multiplicative-congruence relationship.

A number of general rules have been devised for use with the multiplicative method of generating sequences that have a long period length, satisfy tests for randomness, and are suitable for use with digital computers. As summarized by Naylor and others (1966), they are as follows:

1. For binary computers m is chosen as 2^b, where the base 2 is selected because of the binary counting system and b is the number of binary digits in a computer word. For IBM 360 computers, which have a word length of 32 bits (31 bits for accuracy plus 1 for the sign), $m = 2^{31} = 2{,}147{,}483{,}648$ is a possible choice.

2. The value of a is an odd integer of the form $8t \pm 3$, where t is any positive integer. A value of a close to $2^{b/2}$ has been found to

satisfy certain statistical requirements. Thus for $b = 31$, $a = 2^{16} + 3 = 65{,}539$ is a satisfactory choice.

3. The initial value of x, (x_0) is an odd integer smaller than m.

Generally we are not interested in obtaining sequences of integers but instead seek a sequence of decimal fractions that range between 0.0 and 1.0. The sequence of integers can be converted to decimal fractions in this range simply by dividing by m.

A generalized FORTRAN subroutine called SUBROUTINE RANDOM for the generation of pseudorandom numbers by the mixed congruence method as defined in Equation 3-1 is listed in Table 3-3.

TABLE 3-3 FORTRAN Subroutine RANDOM for Generating Pseudorandom Numbers by Mixed Congruential Method

```
SUBROUTINE RANDOM(IX,IA,IC,M,FM,R)
LX=IA*IX+IC
IX=MOD(LX,M)
X=IX
R=X/FM
RETURN
END
```

This subroutine can be used with any digital computer with a FORTRAN compiler. The calling program must contain a statement such as

```
.
.
.
CALL RANDOM(IX,IA,IC,M,FM, R)
.
.
.
```

where IX = initial value of x, a random integer (after the initial call, the value of IX will be a pseudorandom integer generated in the subroutine)
IA = value of constant a, an integer
IC = value of constant c, an integer
M = value of modulus m, an integer
FM = floating-point value of m
R = pseudorandom number in range 0.0 to 1.0

Successive calls of this type will produce a sequence of pseudorandom integers assigned in succession to IX and pseudorandom decimal fractions in the range 0.0 to 1.0 assigned to R. After the initial call, IX need never be redefined, as the value of IX calculated in the previous call is used automatically.

A second subroutine (SUBROUTINE RANDU) is listed in Table 3-4. This is taken from the IBM scientific subroutine package for IBM 360 series computers, which use a word length of 32 binary digits.

TABLE 3-4 FORTRAN Subroutine RANDU for Generating Random Numbers with a Uniform Distribution in the Range 0.0 to 1.0[a]

```
  SUBROUTINE RANDU(IX,IY,R)
  IY=IX*65539
  IF (IY) 5,6,6
5 IY=IY+2147483647+1
6 R=IY
  R=R*.4656613E-9
  RETURN
  END
```

[a]For use with IBM 360 system computers.

RANDU is machine dependent in that it should be used only with computers that have 32-bit words. However, it is more efficient than RANDOM. The arguments for RANDU are

IX = starting integer, which must be an odd number 9 digits long
IY = pseudorandom integer
YFL = pseudorandom number in the range 0.0 to 1.0

After calling RANDU from a program, the new value of IX must be defined as the value of IY before the next call. In other words, the call statement should be followed by a statement in which the value of IY is assigned to IX.

```
.
.
.
CALL RANDU(IX,IY,YFL)
IX=IY
.
.
.
```

Since most computation centers have their own library subroutines for generating random numbers, in practice it may be unnecessary for users to write their own subroutines.

Although congruential methods are simple and have enjoyed wide use, there is growing concern that a single congruential method may not yield sequences of numbers that are sufficiently "random" for some purposes (Poore, 1964; Whittlesey, 1968). An alternative is to mix several simple generating methods together, as has been suggested by Marsaglia and Bray (1968).

The generation of pseudorandom numbers is an exceedingly involved matter, both in terms of mathematical theory and in computing practice. For thorough mathematical treatment we suggest that Jansson (1966) be consulted. Readable, and less detailed, descriptions are provided by Tocher (1963), Naylor and others (1966), Mize and Cox (1968), and Hammersley and Handscomb (1964).

STATISTICAL TESTS OF PSEUDORANDOM NUMBERS

It is clear that pseudorandom numbers are in reality anything but random in that they are completely predetermined. However, if a sequence of pseudorandom numbers can pass statistical tests for randomness, it can be used as if it were a random one. Thus the application of suitable tests for randomness is important. We should point out, however, that it is impossible to ensure absolute randomness because this would require an infinitely long sequence of numbers and an infinite number of tests. Ideally users of pseudorandom numbers should satisfy themselves that the random-number-generation methods they are using are appropriate for their needs. In practice, of course, most persons engaged in digital simulation experiments will rely on others to provide appropriate methods, which, it is to be hoped, have been previously thoroughly checked and found satisfactory.

In making tests of randomness it is possible to test either the individual digits in numbers or the numbers themselves. It can be argued that, if the digits 0 to 9 are truly at random, then the numbers composed of them will necessarily also be random. The tests may be applied to sequences of numbers that are presumed to be either uniformly distributed or, alternatively, to occur in some nonuniform type of distribution, such as the familiar normal, or Gaussian, distribution.

Tests for Uniform Distribution

A popular method for testing the distribution of numbers within a sequence is the *chi-square* test. Suppose that we wish to test a set of numbers whose range is the unit interval 0.0 to 0.1, which in turn can be divided into k equal subintervals. Given these assumptions, the chi-square test may be defined as

$$\chi^2 = \sum_{i=1}^{k} \frac{(m_i - n_i)^2}{n_i} \tag{3-3}$$

where k = number of subintervals
m_i = observed frequency for ith subinterval
n_i = expected frequency for ith subinterval

We wish to see if the pseudorandom numbers are uniformly distributed. Therefore the expected frequency in each subinterval is given by N/k, where N is the total number of pseudorandom numbers in the sequence. If these numbers are uniformly distributed, then the computed values of χ^2 will be close to the theoretical values of χ^2, determined from statistical tables. The null hypothesis that the pseudorandom numbers are uniformly distributed is rejected if the calculated value of χ^2 exceeds the tabulated χ^2 value for $k-1$ degrees of freedom, at some predetermined level of confidence, α. The calculations involved in such a test are illustrated in the example that follows.

For testing purposes we generated a sequence of 200 pseudorandom numbers in the unit interval 0.0 to 1.0 using subroutine RANDU. The unit interval was divided into $k = 10$ subintervals, as illustrated in Figure 3-2. The expected frequency in each subinterval is $200/10 = 20$. Then χ^2 is computed as follows:

$$\begin{aligned}\chi^2 = \tfrac{1}{20}\ &[(14-20)^2 + (18-20)^2 + (26-20)^2 + (18-20)^2 \\ &+ (20-20)^2 + (18-20)^2 + (24-20)^2 + (22-20)^2 + (16-20)^2 \\ &+ (24-20)^2] = 6.8\end{aligned} \tag{3-4}$$

If α has been selected as 0.05, we can consult tables of χ^2 published by Fisher and Yates (1963). For 9 degrees of freedom and the specified value of α, the tables yield a value of 16.92. Since our calculated value of 6.8 is substantially less than the tabulated value, we accept the hypothesis that the distribution from which the numbers were generated is uniform because we do not have sufficient evidence to reject it at the chosen level of significance.

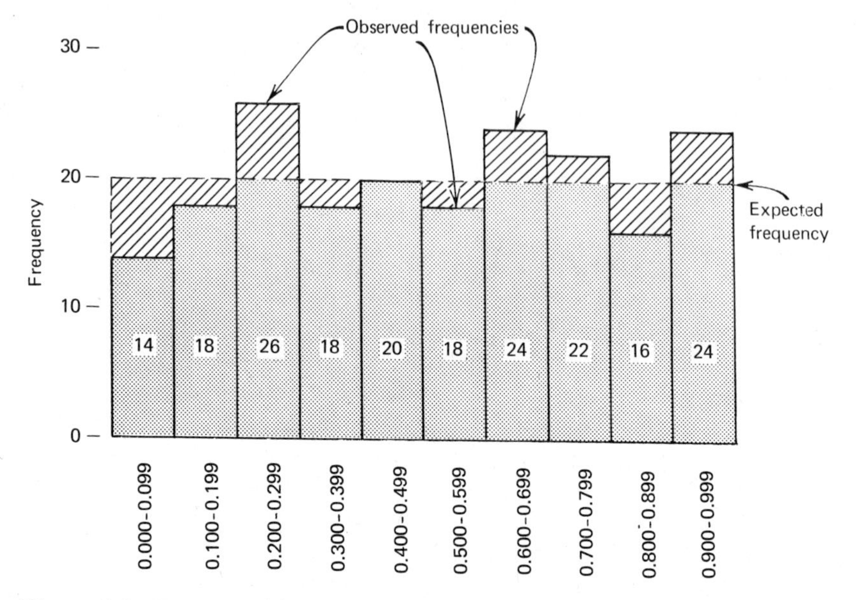

Figure 3-2 Histogram of 10 frequency classes, illustrating values of differences between observed and expected frequencies for a hypothetical uniform distribution of 200 random numbers. Diagonally ruled area denotes size of these differences, reflecting magnitude of χ^2.

Tests for Independence

So far we have considered a test for distribution, but we have not touched on *tests for independence*, which are concerned with whether there is a correlation between numbers in the order of their generation. For example, an ascending sequence of numbers 1, 2, 3, · · ·, 100 would satisfy the tests for a uniform distribution, but the numbers clearly are not at random within this sequence. Consequently tests for independence are needed as well.

The chi-square statistic also may be used in tests of independence, although its application differs somewhat from its use in distribution tests. Let us suppose that we have generated N numbers that we wish to test for randomness in terms of independence. The unit interval 0.0 to 0.1 is again divided into k subintervals. A $k \times k$ matrix is then constructed to record the frequency of numbers that are in the ith subinterval followed by a number in the jth subinterval, i and j being the matrix row and column indices, respectively. These relationships can be visualized by reference to a three-dimensional histogram (Figure 3-3) or stated in the following matrix:

$$
\begin{array}{c|ccccc|}
 & j=1 & 2 & 3 & \cdots & k \\
i=1 & m_{11} & m_{12} & m_{13} & \cdots & m_{1k} \\
2 & m_{21} & m_{22} & m_{23} & \cdots & m_{2k} \\
3 & m_{31} & m_{32} & m_{33} & \cdots & m_{3k} \\
\vdots & \vdots & \vdots & \vdots & & \vdots \\
k & m_{k1} & m_{k2} & m_{k3} & \cdots & m_{kk}
\end{array}
$$

It has been shown by Good (1953) that for large values of N the following statistic is approximately distributed as χ^2 with k^2-k degrees of freedom:

$$
\chi^2 = \sum_{i=1}^{k}\sum_{j=1}^{k}\frac{(m_{ij}-n_{ij})^2}{n_{ij}} - \sum_{i=1}^{k}\frac{\left(\sum_{j=1}^{k} m_{ij} - \sum_{j=1}^{k} n_{ij}\right)^2}{\sum_{j=1}^{k} n_{ij}} \tag{3-5}
$$

where m_{ij} = observed frequency in row i, column j of the above matrix

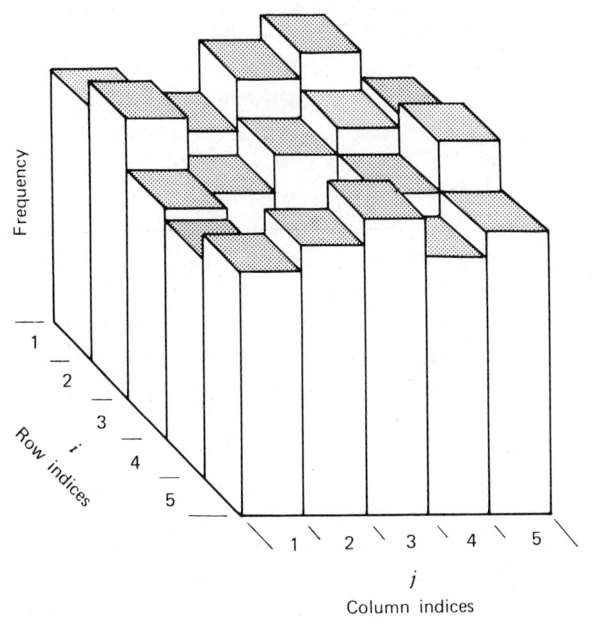

Figure 3-3 Three-dimensional histogram graphically displaying frequencies with which numbers in the ith frequency class are succeeded by those of the jth frequency class, in a 5×5 matrix containing five subintervals or frequency classes.

n_{ij} = expected frequency for numbers in jth subinterval following numbers in the ith subinterval $= (N-1)/k^2$, provided that the subintervals are equal.

The hypothesis of independence is accepted if the calculated value of χ^2 is less than the tabulated value of χ^2 for k^2-k degrees of freedom and some preselected value of α.

The independence of numbers in a sequence can be tested by other methods, which include the lagged-product, the runs, and the gap tests. Naylor and others (1966, pp. 59–62) provide a brief description of these and other tests.

MONTE CARLO SAMPLING FROM KNOWN DISTRIBUTIONS

Our discussion of the methods of generating random numbers and associated statistical tests has thus far been confined to uniform distributions in which any value, within the specified range of the distribution, is equally likely to occur. In nature, however, the frequency distributions of most variables are not uniform. Many actual frequency distributions, if they are continuous, can be graphed as the familiar bell-shaped curve—the Gaussian, or normal, curve. Many geological variables are lognormally distributed; others may be described by the gamma distribution, the exponential distribution, and other distributions discussed by, among others, Krumbein and Graybill (1965), Griffiths (1967), and Vistelius (1967).

In simulation models we often wish to generate random numbers whose distributions are other than uniform. This is usually done by generating uniformly distributed pseudorandom numbers and using these to "draw" random samples from known frequency distributions. This random-sampling process is sometimes known as Monte Carlo simulation and is widely used (e.g., by Hewlett, 1964).

To illustrate this method consider a hypothetical simulation model dealing with the deposition of volcanic ash in the Pleistocene. The model is constructed so that the quantity of ash falling per increment of time (50 years) is an exogenous input to the model. The quantity of ash can be defined as the thickness deposited uniformly over a particular area on the sea floor per increment of time.

In developing the model an initial step would be to gather actual data about the frequency distribution of ash quantities through time. In practice this would be difficult, but we might suppose that we have measured the thicknesses of ash deposited in successive time intervals in a stratigraphic sequence. The availability of close-spaced radiocarbon dates enables us to subdivide the sequence into uniform intervals of time. Let us say that 1000 values are available to us. After we have obtained the ash-fall-thickness values per time unit, we can display the results as a number of discrete intervals in a histogram, which is based on the frequency with which measurements fall into each class of the histogram (Figure 3-4). To simulate the fall of

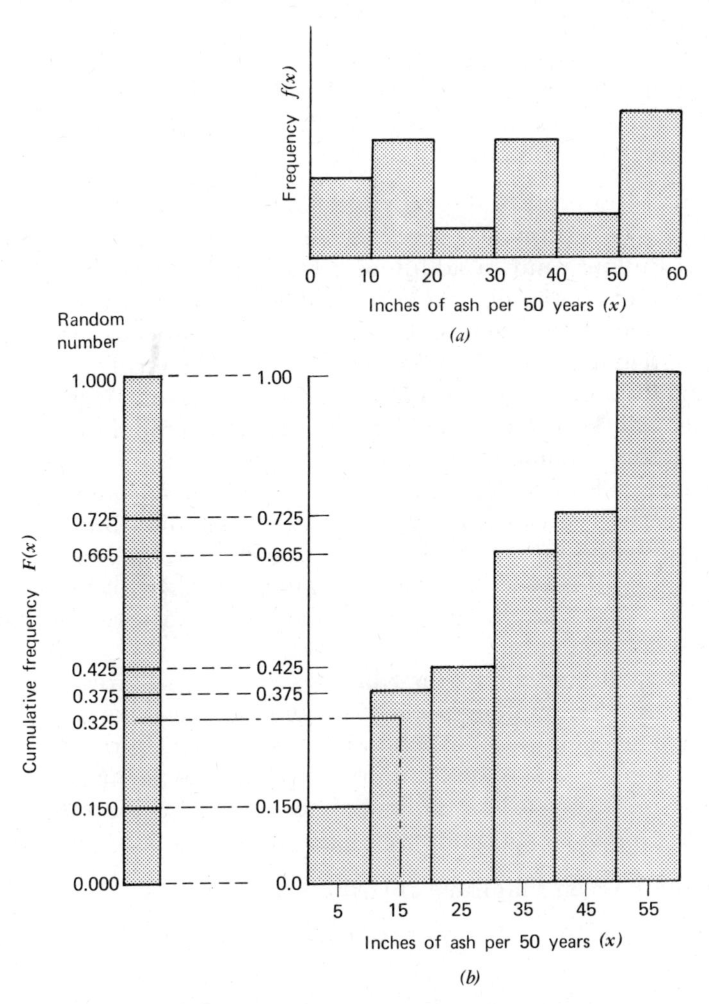

Figure 3-4 Histograms of hypothetical ash-fall frequency in inches per 50 years. (*a*) conventional frequency distribution; (*b*) cumulative frequency distribution based on data in conventional distribution above. Left side of figure illustrates how random numbers with uniform distribution in the range 0.0 to 1.0 are used to sample cumulative distribution.

volcanic ash in a computer model we can sample the discrete, empirical frequency distribution represented by the histogram through the use of a random-number generator that produces a uniform distribution. This sampling can be done as follows:

1. We can draw random samples from the observed, empirical distribution itself.
2. Alternatively, if the frequency distribution can be approximated by a known theoretical distribution, we can draw samples from the theoretical distribution, provided that we can write its function and can estimate its parameters (such as the mean and standard deviation for a normal distribution).

There is no limit to the possible number of theoretical frequency distributions. These can be divided into two general classes, discrete and continuous, and in addition, a few distributions combine both discrete and continuous properties. We do not concern ourselves in this volume with discrete theoretical distributions, although discrete distributions are important in geology (e.g. Griffiths, 1967), and they have been widely used in various nongeological simulation applications. Discrete theoretical distributions include the binomial, the negative binomial, the geometric, the hypergeometric, and the Poisson distribution. Naylor and others (1966) provide an explanation and FORTRAN algorithms for generating these distributions. We are concerned, however, with empirical discrete distributions. These can be based on direct observations or can be dealt with in alternative form—for example, in Markov transition matrices, discussed in Chapter 4.

Theoretical continuous distributions are also of direct concern to us. We shall deal with methods of generating variables with uniform, normal, lognormal, and exponential distributions. Other theoretical distributions that may be of interest include the gamma distribution and the multivariate normal distribution. Naylor and others (1966) provide good explanations of how to simulate these distributions.

Sampling from an Empirical Distribution

In sampling from an empirical distribution the central problem is to transform uniformly distributed random numbers to the new empirical distribution. The first step is to convert the empirical frequency distribution $f(x)$ into the cumulative form $F(x)$. This can be accomplished with the aid of a simple table (Table 3-5) in which the number in each frequency class is divided by the total frequency, and then the transformed values are cumulatively summed and plotted as

TABLE 3-5 Frequency of Occurrence of 1000 Ash-Fall-Thickness Values Pertaining to a Hypothetical Ash-Fall Model, Illustrating the Preparation of a Cumulative Frequency Distribution

Interval J	*Range of Frequency Class in Inches per 50 Years*	*Midrange Value X(J)*	*Frequency of Occurrence*	*Transformed Frequency*	*Cumulative Frequency on Transformed Scale FX(J)*
1	0–10	5	150	0.150	0.150
2	10–20	15	225	0.225	0.375
3	20–30	25	50	0.050	0.425
4	30–40	35	240	0.240	0.665
5	40–50	45	60	0.060	0.725
6	50–60	55	275	0.275	1.000

a cumulative histogram (Figure 3-4). This empirical distribution is, of course, in discrete form, since we would have to represent it by a known continuous function if it were to be represented in continuous form. We shall, however, use a continuous uniform distribution to sample the discrete empirical distribution. In the sampling process it is essential that the range of the cumulative distribution $F(x)$ be the same as that of the uniformly distributed random-number sequence, which is commonly 0.0 to 1.0. To illustrate the use of the cumulative histogram let us then draw a random number whose value is, say, 0.325 and project this number onto the cumulative distribution, as shown in Figure 3-4. If we do this, we obtain an ash-fall value of 15 inches per 50 years. A value of 15 is of course at the midpoint between 10 and 20 inches, and our histogram has been arbitrarily arranged into classes in steps of 10 inches of ash per 50 years. Now, considering the cumulative histogram, the probability of drawing a random number between 0.150 and 0.375 is $(0.375 - 0.150) = .225$ from a uniform distribution whose range is 0.0 to 1.0. But according to our cumulative histogram the probability of drawing a random number from the ash fall distribution that lies between 10 and 20 inches of ash per 50 years is also .225. In effect we have divided the range of uniformly distributed random numbers into the same number of classes as the empirical ash fall distribution.

Computer algorithms to sample from empirical distributions are quite simple. Table 3-6 provides an example, listing subroutine RANEMP, which may be used to obtain samples from any empirical distribution. The number of classes, NCLASS, the array containing

TABLE 3-6 Subroutine RANEMP for Obtaining Random Samples from Any Empirical Distribution, Used in Conjunction with Subroutine RANDU, Which Generates Uniformly Distributed Pseudorandom Numbers in the Range 0.0 to 1.0

```
      SUBROUTINE RANEMP(X,FX,NCLASS,RANDOM,IX)
      DIMENSION X(NCLASS), FX(NCLASS)
      CALL RANDU(IX,IY,R)
      IX=IY
      DO 10 J=1,NCLASS
   10 IF (R.LE.FX(J)) GO TO 20
   20 RANDOM=X(J)
      RETURN
      END
```

the midrange values, X(J), and the array containing the cumulative distribution values in the range 0.0 to 1.0, FX(J), are made available to the subroutine via its argument list. In turn subroutine RANEMP calls subroutine RANDU (Table 3-4), which generates pseudorandom numbers that are uniformly distributed in the range 0.0 to 1.0.

Inverse-Transformation Method

The procedure of transforming a uniform distribution to an empirical distribution is an application of the so-called inverse-transformation method (Naylor and others, 1966, pp. 70–73). This method can be used with any source of random numbers r in the range 0.0 to 1.0 for generating theoretical as well as empirical distributions. A random number of given value r_0 is simply equated with the cumulative frequency distribution, expressed either as a histogram or as a continuous functional relationship (Figure 3-5), and the value of x_0 corresponding to r_0 is obtained. The initial step is to define the cumulative frequency distribution $F(x)$, obtained from the conventional frequency distribution $f(x)$ by summation over each discrete class or

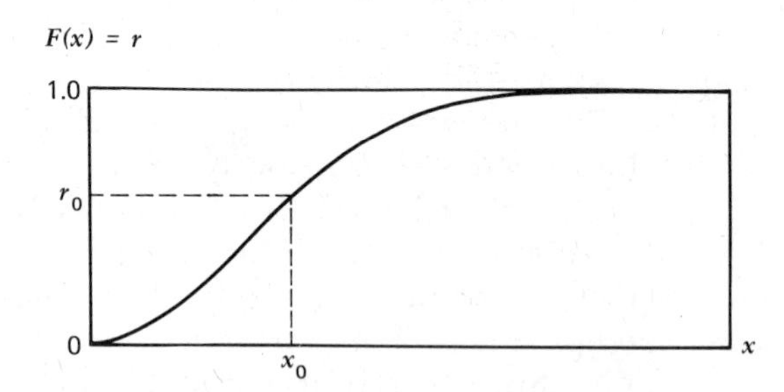

Figure 3-5 Continuous cumulative frequency distribution used to transform given random variable r_0 from continuous uniform distribution to corresponding random variable x_0.

by integration of a continuous distribution, if necessary dividing by the total to scale the range from 0.0 to 1.0.

If both r and $F(x)$ are defined over the range 0.0 to 1.0, we can determine a random value of x that will have the cumulative frequency distribution $F(x)$ from the relationship

$$r = F(x) \tag{3-6}$$

For any particular value of r, say r_0, x_0 can be obtained by the inverse transformation:

$$x_0 = F^{-1}(r_0) \tag{3-7}$$

which simply means that x_0 is solved in terms of r_0.

Sampling from Continuous Theoretical Distributions

Uniform Distribution

Let us illustrate the inverse-transformation method with reference to a uniform, or rectangular, distribution. By uniform distribution we mean a distribution with range a to b, where a and b are the lower and the upper limits respectively. The probability-density function $f(x)$ for a uniform distribution (Figure 3-6) can be defined as

$$f(x) = \frac{1}{b-a} \tag{3-8}$$

The cumulative-distribution function $F(x)$ can be obtained by integration:

$$F(x) = \int_a^x \frac{1}{b-a}\,dt = \frac{x-a}{b-a} \tag{3-9}$$

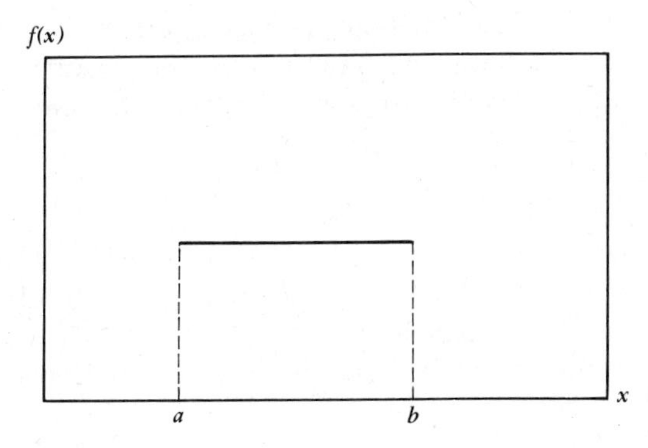

Figure 3-6 Uniform frequency distribution between limits a and b.

Then, equating r with $F(x)$, we obtain

$$r = F(x) = \frac{x-a}{b-a} \tag{3-10}$$

Solving for x yields

$$x = r(b-a)+a \tag{3-11}$$

This says that the variable x is obtained by multiplying the random number r by the range of the uniform distribution $(b-a)$, then adding the lower limit of the range a. This solution is also intuitively obvious. It is assumed, of course, that r has a uniform distribution in the range 0.0 to 1.0.

The FORTRAN statements to generate a random variable X with a uniform distribution that ranges between RLOW and RHIGH can be written as follows:

```
        .
        .
        .
CALL RANDU(IX,IY,R)
IX=IY
X=R * (RHIGH-RLOW)+RLOW
        .
        .
        .
```

The results of a test involving 1000 "draws" using this algorithm are plotted in histogram form in Figure 3-7. In this test the upper limit of the range is 130.0, the lower limit is 20.0, and 11 frequency classes, each progressively increasing by 10.0, are employed. There is some variation in the frequency in each class, but this is due to the moderate size of the sample. If 10,000 samples had been drawn, we would expect the classes to be more equal in size as the observed frequencies more closely approach the expected uniform frequencies over the range.

Exponential Distribution

The exponential distribution also can be sampled by the inverse-transformation method. This distribution is employed in continuous-time Markov models, discussed in Chapter 4. A variable x is exponentially distributed if its probability-density function $f(x)$ is defined as

$$f(x) = \lambda e^{-\lambda x} \tag{3-12}$$

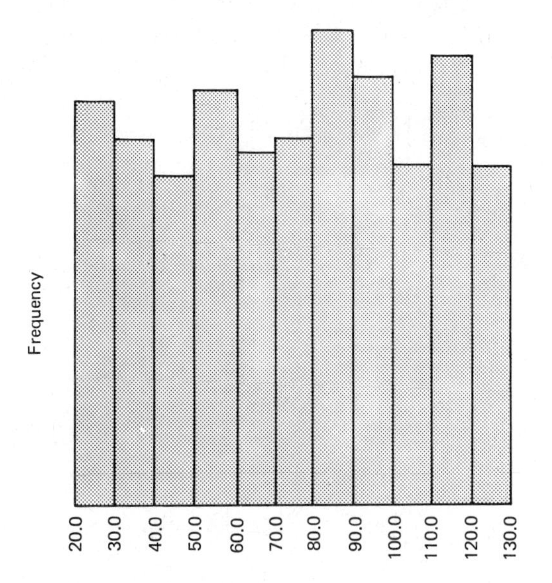

Figure 3-7 Histogram showing results of 1000 draws of uniform random distribution in the range 20.0 to 130.0.

where λ = parameter greater than zero, equivalent to $1/E(x)$, where $E(x)$ is the expected value of x

e = base of natural logarithms $\simeq 2.71828$

The function $f(x)$ can be integrated to give the cumulative-probability-density function $F(x)$:

$$F(x) = \int_0^x \lambda e^{-\lambda t}\, dt = 1 - e^{-\lambda x}$$

Now instead of setting $r = F(x)$ we can set $1 - r = F(x)$ because of the symmetry of r in the range 0.0 to 1.0. Thus $r = 1 - F(x) = e^{-\lambda x}$, and solving for x we obtain

$$x = \frac{-\log_e r}{\lambda} \tag{3-13}$$

Alternatively, if $E(x)$ is given instead of λ,

$$x = -E(x) \log_e r \tag{3-14}$$

We may state this in words as follows: to obtain an exponentially distributed random variable x take the natural logarithm of random number r and multiply it by the negated expected value of x. The following FORTRAN statements may be used to generate an exponentially distributed random variable X given its expected value EX:

```
CALL RANDU(IX,IY,R)
IX=IY
X=-EX*ALOG(R)
```

The results of a test of this algorithm employing 1000 "draws" are shown in histogram form in Figure 3-8.

Normal Distribution

Many geological variables have Gaussian, or normal, frequency distributions. The porosity of some sandstones and the proportions of abundant minerals in rocks provide examples. The probability-density function $f(x)$ for a variable x with a normal distribution is

$$f(x) = \frac{1}{\sigma_x\sqrt{2\pi}} \exp\left[-\frac{1}{2}\left(\frac{x-\mu_x}{\sigma_x}\right)^2\right] \quad (-\infty < x < \infty) \tag{3-15}$$

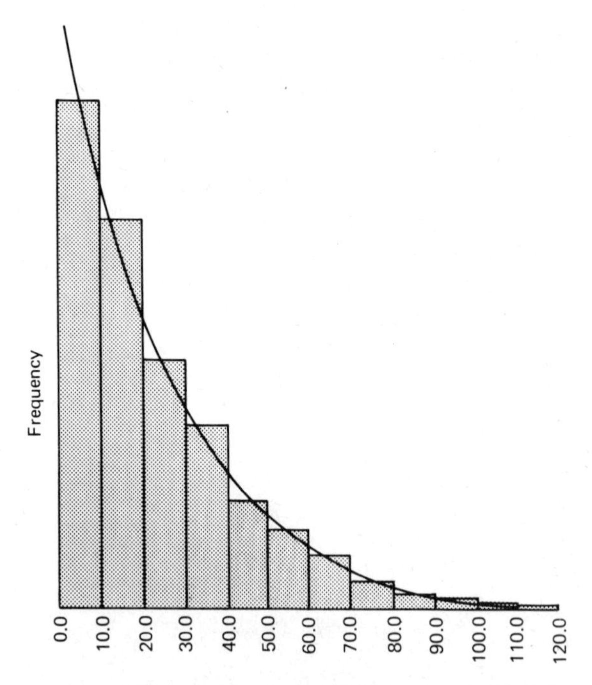

Figure 3-8 Histogram of discrete distribution obtained in test involving 1000 draws of exponentially distributed random variable, whose distribution has an expected value of 30.0. Curve of theoretical continuous distribution is superimposed.

where σ_x = standard deviation of x
μ_x = mean or expected value of x

This formula describes the familiar bell-shaped curve (Figure 3-9) of the normal distribution.

If $\mu = 0$ and $\sigma = 1$, the distribution function is known as the *standard normal distribution*, which is given by

$$f(y) = \frac{1}{\sqrt{2\pi}} e^{-1/2y^2} \qquad (-\infty < y < \infty) \tag{3-16}$$

Any normal distribution can be converted to standard form by the substitution

$$y = \frac{x - \mu_x}{\sigma_x} \tag{3-17}$$

Conversely, if we have a distribution in standard form, we can obtain a normal distribution with any specified mean and standard deviation by the substitution

$$x = y\sigma_x + \mu_x \tag{3-18}$$

Neither $f(x)$ nor $f(y)$ can be directly integrated to give the cumulative distribution $F(x)$. The inverse-transformation method therefore cannot be used unless $F(x)$ is obtained by numerical integration. A much easier way of generating normally distributed random variables is to use the following formula derivable from the central limit theorem (Naylor and others, 1966):

$$y = \frac{\sum_{i=1}^{k} r_i - (k/2)}{\sqrt{k/12}} \tag{3-19A}$$

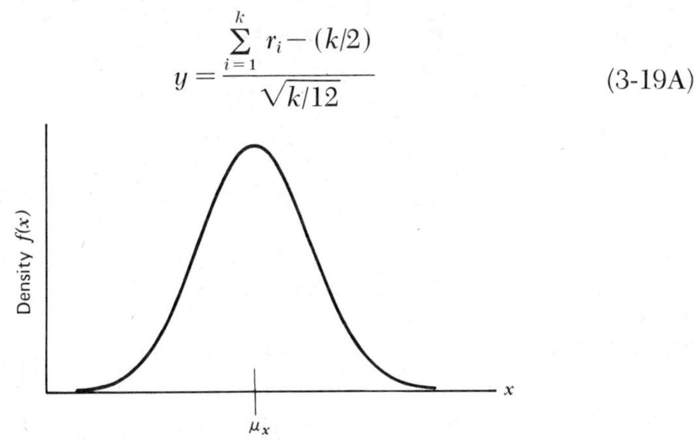

Figure 3-9 Graph of continuous normal, or Gaussian, distribution, in which density is represented on the vertical axis and values of x on the horizontal axis. Mean or expected value is denoted by μ_x.

where y = random variable with standard normal distribution with $\mu = 0, \sigma = 1$
r_i = ith element of sequence of random numbers from a uniform distribution in the range 0.0 to 1.0
k = number of values of r_i to be used

As k approaches infinity, y approaches a true normal distribution. If k is 12, a normal distribution with tails truncated at six standard deviations is produced. For most applications this is adequate. If tail probabilities are critical beyond these points, k must be increased accordingly. With $k = 12$, Equation 3-19A becomes

$$y = \sum_{i=1}^{12} r_i - 6.0 \tag{3-19B}$$

Thus all that is required in order to generate a normally distributed random variable x with mean μ_x and standard deviation σ_x is to sum 12 random numbers drawn from a uniform distribution in the range 0.0 to 1.0, subtract 6, then multiply by σ_x, and add μ_x as indicated by Equation 3-18.

A FORTRAN routine for performing these operations is shown below.

```
        .
        .
        .
        SUM=0.0
        DO 10 I=1,12
        CALL RANDU(IX,IY,R)
        IX=IY
     10 SUM=SUM+R
        X=(SUM-6.0) * STDEV+RMEAN
        .
        .
        .
```

where X = normally distributed random variable x
STDEV = standard deviation of x
RMEAN = mean of x

The results of 1000 "draws" employing the algorithm above are shown in histogram form in Figure 3-10.

An alternative method for generating normally distributed random variables that is slightly more convenient from a computing point of

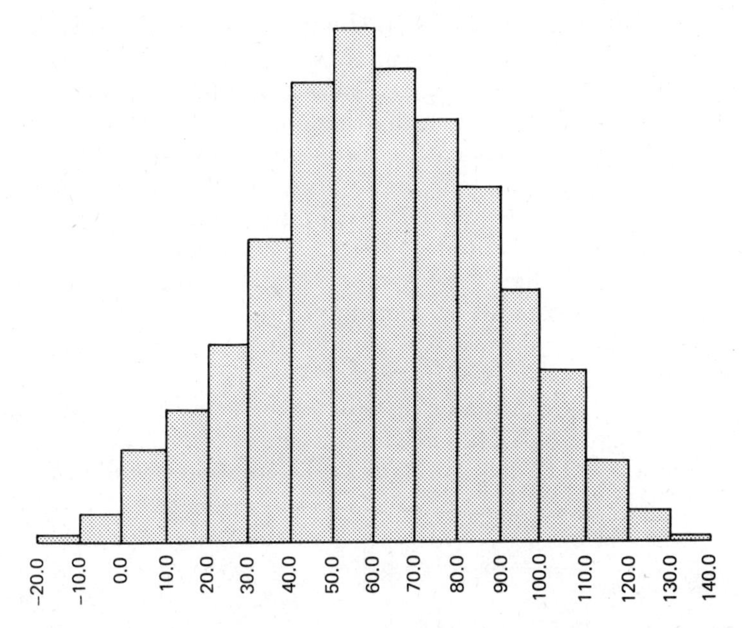

Figure 3-10 Histogram showing results of 1000 draws employing algorithm to produce normal distribution. Distribution has a mean of 60 and a standard deviation of 30.

view has been developed by Box and Muller (1958). Each use of the method requires a pair of independent random numbers from a rectangular distribution as input and yields a pair of numbers from a normally distributed independent random variable with zero mean and unit variance as output. Thus there is one-to-one ratio between rectangularly distributed numbers as input, and the normally distributed numbers produced as output. The defining equations are

$$y_1 = (-2 \log_e r_1)^{1/2} \cos 2\pi r_2 \tag{3-20A}$$

$$y_2 = (-2 \log_e r_1)^{1/2} \sin 2\pi r_2 \tag{3-20B}$$

where r_1 and r_2 = independent random variables from rectangular distribution in range 0.0 to 1.0

y_1 and y_2 = independent normally distributed random variables with $\mu = 0, \sigma = 1$

Lognormal Distribution

Many geological variables, such as the distribution of traces of metals in the earth's crust or the magnitude of oil pools, are more or less lognormally distributed. By this we mean that the logarithms of the variable in question are normally distributed. We can also state the relationship in reverse order. If the logarithm of a random variable has a normal distribution, the random variable itself has a positively skewed distribution known as the lognormal distribution. The lognormal distribution is commonly used to describe random processes that represent the product or combined effect of several lesser independent events. Use of a logarithmic scale has the effect of expanding the low values, and compressing the high values, of the original scale. In general a logarithmic transformation is likely to change any positively skewed distribution into one that is more nearly symmetrical.

Generation of lognormally distributed sets of random numbers involves an adaptation of the formula (Equation 3-19A) for generating sets of numbers with a standard normal distribution, as follows:

$$x = \exp\left\{\mu_y + \sigma_y\left[\frac{\sum_{i=1}^{k} r_i - (k/2)}{\sqrt{k/12}}\right]\right\} \tag{3-21}$$

where $y = \log_e x$

r_i = ith element of sequence of random numbers drawn from a uniform distribution in the range 0.0 to 1.0

k = number of values of r_i to be used

μ_y = mean of y

σ_y = standard deviation of y

If $k = 12$, Equation 3-21 simplifies to

$$x = \exp\left[\mu_y + \sigma_y\left(\sum_{i=1}^{k} r_i - 6\right)\right] \tag{3-22}$$

If we employ Equation 3-22, obtaining a single number from a lognormal distribution involves three steps:

1. Generation of a single value drawn from a standard normal distribution by summing 12 random numbers drawn from a uniform distribution in the range 0.0 to 1.0 and subtracting 6 from the sum.
2. Transformation of this number to one drawn from a normal distribution of specified mean and standard deviation by multiplying by standard deviation σ_y and adding mean μ_y.
3. Transformation of this value to a number drawn from a lognormal distribution by raising e to the power of the value obtained in step 2.

These steps are translated into FORTRAN statements below.

```
      .
      .
      .
      SUM=0.0
      DO 10 I=1,12
      CALL RANDU(IX,IY,R)
      IX=IY
   10 SUM=SUM+R
      X=EXP((SUM-6.0)*STDEVY+RMEANY)
      .
      .
      .
```

where STDEVY = standard deviation of y
RMEANY = mean of y
X = lognormally distributed random variable x

STRATIGRAPHIC-SEQUENCE GENERATOR

To illustrate the use of random variables in a simulation model a simple method of generating an artificial stratigraphic sequence is described below. Let us assume that we have made measurements on an actual stratigraphic section consisting of beds of sandstone and shale. Each unit of sandstone may consist of one or more beds but is terminated at top and bottom by a transition to shale. Likewise each unit of shale is defined by transition to sand at its upper and lower surfaces. Thus the sequence is characterized by a regular alternation between sandstone and shale units of varying thickness.

By recording the thicknesses of sandstone and shale from measured stratigraphic sections, histograms of frequency distributions for both sandstone and shale are constructed. The shape of the sandstone histogram (Table 3-7 and Figure 3-11) suggests that the distribution is approximately normal (which can be ascertained by a χ^2 goodness-of-fit test). The sample mean $\bar{y}$ and sample standard deviation s_y of the original observations are

$$\bar{y} = \frac{\sum_{i=1}^{n} y_i}{n} = 5.08, \qquad s_y = \sqrt{\frac{\sum_{i=1}^{n} (\bar{y} - y_i)^2}{n-1}} = 1.24$$

where n = number of observations = 250. In turn these sample statistics can be used as estimates of the population mean and standard

Table 3-7 Observed Frequencies for Hypothetical Sandstone Unit Thicknesses, Subdivided into 10 Size Classes[a]

Class Number	*Class Range*	*Class Midpoint*	*Observed Frequency*	*Probability Density* f(x)	*Cumulative Probability Density* F(x)
1	0.00–0.99	0.5	1	.004	.004
2	1.00–1.99	1.5	1	.004	.008
3	2.00–2.99	2.5	10	.040	.048
4	3.00–3.99	3.5	34	.136	.184
5	4.00–4.99	4.5	69	.276	.460
6	5.00–5.99	5.5	81	.324	.784
7	6.00–6.99	6.5	41	.164	.948
8	7.00–7.99	7.5	9	.036	.984
9	8.00–8.99	8.5	3	.012	.996
10	9.00–9.99	9.5	1	.004	1.000
			250	1.000	

[a]Probability density and cumulative probability density have been calculated for each size class. Mean and standard deviation of original observations are 5.08 and 1.24, respectively.

deviation. Using pseudorandom numbers drawn from a normal distribution, we can simulate a sequence of any desired length containing units whose mean thickness is 5.08 feet and whose standard deviation is 1.24 feet.

The frequency distribution of shale thicknesses (Table 3-8 and Figure 3-12*a*), however, does not accord with a known theoretical

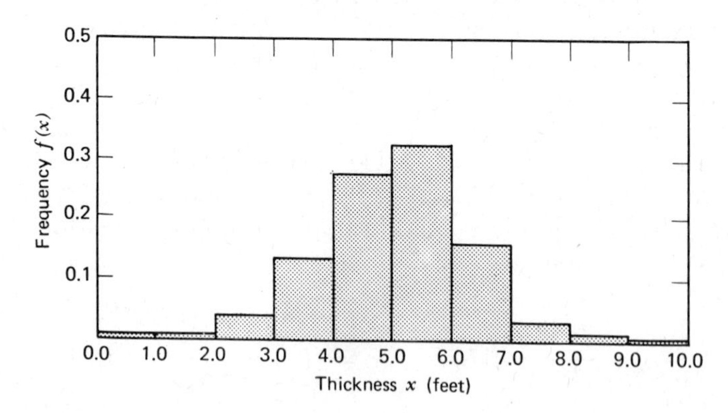

Figure 3-11 Frequency-distribution histogram of observed sandstone thicknesses listed in Table 3-7.

TABLE 3-8 Observed Frequencies of 10 Size Classes of Shale Thicknesses, and Probability Density and Cumulative Probability Density of Each Class

Class Number	*Class Range*	*Class Midpoint*	*Observed Frequency*	*Probability Density* f(x)	*Cumulative Probability Density* F(x)
1	0.00–0.99	0.50	75	.25	.25
2	1.00–1.99	1.50	21	.07	.32
3	2.00–2.99	2.50	39	.13	.45
4	3.00–3.99	3.50	21	.07	.52
5	4.00–4.99	4.50	30	.10	.62
6	5.00–5.99	5.50	15	.05	.67
7	6.00–6.99	6.50	24	.08	.75
8	7.00–7.99	7.50	15	.05	.80
9	8.00–8.99	8.50	21	.07	.87
10	9.00–9.99	9.50	39	.13	1.00
			300	1.00	

frequency distribution. Therefore to generate a random variable with similar characteristics we must sample the empirical distribution after it has been transformed to cumulative form (Figure 3-12*b*).

A FORTRAN program for generating a synthetic sand–shale stratigraphic sequence is listed in Table 3-9. A general flow chart describing the principal operations is shown in Figure 3-13, and examples of input and output from the program are shown in Table 3-10 and Figure 3-14, respectively.

Although the sand and shale thicknesses generated by the program

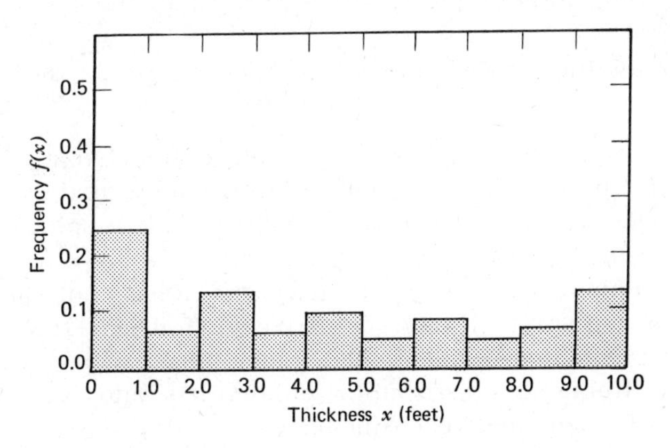

Figure 3-12a Conventional frequency-distribution histogram of thicknesses of shale units.

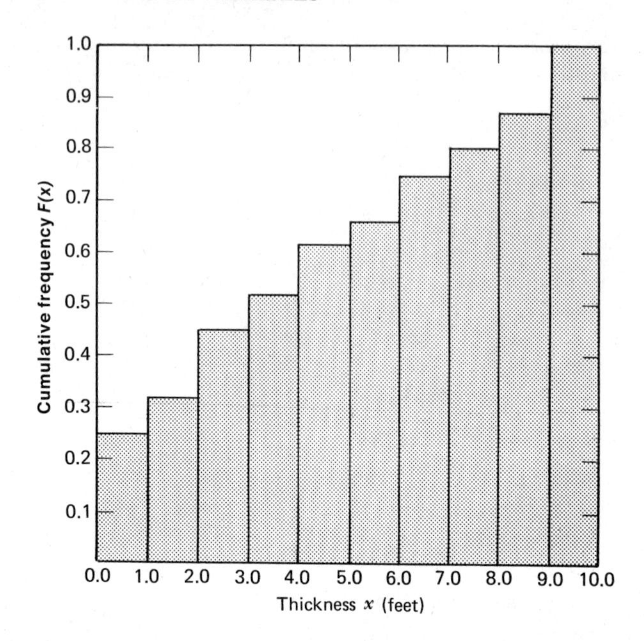

Figure 3-12*b* Cumulative frequency-distribution histogram derived from Figure 3-12a.

are real numbers, they have been truncated to integers for output purposes in Figure 3-14. Each rock unit is thus represented by a whole number of lines of computer output (dots for sandstone and dashes for shale). Rounding of each real number to an integer involves adding 0.5 to the real number and then truncating to an integer, thus rounding either upward or downward to the nearest whole number.

The generating process in the program consists of the following steps:

1. The initial lithology to be generated is selected at random by producing a pseudorandom number R from a uniform distribution in the range 0.0 to 1.0. If R is greater than 0.5, shale is selected; if R is 0.5 or less, sandstone is selected.
2. A random number, SAND, is generated; it has a normal distribution with mean SANDMN and standard deviation SANDEV.
3. A random number, SHALE, is generated with a prescribed empirical frequency distribution subdivided into NCLASS subintervals. The cumulative frequency distribution is stored in array FX(I), and the values of the midpoints in each class are stored in array XMID(I) (index I ranges from 1,2,· · ·, NCLASS).

TABLE 3-9 FORTRAN Program for Generating Synthetic Stratigraphic Sequence of Sandstone and Shale[a]

```
C.....SIMPLE SAND/SHALE STRATIGRAPHIC SEQUENCE GENERATOR
C.....  NLINES          NO. OF LINES TO BE GENERATED
C.....  FTLINE          NO. OF FEET OF SEDIMENT EQUIVALENT TO 1 LINE
C.....  IX              ODD INTEGER 9 DIGITS LONG FOR STARTING RANDU
C.....  SANDMN          MEAN OF SANDST. THICKNESS DISTRIB. (GAUSSIAN)
C.....SANDEV            ST. DEV OF SANDSTONE THICKNESS
C.....  NCLASS          NO. OF FREQUENCY CLASSES USED FOR SHALE THICKNESS
C.....  FX(I)           CUMULATIVE FREQ. IN I-TH SHALE CLASS
C.....  XMID(I)         SHALE THICKNESS AT MID POINT OF I-TH FREQ. CLASS
C.....  ALPHA(I,L)      12-SYMBOL STRING FOR DISPLAY OF I-TH LITHOLOGY
      DIMENSION FX(20), XMID(20), ALPHA(2,3)
      INTEGER STRATA(1000)
    1 FORMAT(3I10, 3F10.2)
    2 FORMAT(16F5.2/16F5.2)
    3 FORMAT(3A4/3A4)
    4 FORMAT(1H1, 'RANDOM STRATIGRAPHIC SEQUENCE')
    5 FORMAT(1H , I5, 2X, 3A4)
C.....READ INPUT PARAMETERS
      READ(5,1) NLINES,NCLASS,IX,FTLINE,SANDMN,SANDEV
      READ(5,2) (FX(I), I=1,NCLASS), (XMID(I), I=1,NCLASS)
      READ(5,3) ((ALPHA(INDEX,L), L=1,3), INDEX=1,2)
      NSTOP=0
      SAND=0.0
      SHALE=0.0
C.....SELECT INITIAL LITHOLOGY AT RANDOM
      CALL RANDU(IX,IY,R)
      IX=IY
      IF (R-0.5) 10, 10, 30
C.....GENERATE SAND THICKNESS (NORMAL DISTRIBUTION)
   10 SUM=0.0
      DO 20 K=1,12
      CALL RANDU(IX,IY,R)
      IX=IY
   20 SUM=SUM+R
      SAND=(SUM-6.0)*SANDEV+SANDMN
C.....GENERATE SHALE THICKNESS (EMPIRICAL DISTRIBUTION)
   30 CALL RANEMP(XMID,FX,NCLASS,SHALE,IX)
      WRITE(6,99) SAND,SHALE
   99 FORMAT(1H , 2F10.3)
C.....STORE IN STRATIGRAPHIC ACCOUNTING ARRAY
      NSAND=SAND/FTLINE+0.5
      IF (NSAND.LT.1) GO TO 50
      NSTART=NSTOP+1
      NSTOP=NSTART+NSAND
      DO 40 K=NSTART,NSTOP
      IF (K.GT.NLINES) GO TO 80
   40 STRATA(K)=1.0
   50 NSHALE=SHALE/FTLINE+0.5
      IF (NSHALE.LT.1) GO TO 70
      NSTART=NSTOP+1
      NSTOP=NSTART+NSHALE
      DO 60 K=NSTART,NSTOP
      IF (K.GT.NLINES) GO TO 80
   60 STRATA(K)=2.0
C.....RETURN TO 10 FOR ANOTHER DRAW
   70 GO TO 10
C.....PRINT OUT STRATIGRAPHIC SEQUENCE
   80 WRITE(6,4)
      DO 90 KK=1,NLINES
      K=NLINES-KK+1
      INDEX=STRATA(K)
   90 WRITE(6,5) K, (ALPHA(INDEX,L), L=1,3)
      RETURN
      END
```

[a]Sandstone thicknesses are drawn from theoretical normal distribution, whereas shale thicknesses are drawn from empirical distribution. Subroutines RANEMP and RANDU, which are included in the program, have been listed earlier in this chapter.

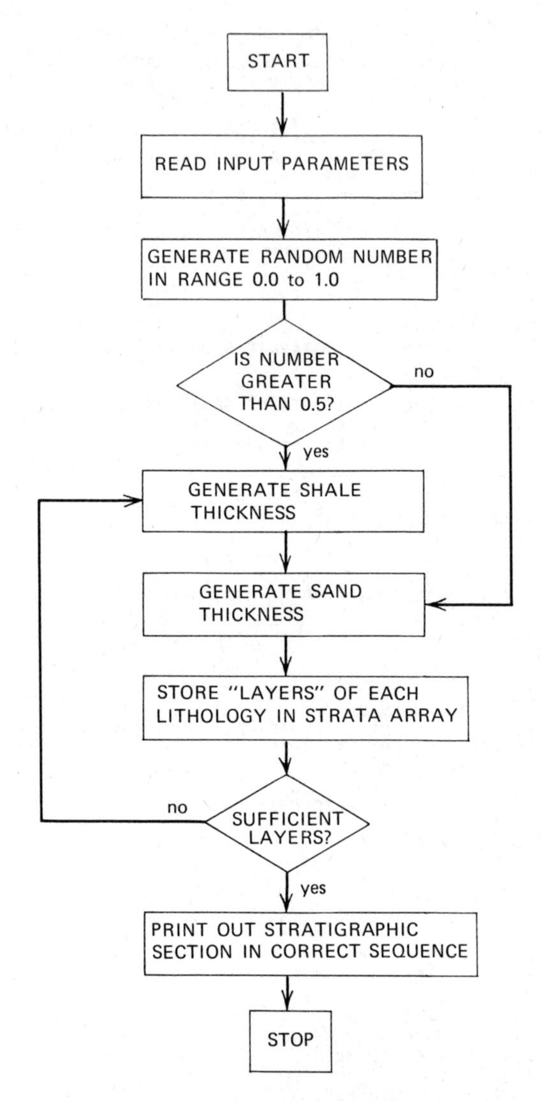

Figure 3-13 Flow chart for generating sand–shale stratigraphic sequence.

4. The real numbers SAND and SHALE are transformed into integers NSAND and NSHALE, respectively. In turn a sequence of real numbers is assigned to array STRATA(K) for plotting purposes. Values of 1.0 and 2.0 are used to represent sandstone and shale, respectively. Each element in STRATA(K) is then regarded as representing one row of symbols portraying a bed in the printed graphic

TABLE 3-10 Data for Trial Run of Stratigraphic-Sequence-Generating Program

```
      100         10 98989879        1.00       5.08        1.24
0.25 0.32 0.45 0.52 0.62 0.67 0.75 0.80 0.87 1.00
0.50 1.50 2.50 3.50 4.50 5.50 6.50 7.50 8.50 9.50
............
--------
```

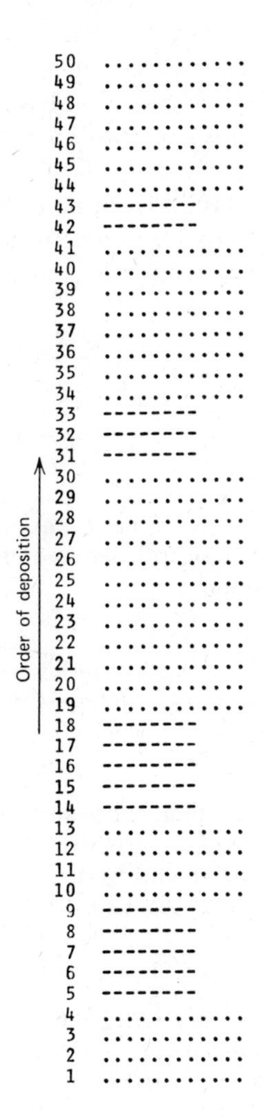

Figure 3-14 Part of computer output showing alternating sandstone (dots) and shale (dashes) in synthetic stratigraphic sequence. Each line represents 1 foot of sediment. Input for run is listed in Table 3-10.

column. Although the vertical scale of the column is arbitrary, it is convenient to have each line represent 1 foot.

5. If the number of feet of sediment generated is greater than NLINES (a control parameter read in at the start), the computations cease and control is transferred to step 6, otherwise computations start again at step 2.

6. The stratigraphic section is printed out starting with STRATA-(NLINES), which is the youngest layer of sediment and last to be generated, sequentially printing out progressively older beds. For each line a row of dots is printed if STRATA(K) = 1.0, to represent sandstone, or a row of dashes is printed if STRATA(K) = 2.0, to represent shale.

Although this model is trivial from a geological standpoint, it provides an example of the steps involved in generating stochastic variables with various statistical properties and applying them in a simple simulation model.

Problems

Problem 3-1

Write a simple FORTRAN program to generate pseudorandom integer numbers by employing the multiplicative-congruence relationship. Make the program machine independent.

Problem 3-2

After the program to produce pseudorandom integers is operating, modify it so as to produce decimal-point numbers that are uniformly distributed within any prescribed range.

Problem 3-3

Write a FORTRAN program that will calculate χ^2 to test for uniform distribution, adapting Equation 3-3. Use the program to test one or more sequences of real numbers produced by the program specified in Problem 3-2, employing 10 frequency classes. Consult tables for values of χ^2, setting α at 0.05.

Problem 3-4

Write a FORTRAN program that will calculate χ^2 to test for independence in sequences of real pseudorandom numbers produced above, adapting Equation 3-5 for this purpose. Set α at 0.05 in using tables for values of χ^2.

Problem 3-5

Write a FORTRAN program to provide pseudorandom numbers drawn from a discrete empirical frequency distribution that you have devised. In turn algebraically combine numbers produced in this way with a sine function to produce a composite function that has a deterministic component and a random component similar to that graphed in Figure 3-1.

Problem 3-6

Either modify the program in Table 3-9 or write an entirely new program to generate a stratigraphic sequence comprising alternations of limestone and shale. Assume that the limestone-bed thicknesses are lognormally distributed and that the natural logarithms of bed thickness have a mean of 5 feet and a standard deviation of 2 feet. Assume that the shale-bed thicknesses are distributed as follows:

Thickness (feet)	*Frequency*
0.00–0.49	10
0.50–0.99	70
1.00–1.49	142
1.50–1.99	39
2.00–2.49	25
2.50–2.99	82
3.00–3.49	76
3.50–3.99	21
4.00–4.49	5
4.50–4.99	1

Annotated Bibliography

Box, G. E. P., and Muller M. E., 1958, "A Note on the Generation of Random Variables", *Annals of Mathematical Statistics*, v. 30, pp. 610–611.

Fisher, R. A., and Yates, F., 1963, *Statistical Tables for Biological, Agricultural, and Medical Research*, 6th edition, Hafner, New York, 145 pp. (published in Britain by Oliver and Boyd, Edinburgh).

Contains tables of χ^2, as well as many other statistics.

Good, I. J., 1953, "The Serial Test for Sampling Numbers and Other Tests for Randomness," *Proceedings of the Cambridge Philosphical Society*, v. 49, pp. 276–284.

Griffiths, J. C., 1967, *Scientific Method in Analysis of Sediments*, McGraw-Hill, New York, 508 pp.

Detailed, authoritative description of the application of statistical methods to the analysis of sediments.

Hammersley, J. M., and Handscomb, D. C., 1964, *Monte Carlo Methods*, Methuen, London, 178 pp.

Chapter 3 provides a readable introduction to the generation of pseudorandom numbers by multiplicative-congruence methods, coupled with a brief discussion of the techniques of random sampling from other distributions.

Hewlett, R. F., 1964, *Application of Simulation in Evaluating Low-Grade Mineral Deposits*, U.S. Bureau of Mines, Report of Investigations 6501, 62 pp.

Uses Monte Carlo methods to generate additional data for evaluating mineral deposits, where only sparse information is available. This overcomes the difficulties encountered in applying evaluation techniques that require large volumes of data.

Jansson, B., 1966, *Random Number Generators*, Victor Pettersons Bokindustri, Stockholm, 205 pp.

This appears to be one of the most thorough treatments of pseudorandom-number-generating methods available in English. The period-length properties of congruential methods are extensively discussed. An additional feature lies in comparisons of actual statistical tests.

Krumbein, W. C., and Graybill, F. A., 1965, *An Introduction to Statistical Models in Geology*, McGraw-Hill, New York, 475 pp.

A detailed authoritative work dealing with a large number of statistical methods and their geological applications.

Lehmer, D. H., 1951, "Mathematical Methods in Large-Scale Computing Units," *Annals of the Computer Laboratory*, Harvard University, v. XXVI, pp. 141–146.

Marsaglia, G., and Bray, T. A., 1968, *One-Line Random Number Generators and Their Use in Combinations*, Mathematical Note No. 551, Mathematical Research Laboratory, Boeing Scientific Research Laboratories, Seattle, 9 pp.

Mize, J. H., and Cox, J. G., 1968, *Essentials of Simulation*, Prentice-Hall, Englewood Cliffs, N.J., 234 pp.

Chapters 3, 4, and 5 provide a readable introduction to types of frequency distributions, pseudorandom-number generation, and Monte Carlo sampling methods.

Naylor, T. H., Balintfy, J. L., Burdick, D. S., and Chu, K., 1966, *Computer Simulation Techniques*, John Wiley and Sons, New York, 352 pp.

Chapter 3 provides a readable, medium-detail introduction to congruential methods of generating pseudorandom numbers, followed by a useful compilation of types of statistical tests applicable to sequences of pseudorandom numbers. Chapter 4 provides a valuable survey of both continuous and discrete probability distributions, coupled with examples of FORTRAN subroutines for their use.

Poore, J. H., Jr., 1964, *Computational Procedures for Generating and Test-*

ing Random Numbers, Division of Business and Economic Research, Louisiana Polytechnic Institute, 36 pp.

This is a well-written discussion of some of the problems that arise in the use of congruential methods of generating pseudorandom numbers. FORTRAN algorithms are listed for the statistical testing of pseudorandom-number sequences.

Tocher, K. D., 1963, *The Art of Simulation*, Van Nostrand, Princeton, N.J. 184 pp.

Chapter 5 provides an excellent exposition of mechanical and electrical random-number generators. Chapter 6 provides a readable and medium-detail discussion of congruential methods for the generation of pseudorandom numbers. Chapter 2 is a useful introduction to sampling from a distribution.

Vistelius, A. B., 1967, *Studies in Mathematical Geology*, Consultants Bureau, New York, 294 pp.

Translation from Russian. Part II (pp. 47–94) discusses the application of frequency distributions to geologic data, with a comprehensive list of references.

Whittlesey, F. R. B., 1968, "A Comparison of the Correlational Behavior of Random Number Generators for the IBM 360," *Communications of the Association for Computing Machinery*, v. 11, pp. 641–644.

The author demonstrates that multiplicative-congruence pseudorandom-number generators tend to produce sequences of numbers that are autocorrelated and therefore nonrandom, particularly when certain multiplicative constants are used. The autocorrelation effect may be present even though sequences of numbers generated pass the usual χ^2 tests for distribution and independence. Furthermore the statistical behavior of generators tends to be markedly influenced by computer-word length.

CHAPTER 4

Markov Chains

Many natural processes that are random in their occurrence also exhibit an effect in which previous events influence, but do not rigidly control, subsequent events. These processes are named *Markov processes* after the work of the Russian mathematician Markov early in the twentieth century. We can regard a Markov process as a process in which the probability of the process being in a given state at a particular time may be deduced from knowledge of the immediately preceding state. One form of a Markov process is a *Markov chain*, which may be regarded as a sequence or chain of discrete states in time (or space) in which the probability of the transition from one state to a given state in the next step in the chain depends on the previous state.

In its general form a Markov chain may be regarded as a series of transitions between different states such that the probabilities associated with each transition depend only on the immediately preceding state, and not on how the process arrived in that state. Furthermore the general form of a Markov chain is such that it contains a finite number of states, and the probabilities associated with the transitions between states are *stationary* (i.e., do not change) with time. If we adhere to this definition, we may regard a Markov chain in its general form as having a very short "memory," extending only for a single step at a time and ceasing beyond a single step. Such a chain can be termed a *first-order* Markov chain. If, however, we extend the definition of a Markov chain so that the probabilities associated with each transition depend on events earlier than the immediately preceding event, the Markov chain has a longer memory and is a higher-order chain. Furthermore, chains may be defined so that the probabilities associated with each transition depend jointly on more than one previous event, or in other words, the chain exhibits multiple dependence relationships.

An important aspect of Markov chains is that they exhibit the *Markov property*—that is, the dependence of the probabilities associated with each transition on the immediately preceding state (or several states if we extend the definition to a higher-order chain). Many geologic processes exhibit the Markov property. Consequently we can make effective use of Markov chains as components in probabilistic dynamic models if we wish to represent the Markov property. Mathematical models that employ Markov chains occupy an intermediate position in the spectrum of dynamic models, ranging from classical, deterministic models at one extreme to purely random models at the other extreme, in which all events are independent. If, for example, a particular process operating through either time or space is represented as a purely deterministic model, the state of the system at any point in space or time can be predicted precisely by evaluating the model. At the other extreme, in a purely random model the state of the system at any point in time or space is wholly independent of previous events. A Markov-chain model is intermediate in that a random component or components are present, but that the state of the system at any point in time or space is not independent of the previous event or events.

The Markov property is illustrated by a number of everyday phenomena. Assume, as an example, that a man lives near a river whose depth fluctuates widely. Every Sunday this man records the depth of flow. Furthermore he classifies the condition of the river into one of three stages: low, normal, or flood. After doing this for several years he becomes aware of the fact that the three stages (low, normal, and flood) are by no means independent of one another. If the river is normal one week, it is usually normal the next week. If the river is at low stage one week, it is very unlikely that it will be in flood stage the following week. From his record of 148 measurements he tabulates the frequency of transition from each type of river stage to every other type of river stage on a weekly basis (Table 4-1).

Being a statistically minded person, he notices that each entry in

TABLE 4-1 Frequencies of Transitions between Three River Stages Tabulated on a Weekly Basis over 148 Weeks

From Stage	To Stage: Low	Normal	Flood	Row Totals
Low	12	6	0	18
Normal	5	80	15	100
Flood	0	14	16	30

this table can also be expressed as a probability by simply dividing that entry by the row total. In other words, the low river stage is followed at each weekly measurement by another low river stage 12 times. But the low stage was only recorded a total of 18 times. So, based on his observations, the probability of a low stage occurring immediately after another low stage is 12/18 = .67. This is of course an estimate of the probability based on the sample data. Thus a table of estimated probability values (Table 4-2) can be substituted for the

TABLE 4-2 Estimated Transition Probabilities Pertaining to River Stages Calculated from Hypothetical Transition Frequencies in Table 4-1

From Stage	To Stage: Low	Normal	Flood	Row Totals
Low	.67	.33	.00	1.00
Normal	.05	.80	.15	1.00
Flood	.00	.47	.53	1.00

table listing the transition frequencies. The probability values are termed *Markov transition probabilities*, and they form a *transition matrix*, in which the number of rows equals the number of columns (thus a square matrix), corresponding to the number of states. Furthermore the probability values in each row sum to 1.0, which says in effect that there is absolute certainty of a transition from each state to some state in the next event in the chain of finite events. Our river example has three states (low, normal, and flood), and thus we can represent its behavior with a 3×3 transition matrix. The matrix in our example thus provides a precise and explicit summary of the Markov property as it pertains to the river's depth fluctuations observed on a weekly basis.

The hypothetical river example concerns a Markov process with discrete states (the three river stages) and discrete time (weeks). In other examples, however, we could distinguish Markov processes that involve continuous states and continuous time. It is convenient to erect the following simple classification of Markov processes:

Discrete Time		*Continuous Time*	
Discrete states	Continuous states	Discrete states	Continuous states

We shall treat discrete time-discrete state Markov processes in most of this chapter, but touch upon continuous-time Markov processes later in the chapter.

Although a wide variety of geological phenomena seem to possess the Markov property, many of these processes are probably not represented by Markov chains of the general form (first-order chains with stationary transition probabilities) but instead involve more complex, multiple-dependence chains with nonstationary transition probabilities. For a more extended introduction to Markov chains the reader is referred to books by Kemeny, Snell, and Thompson (1956), Cox and Miller (1965), Parzen (1962), Pfeiffer (1965), and particularly the book by Kemeny and Snell (1960), which has provided background for parts of this chapter.

MARKOV TRANSITION MATRICES

As we have seen, a matrix of transition probabilities provides a succinct description of the behavior of a Markov chain. Each element in the matrix is the probability of the transition from a particular state (that state pertaining to the particular row in the matrix) to the next state (that state pertaining to the particular column). Thus all possible transitions are provided for, assuming a given number of states. A Markov transition matrix **P** with three states may be written as follows:

$$\mathbf{P} = \begin{array}{c} \\ s_1 \\ s_2 \\ s_3 \end{array} \begin{array}{c} \begin{array}{ccc} s_1 & s_2 & s_3 \end{array} \\ \begin{bmatrix} p_{11} & p_{12} & p_{13} \\ p_{21} & p_{22} & p_{23} \\ p_{31} & p_{32} & p_{33} \end{bmatrix} \end{array} \tag{4-1}$$

where the p's signify the probabilities of transition from one of the states whose rows are denoted by s_1, s_2, and s_3 to the next state, denoted by s_1, s_2, and s_3 written above the matrix. Thus the probability of a transition from state s_1 to state s_1, s_2, or s_3 is given by the values assigned to the p's in the top row, the transitions from state s_2 by the probabilities in the middle row, and so on.

The aspect of memory in a Markov chain involves *conditional probability*, which implies that the probability of a particular event is conditional on some other event. Consider, initially, the definition of probability independent of other events. The probability of statement p, independent of other considerations, may be written $\mathbf{P}r[p]$. Suppose that additional information is received in the form of statement q that modifies the probability. The probability of p upon receipt of q is its conditional probability and is written $\mathbf{P}r[p|q]$, which is read the probability of p, given q. Markov chains are one way of dealing with sequences of events whose probabilities are conditional on other events.

Now consider a matrix of transition probabilities pertaining to a hypothetical stratigraphic sequence. Assume that we are concerned with the transitions in a spatial, rather than in a time, sense and that we observe the transitions from bed to bed, moving upward in the sequence. If there are only three lithologic types (states), a 3×3 transition matrix is capable of probabilistically describing the behavior of the sequence in ascending order. If we let $s_1 =$ sandstone, $s_2 =$ shale, and $s_3 =$ limestone, the values in row 1 of the matrix govern the likelihood of going from sandstone to sandstone, sandstone to shale, or sandstone to limestone. Consider the following hypothetical transition matrix:

$$\begin{array}{lc} & \begin{array}{ccc} s_1 & s_2 & s_3 \end{array} \\ \begin{array}{ll} \text{sandstone} & s_1 \\ \text{shale} & s_2 \\ \text{limestone} & s_3 \end{array} & \begin{bmatrix} 0 & 1 & 0 \\ 0 & \frac{1}{3} & \frac{2}{3} \\ \frac{1}{2} & 0 & \frac{1}{2} \end{bmatrix} \end{array} \tag{4-2}$$

This matrix says in effect that the only possible transition from sandstone (s_1) is to shale (s_2). Thus there is absolute certainty (probability, $p_{12} = 1$) that sandstone will be succeeded by shale. There is also absolute certainty that sandstone will not be succeeded by sandstone ($p_{11} = 0$) nor by limestone ($p_{13} = 0$). When in shale (s_2), however, there is zero probability of an upward transition to sandstone, $\frac{1}{3}$ probability of a transition to shale, and $\frac{2}{3}$ probability of a transition to limestone. Each row sums to 1, reflecting the fact that there is absolute certainty of a transition from any lithology to some lithology. In an actual stratigraphic example the probability values may be based on observed frequencies, as the river-stage transition probabilities.

An alternative means of representing transition probabilities in a Markov matrix is with a transition diagram (Figure 4-1). The arrows from each state indicate the possible states to which a transition may be made, and the values pertain to the probability of each transition.

MARKOVIAN STRATIGRAPHIC SEQUENCES

We would not argue that the transition-probability values in the preceding hypothetical stratigraphic examples are reasonable. It can be shown, however, that some actual stratigraphic sequences exhibit the Markov property. For example, in coal-bearing strata the probability of occurrence of a coal bed at a particular horizon in a sequence of strata may be strongly influenced by the presence of an underclay in the immediately underlying bed. The first stratigraphic application of Markov chains appears to have been made by the Soviet geologist Vistelius (1949). In company with his colleagues, Vistelius has made a number of significant contributions in this field (1965, 1966, 1967). Allegre (1964), Potter and his students (Carr and others, 1966), Agterberg (1966), Krumbein (1967, 1968b), Dowds (1968), and others have applied the Markov theory to geological problems. Carr and others dealt with the Mississippian Chester series of the Illinois

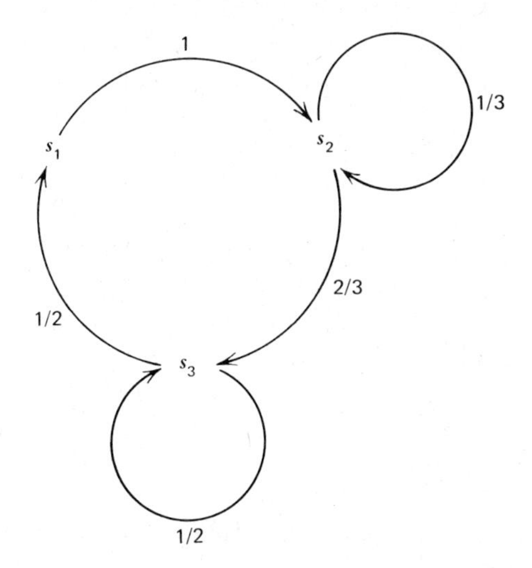

Figure 4-1 Markov transition diagram corresponding to transition-probability matrix 4-2. Numerical values are transition probabilities.

basin. On the basis of observations, they computed a transition matrix for three lithologies in the Chester series, as follows:

$$\begin{array}{lc} & \begin{array}{ccc} s_1 & s_2 & s_3 \end{array} \\ \begin{array}{ll} \text{sandstone} & s_1 \\ \text{shale} & s_2 \\ \text{limestone} & s_3 \end{array} & \begin{bmatrix} .1 & .8 & .1 \\ .4 & .2 & .4 \\ .1 & .8 & .1 \end{bmatrix} \end{array} \tag{4-3}$$

This matrix says in effect that, if the system is in state s_1 (sandstone), the most probable next state is shale (s_2). On the other hand, if the system is in state s_2 (shale), a transition to either sandstone or limestone is equally probable. The transition matrix clearly indicates that aspects of the sequence of lithologic states in the Chester series are strongly Markovian. For example, shale tends to "remember" to succeed sandstone.

A transition matrix based on observations is a useful analytical tool in itself. As we have pointed out, it provides a way of probabilistically describing a succession of events in time or space. An alternative application of a transition matrix, however, is to use it as a regulating mechanism in dynamic probabilistic simulation models. Carr and others (1966) used the Chester series transition matrix to produce simulated, or artificial, stratigraphic sequences, which were then compared with the real stratigraphic sequence. In simulating the sequence they were concerned with variations in the thickness of the different units, as well as with the order of their succession. By tabulating the thicknesses of each stratigraphic unit, frequency distributions of thicknesses for each state were prepared. Given this information, they prepared the synthetic sequences with the following set of instructions:

1. Select an initial state at random. Say it is sandstone. Then draw a thickness at random from the sandstone-thickness distribution.
2. Select the succeeding lithology at random from the first row of the transition matrix. Say it is limestone. Then draw a thickness value at random from the limestone-thickness distribution.
3. Return to the transition matrix and select the next lithology at random from the third row of the matrix. For example, if it is shale, draw a thickness value at random from the shale-thickness frequency distribution.
4. Continue this procedure until a simulated section with some specified number of transitions is obtained.

Thus the simulation process involves use of a transition-probability matrix in selecting each rock type or state and use of thickness frequency distributions to obtain the thickness of each state.

An example of a synthetic sequence generated with the transition-probability values of matrix 4-3 is shown in Figure 4-2. Note that in this sequence lesser sequences consisting of a single lithology (termed multistory stages) have been generated where a given lithology has been succeeded by itself. The occurrence of multistory lithologies is controlled by the probabilities along the diagonal of the transition matrix.

Calculation of Transition Probabilities

Transition probabilities are commonly based on frequency distributions. A frequency distribution of transitions is simply a tabulation of the number of transitions from each state to each other state in the system under consideration. The table of hypothetical river-stage transitions on a week-by-week basis (Table 4-1) is an example. The frequencies are converted to estimates of the probabilities by dividing through by the row totals.

Transition probabilities in stratigraphic sequences can be calculated in two principal ways—that is, for transitions between rock units only or, alternatively, for transitions between intervals of specified thickness. A fictitious stratigraphic sequence developed by Krumbein (1967), shown in Figure 4-3, illustrates the two methods. The transitions on a purely lithologic basis are tabulated in a "tally" matrix in the upper right. Moving upward from the base of the stratigraphic sequence, the transitions are recorded by placing a tally mark in the appropriate box. For example, a transition from A to B is recorded by a tally mark in column B of row A. The next transition is from B to C, which gives a tally mark in column C of row B. If the transition occurs within a multistory unit, as in the change from B to B (about 30 feet above the bottom of the section), this means that B exhibits a transition to itself, by virtue of a prominent bedding plane or conspicuous change in character, and a tally mark is placed in column B of row B. When a sufficient number of transitions have been observed, the tally marks in each row are summed and the tally values in each box are divided by the row sums to compute transition-probability values. The probability values necessarily sum to 1.0 in each row.

Alternatively transition probabilities may be computed by using fixed vertical intervals, as illustrated in the other tally matrices of Figure 4-3. The results are sensitive to the interval employed. A thin interval (such as the 1-foot interval in the hypothetical example) tends to yield a high proportion of multistory stages (i.e., sandstone to

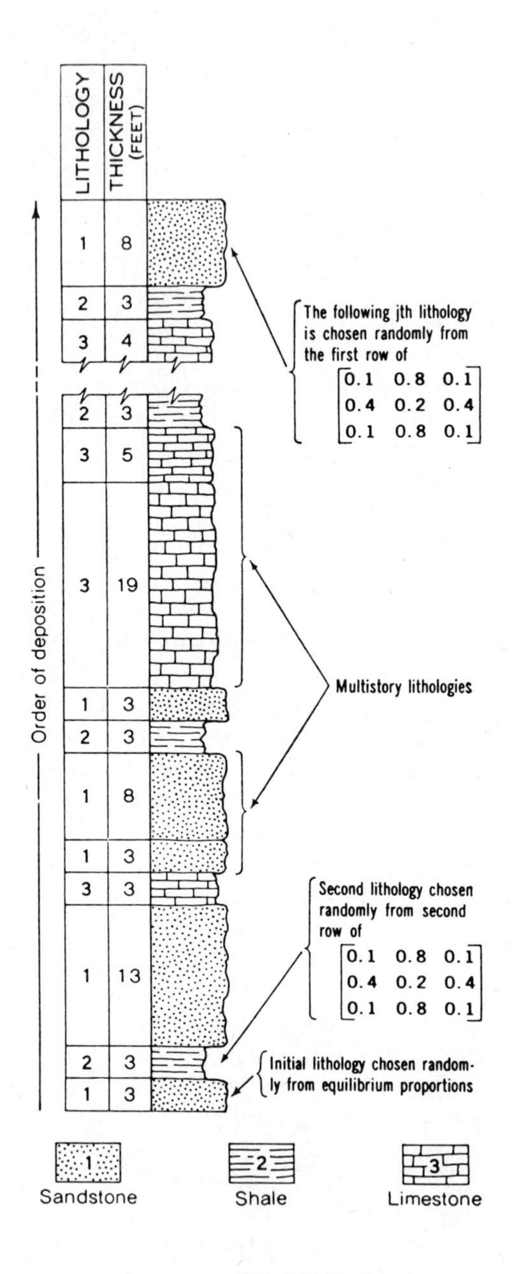

Figure 4-2 Stratigraphic sequence synthesized by employing a Markov transition matrix. From Carr and others (1966); copyright by the American Association for the Advancement of Science.

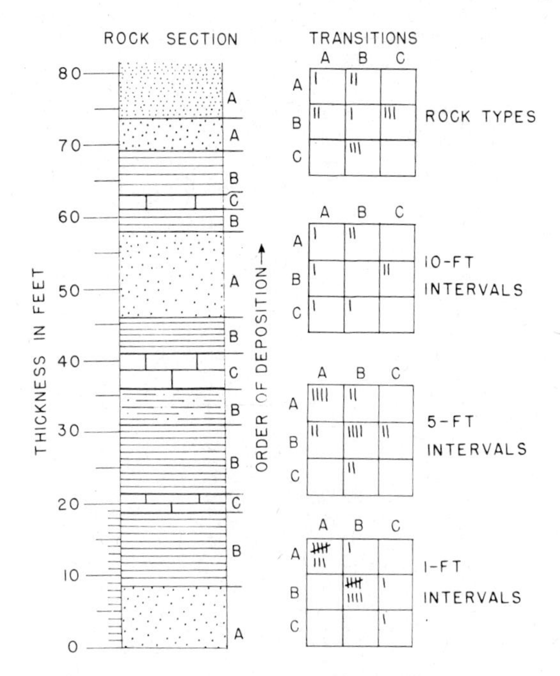

Figure 4-3 Fictitious stratigraphic section illustrating sampling techniques to calculate estimates of Markov transition matrix probability values. "Tally" matrices in right column illustrate results obtained for transitions between rock types and transitions at specified thickness intervals. From Krumbein (1967).

sandstone), whereas a relatively thick interval tends to ignore thin units. For example, a limestone bed with a thickness of 5 feet may be missed if an interval of 10 feet is employed.

As we pointed out earlier, transition probabilities obtained by observation can be used to generate synthetic sequences, which in turn can be compared with the original sequence. Examples are provided by Krumbein (1967), who computed a transition-probability matrix based on transitions between 309 units of equal thickness observed in Chester strata of the Illinois basin (Table 4-3). Given these probability values, a synthetic stratigraphic sequence (Figure 4-4) was produced, and its transition frequencies computed (Table 4-4). The transition-probability values of the synthetic sequence compare favorably with those based on observations of the real sequence, as they should.

A FORTRAN program for computing transition-probability matrices is listed in Table 4-5. Input to the program consists of a sequence of integers, in which the value of each integer denotes a state. For

TABLE 4-3 Tally Matrix and Corresponding Transition-Probability Matrix Based on Observations of 309 Thickness-Unit Lithologic Transitions in the Chester Series[a]

Tally Matrix

		A	B	C	Totals
sand	A	58	18	2	78
shale	B	15	86	39	140
lime	C	5	35	51	91
Totals		78	139	92	309

Transition-Probability Matrix

		A	B	C
sand	A	.74	.23	.03
shale	B	.11	.61	.28
lime	C	.06	.38	.56

[a]Modified after Krumbein (1967).

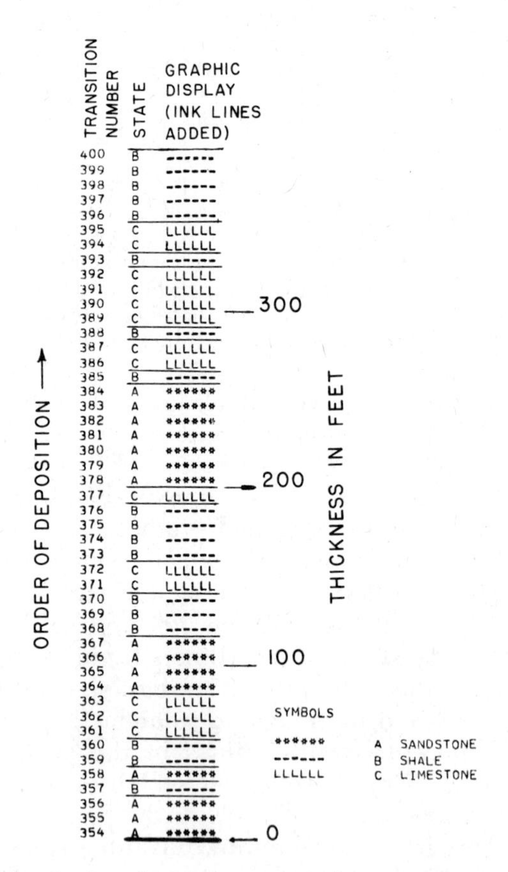

Figure 4-4 Graphic display of synthetic stratigraphic sequence employing transition-probability values of Table 4-3. From Krumbein (1967).

TABLE 4-4 Numerical Results of Simulation Experiment Employing Transition-Probability Values of Table 4-3[a,b]

Rock Type	*Number of Occurrences*	*Thickness (feet)*	*Percentage of Total Thickness*	*Average Bed Thickness (feet)*	*Percentage of Total Beds*
Sandstone	28	944	23.5	33.7	16.5
Shale	75	1736	43.4	23.1	44.1
Limestone	67	1328	33.1	19.8	39.4
Totals	170	4008	100.0		100.0

Relative Frequencies of Transitions Realized in 500-Transition Simulation Experiment

$$\begin{array}{c} \\ A \\ B \\ C \end{array}\begin{array}{c} \begin{array}{ccc} A & B & C \end{array} \\ \begin{bmatrix} .77 & .18 & .05 \\ .08 & .65 & .27 \\ .07 & .32 & .61 \end{bmatrix} \end{array}$$

[a]From Krumbein (1967).
[b]Relative frequencies after simulation run accord relatively well with those in the transition-probability matrix of Table 4-3.

example, if there are five states in the sequence, the integers 1 through 5 could be used.

While extolling the virtues of Markov models we should point out that there are a number of dangers inherent in the indiscriminate application of Markov chains to stratigraphic analysis. Part of the problem lies in the oversimplification.of lithologies used in establishing states for transition matrices. Despite their realism, some artificially generated stratigraphic sequences contain an element of false cyclicity induced by the small number of states present in the system. Furthermore the selected number of states is often too small to accommodate the lithologic variation in a stratigraphic section. This results in the lumping together of dissimilar lithologies. Experimenters with Markov chains should also be careful that (a) the transition probabilities do not vary drastically with position in the sequence (this problem of *stationarity* is treated later in this chapter) and (b) the number of transitions measured is sufficiently large to obtain statistically reliable estimates of the transition probabilities.

Despite these problems, Markov transition matrices do provide a valuable means for describing dependency relationships in sequences of events. For example, Vistelius (1967) derived a transition matrix based on a theoretical model proposed by Kuenen (1950) for turbidite succession. Vistelius then made empirical estimates of transition probabilities with a supposed turbidite sequence from the Paleozoic

TABLE 4-5 FORTRAN Program for Computing Transition-Probability Matrices from Sequences of *N* Integers [a]

```
C.....PROGRAM TO COMPUTE MARKOV TRANSITION MATRICES
C.....FROM SEQUENCES OF DATA CONSISTING OF INTEGERS
C.....IN THE RANGE 1 TO M, WHERE M = NUMBER OF STATES
C.....IN THE CHAIN,AND N = NUMBER OF DATA VALUES
C.....A MAXIMUM OF 50 STATES AMD 10,000 DATA VALUES CAN BE
C.....ACCOMODATED
      DIMENSION P(50,50), IDATA(10000)
    1 FORMAT(16I5)
    2 FORMAT(1H ,13F10.4)
      READ(5,1)N,M
      READ(5,1)(IDATA(I),I=1,N)
C.....CLEAR TALLY ARRAY TO ZERO
      DO 10 I=1,M
      DO 10 J=1,M
   10 P(I,J)=0.0
C.....FILL TALLY MATRIX
      DO 11 K=2,N
      I=IDATA(K-1)
      J=IDATA(K)
   11 P(I,J)=P(I,J)+1.0
C.....CONVERT TALLY MATRIX VALUES TO TRANSITION PROBABILITIES
C.....BY DIVIDING BY ROW SUMS
      DO 13 I=1,M
      SUM=0.0
      DO 12 J=1,M
   12 SUM=SUM+P(I,J)
      DO 13 J=1,M
   13 P(I,J)=P(I,J)/SUM
C.....WRITE TRANSITION MATRIX VALUES
      DO 15 I=1,M
   15 WRITE(6,2) (P(I,J),J=1,M)
      RETURN
      END
```

[a] Integers should be in the range 1 to M. Resulting transition-probability matrix is M × M.

flysch of the Caucasus and matched the estimated transition matrix with the theoretical transition matrix. The theoretical matrix was found to be in close agreement with the empirical matrix, showing that his theory of turbidite origin is consistent with the data.

Algorithm for Generating Synthetic Stratigraphic Sequences

A brief FORTRAN program for generating synthetic stratigraphic sequences is listed in Table 4-6. It is designed to accept transition matrices containing up to 10 states. The transition matrices must be read in as input to the program in cumulative form. In this form the probability values in each row progressively sum to 1.0. For example, if the transition-probability values of Table 4-3 are converted to cumulative form, we obtain

$$\begin{bmatrix} .74 & .97 & 1.00 \\ .11 & .72 & 1.00 \\ .06 & .44 & 1.00 \end{bmatrix} \tag{4-4}$$

TABLE 4-6 Program for Generating Synthetic Stratigraphic Columns, Using First-Order Markov Transition-Probability Matrices

```
C.....PROGRAM TO GENERATE STRATIGRAPHIC SEQUENCES
C.....USING A FIRST ORDER MARKOV TRANSITION PROBABILITY MATRIX
C.....    TITLE(M)              ALPHAMERIC TITLE
C.....    TRANS(I,J)            CUMULATIVE TRANSITION PROBABILITY MATRIX
C.....    OBSVAT(N)             SEQUENCE OF SIMULATED STATES
C.....    ALPHA(N,L)            ALPHAMERIC SYMBOLS FOR PLOTTING STRAT COLUMN
C.....    NSTATE                NUMBER OF STATES
C.....    NTIM                  NO. OF TRANSITIONS TO BE SIMULATED
C.....    IX                    ODD INTEGER  .GT. 8 DIGITS LONG FOR RANDU
C.....    RANDU                 RANDOM NO. GENERATOR, SEE CHAPTER 3
      DIMENSION TRANS(10,10), TITLE(20), ALPHA(10,3)
      INTEGER OBSVAT(50)
    1 FORMAT(20A4)
    2 FORMAT(1H1, 20A4///)
    3 FORMAT(3I10)
    4 FORMAT(10F5.2)
    5 FORMAT(1H0, 10F10.2)
    6 FORMAT(1H , 3A4)
C.....READ INPUT PARAMETERS
      READ(5,1) TITLE
      WRITE(6,2) TITLE
      READ(5,3) NSTATE, NTIM, IX
      DO 10 I=1,NSTATE
   10 READ(5,1) (ALPHA(I,L), L=1,3)
      DO 20 I=1, NSTATE
      READ(5,4) (TRANS(I,J), J=1,NSTATE)
   20 WRITE(6,5) (TRANS(I,J), J=1,NSTATE)
C.....GENERATE INITIAL STATE AT RANDOM
      CALL RANDU(IX, IY, YFL)
      IX=IY
      F=NSTATE
      OBSVAT(1)=F*YFL+0.9999999
C.....GENERATE STATE 2....NTIM
      DO 50 NT=2,NTIM
      I=OBSVAT(NT-1)
      CALL RANDU(IX,IY,YFL)
      IX=IY
      DO 30 J=1,NSTATE
   30 IF (YFL.LE.TRANS(I,J)) GO TO 40
   40 OBSVAT(NT)=J
   50 CONTINUE
C.....PRINT OUT STRAT. SEQUENCE IN RIGHT ORDER
      DO 60 NN=1,NTIM
      NT=NTIM-NN+1
      INDEX=OBSVAT(NT)
   60 WRITE(6,6) (ALPHA(INDEX,L), L=1,3)
      RETURN
      END
```

Conversion to cumulative form is a simple matter of adding each probability value to each succeeding value, moving from left to right within each row.

The cumulative transition matrix is stored in the TRANS array. After the transition matrix and other controlling parameters have been read in, the program generates a sequence of states. The initial state is generated at random, giving each of the states an equal probability of being selected. Subroutine RANDU, which produces uniformly distributed numbers, is used to generate a pseudorandom number, which lies in the range 0.0 to 1.0 and is assigned to identifier YFL. In turn this number is transformed so that it lies within a range of integers extending from 1 to NSTATE, where NSTATE is the total number of states. The resulting integer selected within this range provides the starting state. From then on the program generates subsequent states by sampling the cumulative-probability matrix. To select the state at time t the row of probability values pertaining to the state chosen at time $t-1$ is sampled. This is accomplished by calling RANDU, which produces pseudorandom number YFL between 0.0 and 1.0; YFL is then progressively compared with each element in the appropriate row of the TRANS array, starting with the lowest value, in the left-most column. Comparison of the numbers continues until one is found that is equal to or greater than YFL. The column containing that number identifies the next state. The process is illustrated in Figure 4-5.

Output from the program consists of a sequence of symbols (Figure 4-6) printed so as to imitate a graphic stratigraphic column, similar to that used in Figure 4-4. The program user specifies the symbols to be used. All of the standard alphanumeric characters are available, including letters, numbers, and other characters.

One problem in printing out the resulting stratigraphic column is that if each line is printed in the order in which it is calculated, the sequence will be upside down in the computer printout, with the order of deposition starting at the top instead of at the bottom. To avoid this difficulty the program provides for storage of each state in array OBSVAT(K), which is subsequently used to write out the strati-

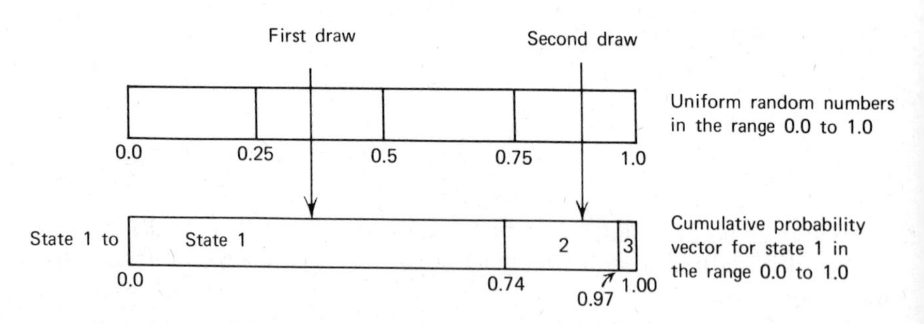

Figure 4-5 Illustration of use of uniform random number source to sample from cumulative probability vector. For example, on first draw random number is 0.35, which is less than 0.74, resulting in the selection of state 1 to succeed state 1. On second draw, random number is 0.88, which is less than 0.97 but greater than 0.74, resulting in the selection of state 2 to succeed state 1.

graphic column in the correct order when the simulation calculations are complete. The synthetic stratigraphic column in Figure 4-6 is based on input listed in Table 4-7, including alphanumeric symbols and a three-state cumulative transition matrix adapted from Krumbein's transition matrix based on observations of Chester strata (Table 4-3).

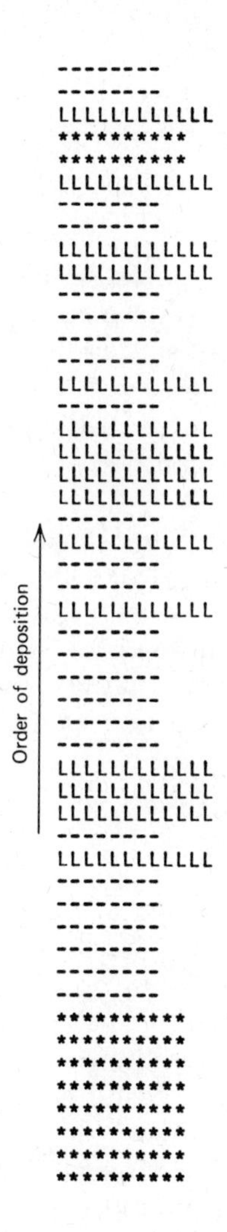

Figure 4-6 Output from FORTRAN program listed in Table 4-6. Probability values of three-state cumulative transition matrix are listed in cumulative form in Table 4-7. Symbols in stratigraphic column denote sandstone (state 1 = *****), shale (state 2 = -----), and limestone (state 3 = LLLLL).

TABLE 4-7 Listing of Input Data to FORTRAN Program to Produce Output Listed in Figure 4-6

```
KRUMBEIN MATRIX
           3                 50 121212121
**********
--------
LLLLLLLLLLLL
   74    97   100
   10    71   100
   05    43   100
```

POWERS OF MARKOV TRANSITION MATRICES

So far we have considered only single steps in Markov chains. The probability values in a transition matrix pertain to a transition from one state to the immediately succeeding state—that is, a single-step transition. What if we are interested in probabilities associated with more than a single step? Fortunately the probability values pertaining to multiple-step transitions can be calculated readily by powering a Markov transition matrix.

Suppose that we begin in state i. What is the probability that after n steps we shall be in state j? We denote this probability by $p_{ij}^{(n)}$. In using this representation we do not mean the nth power of the element p_{ij}. Instead we mean the probability of passing from state i to state j in n steps. We can represent these probability values, as before, in a matrix. A three-state Markov chain involving n steps may be written as follows:

$$\mathbf{P}^{(n)} = \begin{bmatrix} p_{11}^{(n)} & p_{12}^{(n)} & p_{13}^{(n)} \\ p_{21}^{(n)} & p_{22}^{(n)} & p_{23}^{(n)} \\ p_{31}^{(n)} & p_{32}^{(n)} & p_{33}^{(n)} \end{bmatrix} \tag{4-5}$$

To illustrate such a matrix let us construct a "tree" (Figure 4-7) involving a total of three steps showing transitions from state s_1 to all possible subsequent states, employing the single-step transition probabilities of matrix 4-2. State s_1 can pass only to state s_2 (probability of 1). Once in state s_2, however, there is a probability of $\frac{1}{3}$ of passing to state s_2, and $\frac{2}{3}$ probability of passing to state s_3. State s_3 can be reached via two paths. Thus the probability $p_{13}^{(3)}$ of reaching state s_3, beginning in state s_1, in three steps is $(1)(\frac{1}{3})(\frac{2}{3}) + (1)(\frac{2}{3})(\frac{1}{2}) = \frac{5}{9}$. Similarly the probability of reaching state s_2 is $p_{12}^{(3)} = (1)(\frac{1}{3})(\frac{1}{3}) = \frac{1}{9}$, and the probability of reaching state s_1 is $p_{11}^{(3)} = (1)(\frac{2}{3})(\frac{1}{2}) = \frac{3}{9}$. The values for

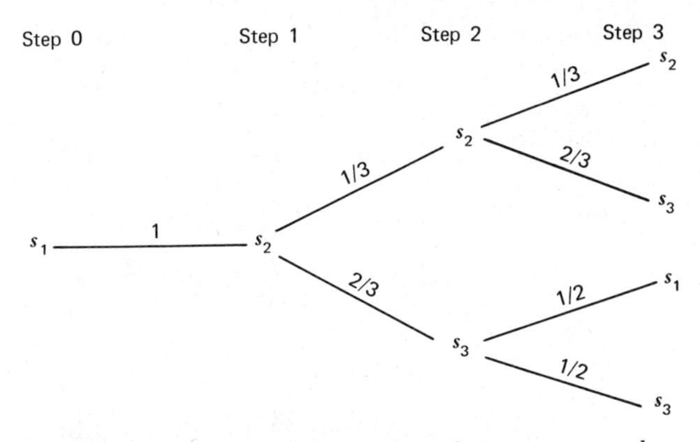

Figure 4-7 Three-step tree, illustrating transitions from state s_1 to subsequent states for transition matrix 4-2.

the other three-step transition probabilities are obtained similarly and may be written in matrix form as

$$\mathbf{P}^{(3)} = \begin{array}{c} \\ s_1 \\ s_2 \\ s_3 \end{array}\begin{array}{c} \begin{array}{ccc} s_1 & s_2 & s_3 \end{array} \\ \begin{bmatrix} \frac{3}{9} & \frac{1}{9} & \frac{5}{9} \\ \frac{3}{9} & \frac{1}{9} & \frac{5}{9} \\ \frac{1}{4} & \frac{2}{4} & \frac{1}{4} \end{bmatrix} \end{array} \tag{4-6}$$

As in the single-step transition matrix, the values in each row sum to 1.

To calculate the probability values after a prescribed number of steps we could of course perform calculations similar to that shown above. This would be tedious, however, and there is a simpler way of doing this via matrix algebra. Let us first write down the individual relationships in normal algebra for two successive steps in a two-state matrix:

$$\begin{aligned} p_{11}^{(2)} &= p_{11}p_{11} + p_{12}p_{21} \\ p_{12}^{(2)} &= p_{11}p_{12} + p_{12}p_{22} \\ p_{21}^{(2)} &= p_{21}p_{11} + p_{22}p_{21} \\ p_{22}^{(2)} &= p_{21}p_{12} + p_{22}p_{22} \end{aligned} \tag{4-7}$$

In matrix algebra this is exactly equivalent to multiplying the one-step transition matrix by itself, or

$$\begin{bmatrix} p_{11}^{(2)} & p_{12}^{(2)} \\ p_{21}^{(2)} & p_{22}^{(2)} \end{bmatrix} = \begin{bmatrix} p_{11} & p_{12} \\ p_{21} & p_{22} \end{bmatrix} \times \begin{bmatrix} p_{11} & p_{12} \\ p_{21} & p_{22} \end{bmatrix} \tag{4-8}$$

In condensed form we may write

$$\mathbf{P}^{(2)} = \mathbf{PP} \tag{4-9}$$

The three-step transition may be written

$$\begin{bmatrix} p_{11}^{(3)} & p_{12}^{(3)} \\ p_{21}^{(3)} & p_{22}^{(3)} \end{bmatrix} = \begin{bmatrix} p_{11}^{(2)} & p_{12}^{(2)} \\ p_{21}^{(2)} & p_{22}^{(2)} \end{bmatrix} \times \begin{bmatrix} p_{11} & p_{12} \\ p_{21} & p_{22} \end{bmatrix} \tag{4-10}$$

or

$$\mathbf{P}^{(3)} = \mathbf{P}^{(2)}\mathbf{P} \tag{4-11}$$

The four-step matrix is obtained by another multiplication,

$$\mathbf{P}^{(4)} = \mathbf{P}^{(3)}\mathbf{P} \tag{4-12}$$

In general for the nth step we may write

$$\mathbf{P}^{(n)} = \mathbf{P}^{(n-1)}\mathbf{P} \tag{4-13}$$

If a matrix of transition probabilities is successively powered with the result that each row (row vector) is the same as every other row (forming a fixed probability vector), the matrix is termed a regular transition matrix. To illustrate, consider a two-state matrix with one-step transition probabilities as follows:

$$\mathbf{P} = \begin{bmatrix} .67 & .33 \\ .50 & .50 \end{bmatrix} \tag{4-14}$$

Raising this to the second power yields

$$\mathbf{P}^{(2)} = \mathbf{PP} = \begin{bmatrix} .67 & .33 \\ .50 & .50 \end{bmatrix} \times \begin{bmatrix} .67 & .33 \\ .50 & .50 \end{bmatrix} = \begin{bmatrix} .61 & .39 \\ .58 & .42 \end{bmatrix} \tag{4-15}$$

Then for the third power

$$\mathbf{P}^{(3)} = \mathbf{P}^{(2)}\mathbf{P} = \begin{bmatrix} .61 & .39 \\ .58 & .42 \end{bmatrix} \times \begin{bmatrix} .67 & .33 \\ .50 & .50 \end{bmatrix} = \begin{bmatrix} .60 & .40 \\ .60 & .40 \end{bmatrix} \tag{4-16}$$

Thus after only three steps, and rounding to two decimal places, the rows in the matrix have converged to the same values. Subsequent powering will not alter the composition of this matrix. These values may be verified by working them out with the aid of the tree diagram in Figure 4-8. The limiting matrix obtained by successive powering may be called the **T**-matrix.

$$\mathbf{T} = \begin{array}{c} \\ s_1 \\ s_2 \end{array}\begin{array}{c} \begin{array}{cc} s_1 & s_2 \end{array} \\ \begin{bmatrix} .60 & .40 \\ .60 & .40 \end{bmatrix} \end{array} \tag{4-17}$$

This matrix says in effect that, when the limit **T** has been reached, the probabilities of passing to either state s_1 or s_2 are independent of the starting state because the two rows of transition probabilities are identical. This outcome accords with experience. For example, the day-to-day weather is strongly Markovian. If today is rainy, tomorrow is more likely to be rainy than it would be if today were a sunny day. But, on a day-to-day basis, the probability of a rainy day succeeding a rainy day 30 days from now is very little influenced by the state of today's weather. In matrix terms, if a matrix of transition probabilities describing day-to-day weather is raised to the 30th power, the probability values will be virtually the same on a row-by-row basis, signifying that the particular starting state has virtually no effect by the time the original matrix has been raised to the 30th power. The limiting matrix **T** has the following property: if we multiply any row vector (labeled **t**) of matrix **T** by the matrix of single-step transition probabilities, we obtain **t**. For example, if

$$\mathbf{P} = \begin{bmatrix} .667 & .333 \\ .500 & .500 \end{bmatrix} \tag{4-18}$$

and

$$\mathbf{t} = [.6 \quad .4] \tag{4-19}$$

then

$$\mathbf{tP} = \mathbf{t} \tag{4-20}$$

For example,

$$[.6 \quad .4] \times \begin{bmatrix} .667 & .333 \\ .500 & .500 \end{bmatrix} = [.6 \quad .4] = \mathbf{t} \tag{4-21}$$

We may also notice that the fixed probability vector **t** expresses the equilibrium proportions of the various states. For example, a two-

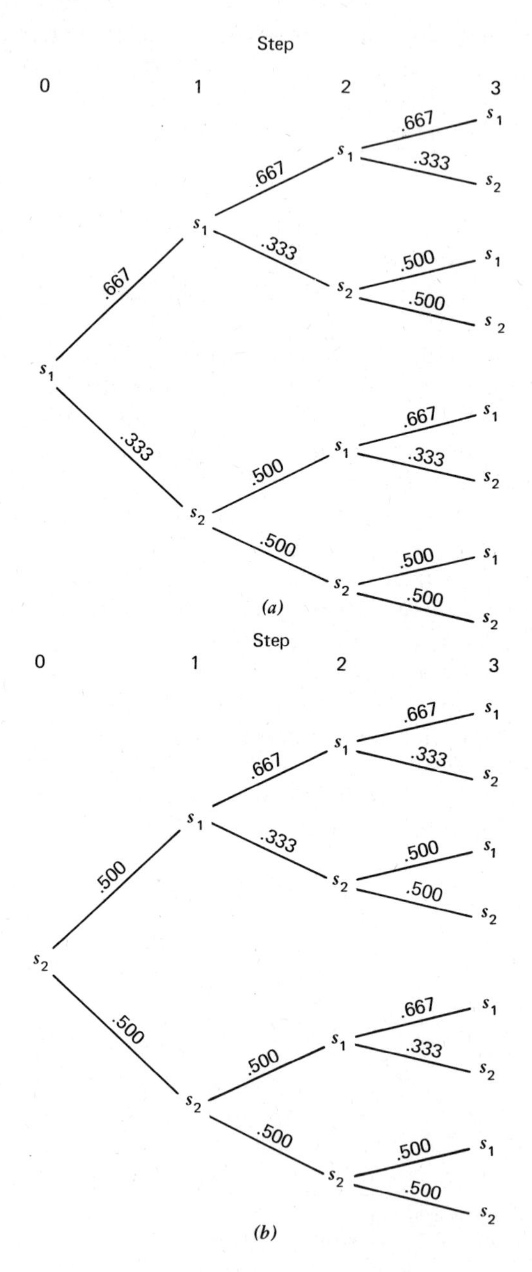

Figure 4-8 Tree diagram for two-state Markov chain to show successive probabilities of reaching states s_1 and s_2 in three steps. (*a*) Tree beginning with state s_1; (*b*) tree beginning with state s_2.

state sedimentary sequence of sand and shale with the transition matrix

$$\begin{array}{c} \\ \text{sand} \\ \text{shale} \end{array} \begin{array}{c} \begin{array}{cc} \text{sand} & \text{shale} \end{array} \\ \begin{bmatrix} .67 & .33 \\ .50 & .50 \end{bmatrix} \end{array} \tag{4-22}$$

gives the fixed probability vector

$$\begin{array}{cc} \text{sand} & \text{shale} \\ [.60 & .40] \end{array} \tag{4-23}$$

This means that, if there are sufficient steps, the probability of making a transition to sand is .60 and to shale .40. In turn this implies that the proportions of sand and shale in the entire sequence are 0.6 and 0.4, respectively.

Since obtaining the successive powers of a transition matrix is simply a matter of repeated matrix multiplication, we can readily devise an algorithm to carry out the operations. Table 4-8 provides a

TABLE 4-8 FORTRAN Program for Computing Probability Values After N Successive steps of First-Order Transition Matrix Containing M States[a]

```
C.....PROGRAM TO COMPUTE PROBABILITIES AFTER N STEPS OF
C.....FIRST-ORDER TRANSITION MATRIX CONTAINING M STATES
      DIMENSION TM(25,25), PM(25,25) , PROD(25,25)
    1 FORMAT  (2I10)
    2 FORMAT(7F10.2)
    3 FORMAT(1H ,7F10.4)
    4 FORMAT(1H ,'STEP ',I2)
      READ(5,1)N,M
      DO 5  I=1,M
    5 READ(5,2)(TM(I,J),J=1,M)
      DO  6 I=1,M
      DO  6 J=1,M
    6 PM(I,J)=TM(I,J)
C.....PRINT OUT TRANSITION MATRIX VALUES
      L=1
      WRITE(6,4)L
      DO 8 I=1,M
    8 WRITE(6,3)(TM(I,J),J=1,M)
C.....FOR EACH STEP, PERFORM MATRIX MULTIPLICATION
      DO 30 L=1,N
      L1=L+1
      WRITE(6,4) L1
      DO 10 I=1,M
      DO 10 J=1,M
      PROD(I,J)=0.
      DO 10 K=1,M
   10 PROD(I,J)=PROD(I,J)+TM(I,K)*PM(K,J)
      DO 30 I=1,M
      DO 20 J=1,M
   20 PM(I,J)=PROD(I,J)
   30 WRITE(6,3) (PM(I,J), J=1,M)
      RETURN
      END
```

[a]Output from program, dealing with a 3×3 matrix through 11 steps, is listed in Table 4-9.

listing of a FORTRAN program for computing the successive powers of a transition matrix. The program is generalized and can be used for matrices of any size.

An example of output from the program is listed in Table 4-9. The initial transition matrix (step 1) is the 3 × 3 transition matrix of Table 4-4 obtained in a stratigraphic simulation experiment. As the matrix is raised to successive powers, the sequence of values in each row approaches more and more closely the sequence in the other rows.

TABLE 4-9 Output from FORTRAN Program in Table 4-8, Listing Probability Values in 3 × 3 Matrix through N Steps[a]

```
STEP  1
     0.7700      0.1800      0.0500
     0.0800      0.6500      0.2700
     0.0700      0.3200      0.6100
STEP  2
     0.6108      0.2716      0.1176
     0.1325      0.5233      0.3442
     0.1222      0.4158      0.4620
STEP  3
     0.5003      0.3241      0.1756
     0.1680      0.4741      0.3579
     0.1597      0.4401      0.4002
STEP  4
     0.4234      0.3569      0.2196
     0.1923      0.4529      0.3547
     0.1862      0.4429      0.3709
STEP  5
     0.3700      0.3785      0.2515
     0.2092      0.4425      0.3483
     0.2048      0.4401      0.3552
STEP  6
     0.3328      0.3931      0.2741
     0.2208      0.4368      0.3424
     0.2177      0.4366      0.3457
STEP  7
     0.3069      0.4031      0.2900
     0.2290      0.4332      0.3378
     0.2268      0.4336      0.3396
STEP  8
     0.2888      0.4101      0.3011
     0.2346      0.4309      0.3345
     0.2331      0.4313      0.3356
STEP  9
     0.2763      0.4149      0.3088
     0.2385      0.4294      0.3321
     0.2375      0.4297      0.3328
STEP 10
     0.2675      0.4182      0.3142
     0.2413      0.4283      0.3304
     0.2405      0.4286      0.3309
STEP 11
     0.2615      0.4206      0.3180
     0.2432      0.4276      0.3293
     0.2427      0.4277      0.3296
```

[a]Initial, or single-step, values of the transition matrix are listed under step 1.

TESTING FOR THE MARKOV PROPERTY

The transitions in any series of observations that involve either discrete-state or discrete-time phenomena can be described by a matrix of transition probabilities, provided of course that there is a finite number of states in the system. For example, suppose we conduct experiments with a statistician's "urn" that contains three red balls, two white balls, and one green ball. If we pick a ball at random, replace it, and repeat the process many times, we can obtain information on which to base a transition matrix. This matrix will describe the probability of drawing a white ball immediately after a red ball, and so on, as follows:

$$\begin{array}{l} \\ \text{red} \\ \text{white} \\ \text{green} \end{array} \begin{array}{c} \text{red white green} \\ \begin{bmatrix} \frac{1}{2} & \frac{1}{3} & \frac{1}{6} \\ \frac{1}{2} & \frac{1}{3} & \frac{1}{6} \\ \frac{1}{2} & \frac{1}{3} & \frac{1}{6} \end{bmatrix} \end{array} \tag{4-24}$$

We note that each row is the same. This tells us that the probability of drawing a particular color does not depend on the color of the ball that has been drawn previously. In other words, there is no Markov property, or memory effect.

A statistical test for the Markov property forms an important component in the experimenter's list of tools. The test distinguishes between the two alternative hypotheses that either the successive events are independent of each other (the null hypothesis) or the events are not independent. If not independent, they could form a first-order Markov chain. The test statistic λ is

$$\lambda = \prod_{i,j}^{m} \left(\frac{p_j}{p_{ij}}\right)^{n_{ij}} \tag{4-25}$$

where $-2 \log_e \lambda$ is distributed asymptotically as χ^2 with $(m-1)^2$ degrees of freedom (Anderson and Goodman, 1957). This expression is equivalent to the more convenient computational equation

$$-2 \log_e \lambda = 2 \sum_{i,j}^{m} n_{ij} \log_e \left(\frac{p_{ij}}{p_j}\right) \tag{4-26}$$

where p_{ij} = probability in cell i, j of the transition-probability matrix

p_j = marginal probabilities for the jth column $\left(= \sum_{i}^{m} n_{ij} / \sum_{i,j}^{m} n_{ij}\right)$

n_{ij} = transition frequency total in cell i, j of the original tally matrix of observed transitions

m = total number of states

Let us illustrate the calculation of this statistic with reference to the data from Table 4-3. The values of p_j are calculated by taking the marginal total for the jth column in the tally matrix and dividing it by the overall total. For the first column $p_1 = 78/309 = .25$, for the second column $p_2 = 140/309 = .45$, and $p_3 = 91/309 = .29$. Then

$$
\begin{aligned}
-2 \log_e \lambda = 2\Big[& 58 \log_e \frac{.74}{.25} + 18 \log_e \frac{.23}{.45} + 2 \log_e \frac{.03}{.29} \\
& + 15 \log_e \frac{.11}{.25} + 86 \log_e \frac{.61}{.45} + 39 \log_e \frac{.28}{.29} \\
& + 5 \log_e \frac{.06}{.25} + 35 \log_e \frac{.38}{.45} + 51 \log_e \frac{.56}{.29}\Big] \\
& = 2(62.941 - 12.081 - 4.537 - 12.315 + 26.162 - 1.369 \\
& \quad - 7.136 - 5.918 + 33.561) \\
& = 158.616
\end{aligned}
$$

The number of degrees of freedom $(m-1)^2 = (3-1)^2 = 4$. If we make the level of significance $\alpha = 0.05$, we can now consult a table of values of the χ^2 distribution (Fisher and Yates, 1963) where the corresponding χ^2 value is 9.49. Our calculated value of $-2 \log \lambda = 158.6$ is much greater. Therefore we can reject the null hypothesis that these transitions are from an independent events process and can accept the alternative hypothesis that the transitions have the Markov property. Anderson and Goodman (1957) describe the details of the testing process, and Krumbein (1967) presents a FORTRAN program for carrying out the calculations.

STATIONARITY OF MARKOV CHAINS

Up to this point we have assumed that the Markov matrices are the result of a process that is stationary in time or space. By stationary we mean that the transition probabilities are constant through time or space. Thus, if we take a very long stratigraphic sequence and break it up into a number of subintervals, we can calculate a separate transition-probability matrix for each (Figure 4-9). For a stationary process these matrices should all be approximately equal to each other. Anderson and Goodman (1957) present some statistics for testing for stationarity.

In a stationary Markov chain p_{ij} is the probability of a transition from state i at time $t-1$ to state j at time t. In a nonstationary Markov chain the transition probabilities vary with time (or space). Thus $p_{ij}(t)$ is the

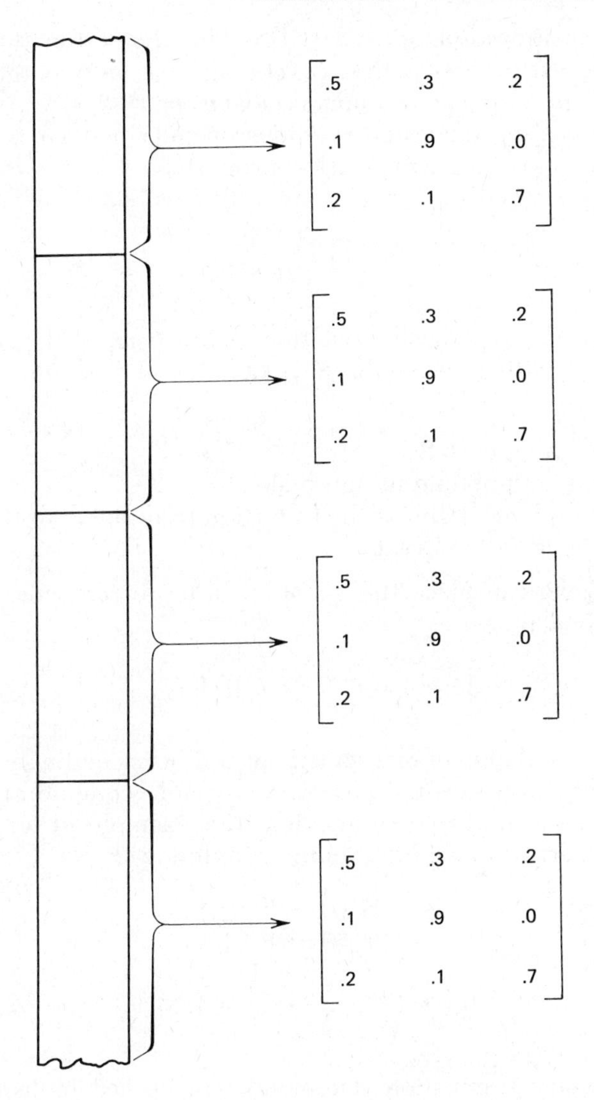

Figure 4-9 Stratigraphic sequence subdivided into intervals for each of which a separate transition matrix is calculated. If matrices are nearly identical, Markov process can be considered to be stationary.

probability of a transition from state i to state j and is a function of time. The null hypothesis that we wish to test is that $p_{ij}(t) = p_{ij}$ for $(t = 1, \cdots, T)$. In other words, the test is to see if the transition probabilities calculated from each subinterval of time are equal to the lumped transition-probability matrix obtained by estimation over the whole interval (null hypothesis). The alternative hypothesis is that the Markov process is nonstationary; that is, $p_{ij}(t) \neq p_{ij}$.

The test statistic suggested by Anderson and Goodman is somewhat similar to that for testing for Markov property:

$$\lambda = \prod_{t}^{T} \prod_{i,j}^{m} \left[\frac{p_{ij}}{p_{ij}(t)} \right]^{n_{ij}(t)} \tag{4-27}$$

When the null hypothesis of stationarity is true, $-2 \log_e \lambda$ is distributed at χ^2 with $(T-1)[m(m-1)]$ degrees of freedom. The other symbols are

m = number of states
T = number of time subintervals
$n_{ij}(t)$ = frequency tally for the transition from state i to state j in the the tth subinterval

As before, we can place the equation in more convenient form for computational purposes:

$$-2 \log_e \lambda = 2 \sum_{t}^{T} \sum_{i,j}^{m} n_{ij}(t) \log_e \left[\frac{p_{ij}(t)}{p_{ij}} \right] \tag{4-28}$$

Manual calculation of this statistic would be exceedingly tedious. It is a straightforward matter, however, to write a computer program to accomplish the multiple summation. For example, if we name the FORTRAN arrays and other identifiers as follows:

$n_{ij}(t) \rightarrow$ N(I,J,K)
$p_{ij}(t) \rightarrow$ P(I,J,K)
$p_{ij} \rightarrow$ PP(I,J)
$i, j = 1 \cdots m \rightarrow$ I,J = 1, M
$t = 1 \cdots T \rightarrow$ K=1,NT

The necessary summation statements can be accomplished by the following nested DO-loops:

```
      .
      .
      .
      SUM=0.0
      DO 10 K=1,NT
      DO 10 I=1,M
```

```
      DO 10 J=1,M
   10 SUM=SUM+N(I,J,K)*ALOG(P(I,J,K)/PP(I,J))
      TESTAT=2.0*SUM
      NDEGFR=(NT-1)*M*(M-1)
      WRITE(6, 1)TESTAT,NDEGFR
    1 FORMAT(1H , F15.5,15)
      .
      .
      .
```

If the null hypothesis of stationarity is to be accepted, then the calculated value of χ^2 must be less than the tabulated value at some preselected level of significance for that number of degrees of freedom.

Where a sedimentary sequence is shown to be nonstationary, we might hypothesize, among various alternative explanations, that more than one regime of sedimentary environmental conditions had influenced the sequence of deposits. It would be instructive, for example, to compare Markov transition matrices for different types of sedimentary environments. It might be possible to show that certain transition matrices are highly characteristic for certain sedimentary regimes. If so, this fact would be useful for simulation, as well as for more conventional analytical studies, perhaps providing an additional tool for discriminating between various ancient sedimentary environments.

MARKOV CHAINS CLASSIFIED BY DEPENDENCE, ORDER, AND STEP LENGTH

Markov chains may be classified according to dependence, order, and step length. Thus far we have considered chains that are dependent only upon the immediately preceding state. These chains may be defined as being single-dependence chains because only a single preceding state is involved. They are also first-order chains because the preceding state is the *immediately* preceding state. Furthermore, the transition involves a single step, of unit length.

We can define chains, however, whose dependency relationships involve more than one preceding state. A double-dependence chain, for example, is dependent on two preceding states. If these two states are the two immediately preceding states, the chain is a second-order chain, each step of unit length. However, if the two steps on which the chain is dependent are not the immediately preceding two states, the chain is of higher order than a second-order chain, and either or both of the steps involved may be greater than a unit step.

The significance of a multiple-dependence chain may be clarified if we consider the simplest possible example, a two-state, second-order chain. In such a chain the probability of a transition to either of the states is influenced not only by both the preceding states but also by the sequence in which the two states occurred. We could represent these relationships with the following transition matrix:

$$
\begin{array}{cc|cc}
\multicolumn{2}{c|}{\text{preceding states}} & \multicolumn{2}{c}{\text{succeeding state at } t} \\
t-2 & t-1 & s_1 & s_2 \\
s_1 & s_1 & p_{11} & p_{12} \\
s_1 & s_2 & p_{21} & p_{22} \\
s_2 & s_1 & p_{31} & p_{32} \\
s_2 & s_2 & p_{41} & p_{42}
\end{array}
\qquad (4\text{-}29)
$$

Although transition matrices are not ordinarily written in this fashion, this matrix makes it clear that the states at both $t-2$ and $t-1$ influence the transition from $t-1$ to t. A triple-dependence, two-state transition matrix written in this fashion contains double the number of elements, the number of rows increasing by a power of 2. These simple examples point out that multiple-dependence Markov chains are exceedingly complex if the number of states is very great. For example, a transition matrix for a three-state, double-dependence chain, if written in the form of matrices 4-29 and 4-30, has 9 rows, and a three-state, triple-dependence transition matrix has 27 rows.

$$
\begin{array}{ccc|cc}
\multicolumn{3}{c|}{\text{preceding states}} & \multicolumn{2}{c}{\text{succeeding state at } t} \\
t-3 & t-2 & t-1 & s_1 & s_2 \\
s_1 & s_1 & s_1 & p_{11} & p_{12} \\
s_1 & s_1 & s_2 & p_{21} & p_{22} \\
s_1 & s_2 & s_1 & p_{31} & p_{32} \\
s_1 & s_2 & s_2 & p_{41} & p_{42} \\
s_2 & s_1 & s_1 & p_{51} & p_{52} \\
s_2 & s_1 & s_2 & p_{61} & p_{62} \\
s_2 & s_2 & s_1 & p_{71} & p_{72} \\
s_2 & s_2 & s_2 & p_{81} & p_{82}
\end{array}
\qquad (4\text{-}30)
$$

For transition matrices written in this form, the general rule is that the number of rows is equal to the number of states, raised to the power equal to the dependence number (single, double, triple, etc.) of the chain. It is difficult to deal with Markov chains higher than quadruple because of their complexity.

The relationships between dependence, order, and step length may be clarified if we think of a Markov chain as a series of steps on a ladder in which the rungs are equally spaced. We can think of the chain as advancing forward through time in equal increments, each increment (or rung) being indexed numerically, as for example $\cdots\cdots, t-2, t-1, t, t+1, t+2, \cdots$. The transition from $t-2$ to $t-1$, etc., constitutes a step of unit length. A transition in a single step from $t-3$ to $t-1$ would involve two unit lengths, and so on. Figure 4-10 contains example "ladder diagrams" illustrating various combinations of dependence, order, and step length. It should be noted that the terms "order" and "dependence" have not been used consistently in the literature. In the geological literature, "order" has commonly been used in the sense that we have used "dependence". While we have attempted to adhere to the distinction between the terms, "order" has been used within the FORTRAN programs of Tables 4-10 and 4-12, whereas "dependence" would be more correct.

Although it is possible to represent multiple-order Markov chains by elongate, rectangular, two-dimensional transition matrices (matrices 4-29 and 4-30), it is more convenient for most algebraic and computational purposes to use multidimensional, "square" matrices or arrays. For example, a single-dependence chain can be represented by a square two-dimensional array (Figure 4-11*a*), a double-dependence chain can be represented by a transition matrix in the form of a three-dimensional array forming a cube (Figure 4-11*b*), and a triple-dependence chain, in turn, can be represented by a four-dimensional array that forms a "hypercube."

The elements in transition matrices are identified by subscripts. Elements in a two-dimensional matrix (single-dependence chain) are specified by two subscripts; i and j are commonly used as row and column subscripts, respectively (Figure 4-12*a*). Elements in a double-dependence chain's transition matrix require a third subscript, i commonly being used as index to the "layers" in the cube representing the matrix (Figure 4-12*b*). Elements in a triple-dependence transition matrix require four subscripts, for example, p_{ijkl}.

In a single-dependence chain state j at time t is conditional on state i at time $t-\mu$. In a double-dependence chain state k at time t is conditional on state j at time $t-\mu$ and also on state i at time $t-\mu-\tau$. In a

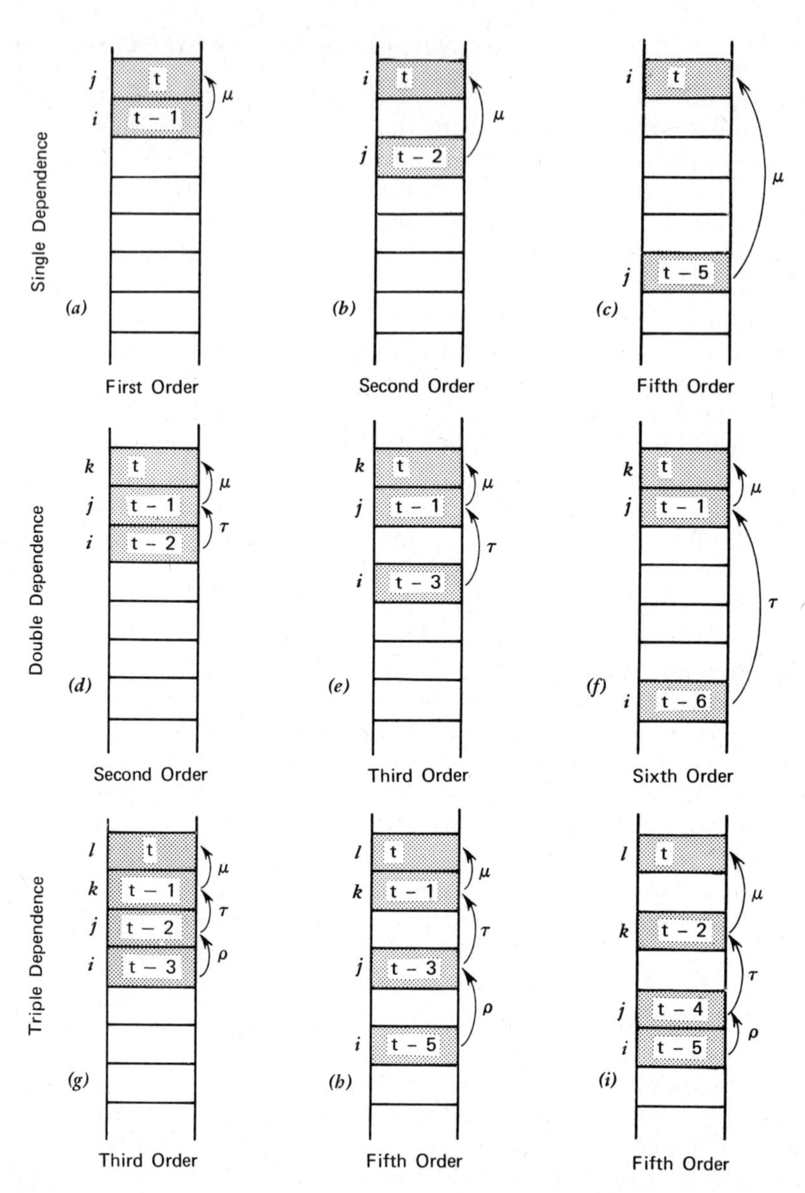

Figure 4-10 "Ladder" diagrams to illustrate classification of Markov chains according to various combinations of dependence, order, and step length. Distance between each rung in ladders is of unit length. Upper row (*a*, *b*, and *c*) illustrates chains involving a single dependence. Chain *a* is first order because state *j* at time *t* is dependent upon immediately preceding state *i* at time *t*-1. Chains *b* and *c* are higher order because they are dependent upon earlier states, even though only a single dependence is involved; μ denotes step length. Examples involving double dependence (*d*, *e* and *f*) are shown in middle row; τ denotes step length for earlier of the two steps involved in the dependence relationship. Examples (*g*, *h*, and *i*) involving triple dependence are shown in third row, where ρ denotes length of initial step in dependency relationship.

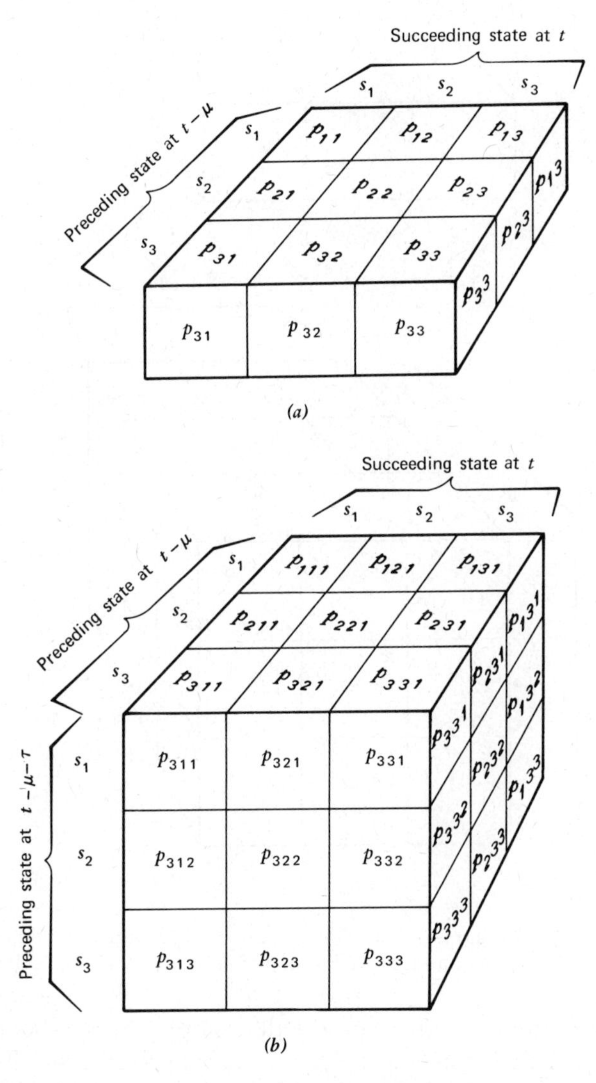

Figure 4-11 Block diagrams to illustrate organization of three-state transition matrices for (*a*) single-dependence Markov chain and (*b*) double-dependence Markov chain. Probability values in elements in each row in (*a*) sum to 1.0. In (*b*) each tier of elements defined by an individual row and layer sums to 1.0. Use of i, j, and k to represent subscript values is shown in Figure 4-12.

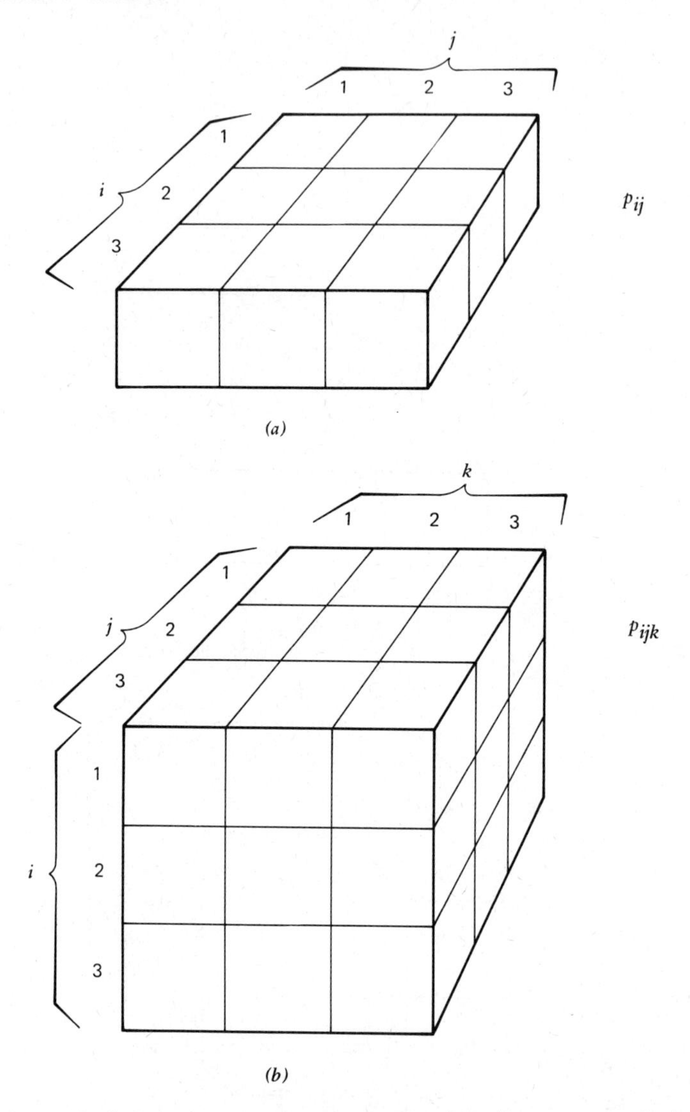

Figure 4-12 Block diagrams to accompany Figure 4-11, illustrating method of indexing (*a*) single-dependence and (*b*) double-dependence matrices with subscripts.

triple-dependence chain state l is dependent on state k at $t-\mu$, state j at $t-\mu-\tau$, and state i at $t-\mu-\tau-\rho$, where ρ is the length of the third-dependence step. The use of subscripts with multidimensional arrays makes it feasible to manipulate double and higher-dependency transition matrices.

Many natural phenomena have more than a first-order memory. Some sedimentary sequences appear to have major cycles superimposed on minor cycles. It is possible to model these complex rhythmic sequences with Markov chains greater than single-dependence. For example, Harbaugh (1966) employed triple-dependence, third-order Markov chains to simulate the stratigraphic succession of Pennsylvanian marine bank-building organisms in southeastern Kansas.

Schwarzacher (1967) examined a Pennsylvanian rock sequence in Kansas by Markov methods. He sampled the four-state (rock type) system at 5-foot intervals. By applying the maximum-likelihood criterion $-2\log_e \lambda$ discussed previously, he found that the first-order transition-probability matrix had a significant Markov memory. He also tested for stationarity, employing a "runs" test (not the stationarity test outlined above) on the mean sandstone values, averaged over 50-foot intervals. He found only small deviations from stationarity, which were deemed unimportant. Among other tests, Schwarzacher tested his sequence of transitions for double-dependence Markov memory, using another form of the maximum-likelihood criterion proposed by Anderson and Goodman (1957).

Below we discuss methods for generating double-dependence Markov sequences and for testing the sequences for a double-dependence memory.

Algorithm for Simulating Double-Dependence Chains

Table 4-10 lists a FORTRAN program for simulating double-dependence Markov chains, with output resembling a stratigraphic sequence (Figure 4-13). The principles are identical to those used in generating single-dependence chains. Input to the program (Table 4-11) includes the number of states (NSTATE), the length of the second step in whole-number multiples of the unit step (NTAU), the number of transitions to be generated (NTIM), an odd, nine-digit integer random number supplied as the initial value to subroutine RANDU, and the transition probabilities. Several two-dimensional double-dependence transition-probability matrices are read in and stored in a three-dimensional array PROB(I,J,K). They are then automatically transformed to cumulative-probability form. The first (NTAU+1) states are generated at random, and then subsequent states are generated by sampling from the cumulative transition matrices.

For time t state k is conditional on both the state j that was chosen at $t-1$ and state i at $t-1-\tau$. Selection of the probability value in the kth column in the jth row and ith layer (Figure 4-12b) is made with a

TABLE 4-10 FORTRAN Program for Generating Stratigraphic Sequences with Double-Dependence Memory with Variable Length Second Step

```
C.....      NSTATE          NO. OF STATES
C.....      NTAU            LENGTH OF SECOND ORDER STEP
C.....      NTIM            NO. OF TIME STEPS IN SIMULATION
C.....      IX              ODD INTEGER TO START RANDOM NUMBER GENERATOR
C.....      PROB(I,J,K)     TRANSITION PROBABILITY MATRICES
C.....      I               STATE AT TIME (T-NTAU-1)
C.....      J               STATE AT TIME (T-1)
C.....      K               STATE AT TIME (T)
      DIMENSION PROB(10,10,10), CMPROB(10,10,10), SYMBOL(10,3)
      INTEGER OBSVAT(1000)
    1 FORMAT(4I10)
    2 FORMAT(3A4)
    3 FORMAT(10F5.2)
    4 FORMAT(1H1, 'SECOND ORDER MARKOVIAN SEQUENCE'/ 1X, 'SECOND ORDER S
     1TEP LENGTH =', I5//)
    5 FORMAT(1H ,10X, 3A4)
      READ(5,1) NSTATE, NTAU, NTIM, IX
      DO 9 II=1,NSTATE
    9 READ(5,2) (SYMBOL(II,M), M=1,3)
      RSTATE=NSTATE
C.....READ IN TRANSITION PROBABILITY MATRICES
      DO 10 I=1,NSTATE
      DO 10 J=1,NSTATE
   10 READ(5,3) (PROB(I,J,K), K=1,NSTATE)
C.....CALCULATE CUMULATIVE TRANSITION PROBABILITY MATRICES
      NSL1=NSTATE-1
      DO 30 I=1,NSTATE
      DO 30 J=1,NSTATE
      CMPROB(I,J,1)=PROB(I,J,1)
      IF (NSL1.LT.2) GO TO 30
      DO 20 K=2,NSL1
   20 CMPROB(I,J,K)=CMPROB(I,J,K-1)+PROB(I,J,K)
   30 CMPROB(I,J,NSTATE)=1.0000
C.....BEGIN SIMULATION - FIRST SELECT INITIAL (NTAU+1) STATES AT RANDOM
      NSTOP=NTAU+1
      DO 40 N=1,NSTOP
      CALL RANDU(IX,IY,YFL)
      IX=IY
      INDEX=(YFL*RSTATE)+0.99999
   40 OBSVAT(N)=INDEX
      NSTART=NSTOP+1
C.....BEGIN SAMPLING FROM CUMULATIVE TRANSITION MATRICES
      DO 60 N=NSTART, NTIM
      I=OBSVAT(N-NTAU-1)
      J=OBSVAT(N-1)
      CALL RANDU(IX,IY,YFL)
      IX=IY
      DO 50 K=1,NSTATE
   50 IF (YFL.LE.CMPROB(I,J,K)) GO TO 59
   59 OBSVAT(N)=K
   60 CONTINUE
C.....WRITE OUT STRATIGRAPHIC SEQUENCE
      WRITE(6,4) NTAU
      DO 70 NN=1,NTIM
      N=NTIM-NN+1
      LL=OBSVAT(N)
   70 WRITE(6,5) (SYMBOL(LL,M), M=1,3)
      RETURN
      END
```

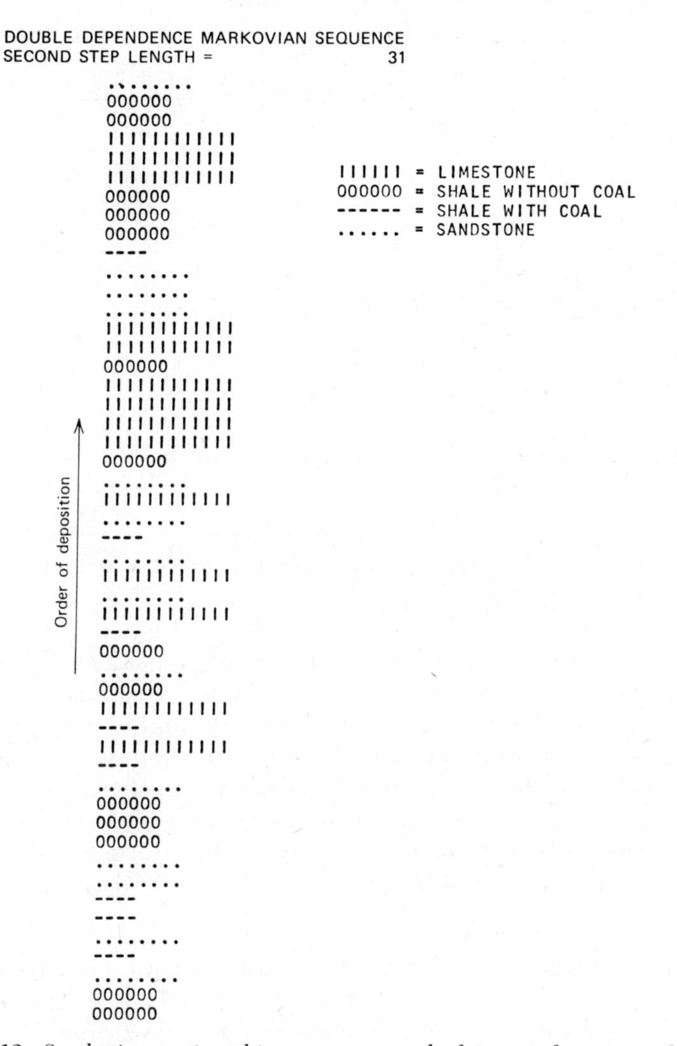

Figure 4-13 Synthetic stratigraphic sequence embodying a four-state, double-dependence Markov chain with earlier step length (τ) of 31 unit steps. Sequence is based on input data listed in Table 4-11, used with program of Table 4-10. Transition-probability values employed were calculated by Schwarzacher (1967) from observations of cyclically bedded strata on Pennsylvanian age in eastern Kansas.

random number drawn from a uniform distribution in the range 0.0 to 1.0. After state k has been chosen in this manner, the sequence is advanced by a unit step, and the process is repeated.

The computer program stores the generated sequence until all the simulation calculations are complete, and then the sequence is

TABLE 4-11 Example of Input Data to Program Listed in Table 4-10, Which Produced Output Listed in Table 4-13[a]

```
          4          31              50 436054367
........
----
||||||||||||
000000
   71    10    14    05 ┐
   00    00    00   100 │ i = 1
   05    00    55    41 │
   00    00   100    00 ┘
  100    00    00    00 ┐
   25    25    25    25 │ i = 2
   38    12    38    12 │
   00    00   100    00 ┘
   22    33    33    11 ┐
   10    30    50    10 │ i = 3
   13    00    73    14 │
   00    03    29    70 ┘
   88    06    00    06 ┐
   00    50    00    50 │ i = 4
   13    00    82    05 │
   19    00    25    56 ┘
```

[a]Lines below fifth line contain transition-probability values calculated by Schwarzacher (1967).

printed out so that the sequence is presented in advancing order upward, so as to accord with the law of stratigraphic succession. It should be emphasized, however, that the program can be used to represent any double-dependence chain. Since the user supplies the symbols to be printed out, he can employ symbols to represent his particular needs, whether he deals with sequences in time or space. To illustrate use of the program two additional synthetic stratigraphic sequences involving double-dependence chains are described below.

Stratigraphic Example 1

Using Schwarzacher's (1967) transition-probability matrices, based on an observed sequence of strata of Pennsylvanian age in eastern Kansas, an artificial stratigraphic column has been produced. The input data are listed in Table 4-11. Here NTAU = 31 and the number of states is four. Figure 4-13 is a short segment of output in an experimental run that includes 1000 transitions. Although the resulting stratigraphic column is relatively complex, the influence of the 31-unit second-dependent step is apparent when a long sequence is examined.

Stratigraphic Example 2

The second example is artifically contrived for illustrative purposes and employs hypothetical transition-probability matrices. For a three-state system consisting of sandstone, shale, and limestone, suppose that we wish to produce a sequence containing limestone marker beds that recur persistently every sixth unit time step. Our example consists of a chain whose second step length is five unit steps ($\tau = 5$). Suppose that between the limestone marker beds shale and sandstone beds alternate at random so that there is no first-order memory (although we could also include a first-order memory effect between markers if we so desired). The three two-dimensional transition matrices illustrated in Figure 4-14 can be used to produce such a sequence. Notice that, whenever a limestone is encountered in state $t-1-\tau$ $(t-6)$, another limestone is generated at time t, irrespective of the state at $t-1$. If, however, either shale or sandstone is encountered at time $t-6$, either shale or sandstone can be generated at time t.

Figure 4-15 illustrates the output produced with these transition matrices and a value of NTAU = 5. Notice that a limestone occurs at every sixth layer. Although the transition-matrix values are extreme, it would be possible to change them to produce a more realistic stratigraphic sequence.

Algorithm for Testing for Double-Dependence Markov Property

At this point it is pertinent to consider methods of estimating transition-probability values of a double-dependence chain. If the data pertaining to the chain are already arranged in a discrete sequence of unit steps, the only decision to be made concerns the length of the earlier or second-dependence step. If, on the other hand, we wish to analyze a continuous sequence, such as a stratigraphic one, we must also select the unit-step length at which the continuous sequence will be sampled. Then, starting with time period t, the states for time t, $t-\mu$, and $t-\mu-\tau$ are examined, and a tally mark is made for the appropriate transition during each unit step. Time t is then advanced by μ and another tally mark is made, and so on. Clearly a large number of transitions are required in order to make reliable estimates of the transition probabilities.

A FORTRAN program for analyzing and testing for the double-dependence Markov property is listed in Table 4-12. Input to the program includes a sequence of states stored in array OBSVAT. For each value of NTAU (length of memory step τ) between LTAU and LIMTAU the program calculates a transition-frequency matrix, which in turn is transformed into a transition-probability matrix. In addition, the program calculates the maximum-likelihood-criterion test statistic

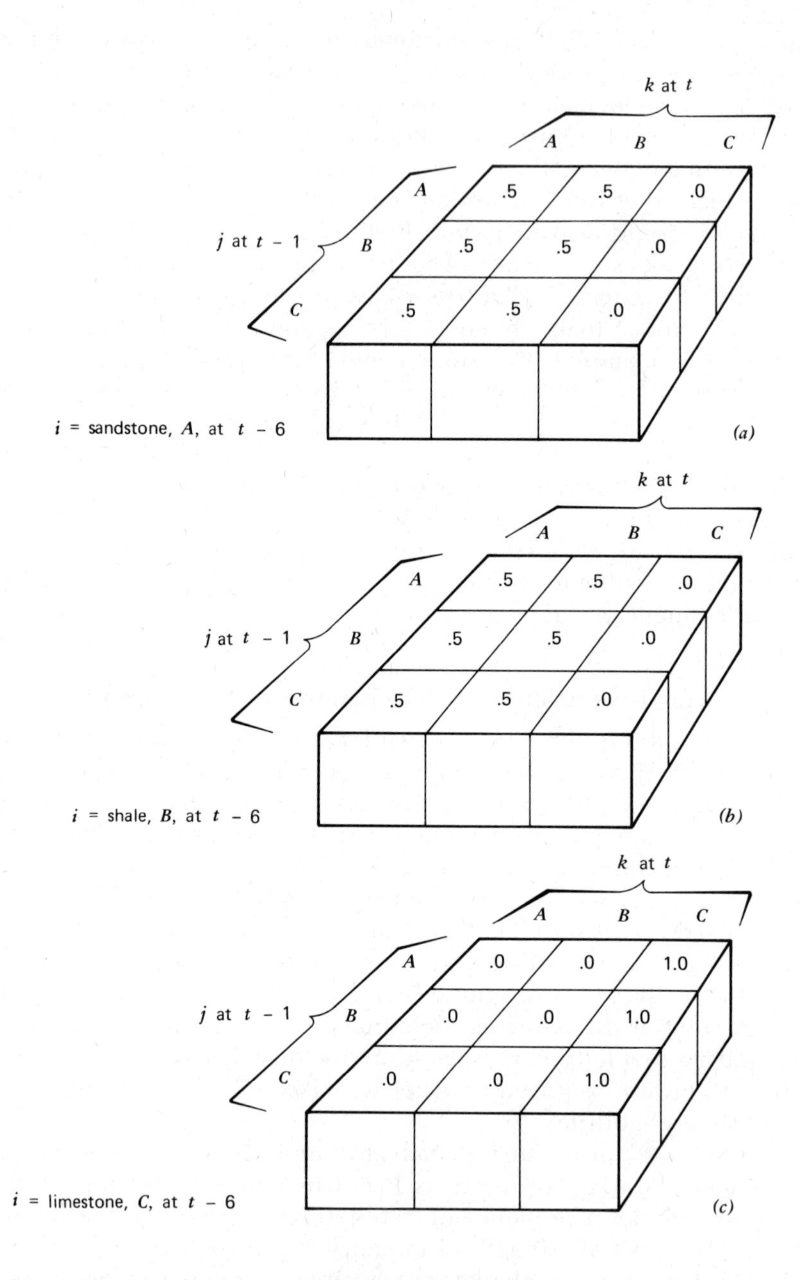

Figure 4-14 Block diagrams illustrating transition matrix for double-dependence, three-state Markov chain pertaining to the synthetic stratigraphic sequence of Figure 4-15. A = sandstone, *B* = shale, and *C* = limestone.

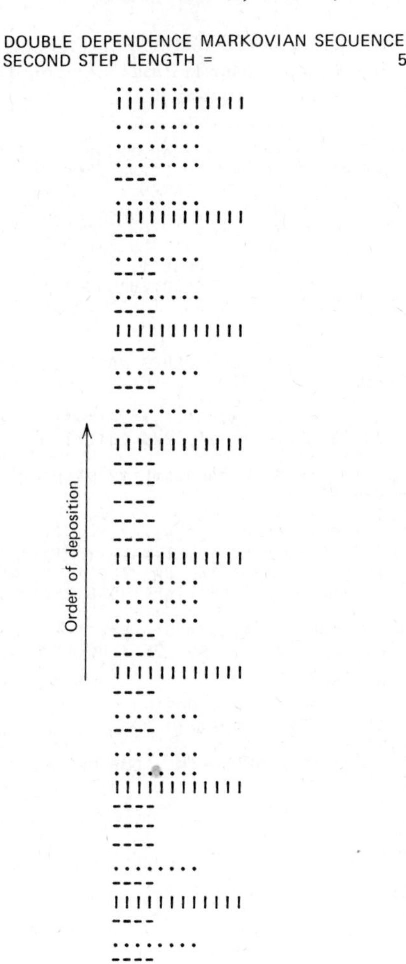

Figure 4-15 Synthetic stratigraphic sequence incorporating double-dependence Markov chain employing transition matrix illustrated in Figure 4-14, and earlier step length (τ) of 5. Effect is such that limestone unit tends to be repeated every six time steps. Rows of dashes represent shale, dots sandstone, and I's represent limestone.

of Anderson and Goodman (1957), which is similar to that for first-order Markov chains (Equation 4-25) and may be written

$$\lambda = \prod_{i,j,k}^{m} \left(\frac{p_{jk}}{p_{ijk}} \right)^{n_{ijk}} \tag{4-31}$$

TABLE 4-12 FORTRAN Program for Calculating Transition Probabilities of Double-Dependence Markov Chain and for Testing for Double-Dependence Markov Property

```
C.....ORIGINAL SERIES OF OBSERVATIONS READ IN AS INPUT (OBSVAT(1000)
C.....STATES CODED NUMERICALLY USING INTEGERS FROM 1 ON UP
C.....    NSTATE               NO. OF STATES
C.....    LTAU                FIRST VALUE OF NTAU (SECOND ORDER STEP)
C.....    LIMTAU               MAX. LENGTH OF SECOND ORDER STEP
C.....    NOBSV                NO. OF OBSVERVATIONS
      DIMENSION NZERO(50), TEST(50), PROB(10,10,10),
     1  PROBI(10,10),  SUMIK(10), SUMK(10,10),SUMI(10,10)
      INTEGER OBSVAT(1000), DGFREE, DGF, TALLY(10,10,10)
    1 FORMAT(4I5)
    2 FORMAT(16I5)
    3 FORMAT(1H1, 'TESTING FOR SECOND ORDER MARKOV PROPERTY'/ 1X,
     1  'LENGTH OF SECOND ORDER STEP =', I5//5X, 'FREQUENCY TALLY MATRIC
     2ES'/5X,'-----------------------')
    4 FORMAT(1H0///' MATRIX FOR STATE I=',I2,'(T-',I2,')'//25X,
     1 'STATE K(T)'/ 8X,'STATE J(T-1)', 2X, 10I8/)
    5 FORMAT(1H / 19X, I3, 11I8)
    6 FORMAT(1H0///5X, 'TRANSITION PROBABILITY MATRICES'/5X,'-----------
     1--------------------')
    7 FORMAT(1H0/19X,I3,11F8.2)
    8 FORMAT(1H0///5X,'SIGNIFICANCE TESTS'/5X,'------------------'// 8X,
     1   'TOTAL NO. OF ZEROES IN P(I,J,K) MATRIX =',I8/ 8X,'TOTAL NO. OF
     2 ZEROES ALONG DIAGONAL      =',I8/ 8X, 'TOTAL DEGREES OF FREEDOM LE
     3SS ZEROES    =',I8/8X, 'VALUE OF LIKELIHOOD RATIO STATISTIC    =',
     4  F8.2///)
    9 FORMAT(1H1//5X, 'SUMMARY STATISTICS'/ 5X, '------------------'/
     1 8X, 'SECOND ORDER STEP', 3X, 'NO. OF ZEROES', 5X, 'STATISTIC'/)
   10 FORMAT(1H , 15X, I3, 16X, I3,  6X, F10.3)
C.....READ INPUT PARAMETERS
      READ(5,1) NSTATE, LTAU, LIMTAU, NOBSV
      READ(5,2) (OBSVAT(N), N=1,NOBSV)
      DGFREE=NSTATE*(NSTATE-1)**2
C.....BEGIN MAJOR DO LOOP, ONCE THRU PER VALUE OF TAU
      DO 90 NTAU=LTAU,LIMTAU
C.....SET ARRAYS TO ZERO
      DO 20 J=1,NSTATE
      SUMIK(J)=0.0
      DO 20 I=1,NSTATE
      SUMK(I,J)=0.0
      DO 20 K=1,NSTATE
      SUMI(J,K)=0.0
      PROBI(J,K)=0.0
      PROB(I,J,K)=0.0
   20 TALLY(I,J,K)=0
C.....CALCULATE TALLY FREQUENCIES
      NSTART=NTAU+2
      INDEX=NTAU+1
      DO 30 N=NSTART,NOBSV
      I=OBSVAT(N-INDEX)
      J=OBSVAT(N-1)
      K=OBSVAT(N)
      TALLY(I,J,K)=TALLY(I,J,K)+1
      SUMK(I,J)=SUMK(I,J)+1.0
      SUMI(J,K)=SUMI(J,K)+1.0
   30 SUMIK(J)=SUMIK(J)+1.0
C.....WRITE OUT TALLY TOTALS
      WRITE(6,3)NTAU
      DO 40 I=1,NSTATE
      WRITE(6,4) I, INDEX,   (K, K=1,NSTATE)
      DO 40 J=1,NSTATE
      NUM=SUMK(I.J)
```

Table 4-12 *contd.*

```
   40 WRITE(6,5) J, (TALLY(I,J,K), K=1,NSTATE), NUM
C.....CALCULATE TRANSITION PROBABILITIES
      NZERO(NTAU)=0
      NDIAG=0
      DO 50 K=1,NSTATE
      DO 50 J=1,NSTATE
      IF (SUMIK(J).LT.0.1) GO TO 42
      PROBI(J,K)=SUMI(J,K)/SUMIK(J)
   42 DO 50 I=1,NSTATE
      IF (SUMK(I,J).LT.0.1) GO TO 49
      DUM=TALLY(I,J,K)
      PROB(I,J,K)=DUM/SUMK(I,J)
   49 NZERO(NTAU)=NZERO(NTAU)+1
   50 IF (I.EQ.J.AND.TALLY(I,J,K).LT.1)      NDIAG=NDIAG+1
C.....WRITE OUT TRANSITION PROBABILITIES
      WRITE(6,6)
      DO 60 I=1,NSTATE
      WRITE(6,4) I, INDEX,   (K, K=1,NSTATE)
      DO 60 J=1,NSTATE
   60 WRITE(6,7) J, (PROB(I,J,K), K=1,NSTATE)
C.....CALCULATE TEST STATISTIC
      NZERO(NTAU)=0
      SUM=0.0
      DO 80 I=1,NSTATE
      DO 80 J=1,NSTATE
      DO 80 K=1,NSTATE
      IF (PROB(I,J,K).LT.0.00001) GO TO 80
      SUM=SUM+TALLY(I,J,K)*ALOG(PROB(I,J,K)/PROBI(J,K))
   80 CONTINUE
      TEST(NTAU)=2.0*SUM
      DGF=DGFREE-NZERO(NTAU)
      WRITE(6,8) NZERO(NTAU), NDIAG, DGF, TEST(NTAU)
   90 CONTINUE
C.....TABULATE TEST STATISTIC RESULTS
      WRITE(6,9)
      DO 100 NTAU=LTAU, LIMTAU
  100 WRITE(6,10) NTAU, NZERO(NTAU), TEST(NTAU)
      RETURN
      END
```

For computational purposes it is convenient to write

$$-2\log_e\lambda = 2\sum_{i,j,k}^{m} n_{ijk}\log_e\left(\frac{p_{ijk}}{p_{jk}}\right), \tag{4-32}$$

where m = number of states

p_{jk} = marginal probability for j–kth array, or

$$\frac{\sum_{i}^{m} n_{ijk}}{\sum_{i,k}^{m} n_{ijk}}$$

and the other quantities have been defined previously in connection with Equation 4-26. The null hypothesis under test is that the chain is singly dependent against the alternative that it is doubly dependent. In other words, if $p_{1jk} = p_{2jk} = \cdots p_{mjk} = p_{jk}$, the null hypothesis is true, in which case there is no significant double dependence memory effect. If the null hypothesis is true, $-2 \log_e \lambda$ is distributed as χ^2 with $m(m-1)^2$ degrees of freedom (Anderson and Goodman, 1957). The program calculates this test statistic for each value of the second memory step length.

To illustrate use of the program 1000 lithologic transitions were generated with the program in Table 4-10, employing Schwarzacher's double-dependent transition-matrix values. These lithologic transitions were then coded in integer form (states 1 to 4) and entered as input to the program in Table 4-12. A tabulation of frequency tally matrices and transition-probability matrices forms the bulk of the output from this program. In addition the values of the test statistic, $-2 \log_e \lambda$, are tabulated for each value of τ (Table 4-13). These data have also been plotted as a graph in Figure 4-16, with the tabulated value of χ^2 for $4(4-1)^2 = 36$ degrees of freedom at the 0.001 level of confidence shown as a dashed line. The points that fall above this line indicate values of τ for which the null hypothesis is rejected and the alternative of a double-dependence Markov property accepted. Notice that the value of $-2 \log_e \lambda$ is highly significant for $\tau = 31$. This parallels Schwarzacher's original findings, of course, although the values of $-2 \log_e \lambda$ differ somewhat from those he reported. This is because his values pertain to raw data, whereas those reported here pertain to a simulated sequence.

DISCRETE-STATE CONTINUOUS-TIME MODELS

In the conventional discrete-time Markov chain the system advances in a series of discrete time steps, with transitions occurring with "each tick of a conceptual Markovian clock" (Krumbein, 1968*b*). It has been assumed in the preceding discussions on generating stratigraphic sequences that time and thickness are interchangeable. Instead of using a unit time interval Δt, a unit thickness Δs has been employed. With each transition a layer of "sediment" of unit thickness is deposited, followed by succeeding layers, much like stacking strips of cardboard of unit thickness on one another. Variations in the thickness of a certain lithology may be represented by several unit layers of the same composition occurring in succession. The diagonal elements in the matrix pertain to the probability of a particular state succeeding itself and thus bear indirectly on thickness variations. Potter and Blakely (1967), however, approached the problem differ-

TABLE 4-13 Part of Output from Experimental Run of Program (Table 4-12) for Testing for Double-Dependence Markov Property[a]

SUMMARY STATISTICS

SECOND STEP	STATISTIC
1	44.068
2	31.847
3	37.960
4	29.436
5	30.639
6	64.615
7	24.334
8	29.319
9	39.299
10	43.319
11	64.849
12	54.924
13	40.810
14	53.590
15	46.349
16	44.849
17	23.078
18	26.478
19	21.860
20	37.366
21	45.592
22	36.746
23	44.477
24	43.192
25	28.971
26	34.846
27	38.901
28	25.624
29	46.451
30	70.449
31	353.081
32	77.288
33	42.610
34	36.173
35	43.637

[a]Test data consist of a synthetically generated sequence of 1000 transitions. Statistic in right column is $-2 \log_e \lambda$. Notice strong peak at $\tau = 31$. Test statistic is graphed in Figure 4-16.

ently. They used zero values in the diagonal and determined the thickness of each state appearing in the sequence by sampling from an "embedded" empirical frequency distribution pertaining to the thickness for the particular state (lithology). Figure 4-17 attempts to clarify the differences between these two methods.

In the continuous-time model (stretching a point, we might assume that time is interchangeable with thickness in a stratigraphic sequence), a matrix of *transition rates* q_{ij} is employed. These not only indicate the probability of one state succeeding another but also establish the expected waiting time in each state, or, in other words, the expected thickness of each lithology. This example is shown in

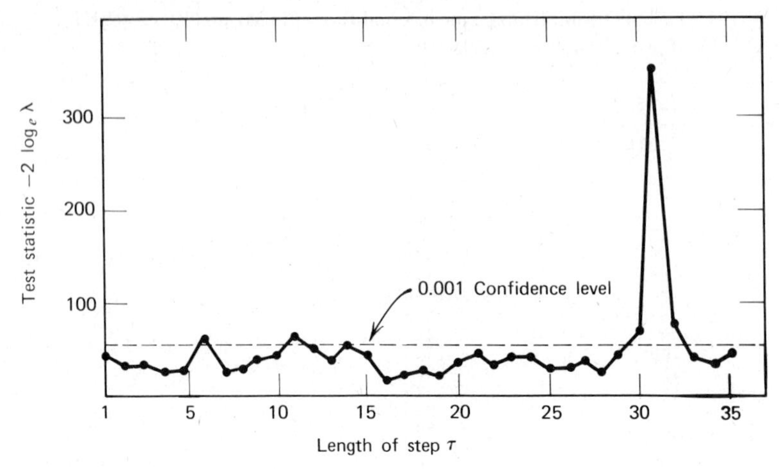

Figure 4-16 Graph of test statistic $-2 \log_e \lambda$ versus length of earlier step τ for sequence of 1000 transitions generated with Schwarzacher's (1967) transition matrices for Kansas rock column. Note highly significant double-dependence memory at $\tau = 31$.

Figure 4-17*c*. In practice the three methods tend to yield graphic columns that are similar in appearance. This is because the synthetic stratigraphic sections produced on a line printer employ a whole number of lines to represent a lithologic unit. Thus units in types (*b*) and (*c*), which have continuous thicknesses, are first rounded to the nearest whole number of lines and then printed. Although the results appear to be similar, the rationale underlying the generating mechanisms differs from type to type.

Krumbein (1968*b*) presents a discussion of the continuous-time model for generating stratigraphic sequences. Much of the following treatment is derived from his paper. In the continuous-time model we are concerned not with the probability of a state change, p_{ij} (where $i \neq j$) but rather with the rate of transfer q_{ij} from state i to state j. The transition rates may be obtained by direct observation of sedimentation rates or alternatively they can be derived by algebraic transformation of the p_{ij}-matrix. The following equations relate transition probabilities to transition rates:

$$p_{ii} = e^{-M_i \Delta t} \tag{4-33}$$

$$p_{ij} = \frac{q_{ij}}{M_i} p_i \qquad (i \neq j) \tag{4-34}$$

$$q_{ij} = \frac{p_{ij}}{p_i} M_i \qquad (i \neq j) \tag{4-35}$$

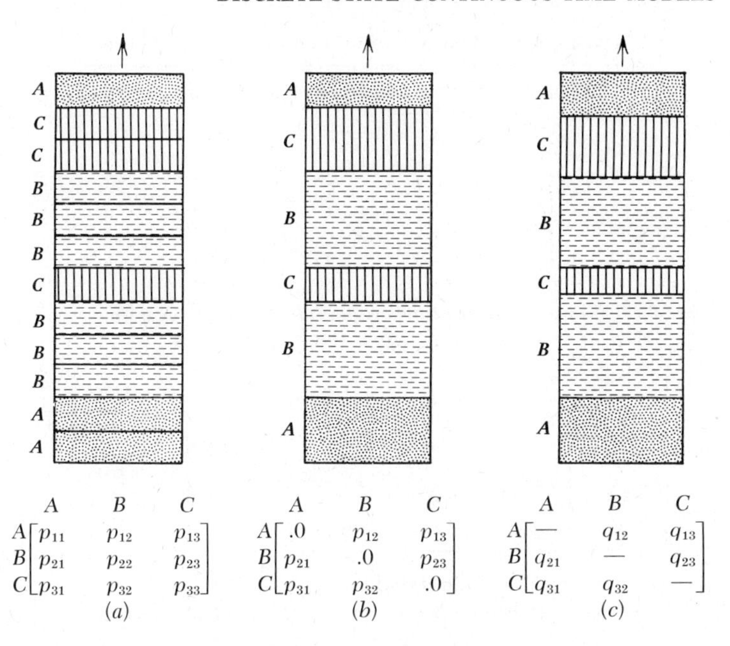

$$A\begin{bmatrix} p_{11} & p_{12} & p_{13} \\ p_{21} & p_{22} & p_{23} \\ p_{31} & p_{32} & p_{33} \end{bmatrix} \qquad A\begin{bmatrix} .0 & p_{12} & p_{13} \\ p_{21} & .0 & p_{23} \\ p_{31} & p_{32} & .0 \end{bmatrix} \qquad A\begin{bmatrix} — & q_{12} & q_{13} \\ q_{21} & — & q_{23} \\ q_{31} & q_{32} & — \end{bmatrix}$$

Figure 4-17 Three ways of representing thickness variation in stratigraphic sequences generated by Markov chains: (*a*) conventional discrete-time (or discrete-thickness) method, using P_{ij}-matrix with nonzero diagonal elements; (*b*) method used by Potter and Blakely (1967), employing zero diagonal elements in P_{ij}-matrix and embedded frequency distributions to determine the thickness of lithologies; (*c*) method of Krumbein (1968*b*), employing continuous time (or continuous thickness), using transition-rate matrix q_{ij} where thickness variations are exponentially distributed.

where M_i = sum of the off-diagonal elements in the ith row of the q_{ij}-matrix $\left(\sum_{\substack{j=1 \\ j\neq i}}^{k} q_{ij}\right)$, where k is the number of states

p_i = sum of the off-diagonal elements in the ith row of the p_{ij}-matrix $\left(\sum_{\substack{j=1 \\ j\neq i}}^{k} p_{ij}\right)$ or $[1-p_{ij}]$, where p_{ii} is the diagonal element

Rearranging Equation 4-33, we can solve for M_i:

$$M_i = \frac{-\log_e p_{ii}}{\Delta t} \tag{4-36}$$

To illustrate the steps involved in transforming a p_{ij}-matrix to a q_{ij}-matrix the numerical example in Table 4-14 is used. The values of

TABLE 4-14 Transformation of p_{ij}-Matrix to q_{ij}-Matrix[a]

Transition-Probability Matrix with $\Delta s = 5.0$ *Feet*[b]

	A	B	C	D	p_i
A	.62	.14	.14	.10	.38
B	.06	.63	.20	.11	.37
C	.19	.31	.33	.17	.67
D	.21	.48	.17	.14	.86

Transition-Rate Matrix Calculated from Above with $\Delta t = 1.0$[b]

	A	B	C	D	M_i
A	—	0.176	0.176	0.126	0.478
B	0.075	—	0.250	0.137	0.462
C	0.315	0.512	—	0.282	1.109
D	0.480	1.098	0.389	—	1.967

[a]From Krumbein (1968*b*).
[b]A = sandstone, B = shale, C = siltstone, D = lignite.

M_i are computed for each row of the p_{ij}-matrix by using Equation 4-36. The value of Δt is assumed to be 1.0. The sampling interval Δs, which defines the thickness equivalent to Δt, is assumed to be 5.0 feet. Thus we imply that 5.0 feet of sediment is deposited during 1.0 unit of time. Compaction is not considered. We note that element p_{11} in the first row of the p_{ij}-matrix in Table 4-14 is .62. Therefore

$$M_1 = \frac{-\log_e .62}{1.0} = 0.478$$

In row 2, $p_{22} = .63$; thus

$$M_2 = \frac{-\log_e .63}{1.0} = 0.462$$

Likewise $M_3 = 1.109$ and $M_4 = 1.967$.

These values of M represent the total rates for leaving the designated states. They have the dimensions T^{-1}, which can be interpreted as the reciprocal of time (thickness of deposit) required to leave the state. Thus the expected waiting time for the system to remain in state A is $1/0.478 = 2.09$ time units, which at this rate is equivalent to $2.09 \times 5 = 10.45$ feet of sediment. Next, the values of p_i are calculated by summing the values in each row of the p_{ij}-matrix, omitting the diagonal elements.

Thus

$$p_1 = .14 + .14 + .10 = .38$$
$$p_2 = .06 + .20 + .11 = .37$$

and similarly $p_3 = .67$ and $p_4 = .86$. The elements q_{ij} can now be calculated by using Equation 4-35. There are no diagonal elements, as the rate at which one state moves to itself is meaningless. Thus

$$q_{12} = \left(\frac{0.14}{0.38}\right)(0.478) = 0.176$$
$$q_{13} = \left(\frac{0.14}{0.38}\right)(0.478) = 0.176$$
$$q_{14} = \left(\frac{0.10}{0.38}\right)(0.478) = 0.126$$

and so on for rows 2 through 4 of the q_{ij}-matrix.

The reverse transformation from the q_{ij}-matrix to the p_{ij}-matrix can be made by using the same equations. In this case the values of q_{ij} are known, and values of M_i are calculated by summing each row of the q_{ij}-matrix. The values of p_{ii} are determined with Equation 4-33. Then, given that $p_i = 1 - p_{ii}$, Equation 4-34 is used to determine values of p_{ij}. Krumbein's (1968*b*) paper presents a FORTRAN algorithm for accomplishing these transformations in either direction.

Generating Synthetic Sequences with the q_{ij}-Matrix

The generating process consists of choosing the sequence of states and in turn of determining the length of time (and hence thickness) spent in each state. The succession of time intervals t during which the system is in any given state i forms an exponential distribution described by the parameter M_i, (the expected value of t is $1/M_i$). The distribution may be written

$$f(t) = M_i e^{-M_i t} \tag{4-37}$$

where $f(t)$ is the probability-density function of t. The simulation process involves generating a random variable with the same distribution as $f(t)$. As described in Chapter 3, this is done by obtaining the cumulative-probability-density function $F(t)$, which is also defined so that it lies in the range 0.0 to 1.0 and then equating $F(t)$ with a random number r drawn from a uniform distribution in the range 0.0 to 1.0:

$$F(t) = \int_0^t M_i e^{-M_i t}\, dt = 1 - e^{-M_i t} \tag{4-38}$$

Because of the symmetry of the uniform distribution of r between 0.0 and 1.0, instead of setting $r = 1 - e^{-M_i t}$, we can set

$$r = e^{-M_i t} \tag{4-39}$$

To solve for t we take the logarithm to base e of both sides:

$$\log_e r = -M_i t \tag{4-40}$$

Then

$$t = -\frac{\log_e r}{M_i} \tag{4-41}$$

This relationship can be numerically illustrated by employing the q_{ij}-matrix from Table 4-14. Let us suppose that the initial state is A (sandstone). We draw a random number r and find it to be, say, 0.6623. Given $M_i = 0.478$, we can solve for t as follows:

$$\begin{aligned} t &= -\frac{\log_e 0.6623}{0.478} \\ &= 0.862 \text{ time units} \\ &= 0.862 \times 5 \\ &= 4.310 \text{ feet thick} \end{aligned}$$

If the random number r is close to 1.0 or 0.0, the corresponding extremes of the exponential distribution yield thickness values that are either infinitesimally small or extremely thick. Therefore it is expedient to truncate the uniform distribution so that the range of r is reduced to, say, 0.1 to 0.9.

The next step in the simulation process is to choose the state to enter after the first state. To make this decision we need to know the probability of going from each state to each other state. In other words, we need a probability matrix, similar to p_{ij} but with zero elements in the diagonal. We can calculate such a matrix by using the relation

$$p'_{ij} = \frac{q_{ij}}{M_i} \qquad (i \neq j) \tag{4-42}$$

where p'_{ij} is the probability of moving from state i to state j. For example, the probability of moving from state A $(i = 1)$ to state B $(i = 2)$ may be calculated as follows:

$$p'_{12} = \frac{q_{12}}{M_1} = \frac{.176}{.478} = .3682$$

The full p'_{ij}-matrix obtained in this manner from the q_{ij}-matrix (Table 4-14) is listed in Table 4-15.

TABLE 4-15 Transition-Probability Matrix p'_{ij}, obtained from the q_{ij}-Matrix in Table 4-14, and Cumulative Matrix p''_{ij}

Matrix in Conventional Form (p'_{ij})

	A	B	C	D
A	.000	.368	.368	.264
B	.162	.000	.541	.297
C	.284	.462	.000	.254
D	.244	.558	.198	.000

Matrix in Cumulative Form (p''_{ij})

	A	B	C	D
A	.000	.368	.736	1.000
B	.162	.162	.703	1.000
C	.284	.746	.746	1.000
D	.244	.802	1.000	1.000

The cumulative form of the p'_{ij}-matrix (p''_{ij}) is used to determine transitions from state to state, employing uniformly distributed random numbers in the range 0.0 to 1.0. For example, assume that the system starts in state *A*. We generate a random number, say $r = 0.641$, which lies between the values of p''_{12} and p''_{13}. State *A* is therefore succeeded by state *C* ($i = 3$). Subsequent states are chosen similarly. A state can never be succeeded by itself, however, as the diagonal elements of the p'_{ij}-matrix are zero.

A FORTRAN program that generates a stratigraphic sequence by using a transition-rate matrix is listed in Table 4-16. Output from an experimental run with the program using the matrix of Table 4-14 (input data are listed in Table 4-17) is shown in Figure 4-18.

To conclude, we may comment on the comparison between discrete-time and continuous-time Markov models. Superficially there seems to be little advantage in using a q_{ij} rate matrix in preference to a p_{ij} probability matrix. Krumbein (1968*b*) points out, however, that the continuous-time model focuses more strongly on the relationship between time and thickness in stratigraphy. Ultimately the development of rate matrices may be a better way of linking observations on sedimentary processes with the stratigraphic record. The analysis of stratigraphic sequences in terms of p_{ij}- or q_{ij}-matrices is, however,

TABLE 4-16 FORTRAN Program for Generating Synthetic Stratigraphic Sequences Given q_{ij}-Matrix as Input

```
C.....ALGORITHM FOR CONTINUOUS-TIME DISCRETE-STATE MARKOV MODELING
C.....   QRATE(I,J)      TRANSITION RATE MATRIX
C.....   M(I)            SUM OF OFF-DIAGONAL ELEMENTS OF QRATE() MATRIX
C.....   NSTATE          NO. OF STATES
C.....   NLINES          MAX NO OF LINES OF PRINTED OUTPUT
C.....   DS              THICKNESS EQUIVALENT TO UNIT OF TIME
C.....   FTLINE          NO OF FEET TO BE PRINTED PER LINE
C.....   ALPHA(I,J)      ALPHAMERIC SYMBOLS TO BE USED FOR I-TH STATE
      DIMENSION QRATE(10,10), CMPROB(10,10), ALPHA(10,3)
      INTEGER OBSVAT(1000)
      REAL M(10)
    1 FORMAT(3I10, 2F5.2)
    2 FORMAT(3A4)
    3 FORMAT(10F10.3)
    4 FORMAT(1H1, 'TRANSITION RATE MATRIX'/ 6X, 10I10/)
    5 FORMAT(1H , I5,11F10.3)
    6 FORMAT(1H1)
    7 FORMAT(1H , 3A4)
C.....READ INPUT PARAMETERS
      READ(5,1) NSTATE, NLINES, IX, DS, FTLINE
      DO 10 I=1,NSTATE
   10 READ(5,2) (ALPHA(I,J), J=1,3)
      DO 15 I=1,NSTATE
   15 READ(5,3) (QRATE(I,J), J=1,NSTATE)
C.....CALCULATE M(I) BY SUMMING EACH ROW
      DO 20 I=1,NSTATE
      M(I)=0.0
      DO 20 J=1,NSTATE
   20 M(I)=M(I)+QRATE(I,J)
C.....WRITE QRATE MATRIX
      WRITE(6,4) (J, J=1,NSTATE)
      DO 25 I=1,NSTATE
   25 WRITE(6,5) I, (QRATE(I,J), J=1,NSTATE), M(I)
C.....CALCULATE CUMULATIVE TRANSITION PROBABILITY MATRIX
      DO 31 I=1,NSTATE
      DUM=0.0
      DO 30 J=1,NSTATE
      A=0.0
      IF (I.NE.J) A=QRATE(I,J)/M(I)
      DUM=DUM+A
   30 CMPROB(I,J)=DUM
   31 CONTINUE
C.....DETERMINE INITIAL STATE AT RANDOM
      CALL RANDU(IX, IY, YFL)
      I=FLOAT(NSTATE)*YFL+0.999999
C.....BEGIN GENERATING SEQUENCE
      NSTOP=0
      TOTHIK=0.0
C.....CALCULATE THICKNESS OF STATE I
   35 CALL RANDU(IX,IY,YFL)
      IX=IY
C......CONVERT RANGE TO 0.1 TO 0.9
      R=YFL*0.8+0.1
      THICK=-ALOG(R)/M(I)*DS
      TOTHIK=TOTHIK+THICK
      NSTART=NSTOP+1
      NSTOP=TOTHIK/FTLINE+0.5
      IF (NSTOP.LT.NSTART) GO TO 50
      DO 40 N=NSTART,NSTOP
      IF (N.GT.NLINES) GO TO 80
   40 OBSVAT(N)=I
```

Table 4-16 *contd.*

```
C.....CHOOSE NEXT STATE
   50 CALL RANDU(IX,IY,YFL)
      IX=IY
      DO 60 J=1,NSTATE
   60 IF (YFL.LE.CMPROB(I,J)) GO TO 70
   70 I=J
      GO TO 35
C.....PRINT OUT STRATIGRAPHIC SEQUENCE
   80 WRITE(6,6)
      DO 90 NN=1,NLINES
      N=NLINES-NN+1
      INDEX=OBSVAT(N)
   90 WRITE(6,7) (ALPHA(INDEX,K), K=1,3)
      RETURN
      END
```

Table 4-17 Data for Input in Trial Run with FORTRAN Program of Table 4-16 for the Continuous-Time Model[a]

```
         4          50 121165379 5.00 1.00
..........
------
-.-.-.-.
cccccccccccc
       000        176        176        126
       075        000        250        137
       315        512        000        282
       480       1098        389        000
```

[a]Transition-rate matrix (Krumbein 1968*b*) is represented in last four lines (decimal points have been omitted).

strictly an empirical approach. It would be more illuminating to start with observations on a sedimentary process itself, develop the rate matrix, and then generate a synthetic sequence. This would also be empirical, but it should provide a better understanding of the processes involved in the deposition of stratigraphic sequences.

Two further points raised by Krumbein are worth mentioning. First, the continuous-time model could incorporate a means for eroding previously deposited layers by simply turning back the "clock" and wiping out these records concerned. In this way the disconformities so common in depositional sequences could be modeled. Second, the exponential waiting times in each state of the continuous-time model may not correspond to the distribution of bed thicknesses in stratigraphic sections because the bed thicknesses tend to be lognormally, rather than exponentially, distributed. Potter and Blakely (1967) use an embedded chain that employs thickness frequency distributions obtained by direct observation. In this respect their method provides a more satisfactory match with actual bed thickness than through use of a rate matrix. A rate matrix, however, more compactly represents the information, incorporating data pertaining both to transitions and to thicknesses.

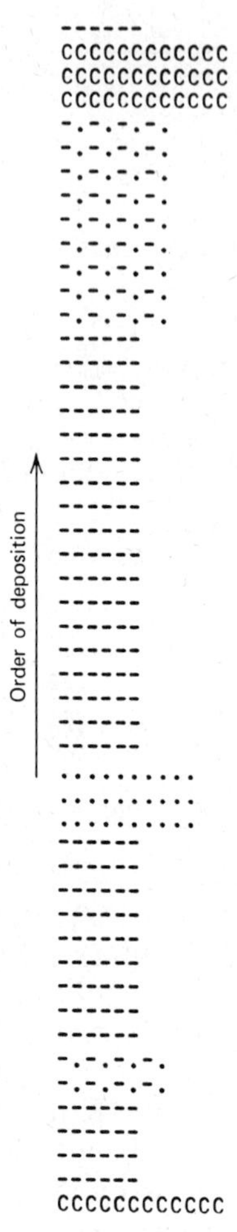

Figure 4-18 Synthetic stratigraphic section generated by program in Table 4-16, using data listed in Table 4-17. Symbols are arbitrary. In example C's represent coal, dashes shale, dots sand, and mixed dots and dashes represent silty sandstone.

GENERATING TWO-DIMENSIONAL STRATIGRAPHIC SECTIONS

Stratigraphic applications of Markov chains are normally confined to the analysis of one-dimensional sedimentary sequences. One might initially suppose that a two-dimensional vertical stratigraphic section could be modeled by independently generating several one-dimensional columns of sediment and later placing them next to each other. This would be unrealistic, however, as the resulting section would lack lateral continuity. Although such an approach would include vertical dependency relationships in a given column, it ignores lateral dependency relationships between columns. Clearly, neighboring columns in an actual stratigraphic section are not independent of one another.

Krumbein's (1968*b*) model of transgressive-regressive strand-line deposits avoids these difficulties. It assumes that each time unit includes a beach sand of fixed width, with marine deposits on one side and nonmarine deposits on the other. The model uses a transition matrix to control the lateral shifting of the beach sand from column to column, rather than the change from one lithology to another in a single vertical sequence. Each column corresponds to a state in the transition matrix.

The principles of the method used with a discrete-time model are illustrated in Figures 4-19 and 4-20. In Figure 4-19 the lowest posi-

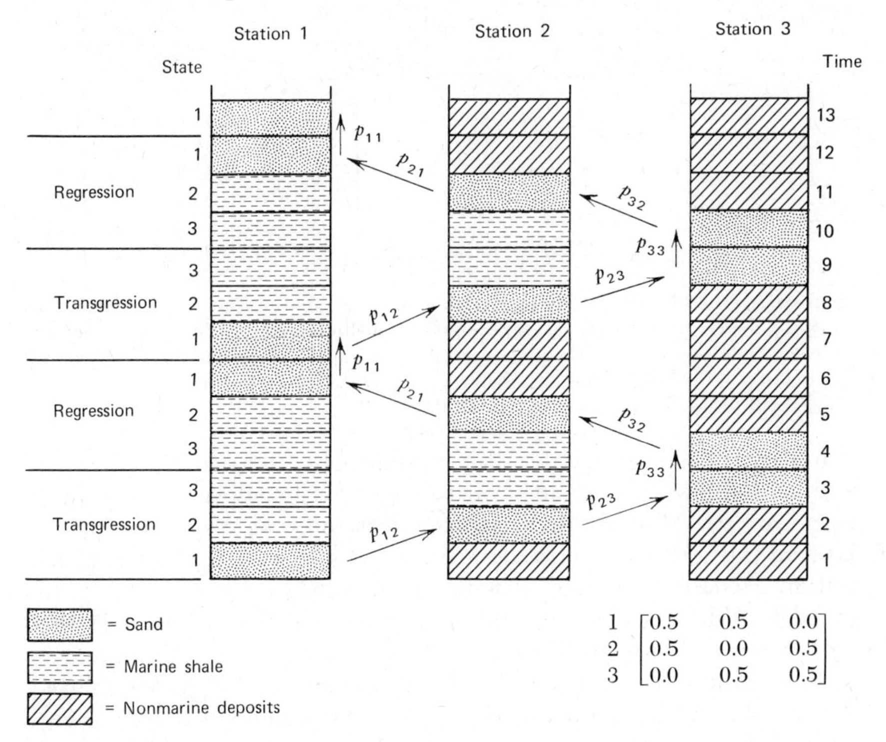

Figure 4-19 Representation of lateral migration of beach sand from station to station through time, employing three-state discrete-time transition-probability matrix. In example beach sand has constant width at each time step and is flanked by marine shale on left and nonmarine deposits on right.

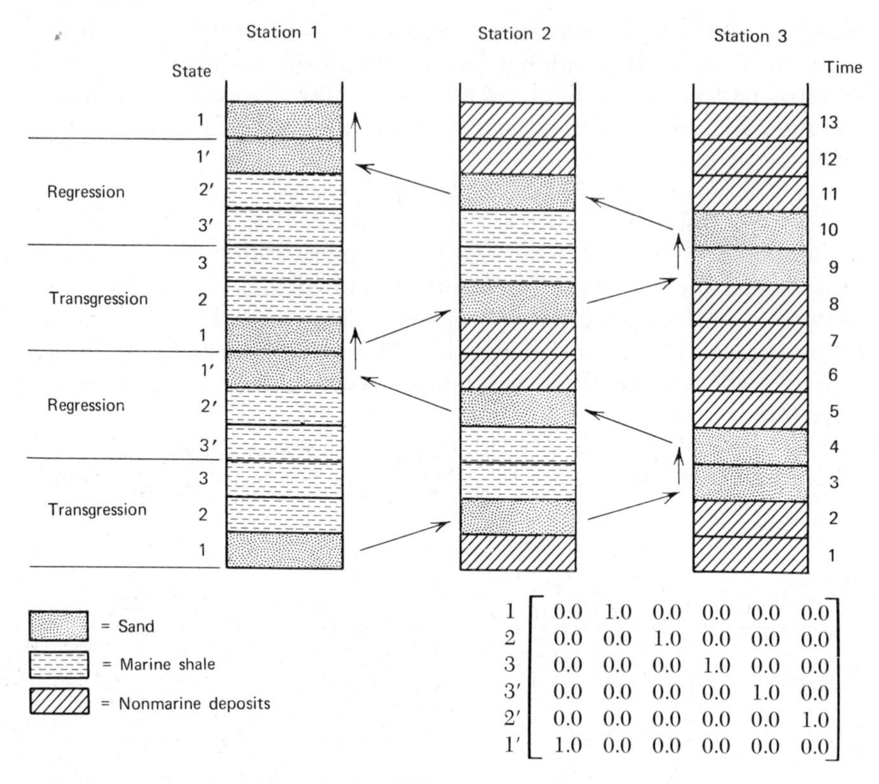

Figure 4-20 Beach-sand lateral-migration diagram similar to Figure 4-19, except that transition matrix contains six states. One set of transition probabilities denotes states in the regressive phase of beach sand, and other set denotes states in the transgressive phase. Transition probabilities thus define a completely deterministic cycle, as indicated by probability values that are either .0 or 1.0.

tion of the beach sand in the stratigraphic sequence (time 1) coincides with station 1. On the shoreward side of the beach nonmarine sediments are deposited in the corresponding interval at stations 2 and 3. During the next time increment the process moves from station 1 to station 2 (state 1 to state 2) with probability p_{12}, which is .5 in the example. When the beach sand is positioned at station 2, it is flanked with marine shales on the seaward side in station 1 and nonmarine sediments on the shoreward side in station 3. The system is then advanced another time step and in our example moves to state 3 with $p_{23} = .5$. Each transition implies a lateral shift of the strand line.

It is clear that the transition-probability matrix shown in Figure 4-19 would be unlikely to produce the regular transgressive and regressive cycles shown in the stratigraphic sequence. The strand line would be more likely to oscillate back and forth in less regular fashion. To control the transgression–regression cycle more closely the number of states in the system can be doubled (Figure 4-20). Each station now has two possible states, one for a regressive beach sand, the other for a transgressive beach sand. In the p_{ij}-matrix shown in Figure 4-20 a completely deterministic transgressive–regressive cycle is represented. For example, at the end of time 1 the system moves from state 1 to state 2 with probability 1.0. Similarly at the end of time 2 the system moves from state 2 to state 3 with $p_{23} = 1.0$, and so on. Elements with values other than 0.0 and 1.0 can be introduced into this matrix to cause the chain to behave in a stochastic rather than a completely deterministic manner.

Continuous-time models also can be effectively used to represent the lateral shifts of a strand line (Krumbein, 1968*b*). In fact the transition-rate matrix q_{ij} provides an effective means for denoting variable transition rates, giving greater flexibility than the transition-probability matrix p_{ij}. For example, the q_{ij} rate matrix of Table 4-18 parallels the p_{ij}-matrix of Figure 4-20, but the rate during the transgressive phase is four times the rate of the regressive phase.

TABLE 4-18 Transition-Rate Matrix q_{ij}, Characterizing Completely Deterministic Transgressive–Regressive Cycle with Transgression Occurring Four Times Faster than Regression[a]

		1	2	3	3′	2′	1′	M_i
Transgression	1	—	4.0	0.0	0.0	0.0	0.0	4.0
	2	0.0	—	4.0	0.0	0.0	0.0	4.0
	3	0.0	0.0	—	4.0	0.0	0.0	4.0
Regression	3′	0.0	0.0	0.0	—	1.0	0.0	1.0
	2′	0.0	0.0	0.0	0.0	—	1.0	1.0
	1′	1.0	0.0	0.0	0.0	0.0	—	1.0

[a]From Krumbein (1968*b*).

SPECIAL FEATURES OF MARKOV CHAINS

To conclude this chapter some special features of finite Markov chains are discussed and are classified. The classification applies to single-dependence chains, although the concepts are applicable to multiple-dependence chains. Such a classification is useful for demonstrating the flexibility of transition matrices and for suggesting the application of Markov chains in modeling such phenomena as regressive–transgressive sedimentation cycles, stochastic diffusion processes (Hagerstrand, 1968), random-walk drainage networks (Schenck, 1965), and other phenomena of interest to geologists.

First let us consider a chain containing *absorbing* states. By an absorbing state we mean a state from which, once it is entered, there is no departure. The state thus absorbs the Markov process. Matrix 4-43 is an example of the transition matrix of a chain containing two absorbing states.

$$
\begin{array}{c}
\\
\mathbf{P} = \begin{array}{c} s_1 \\ s_2 \\ s_3 \\ s_4 \\ s_5 \end{array}
\end{array}
\begin{array}{c}
\begin{array}{ccccc} s_1 & s_2 & s_3 & s_4 & s_5 \end{array} \\
\begin{bmatrix}
1 & 0 & 0 & 0 & 0 \\
\frac{1}{2} & 0 & \frac{1}{2} & 0 & 0 \\
0 & \frac{1}{3} & 0 & \frac{2}{3} & 0 \\
0 & 0 & \frac{1}{4} & 0 & \frac{3}{4} \\
0 & 0 & 0 & 0 & 1
\end{bmatrix}
\end{array}
\qquad (4\text{-}43)
$$

This chain consists of five states. States s_2, s_3, and s_4 form a set of transient states in that they give rise to transitions to subsequent states, including themselves, and to states s_1 and s_5. States s_1 and s_5, however, are absorbing states because, once they are entered, there is no return to the transient set. We may generalize that an absorbing state is represented by probability value of 1 in the principal diagonal and zeros in the remainder of the row. The matrix above may be written in more generalized form as

$$
\begin{array}{c}
\\
\mathbf{P} = \begin{array}{c} s_1 \\ s_2 \\ s_3 \\ s_4 \\ s_5 \end{array}
\end{array}
\begin{array}{c}
\begin{array}{ccccc} s_1 & s_2 & s_3 & s_4 & s_5 \end{array} \\
\begin{bmatrix}
1 & 0 & 0 & 0 & 0 \\
q & 0 & p & 0 & 0 \\
0 & q & 0 & p & 0 \\
0 & 0 & q & 0 & p \\
0 & 0 & 0 & 0 & 1
\end{bmatrix}
\end{array}
\qquad (4\text{-}44)
$$

where q and p represent two probability values, each of which is greater than zero and less than unity, and both sum to 1. A tree diagram representing these properties is shown in Figure 4-21.

Another property of some Markov chains is that of "reflection."

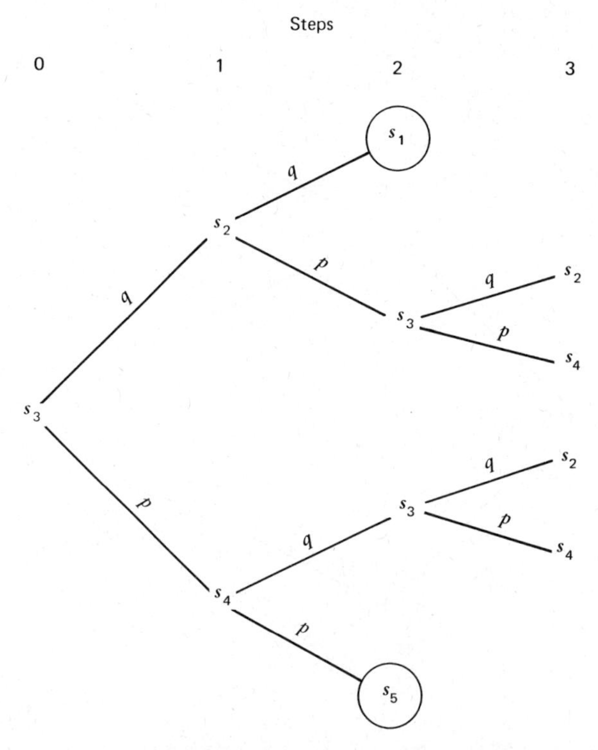

Figure 4-21 Tree diagram of five-state Markov chain for transition matrix 4-43. States s_1 and s_5 are absorbing states. Chain arbitrarily begins (step 0) in state s_3.

Consider a five-state chain (Figure 4-22) whose transition matrix is

$$\mathbf{P} = \begin{array}{c} \\ s_1 \\ s_2 \\ s_3 \\ s_4 \\ s_5 \end{array} \begin{array}{c} \begin{array}{ccccc} s_1 & s_2 & s_3 & s_4 & s_5 \end{array} \\ \begin{bmatrix} 0 & 1 & 0 & 0 & 0 \\ q & 0 & p & 0 & 0 \\ 0 & q & 0 & p & 0 \\ 0 & 0 & q & 0 & p \\ 0 & 0 & 0 & 1 & 0 \end{bmatrix} \end{array} \tag{4-45}$$

In this chain the only entry to state s_1 is via state s_2, from which there is probability q that state s_1 will be entered next and probability p that state s_3 will be entered next. If state s_1 is entered, the subsequent transition is back to state s_2 because element p_{12} in the transition matrix has a value of 1, whereas all remaining elements in that row are 0. A transition into state s_5 is likewise reflecting, the transition into state s_5 is possible only from state s_4, and the reflection is back to state s_4.

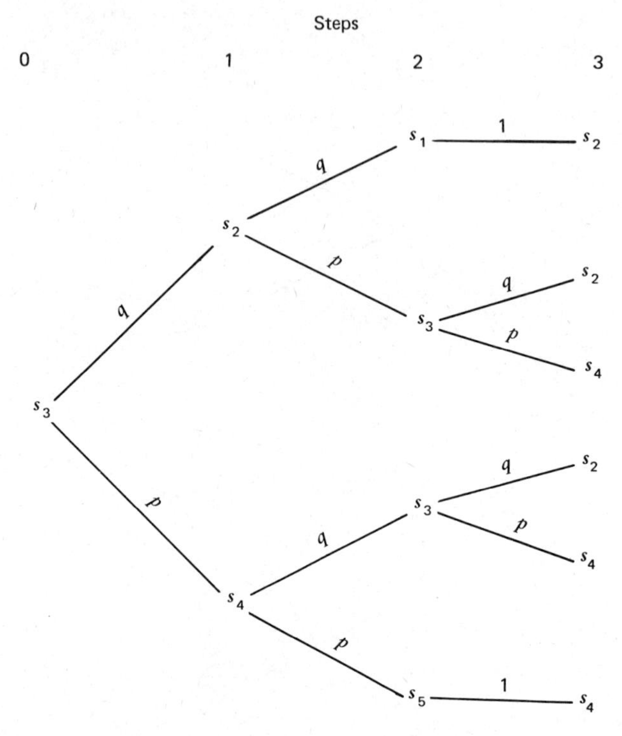

Figure 4-22 Tree diagram of five-state Markov chain for transition matrix 4-45. States s_1 and s_5 are reflecting states in that they reflect back to the immediately preceding state. Tree arbitrarily starts in state s_3.

Now let us consider a Markov chain in which there are transitions from boundary states s_1 and s_5 to the center state s_3, as governed by the following transition matrix:

$$\mathbf{P} = \begin{array}{c} \\ s_1 \\ s_2 \\ s_3 \\ s_4 \\ s_5 \end{array} \begin{array}{c} \begin{array}{ccccc} s_1 & s_2 & s_3 & s_4 & s_5 \end{array} \\ \begin{bmatrix} 0 & 0 & 1 & 0 & 0 \\ q & 0 & p & 0 & 0 \\ 0 & q & 0 & p & 0 \\ 0 & 0 & q & 0 & p \\ 0 & 0 & 1 & 0 & 0 \end{bmatrix} \end{array} \qquad (4\text{-}46)$$

The transition from state s_3 is to either state s_2 or s_4. If the transition is to s_2, the subsequent transition is either to state s_1 (where it will be immediately followed by a transition to state s_3) or directly to s_3.

A special case (matrix 4-47) occurs when the transition steps in Markov chains reach the point where they are confined to two boundary states. Once a boundary state has been reached, the transition stays at this state with some probability, say $\frac{1}{2}$, or moves to the other boundary state with a probability of $\frac{1}{2}$.

$$\mathbf{P} = \begin{matrix} & s_1 & s_2 & s_3 & s_4 & s_5 \\ s_1 & \frac{1}{2} & 0 & 0 & 0 & \frac{1}{2} \\ s_2 & q & 0 & p & 0 & 0 \\ s_3 & 0 & q & 0 & p & 0 \\ s_4 & 0 & 0 & q & 0 & p \\ s_5 & \frac{1}{2} & 0 & 0 & 0 & \frac{1}{2} \end{matrix} \tag{4-47}$$

Another special case is that in which transitions are confined to two boundary states each of which, once it has been entered, oscillates with complete regularity to the other boundary state:

$$\mathbf{P} = \begin{matrix} & s_1 & s_2 & s_3 & s_4 & s_5 \\ s_1 & 0 & 0 & 0 & 0 & 1 \\ s_2 & q & 0 & p & 0 & 0 \\ s_3 & 0 & q & 0 & p & 0 \\ s_4 & 0 & 0 & q & 0 & p \\ s_5 & 1 & 0 & 0 & 0 & 0 \end{matrix} \tag{4-48}$$

CLASSIFICATION OF STATES AND CHAINS

Having explored some of the ideas concerning finite Markov chains, we find it appropriate for our purposes to erect a classification of the types of sets of states that occur in Markov chains. We can distinguish two principal kinds of sets, *transient* and *closed*. A composite Markov chain, for example, might be composed of several transient sets and several closed sets.

First let us consider closed sets. A single closed set is a set of one or more states that have the property of confining a Markov process once it enters. If a closed set contains more than one state in which communication between states is possible, the Markov process can move from state to state within the set, but it can never leave. If more than one closed set is present, the Markov process will eventually be confined within one of the closed sets, and will therefore fail to reach the other closed set or sets. If only one state is present within a closed set, that state is necessarily an absorbing one.

A transient set, as its name implies, involves states that are transient, leading the Markov process to a closed set. Not all chains have a transient set or sets. A matrix containing more than one closed set, but lacking a transient set, in effect contains two or more unrelated Markov chains that have been lumped together. Without a transient set there can be no connection between the closed sets. Under these circumstances the chains represented by the different closed sets may be studied separately.

The distinctions between closed and transient sets can be made clearer if we arrange the elements in transition matrices in *canonical form* so that the states are placed into classes containing either transient or closed sets. The canonical form in an aggregated version is as follows:

$$\mathbf{P} = \begin{array}{c} \begin{array}{cc} r-s & s \end{array} \\ \left[\begin{array}{c:c} \mathbf{S} & \mathbf{0} \\ \hdashline \mathbf{R} & \mathbf{Q} \end{array}\right] \end{array} \begin{array}{l} \\ r-s \\ s \end{array} \tag{4-49}$$

where s = number of transient states

$r-s$ = number of closed states

$\mathbf{Q}$ = square submatrix, containing $s \times s$ elements, which pertains to transitions directly between transient states

$\mathbf{R}$ = rectangular submatrix, containing $s \times (r-s)$ elements, which pertains to transitions from transient set to closed set or sets

$\mathbf{S}$ = square submatrix, containing $(r-s) \times (r-s)$ elements, which forms a class of closed set or sets from which there is no exit after entry

$\mathbf{0}$ = submatrix, containing $s \times (r-s)$ elements, all of which are zero and therefore represent no transition.

Let us now consider some examples of chains in aggregated canonical form. A six-state chain containing a transient set and a closed class is shown below in this form:

$$\mathbf{P} = \begin{array}{c} \\ s_1 \\ s_2 \\ s_3 \\ \\ s_4 \\ s_5 \\ s_6 \end{array} \begin{array}{c} \begin{array}{cccccc} s_1 & s_2 & s_3 & s_4 & s_5 & s_6 \end{array} \\ \left[\begin{array}{ccc:ccc} 1 & 0 & 0 & 0 & 0 & 0 \\ 0 & 1 & 0 & 0 & 0 & 0 \\ 0 & 0 & 1 & 0 & 0 & 0 \\ \hdashline 0 & 0 & 0 & 0 & \frac{1}{2} & \frac{1}{2} \\ \frac{1}{4} & \frac{1}{4} & 0 & 0 & 0 & \frac{1}{2} \\ \frac{1}{4} & 0 & \frac{1}{4} & 0 & \frac{1}{2} & 0 \end{array}\right] \end{array} \tag{4-50}$$

The closed class **S** is composed of three absorbing states (1's in diagonal) and is in itself an identity matrix. Each absorbing state is of course independent of the other absorbing states. In effect we have three separate closed sets. With the transition probabilities as given, the chain would soon enter an absorbing state, regardless of the starting state. If we are not concerned with the particular state by which the closed class is entered, we may simplify the transition matrix so that only one absorbing state remains:

$$\mathbf{P} = \begin{bmatrix} 1 & \cdot & 0 & 0 & 0 \\ \cdot & \cdot & \cdot & \cdot & \cdot \\ 0 & \cdot & 0 & \frac{1}{2} & \frac{1}{2} \\ \frac{1}{2} & \cdot & 0 & 0 & \frac{1}{2} \\ \frac{1}{2} & \cdot & 0 & \frac{1}{2} & 0 \end{bmatrix} \tag{4-51}$$

This matrix preserves the essential features of the preceding one. We have, however, lumped the three original absorbing states s_1, s_2, and s_3 into a single state and have likewise reduced the number of elements in **R** by two-thirds.

At this point it is convenient to present a classification of Markov chains in Table 4-19, closely following the classification of Kemeny and Snell (1960). The primary division in the classification is the (I) absence or (II) presence of transient sets. In turn chains lacking transient sets (I) can be subdivided into (*a*) regular chains and (*b*) cyclic chains.

Regular chains (type I*a*) may be defined as chains that have no transient sets. In a regular Markov chain it is possible to be in any state after a specified number of steps n, regardless of the starting state. Thus, if the matrix **P** is raised to successive powers, eventually there will be no zero elements, even though the single-step transition-probability matrix contains one or more zero elements. Furthermore each row of the limiting matrix **T** is identical to every other row. For example, the following transition matrix is regular:

$$\mathbf{P} = \begin{array}{c} \\ s_1 \\ s_2 \\ s_3 \end{array} \begin{bmatrix} s_1 & s_2 & s_3 \\ \frac{1}{2} & \frac{1}{4} & \frac{1}{4} \\ \frac{1}{2} & 0 & \frac{1}{2} \\ \frac{1}{4} & \frac{1}{4} & \frac{1}{2} \end{bmatrix} \tag{4-52}$$

TABLE 4-19 Classification of Finite Markov Chains

I. ***Chains without Transient Sets*** Chain consists of one or more closed sets. If more than one closed set is present, the chain may be regarded as consisting of unrelated Markov chains.		II. ***Chains with Transient Sets*** Chain contains both closed and transient sets of states. Process tends toward a closed set or sets, with probability of 1 given sufficient time. There is no escape from a closed set once it is entered.			
I*a*. *Regular chains:* Regardless of where the process starts, after sufficient time it can be in any state	I*b*. *Cyclic chains:* Chain has period length *d*; from specified starting subset, the process moves through other subsets in definite order, returning to starting subset in *d* steps	II*a*. *Absorbing chains*: all closed sets consist of absorbing states	II*b*. All closed sets present are regular	II*c*. All closed sets are cyclic	II*d*. Both cyclic and regular closed sets are present

After successive powering it approaches the limiting matrix **T**:

$$\mathbf{T} = \begin{array}{c} \\ s_1 \\ s_2 \\ s_3 \end{array} \begin{array}{c} \begin{array}{ccc} s_1 & s_2 & s_3 \end{array} \\ \begin{bmatrix} .4 & .2 & .4 \\ .4 & .2 & .4 \\ .4 & .2 & .4 \end{bmatrix} \end{array} \tag{4-53}$$

Cyclic closed chains (type Ib) that lack transient sets differ from regular chains in that they consist of d cyclic classes of states. A cyclic chain has a period length d. For a given starting position it will move through the sets in definite order, returning to the set of the starting state after d steps. The simplest possible example of a cyclic chain consists of a two-state chain in which the chain oscillates between the two states with perfect regularity:

$$\mathbf{P} = \begin{bmatrix} 0 & 1 \\ 1 & 0 \end{bmatrix} \tag{4-54}$$

If a cyclic chain consists of a single class, it is possible to go from every state to every other state. However, if the number of cyclic classes is greater than one, the transition from any state to any other state is possible only for special values of n, where n is the value of the power of the matrix. Different powers of the matrix will have zeros in different positions, and these zeros will change cyclically for the powers. Thus $\mathbf{P}^{(n)}$ cannot converge to a matrix **T** with identical rows, as can a regular Markov chain. Failure to converge is an important difference between cyclic and regular chains.

Chains with transient sets form the other major division. We do not discuss these at length but wish to point out that the classification, as given in Table 4-19, places them into four subdivisions on the basis of whether the closed sets are (IIa) absorbing, (IIb) regular, (IIc) cyclic, or (IId) a mixture of cyclic and regular sets.

Illustration with Synthetic Stratigraphic Sequences

The distinctions between the different types of Markov states and chains may be more meaningful if illustrated with synthetic stratigraphic sequences. Four first-order transition matrices are listed below. The first example, matrix 4-55, represents a regular Markov chain (i.e., without transient sets), type Ia of Table 4-19, which has been used to produce the stratigraphic column of Figure 4-23a. In this example it is possible to go directly (in one step) from any state to any

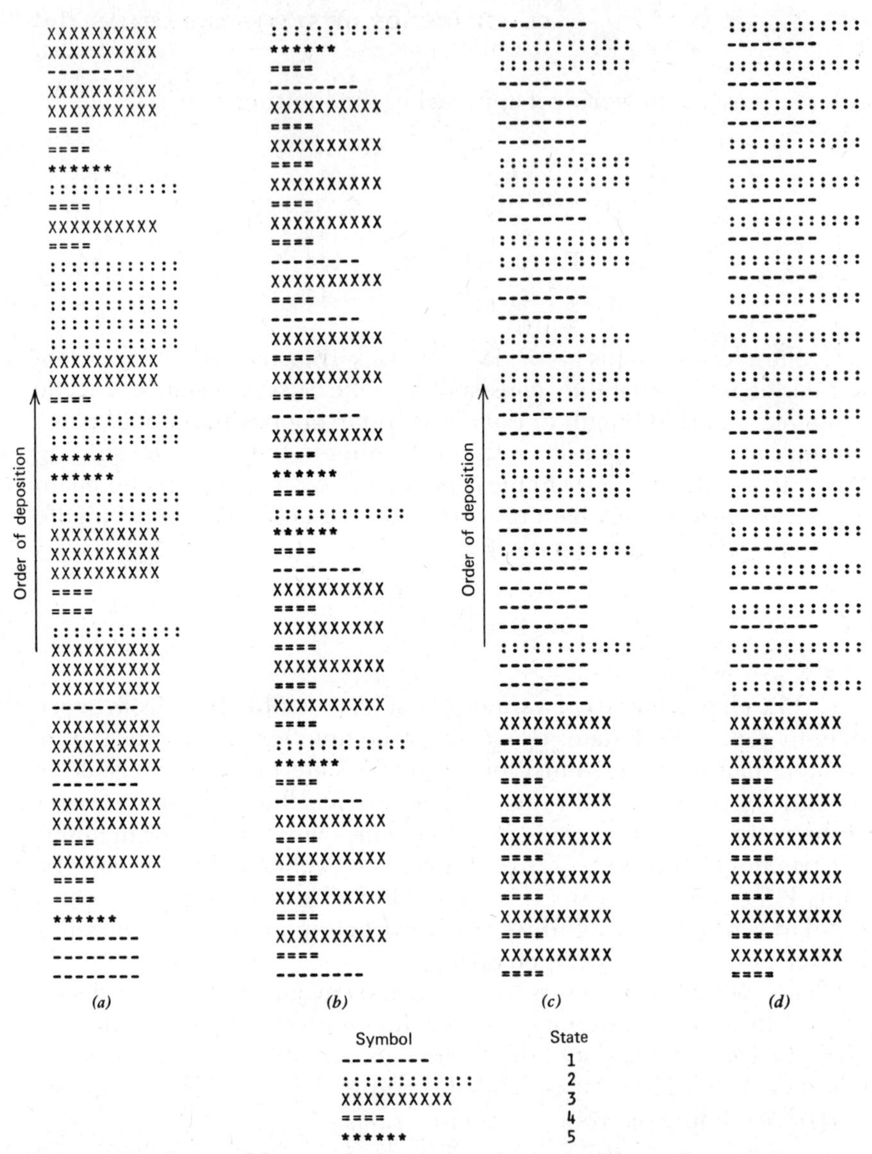

Figure 4-23 Series of four synthetic stratigraphic sequences to illustrate responses produced with different first-order Markov transition matrices (4-55 to 4-58). Sequences have been produced with FORTRAN program listed in Table 4-6. Sequences (*a*) and (*b*) (matrices 4-55 and 4-56) are regular Markov chains without transient sets. Sequence (*b*), however, is reflecting, whereas (*a*) is not. Sequences (*c*) and (*d*) each contain a closed set and a transient set. Closed set in (*c*) is regular, and it is cyclic in (*d*). Chains in both (*c*) and (*d*) were trapped within closed set at 15th step (15th layer above base of columns) in these particular runs. States and their respective symbols are identified at bottom of diagram.

other state. The second example (matrix 4-56 and Figure 4-23*b*) has the property of reflection. When the process emerges from either state 1 or state 2, it is reflected to state 4.

The third and fourth examples are chains consisting of a closed and a transient set. The matrices are shown in aggregated canonical form. In the third example (matrix 4-57 and Figure 4-23*c*) the closed set is regular (type II*b*), whereas in the fourth example (matrix 4-58 and Figure 4-23*d*) the closed set is cyclic (type II*c*). We have shown these brief examples to illustrate the flexibility and simplicity by which finite Markov chains may be used to represent complex stochastic processes. As these examples demonstrate, the properties of a Markov chain of a prescribed number of states are regulated by the values and arrangement of the probabilities in its transition matrix.

$$\mathbf{P} = \begin{array}{c} \\ s_1 \\ s_2 \\ s_3 \\ s_4 \\ s_5 \end{array} \begin{array}{c} \begin{array}{ccccc} s_1 & s_2 & s_3 & s_4 & s_5 \end{array} \\ \begin{bmatrix} .45 & .05 & .25 & .15 & .10 \\ .01 & .65 & .04 & .10 & .20 \\ .20 & .10 & .40 & .20 & .10 \\ .05 & .10 & .30 & .40 & .15 \\ .03 & .15 & .07 & .30 & .45 \end{bmatrix} \end{array} \tag{4-55}$$

$$\mathbf{P} = \begin{array}{c} \\ s_1 \\ s_2 \\ s_3 \\ s_4 \\ s_5 \end{array} \begin{array}{c} \begin{array}{ccccc} s_1 & s_2 & s_3 & s_4 & s_5 \end{array} \\ \begin{bmatrix} .00 & .00 & .00 & 1.00 & .00 \\ .00 & .00 & .00 & 1.00 & .00 \\ .50 & .00 & .00 & .50 & .00 \\ .00 & .00 & .50 & .00 & .50 \\ .00 & .50 & .00 & .50 & .00 \end{bmatrix} \end{array} \tag{4-56}$$

$$\mathbf{P} = \begin{array}{c} \\ s_1 \\ s_2 \\ \\ s_3 \\ s_4 \\ s_5 \end{array} \begin{array}{c} \begin{array}{ccccc} s_1 & s_2 & s_3 & s_4 & s_5 \end{array} \\ \left[\begin{array}{cc:ccc} .50 & .50 & .00 & .00 & .00 \\ .50 & .50 & .00 & .00 & .00 \\ \hdashline .05 & .00 & .00 & .95 & .00 \\ .00 & .00 & .50 & .00 & .50 \\ .00 & .05 & .00 & .95 & .00 \end{array}\right] \end{array} \tag{4-57}$$

$$\mathbf{P} = \begin{matrix} & s_1 & s_2 & & s_3 & s_4 & s_5 \\ s_1 & .00 & 1.00 & \cdot & .00 & .00 & .00 \\ s_2 & 1.00 & .00 & \cdot & .00 & .00 & .00 \\ & \cdot & \cdot & \cdot & \cdot & \cdot & \cdot \\ s_3 & .05 & .00 & \cdot & .00 & .95 & .00 \\ s_4 & .00 & .00 & \cdot & .50 & .00 & .50 \\ s_5 & .00 & .05 & \cdot & .00 & .95 & .00 \end{matrix} \tag{4-58}$$

Problems

Problem 4-1

Analyze an actual stratigraphic sequence that is 1000 ft or thicker, containing at least four, and not more than six, lithologic states. Ideally the lithologic states should be recorded at 1-foot intervals.

a. Construct three tally matrices and the corresponding transition-probability matrices employing sampling intervals of 1, 3, and 5 feet, respectively.

b. Raise the three transition-probability matrices to successively higher powers, until fixed probability vectors describing the equilibrium proportions of lithologies have been reached. Compare the transition matrices as they are raised to successive powers.

c. Determine which sampling interval (1, 3, and 5 feet) yields the most significant first-order Markov property.

d. Use the transition-probability matrix with the most significant memory effect to generate an artificial stratigraphic section. Check the validity of the results by estimating a new transition matrix from the simulated sequence.

e. Check the original stratigraphic sequence for stationarity.

f. Test the original sequence for double-dependence memory effects. Employ the sampling interval found to give the most significant first-order memory. Employing this interval determine the second step length that gives the most significant Markov property.

g. Generate an artificial stratigraphic sequence by employing the double-dependence transition matrix with the most significant second step length.

h. Transform the first-order transition matrix p_{ij} used in (*d*) to a transition-rate matrix q_{ij}. Use this q_{ij}-matrix to generate an artificial stratigraphic sequence. What are the relative merits of generating sequences by using continuous-time versus discrete-time models?

Problem 4-2

Select a geological map on which rock types or sedimentary facies have been plotted. Draw straight lines with various azimuth directions across the map. Select a suitable sampling interval, then take each line and construct a tally matrix of transitions from one lithology to another along a horizontal traverse. Construct transition-probability matrices for each compass direction and test each for the Markov property. If significant differences occur, can they be explained geologically?

Annotated Bibliography

Agterberg, F. P., 1966, "Markov Schemes for Multivariate Well Data," Mineral Industries Experiment Station, Pennsylvania State University, Special Publication 2-65, pp. Y1–Y18.

Application of Markov theory to multivariate series.

Allegre, C., 1964, "Vers une logique mathematique des series sedimentaires," *Bulletin de Societé Geologique de France*, Series 7, v. 6, pp. 214–218.

An early application of Markov principles to problem of stratigraphic succession.

Anderson, T. W., and Goodman, L. A., 1957, "Statistical Inference about Markov Chains," *Annals of Mathematical Statistics*, v. 28, pp. 89–110.

A very useful source of information on statistical tests for single and double-dependence Markov properties, and stationarity of Markov chains.

Carr, D. D., Horowitz, A., Hrabar, S. V., Ridge, K. F., Rooney, R., Straw, W. T., Webb, W., and Potter, P. E., 1966, "Stratigraphic Sections, Bedding Sequences, and Random Processes," *Science*, v. 154, pp. 1162–1164.

Uses first-order transition-probability matrices to model the Mississippian Chester series of the Illinois basin.

Cox, D. R., and Miller, H. D., 1965, *The Theory of Stochastic Processes*, John Wiley and Sons, New York, 398 pp.

Chapters 3, 4, and 5 deal with Markov chains and processes.

Dowds, J. P., 1968, "Oil, Rocks, Information Theory, Markov Chains, Entropy," in *Symposium on Operations Research and Computer Applications in the Mineral Industries*, Colorado School of Mines, 23 pp.

Discusses concept of entropy as applied to first-order Markov chains.

Fisher, R. A., and Yates, F., 1963, *Statistical Tables for Biological, Agricultural, and Medical Research*, Hafner, New York, 6th edition, 145 pp. (published in Britain by Oliver and Boyd, Edinburgh).

Hagerstrand, T., 1968, "A Monte Carlo Approach to Diffusion," in *Spatial Analysis: A Reader in Statistical Geography*, B. J. Berry and D. F. Marble, eds., Prentice-Hall, Englewood Cliffs, N.J., pp. 368–384.

Employs random-walk process for describing spatial diffusion of ideas. Space is subdivided by cellular grid. Each cell can be represented by state in Markov chain. Ideas "move" through grid, their paths governed by transition probabilities.

Harbaugh, J. W., 1966, "Mathematical Simulation of Marine Sedimentation with IBM 7090/7094 Computers," Computer Contribution 1, Kansas Geological Survey, 52 pp.

Complex simulation model that embodies triple-dependence, third-order Markov chains with transition probabilities and treats dependency relationships in a spatial sense as well as through time.

Kemeny, J. C., Snell, J. L., and Thompson, G. L., 1956, *Introduction to Finite Mathematics*, Prentice-Hall, Englewood Cliffs, N.J., 465 pp.

Important general reference on finite Markov chains. Chapter 2, which is of particular interest, deals with the classification of states and chains.

Kemeny, J. C., Snell, J. L., and Thompson, G. L., 1966, *Introduction to Finite Mathematics*, Prentice-Hall, Englewood Cliffs, N.J., 465 pp.

Includes an excellent introduction to basic ideas of probability, Markov chains, and matrix algebra.

Krumbein, W. C., 1967, "FORTRAN IV Computer Programs for Markov Chain Experiments in Geology," Computer Contribution 13, Kansas Geological Survey, 38 pp.

A useful paper that discusses estimation of transition probabilities and tests for first-order Markov property, as well as lists a number of useful programs.

Krumbein, W. C., 1968*a*, "Statistical Models in Sedimentology," *Sedimentology*, v. 10, pp. 7–23.

An informative, readable discussion of the types of mathematical models in geology, with emphasis on the role of Markov chains in probabilistic models.

Krumbein, W. C., 1968*b*, "FORTRAN IV Computer Program for Simulation of Transgression and Regression with Continuous Time Markov Models," Computer Contribution 26, Kansas Geological Survey, 38 pp.

An important paper that introduces the theory of continuous-time models, as well as proposes a model for lateral transitions of a transgressive–regressive strand-line deposit.

Kuenen, P. H., 1950, "Turbidity Currents of High Density," in *Report on the 18th Session of the International Geological Congress*, Great Britain, 1948, Part VIII, pp. 44–52.

Matalas, N. C., 1967, "Some Distribution Problems in Time Series Simulation," Computer Contribution 18, Kansas Geological Survey, pp. 37–40.

Discusses the problem of matching empirical and synthetic Markov sequences generated by simulation methods.

Parzen, E., 1962, *Stochastic Processes*, Holden-Day, San Francisco, 324 pp.

Chapters 6 and 7 deal with Markov chains.

Pattison, A., 1965, "Synthesis of Hourly Rainfall Data," *Water Resources Research*, v. 1, pp. 489–498.

Finds that hourly rainfall records possess significant sixth-order Markov property.

Pfeiffer, P. E., 1965, *Concepts of Probability Theory*, McGraw-Hill, New York, 399 pp.

Chapter 7 includes an elementary discussion of Markov chains.

Potter, P. E., and Blakely, R. F., 1967, "Generation of a Synthetic Vertical Profile of a Fluvial Sandstone Body," *Journal of the Society of Petroleum Engineers of AIME*, pp. 243–251.

Introduction to uses first-order Markov transition-probability matrices with zero elements in the diagonal. Thicknesses of states obtained by sampling from lognormal frequency distributions.

Schenck, H., 1965, "Simulation of the Evolution of Drainage Basin Network with a Digital Computer," *Journal of Geophysical Research*, v. 68, pp. 5739–5745.

A random-walk technique is employed to generate artificial drainage patterns.

Schwarzacher, W., 1967, "Some Experiments To Simulate the Pennsylvanian Rock Sequence of Kansas," Computer Contribution 18, Kansas Geological Survey, pp. 5–14.

An important paper that deals in part with double-dependence Markov chains applied to stratigraphic sequences.

Schwarzacher, W., 1968, "Experiments with Variable Sedimentation Rates," Computer Contribution 22, Kansas Geological Survey, pp. 19–21.

Demonstrates that a stratigraphic sequence may be generated by a wide variety of time processes. Develops a number of Markov models for representing variation in rate of sedimentation.

Vistelius, A. B., 1949, "On the Question of the Mechanism of Formation of Strata," *Doklady Akademii Nauk SSSR*, v. 65, pp. 191–194.

The earliest published work on the geological use of Markov chains known to us.

Vistelius, A. B., 1966, "Genesis of the Mt. Belaya Granite (an Experiment in Stochastic Modeling)," *Doklady Akademii Nauk SSSR*, v. 167, pp. 48–50.

Estimates transitions from one mineral to another in traverses across thin sections of granite. Uses χ^2-tests for stationarity. Finds that estimated transition probabilities are statistically similar to theoretically derived transition probabilities in which a magmatic origin for the granites is assumed.

Vistelius, A. B., 1967, *Studies in Mathematical Geology*, Consultants Bureau, New York, 294 pp.

Translation from Russian. Part IV ("Analysis of Geologic Sections," pp. 252–259) deal with the use of Markov chains for modeling stratigraphic sequences. Shows how Markov transition probabilities can be derived from theory and the theory can be checked by matching a theoretical transition matrix with an empirical matrix obtained from observations.

Vistelius, A. B., and Faas, A. V., 1965, "The Mode of Alternation of Strata in Certain Sedimentary Rock Sections," *Doklady Akademii Nauk SSSR*, v. 164, pp. 40–42.

Uses a χ^2-test for stationarity of first-order transition probabilities in two stratigraphic sections in Russia. Also examines the length and duration of the memory effect in a transition matrix raised to successive powers.

CHAPTER 5

Some Numerical Methods for Solving Equations

The bulk of this chapter deals with certain numerical methods for obtaining solutions to linear differential equations. Clearly it is impossible to treat this topic in any depth in the space of a single chapter, because solution of differential equations encompasses a sizeable segment of both classical mathematics and modern numerical analysis. However, since many simulation applications entail numerical solutions of linear differential equations—for example, in fluid flow and in many rate processes—we present a brief introduction to some methods that are useful for obtaining numerical solutions with computers, treating both linear ordinary and linear partial-differential equations. These methods involve translating the equations into sets of simultaneous linear equations in finite-difference form and then obtaining iterative numerical solutions to the sets of equations.

At this point it is convenient to take an overview of the various types of equations and the relative facility with which they may be solved. Table 5-1 is a classification of equations organized according to whether they are linear or nonlinear and according to whether they are algebraic, ordinary differential, or partial-differential equations. In turn the equations are classed according to the relative degree of difficulty that they present if analytical solutions are to be obtained.

TABLE 5-1 Classification of Equations According to the Relative Degree of Difficulty in Obtaining Solutions by Analytical Methods[a]

	Linear Equations			*Nonlinear Equations*		
Equation	*One Equation*	*Several Equations*	*Many Equations*	*One Equation*	*Several Equations*	*Many Equations*
Algebraic	Trivial	Easy	Essentially impossible	Very difficult	Very difficult	Impossible
Ordinary differential	Easy	Difficult	Essentially impossible	Very difficult	Impossible	Impossible
Partial differential	Difficult	Essentially impossible	Impossible	Impossible	Impossible	Impossible

[a]After Franks (1967).

Note: equations to the right of the heavy bar are not amenable to treatment by strictly analytical methods.

Surprisingly, many of the classes are very difficult or impossible to solve analytically, although numerical solutions by means of computers are feasible for many of the categories. Numerical solution of simultaneous linear equations is briefly treated in the Appendix.

By an analytical solution we mean a solution that is both exact and obtainable by algebraic manipulation. For example, the linear algebraic equation

$$2x+3=0 \tag{5-1}$$

can easily be solved by algebraic manipulation

$$x=-\tfrac{3}{2}=-1.5 \tag{5-2}$$

yielding an exact value for the variable x. Another example is the solution of quadratic equations of the type

$$ax^2+bx+c=0 \tag{5-3}$$

by employing the formula

$$x=\frac{-b\pm\sqrt{b^2-4ac}}{2a} \tag{5-4}$$

which can be derived by algebraic manipulation.

ITERATIVE SOLUTIONS

Equations that cannot be solved precisely by analytical means must be solved approximately by numerical methods. Many numerical techniques, which generally require desk calculators or preferably computers, involve the process of iteration. An iterative method is a numerical method that gradually converges toward a stable solution, repeatedly passing through the same steps of calculation, with a small improvement at each step. To illustrate this important point let us consider an example of a numerical solution involving iteration of a nonlinear algebraic equation.

The equation for calculating the wavelength of a water wave as it enters shallow water is derived from the theory of oscillatory waves and may be written

$$l=L\tanh\left(\frac{2\pi h}{l}\right) \tag{5-5}$$

where l = wavelength
L = wavelength in deep water
tanh = hyperbolic tangent
h = water depth.

Although this is a simple expression, we note that it is nonlinear and that l occurs on both sides of the equation. If we are given L and h, we still cannot solve for l analytically. We can obtain an approximate solution by one of several numerical methods, and in this case we shall use an iterative method. We do this by guessing at the value of l, call it l_0, inserting l_0 on the right side of the equation and solving for l_1, as follows:

$$l_1 = L \tanh\left(\frac{2\pi h}{l_0}\right) \tag{5-6}$$

Then l_1 is inserted on the right side of the equation, and a value for l_2 is obtained:

$$l_2 = L \tanh\left(\frac{2\pi h}{l_1}\right) \tag{5-7}$$

By repeating this iterative process we find that the value of l converges to its true value. In theory we may achieve any degree of accuracy simply by performing a sufficient number of iterations. In practice rounding errors in the calculations set a limit on the attained accuracy.

A short FORTRAN algorithm for solving the wavelength equation is listed in Table 5-2. The iterations are repeated until the value of

TABLE 5-2 FORTRAN Program for the Numerical Solution of the Wavelength Algebraic Equation

```
C.....ITERATIVE SOLUTION TO ALGEBRAIC EQUATION
C.....CALCULATION OF WAVE LENGTH FOR SHOALING OSCILLATORY WAVES
C.....    DEPTH       DEPTH OF WATER
C.....    DPWWVL      WAVE LENGTH IN DEEP WATER
C.....    WVLENG      WAVE LENGTH
C.....    TOL         TOLERABLE ERROR IN SOLUTION
    1 FORMAT(3F10.4)
    2 FORMAT(1H , 3F10.5)
      READ(5,1) DEPTH,  DPWWVL, TOL
      DUMMY=DPWWVL
      TWOPID=2.0*3.14159625*DEPTH
C.....BEGIN ITERATIVE SOLUTION
   10 WVLENG=DUMMY
      DUMMY=DPWWVL*TANH(TWOPID/WVLENG)
      DIFF=ABS(WVLENG-DUMMY)
      WRITE(6,2) WVLENG, DUMMY, DIFF
      IF (DIFF.GT.TOL) GO TO 10
      RETURN
      END
```

WVLENG($= l$) does not change by more than some tolerable error TOL. Table 5-3 lists the output from a computer run with a deep-water wavelength L of 56 feet, a water depth h of 5 feet, and a tolerable error of 0.0001 foot. A total of 31 iterations were needed to reduce the error to the tolerable limit. For the initial value of l the deep-water value of 56 feet was used.

The first column of Table 5-3 lists the trial value of l for each iteration. The second column lists the solution for l at each iteration. The third column lists the difference between the first two columns. When the value in the third column becomes less than 0.0001, the iterations are stopped. The converging nature of this solution is further illustrated in Figure 5-1. We should point out that not all equations will converge to a stable solution when iterative solutions are attempted. Some will diverge from the true solution, and others will oscillate without either converging or diverging.

TABLE 5-3 Output from Program Listed in Table 5-2, Illustrating Iterative Solution of the Wavelength Equation, Using a Water Depth of 5 Feet[a]

WVLENG	DUMMY	DIFF
56.00000	28.48822	27.51178
28.48822	44.88353	16.39531
44.88353	33.84259	11.04094
33.84259	40.86871	7.02612
40.86871	36.18581	4.68291
36.18581	39.22522	3.03941
39.22522	37.21458	2.01064
37.21458	38.52904	1.31445
38.52904	37.66277	0.86627
37.66277	38.23073	0.56796
38.23073	37.85707	0.37366
37.85707	38.10234	0.24527
38.10234	37.94110	0.16124
37.94110	38.04700	0.10590
38.04700	37.97740	0.06960
37.97740	38.02312	0.04572
38.02312	37.99309	0.03003
37.99309	38.01282	0.01973
38.01282	37.99985	0.01297
37.99985	38.00838	0.00853
38.00838	38.00276	0.00562
38.00276	38.00645	0.00369
38.00645	38.00403	0.00243
38.00403	38.00563	0.00160
38.00563	38.00458	0.00105
38.00458	38.00526	0.00069
38.00526	38.00481	0.00046
38.00481	38.00511	0.00031
38.00511	38.00491	0.00020
38.00491	38.00505	0.00014
38.00505	38.00496	0.00009

[a]Note that wavelength (WVLENG) converges under these conditions to a stable value at 38.005 feet.

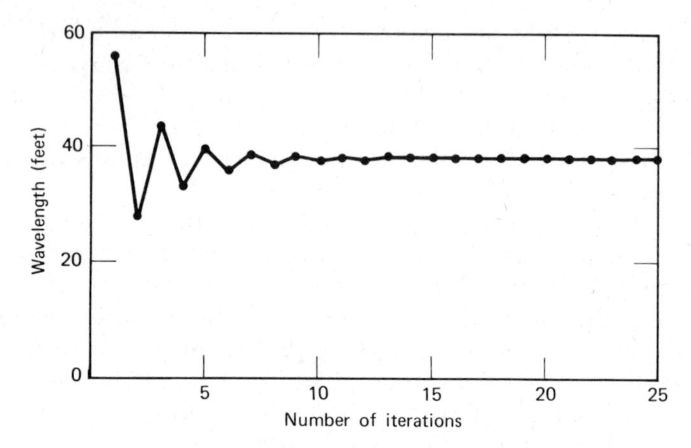

Figure 5-1 Graph of numerical solution of wavelength equation, relating wavelength to number of iterations. Note rapid convergence to solution at about 38 feet.

NUMERICAL SOLUTIONS OF DIFFERENTIAL EQUATIONS

Probably the most effective method for solving complicated differential equations numerically is by means of an analog computer. Because of the continuous nature of differential equations, analog computers that employ continuous means of representing numerical values are able to integrate any expression virtually instantaneously. Most analog computers have no memory capability, however, and are thus strictly limited for many numerical calculations that require an accounting system. Because digital computers tend to be less well suited for solving differential equations, hybrid computers with both analog and digital components will probably come into more widespread use in the future. The user who wishes to solve a differential equation during the course of various other operations will employ digital memory and probably digital logic much of the time but will use an analog integrator, possibly called as a special type of subroutine in the source program, for solving differential equations.

In order to solve a differential equation digitally it is necessary to transform the continuous derivatives to finite differences. Digital computers can calculate derivatives to any desired accuracy by making the finite-difference intervals small enough. Clearly the factors of accuracy and computing time must be weighed against each other in deciding on the finite-difference interval.

Many numerical methods for solving differential equations are available, such as those described by McCalla (1967). For the remainder of this chapter we discuss some of the most simple finite-difference methods, but the reader should be aware that more efficient and accurate methods may be used.

An analytical solution of a differential equation consists of a functional relationship devoid of differentials or derivatives. If the coefficients pertaining to the solution or solutions are known, and if we are given a succession of values for the independent variables, we can evaluate the equation, obtaining numerical values for the dependent variable. A graph relating the variables is thus one means of representing the solution. A numerical solution to a differential equation, on the other hand, can be regarded as a succession of numerical values for the dependent variable and the corresponding independent variable or variables. These values also can be graphed, providing a link between the numerical solution of the equation and its analytical solution.

Translating Derivatives to Finite-Difference Quotients

The numerical solution of differential equations by finite-difference methods entails the transformation of continuous derivatives into finite-difference quotients. Let us begin by considering a very simple example, which we employ for illustrative purposes only. Suppose we wish to solve the equation that describes the growth or decay of populations in which

$$\frac{dN}{dT} = rN \tag{5-8}$$

where N = number of individuals in a population
T = time
r = constant.

This equation has many applications, ranging from growth in organism communities under optimum conditions to radioactive decay and calculation of bank interest. It is a first-order linear differential equation. It says that the rate of change in the size of a population is a proportion of the size of the population. For this simple case we can solve the equation analytically by applying integration rules that lead to the exponential form

$$N_T = N_0 e^{rT} \tag{5-9}$$

This equation, representing the solution, permits us to calculate the value of N for each given time T, provided that we are also given r and an initial condition N_0 at $T = 0$. Table 5-4 gives values for T and N, with $r = 0.5$, and Figure 5-2 is a graph of these relationships.

TABLE 5-4 Values Calculated with the Population-Growth Equation (Equation 5-9), Setting $r = 0.5$

T	N
0	1.00
1	1.65
2	2.72
3	4.48
4	7.39
5	12.19
6	20.09
7	33.12
8	54.60
9	90.00
10	148.40

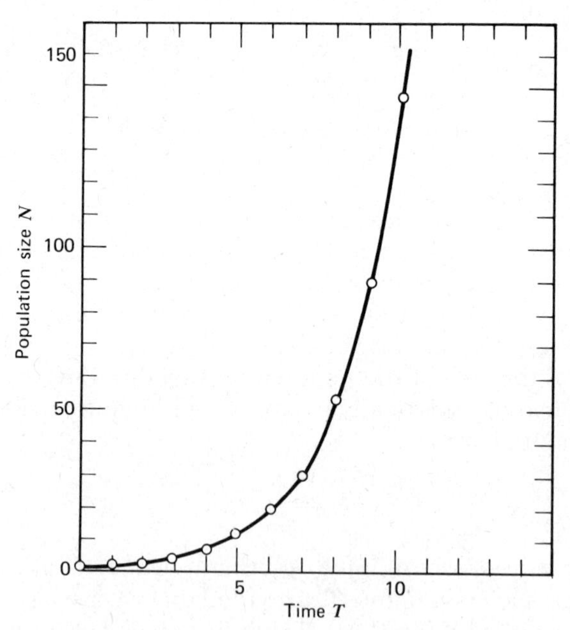

Figure 5-2 Graph of $N = N_0e^{rT}$, setting $r = 0.5$. If N is plotted on logarithmic instead of linear scale, a straight line results.

If for some reason we were unable to integrate Equation 5-8 analytically, we might wish to translate the equation into finite-difference form and proceed in a different fashion. We know that the derivative dN/dT can be regarded as the slope of the line at any time T. Expressing this in another way, we can say that

$$\frac{\Delta N}{\Delta T} \simeq \frac{dN}{dT} \tag{5-10}$$

where ΔN and ΔT are small but finite and dN and dT are infinitesimally small. The approximation in Equation 5-10 holds provided that ΔN and ΔT are small. From here on we shall use Δ to represent the finite-difference operator. The finite-difference quotient $\Delta N/\Delta T$ is thus the slope of a line expressed in finite-difference form.

If we now select a time step ΔT and divide the T-axis of our exponential growth curve into units of ΔT, labeling the particular values of T as $\cdots, t-3, t-2, t-1, t, t+1, t+2, \cdots$ (Figure 5-3), we can calculate the slope of the curve at time t by using the expression

$$\frac{dN}{dT} \simeq \frac{N_{t+1} - N_{t-1}}{2\Delta T} \tag{5-11}$$

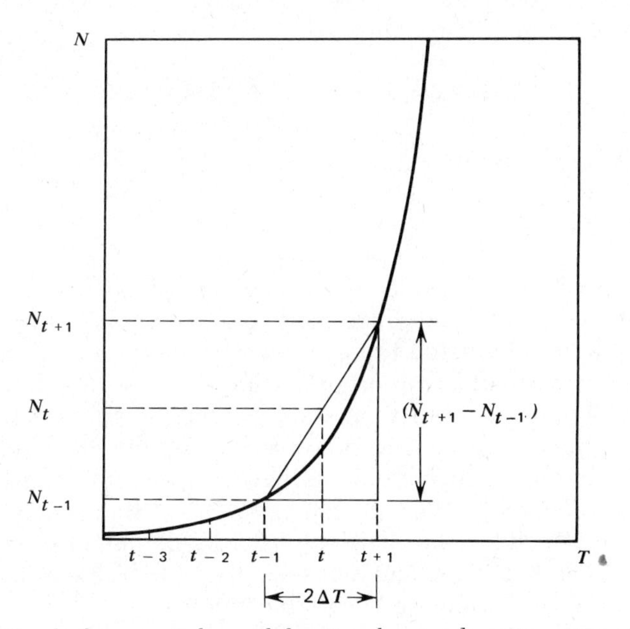

Figure 5-3 Graph of exponential growth function showing derivative at time t approximated by finite-difference quotient $(N_{t+1} - N_{t-1})/2\Delta T$.

If we now combine Equations 5-8 and 5-11 so as to eliminate dN/dT, we obtain

$$\frac{N_{t+1} - N_{t-1}}{2\Delta T} = rN_t \tag{5-12}$$

With rearrangement this becomes

$$N_{t+1} = 2\Delta T r N_t + N_{t-1} \tag{5-13}$$

Given ΔT, r, N_{t-1}, and N_t we can calculate successive values of N. Having obtained N_{t+1} from Equation 5-13, we can then calculate N_{t+2}:

$$N_{t+2} = 2\Delta T r N_{t+1} + N_t \tag{5-14}$$

But N_{t+1} is given by Equation 5-13, and this permits us to write

$$N_{t+2} = 2\Delta T r(2\Delta T r N_t + N_{t-1}) + N_t \tag{5-15}$$

and so on for subsequent values of N. For example, let us set $\Delta T = 1$, $r = 0.5$, $N_1 = 1.00$, and $N_2 = 1.65$, dropping the t subscript from the values of N for simplicity. Then

$$N_3 = (2)(1)(0.5)N_2 + N_1 = N_2 + N_1 = 2.65$$

Similarly it follows that

$$N_4 = N_3 + N_2 = 4.30$$

Table 5-5 lists the results of a succession of calculations made by this particular finite-difference method and compares them with the values obtained by using the analytical expression in Equation 5-9. At the start a small error occurs, which grows cumulatively by a factor of about 2 in each time step, producing an error that ranges from less than 1 percent at the outset to nearly 20 percent of the correct value after 10 increments have elapsed. Obviously such an error is inacceptable. We can reduce the error by reducing the time step ΔT to a smaller value. The values of N, obtained by using a value of $\Delta T = 0.1$ in making calculations are listed in Table 5-6. The error has been reduced to a more acceptable level.

We should point out that the method we have demonstrated here (Nystrom's method) is but one of many finite-difference methods. For

TABLE 5-5 Comparison of Numerical Values Obtained by Finite-Difference Method with Those Obtained by Evaluating the Analytical Expression of Equation 5-9[a]

Time T	*Obtained by Finite-Difference Method* N	*Obtained by Evaluation of Analytical Expression* N*	*Error* N* − N
0	1.00	1.00	0.00
1	1.65	1.65	0.00
2	2.65	2.72	0.07
3	4.30	4.48	0.18
4	6.95	7.39	0.44
5	11.25	12.19	0.94
6	18.20	20.09	1.89
7	29.45	33.12	3.67
8	47.65	54.60	6.95
9	77.10	90.00	12.90
10	124.75	148.40	23.65

[a]Difference between methods, $N^* - N$, represents error resulting from particular finite-difference method. Increment $\Delta T = 1.0$.

TABLE 5-6 Comparison Similar to That of Table 5-5, Except That $\Delta T = 0.1$[a]

T	N	N^*	$N^* - N$
0	1.00	1.00	0.00
1	1.65	1.65	0.00
2	2.72	2.72	0.00
3	4.48	4.48	0.00
4	7.38	7.39	0.01
5	12.17	12.19	0.02
6	20.06	20.09	0.03
7	33.06	33.12	0.06
8	54.50	54.60	0.10
9	89.84	90.00	0.16
10	148.09	148.40	0.31

[a]Only every tenth value of N that has been calculated is shown.

example, there are a number of varieties of the Runge-Kutta method, which will give results superior to those obtained by Nystrom's method. In particular, the predictor-corrector (Adam's) method is to be recommended for superior accuracy. Our purpose in this example is not to dwell on the merits of various methods but simply to illustrate a relatively simple numerical method and to show the effect of changing the size of ΔT.

Furthermore the finite-difference interval (ΔT in this case) is by no means the only source of error. Truncation error, caused by the limit in the number of significant digits that can be stored by the computer, is an important source. The effect of computer truncation error can be roughly checked by comparing the results obtained with single-precision arithmetic with those obtained by double-precision arithmetic, which retains twice the number of significant digits at each step in the calculations.

It is suggested that the reader consult books by McCalla (1967), Stanton (1961), Ralston (1965), and Hamming (1962) for background in various numerical methods.

Algebraic Derivation of Finite-Difference Expressions

The preceding section showed that it is relatively simple to obtain the equivalent of a first derivative in the form of a difference quotient. The transformation is obvious, both geometrically and intuitively. The process of obtaining the difference equivalents of higher-order derivatives involves more algebraic manipulation, but the concepts involved are similar. Shaw (1953) presents a clear description of the transformation. To begin let us review some calculus notation. If y is a function of x, $y=f(x)$, then the first derivative of y with respect to x can be written

$$\frac{dy}{dx} \equiv \frac{d[f(x)]}{dx} \equiv f'(x) \tag{5-16}$$

the second derivative of y with respect to x is

$$\frac{d}{dx}\left(\frac{dy}{dx}\right) \equiv \frac{d^2y}{dx^2} \equiv \frac{d^2[f(x)]}{dx^2} \equiv f''(x) \tag{5-17}$$

and so on for third and higher derivatives.

Consider now the function $y=f(x)$, which is graphed in Figure 5-4. The first derivative, $dy/dx = f'(x)$, at any point on the curve can be defined as

$$\frac{dy}{dx} = \lim_{\Delta x \to 0} \frac{f(x+\Delta x)-f(x)}{\Delta x} \tag{5-18}$$

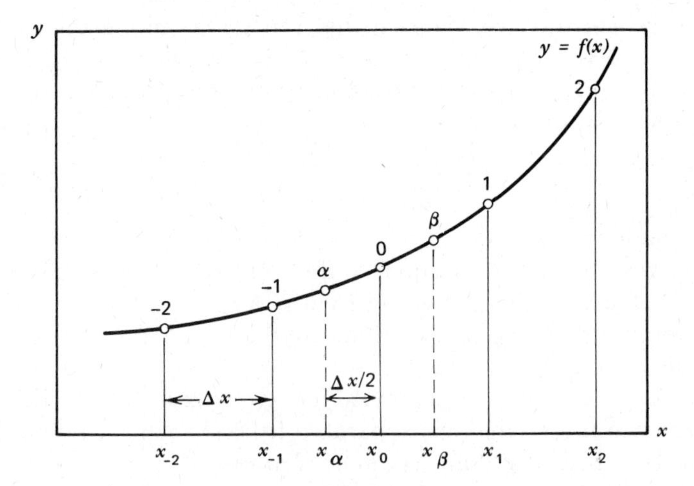

Figure 5-4 Graph of hypothetical function $y = f(x)$, in which the x-axis has been divided into regular units each Δx in width; α and β represent points on graph which are separated from point 0 in the x-direction by $\Delta x/2$.

If we wish to specify the derivative at a particular point, say point $x = \beta$, we can write

$$\Delta x\left(\frac{dy}{dx}\right)_\beta = f(x_0 + \Delta x) - f(x_0) + \epsilon\,\Delta x \tag{5-19}$$

where dy/dx_β = the value of dy/dx at point β
ϵ = an error term, whose value approaches zero as Δx approaches zero

The significance of this equation might be obscure at first glance. However, its meaning should become clear if we keep in mind that the difference in the value of y at two points on the curve, separated by Δx in the x-direction, is approximately equal to Δx times the derivative. The error, or value of ϵ times Δx, becomes vanishingly small as Δx approaches zero. If we drop the error term as Δx becomes very small, we can simplify Equation 5-19, writing

$$\Delta x\left(\frac{dy}{dx}\right)_\beta \simeq f(x_0 + \Delta x) - f(x_0) \tag{5-20}$$

or dividing through by Δx, we obtain

$$\left(\frac{dy}{dx}\right)_\beta \simeq \frac{f(x_0 + \Delta x) - f(x_0)}{\Delta x} \tag{5-21}$$

We can revise this equation so that the points along the x-axis are specified more directly (e.g., $x_1 = x_0 + \Delta x$, $x_2 = x_0 + 2\Delta x$). Likewise $f(x_1)$ represents the value of the function at x_1, and so on. This permits us to write

$$\left(\frac{dy}{dx}\right)_\beta \simeq \frac{f(x_1) - f(x_0)}{\Delta x} \tag{5-22}$$

This equation says in effect that the slope of the line at point β, halfway between x_0 and x_1, is equal to the difference in the values of the function at x_0 and at x_1, divided by Δx. As Δx becomes vanishingly small, the finite-difference quotient (on the right side) becomes equal to the derivative.

It is more common to express the derivative as the difference between two points that are separated by $2\Delta x$. Thus the first derivative for the function at point x_0 can be written

$$\left(\frac{dy}{dx}\right)_0 \simeq \frac{f(x_1) - f(x_{-1})}{2\Delta x} \tag{5-23}$$

The finite-difference approximations to higher derivatives can be obtained in an analogous manner. To develop the approximation to the second derivative let us bring in an additional point on the curve, α. The approximation to the first derivative at point α is

$$\left(\frac{dy}{dx}\right)_\alpha \simeq \frac{f(x_0) - f(x_{-1})}{\Delta x} \tag{5-24}$$

Now a second derivative can be defined as the derivative of a first derivative (i.e., the rate of change of the rate of change). We can write

$$\frac{d^2y}{dx^2} = \frac{d}{dx}\left(\frac{dy}{dx}\right) \tag{5-25}$$

The second derivative at point 0 on the x-axis thus can be written as an approximation to the difference between the first derivatives at points α and β.

$$\left(\frac{d^2y}{dx^2}\right)_0 \simeq \frac{(dy/dx)_\beta - (dy/dx)_\alpha}{\Delta x} \tag{5-26}$$

We cannot make direct use of the second derivative expressed in this form, since the right side contains derivatives. We can simply exchange the definitions of the first derivatives (Equations 5-22 and 5-24), obtaining

$$\left(\frac{d^2y}{dx^2}\right)_0 \simeq \frac{\{[f(x_1) - f(x_0)]/\Delta x\} - \{[f(x_0) - f(x_{-1})]/\Delta x\}}{\Delta x} \tag{5-27}$$

In turn this simplifies to

$$\left(\frac{d^2y}{dx^2}\right)_0 \simeq \frac{f(x_1) - 2f(x_0) + f(x_{-1})}{(\Delta x)^2} \tag{5-28}$$

The approximation to the third derivative can be derived in a similar manner. It is necessary, however, to employ a span along the x-axis that is $2\Delta x$ in width on each side of the point in question, for a total span of $4\Delta x$. If we are to approximate the third derivative at point x_0, we need to obtain the approximations to the second derivatives at points x_1 and x_{-1}. The approximations to the derivatives at these points can be defined, in turn, in terms of the difference between x_2 and x_0, and between x_0 and x_{-2}, respectively. The third derivative is defined, of course, as the derivative of the second derivative

$$\frac{d^3y}{dx^3} = \frac{d}{dx}\left(\frac{d^2y}{dx^2}\right) \tag{5-29}$$

The approximation to the third derivative at point zero can be written

$$\left(\frac{d^3y}{dx^3}\right)_0 \simeq \frac{d}{dx}\left[\frac{(d^2y/dx^2)_1 - (d^2y/dx^2)_{-1}}{2\Delta x}\right] \tag{5-30}$$

employing a modification of Equation 5-28 as the approximation of the second derivative, and employing $2\Delta x$ instead of Δx, since the span is from x_1 to x_{-1}, rather than x_α to x_β (Figure 5-4). Substituting the approximation to the second derivative (Equation 5-28), we obtain

$$\left(\frac{d^3y}{dx^3}\right)_0 \simeq \frac{\{[f(x_2) - 2f(x_1) + f(x_0)]/(\Delta x)^2\} - \{[f(x_{-2}) - 2f(x_{-1}) + f(x_0)]/(\Delta x)^2\}}{2\Delta x} \tag{5-31}$$

which simplifies to

$$\left(\frac{d^3y}{dx^3}\right)_0 \simeq \frac{f(x_2) - 2f(x_1) + 2f(x_{-1}) - f(x_{-2})}{2(\Delta x)^3} \tag{5-32}$$

The finite-difference approximations to the fourth- and higher-order derivatives are derived in a similar manner. The approximation of the fourth derivative at point 0 is written

$$\left(\frac{d^4y}{dx^4}\right)_0 \simeq \frac{f(x_2) - 4f(x_1) + 6f(x_0) - 4f(x_{-1}) + f(x_{-2})}{(\Delta x)^4} \tag{5-33}$$

Although approximations for the first four derivatives would be needed for finite-difference solutions of fourth-order differential equations, our work involves only the approximations to the first and second derivatives. We list the first four for the sake of completeness, however.

To write the finite-difference approximations in a more general form we may drop the $y = f(x)$ notation and simply use the values of y themselves. Then at point j

$$\left(\frac{dy}{dx}\right)_j \simeq \frac{1}{2\Delta x}(y_{j+1} - y_{j-1}) \tag{5-34}$$

$$\left(\frac{d^2y}{dx^2}\right)_j \simeq \frac{1}{(\Delta x)^2}(y_{j+1} - 2y_j + y_{j-1}) \tag{5-35}$$

$$\left(\frac{d^3y}{dx^3}\right)_j \simeq \frac{1}{2(\Delta x)^3}(y_{j+2} - 2y_{j+1} + 2y_{j-1} - y_{j-2}) \tag{5-36}$$

$$\left(\frac{d^4y}{dx^4}\right)_j \simeq \frac{1}{(\Delta x)^4}(y_{j+2} - 4y_{j+1} + 6y_j - 4y_{j-1} + y_{j-2}) \tag{5-37}$$

Numerical Solution in Finite-Difference Form

There are various methods for obtaining solutions in finite-difference form. The method that we illustrate involves three principal steps:

1. Translation of derivatives to finite-difference quotients.
2. Formulation of a set of simultaneous linear equations. The unknowns in these equations are the numerical values of the dependent variables, one for each increment of the independent variable (usually time or distance). The number of equations equals the number of unknowns, but the form of each equation is similar.
3. Iterative solution of the simultaneous linear equations.

To illustrate the translation of a differential equation to finite-difference form consider the linear differential equation

$$\frac{d^2y}{dx^2} - 3\frac{dy}{dx} - 10y = 10x \tag{5-38}$$

Substitution of the finite-difference approximations (Equations 5-34 and 5-35) for point x_j yields

$$\frac{y_{j+1} - 2y_j + y_{j-1}}{(\Delta x)^2} - \frac{3(y_{j+1} - y_{j-1})}{2\Delta x} - 10y_j = 10x_j$$

which simplifies to

$$y_{j-1}(2+3\Delta x)-y_j[4+20(\Delta x)^2]+y_{j+1}(2-3\Delta x)=20x_j(\Delta x)^2 \tag{5-39}$$

Now our problem is to evaluate this equation over a range of values of x. Suppose that we arbitrarily select a range in which the value of x ranges from 0.0 to 1.0 in increments of $\Delta x = 0.1$. We shall thus have 11 values of x, and these can be indexed with j, as follows:

j	x	j	x
0	0.0	6	0.6
1	0.1	7	0.7
2	0.2	8	0.8
3	0.3	9	0.9
4	0.4	10	1.0
5	0.5		

Given these values, it is a mechanical matter to insert them successively into Equation 5-39 to yield a sequence of nine linear equations in which the value of j is incremented in steps of 1, from 1 to 9:

$$\begin{aligned}
2.3y_0-4.2y_1+1.7y_2 &= 0.02\\
2.3y_1-4.2y_2+1.7y_3 &= 0.04\\
2.3y_2-4.2y_3+1.7y_4 &= 0.06\\
\cdot \quad \cdot \quad \cdot \quad &\cdot\\
\cdot \quad \cdot \quad \cdot \quad &\cdot\\
\cdot \quad \cdot \quad \cdot \quad &\cdot\\
2.3y_8-4.2y_9+1.7y_{10} &= 0.18
\end{aligned} \tag{5-40}$$

The reason that we generate only nine equations is that each equation involves three values of x. Our problem is to find the values of y_0 to y_{10} that satisfy the set of equations. But, since there are only nine equations, we can solve for only nine unknowns. Two of the eleven values must be supplied in advance. We can supply, for example, the first two, setting $y_0 = 0.0$ and y_1 at some arbitrarily guessed value, say 1.0. Then we can work forward, calculating values for the successive values of y from y_2 to y_{10}:

$$y_2 = \frac{0.0+4.2+0.02}{1.7} = 2.48$$

$$y_3 = \frac{-2.3 + (4.2)(2.48) + 0.04}{1.7} = 4.7976$$

and so on, until all have been calculated. The method has obvious drawbacks and is not normally attempted in practice. First, the likelihood of obtaining the correct value of y_1 is remote; second, errors are cumulative in the process, so that over a long sequence of operations gross inaccuracies result. It is immediately obvious that it would be wiser to supply values of y at y_0 and y_{10}. Let us assume that $y_0 = 0.0$ and $y_{10} = 100.0$. These are the *boundary* values. Then we can write nine simultaneous equations that have nine unknowns, y_1 through y_9. The equations are the same as those in Equation 5-40 except that $y_0 = 0.0$ is substituted into the first and $y_{10} = 100.0$ is substituted into the last, as follows:

$$\begin{aligned} -4.2y_1 + 1.7y_2 &= 0.02 \\ 2.3y_1 - 4.2y_2 + 1.7y_3 &= 0.04 \\ 2.3y_2 - 4.2y_3 + 1.7y_4 &= 0.06 \\ \cdot \quad \cdot \quad \cdot \quad & \cdot \\ \cdot \quad \cdot \quad \cdot \quad & \cdot \\ \cdot \quad \cdot \quad \cdot \quad & \cdot \\ 2.3y_8 - 4.2y_9 \qquad\qquad &= -169.82 \end{aligned} \tag{5-41}$$

We can now solve these equations iteratively by the Gauss–Seidel method (see Appendix). To do this the equations are first rearranged to give

$$y_1 = \frac{(1.7y_2 - 0.02)}{4.2}$$

$$y_2 = \frac{(2.3y_1 + 1.7y_3 - 0.04)}{4.2}$$

$$y_3 = \frac{(2.3y_2 + 1.7y_4 - 0.06)}{4.2}$$

$$\begin{matrix} \cdot & \cdot & \cdot & \cdot \\ \cdot & \cdot & \cdot & \cdot \\ \cdot & \cdot & \cdot & \cdot \end{matrix}$$

$$y_9 = \frac{(2.3y_8 + 169.82)}{4.2}$$

The general form of these equations for the jth point can be written

$$y_j = \frac{(2.3y_{j-1} + 1.7y_{j+1} - 0.02j)}{4.2} \tag{5-42}$$

Equation 5-42 holds for all of the equations on the first iteration except the first and last, which contain the boundary values y_0 and y_{10}.

The Gauss–Seidel method entails making a guess at the value of the unknowns, using these guesses to obtain approximate solutions to each unknown in turn, then substituting the approximate solutions to the unknowns on the second and subsequent passes through the equations.

On the first pass let us assume that the values of the unknowns are all zero. Then

$$y_1 = \frac{(1.7y_2 - 0.02)}{4.2} = \frac{[(1.7)(0.0) - 0.02]}{4.2} = -0.004761$$

Using this value in the next equation,

$$y_2 = \frac{(2.3y_1 + 1.7y_3 - 0.04)}{4.2} = \frac{[(2.3)(-0.004761) + (1.7)(0.0) - 0.04]}{4.2}$$

$$= -0.012131$$

and

$$y_3 = \frac{(2.3y_2 + 1.7y_4 - 0.06)}{4.2} = \frac{[(2.3)(-0.012131) + (1.7)(0.0) - 0.06]}{4.2}$$

$$= -0.02092857$$

$y_4 \cdots y_9$ are solved similarly. Then the equations are passed through a second time, using the new approximate solutions, to yield revised values for y_1 to y_9.

$$y_1 = \frac{(1.7y_2 - 0.02)}{4.2} = \frac{[(1.7)(-0.012131) - 0.02]}{4.2} = -0.0096721$$

$$y_2 = \frac{(2.3y_1 + 1.7y_3 - 0.04)}{4.2}$$

$$= \frac{[(2.3)(-0.0096721) + (1.7)(-0.02092857) - 0.04]}{4.2} = 0.023291$$

and so on. The method is self-correcting. When y_9 is being calculated, the upper boundary value y_{10} is present and influences the value of y_9. Likewise, when y_8 is calculated, it is influenced by the current value of y_9, and so on down the line. The effect is that the boundary values influence the values y_1 to y_9. As successive iterations of the entire set are calculated, the values of y_1 to y_9 tend to converge toward stable values that are appropriate for the boundary values that have been assigned.

It is obvious that the numerical solution by this method of even a simple linear differential equation involves a prodigious amount of arithmetic. In fact the labor is so great that numerical solutions would not be possible in many applications if the calculations had to be carried out manually. If we write a short FORTRAN program to carry out these operations, we find that very few statements are required due to the repetitive nature of the method of solution (Table 5-7). The general Equation 5-42 is the only equation required. The DO-loop then successively furnishes the correct index values. In the program the sequence $y_0 \cdots y_{10}$ has been shifted to Y(1) $\cdots$ Y(11), inasmuch as zero subscripts are not permitted in FORTRAN. Table 5-8 lists the output from the program. Each row of numbers represents calculations made during a single iteration, here truncated to three decimal places for listing purposes. Thirty iterations were carried out, and the values in each column are seen to converge toward a relatively stable result. If rounding or truncation errors were not present, the greater the number of iterations, the greater would be the accuracy. Rounding is a problem, however, due to limitations on the number of significant digits per number that are imposed by word length of the computer. Also, the choice of the increment Δx is another source of error. A smaller value for Δx, coupled with more iterations, would tend to improve the solution, although a lower limit to the smallness of Δx is set by rounding errors, as well as investment in computing time.

TABLE 5-7 FORTRAN Program for Obtaining the Numerical Solution of Equation 5-38 by the Gauss–Seidel Method

```
C.....SOLVE SIMULTANEOUS EQUATIONS BY GAUSS-SEIDEL METHOD
      DIMENSION Y(11)
    1 FORMAT(1H , I5, 11F10.3)
C.....SET BOUNDARY VALUES
      Y(1)=0.0
      Y(11)=100.0
C.....SET INITIAL GUESSES TO ZERO
      DO 10 I=2,10
   10 Y(I)=0.0
C.....BEGIN ITERATIVE SOLUTION
      DO 30 K=1,30
      DO 20 I=2,10
      RI=I-1
   20 Y(I)=(2.3*Y(I-1)+1.7*Y(I+1)-0.02*RI)/4.2
   30 WRITE(6,1) K, (Y(I), I=1,11)
      RETURN
      END
```

TABLE 5-8 Output from the FORTRAN Program of Table 5-7[a]

Iterations	$x = 0.1$	$x = 0.2$	$x = 0.3$	$x = 0.4$	$x = 0.5$	$x = 0.6$	$x = 0.7$	$x = 0.8$	$x = 0.9$
	y_1	y_2	y_3	y_4	y_5	y_6	y_7	y_8	y_9
1	-0.005	-0.012	-0.021	-0.031	-0.041	-0.051	-0.061	-0.072	40.394
2	-0.010	-0.023	-0.039	-0.057	-0.076	-0.095	-0.114	16.249	49.332
3	-0.014	-0.033	-0.056	-0.080	-0.106	-0.133	6.471	23.473	53.288
4	-0.018	-0.042	-0.070	-0.100	-0.132	2.518	10.847	27.471	55.477
5	-0.022	-0.050	-0.082	-0.118	0.931	4.872	13.754	29.948	56.834
6	-0.025	-0.056	-0.093	0.307	2.116	6.697	15.756	31.594	57.735
7	-0.028	-0.062	0.076	0.879	3.168	8.084	17.182	32.740	58.362
8	-0.030	0.005	0.344	1.452	4.043	9.140	18.224	33.564	58.814
9	-0.003	0.128	0.644	1.970	4.755	9.951	19.002	34.173	59.147
10	0.047	0.277	0.935	2.417	5.328	10.580	19.593	34.632	59.398
11	0.107	0.428	1.198	2.794	5.789	11.072	20.047	34.982	59.590
12	0.168	0.568	1.427	3.106	6.158	11.458	20.401	35.254	59.739
13	0.225	0.691	1.621	3.361	6.455	11.764	20.678	35.466	59.855
14	0.275	0.797	1.783	3.570	6.693	12.006	20.897	35.632	59.946
15	0.318	0.886	1.916	3.739	6.883	12.199	21.070	35.764	60.018
16	0.354	0.960	2.025	3.876	7.036	12.353	21.207	35.869	60.076
17	0.384	1.020	2.113	3.986	7.159	12.476	21.317	35.952	60.121
18	0.408	1.069	2.185	4.075	7.258	12.574	21.404	36.018	60.158
19	0.428	1.109	2.243	4.147	7.336	12.653	21.474	36.071	60.187
20	0.444	1.141	2.289	4.204	7.400	12.716	21.530	36.113	60.210
21	0.457	1.167	2.327	4.250	7.450	12.766	21.575	36.147	60.228
22	0.468	1.188	2.357	4.287	7.491	12.806	21.611	36.174	60.243
23	0.476	1.205	2.381	4.317	7.524	12.839	21.639	36.196	60.255
24	0.483	1.219	2.400	4.341	7.550	12.865	21.663	36.214	60.265
25	0.489	1.230	2.416	4.360	7.571	12.886	21.681	36.228	60.272
26	0.493	1.238	2.429	4.375	7.588	12.902	21.696	36.239	60.278
27	0.496	1.245	2.439	4.388	7.601	12.916	21.708	36.248	60.283
28	0.499	1.251	2.447	4.398	7.612	12.926	21.717	36.255	60.287
29	0.502	1.256	2.453	4.405	7.621	12.935	21.725	36.261	60.290
30	0.503	1.259	2.458	4.412	7.628	12.942	21.731	36.265	60.293

[a]Numbers in the main body of the table represent values of y that correspond with different values of x. Columns contain values of y_1 to y_9 for specified values of x, which range from 0.1 to 0.9. Each row represents a successive iteration. If the column for y_0 were listed, it would contain all zeros. Similarly column y_{10} would contain all 1000.0's.

It is convenient to express the results of the numerical solution of Equation 5-38 in graphic form. Figure 5-5 shows values of y that pertain to the first and thirtieth iterations. The values of y_1 to y_9 change progressively, of course, whereas the boundary values y_0 and y_{10} remain constant, having been defined as 0.0 and 100.0, respectively.

NUMERICAL SOLUTION OF PARTIAL-DIFFERENTIAL EQUATIONS

Two Independent Variables

So far we have considered the finite-difference form of differential equations involving only a single independent variable. In order to deal with applications that involve two or three dimensions, such as flow networks, we need differential equations involving two or more independent variables. Equation 5-50 is an example of a differential equation involving two independent variables and containing partial derivatives. We need to make some changes in notation at this point. In dealing with two independent variables we use z to represent the dependent variable, and x and y to represent the independent variables.

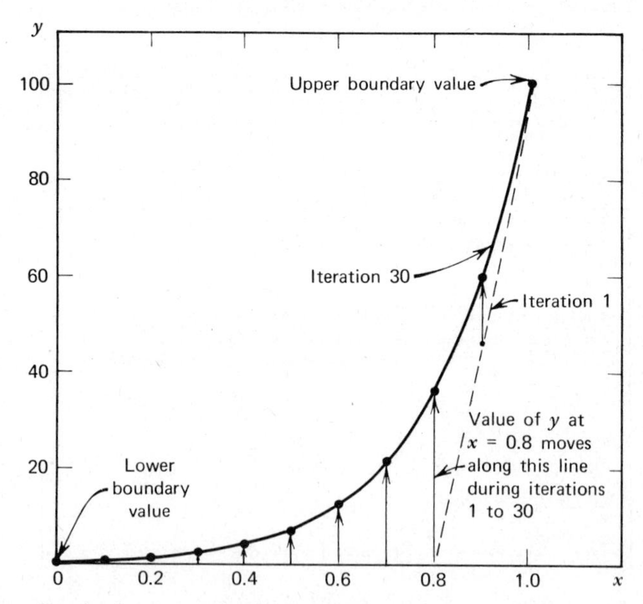

Figure 5-5 Graph of numerical solution of Equation 5-38. Solid line represents 30th iteration, toward which earlier iterations converge.

Our immediate problem is to develop the relationship between partial derivatives and their finite-difference equivalents. Consider a plane representing two independent variables, x and y, which has been subdivided into square cells (Figure 5-6). Each cell contains a point marking its center and each is indexed with a row number and a column number, using i as the row, or y-direction, index and j as the column, or x-direction, index. If we now add a third axis, the dependent variable z, we are capable of representing, in three-dimensional space, the function $z = f(x, y)$ as a surface that involves the two independent variables (Figure 5-7). Derivatives of z will be partial derivatives, with respect to either x or y. First partial derivatives are written

$$\frac{\partial z}{\partial x} \quad \text{and} \quad \frac{\partial z}{\partial y}$$

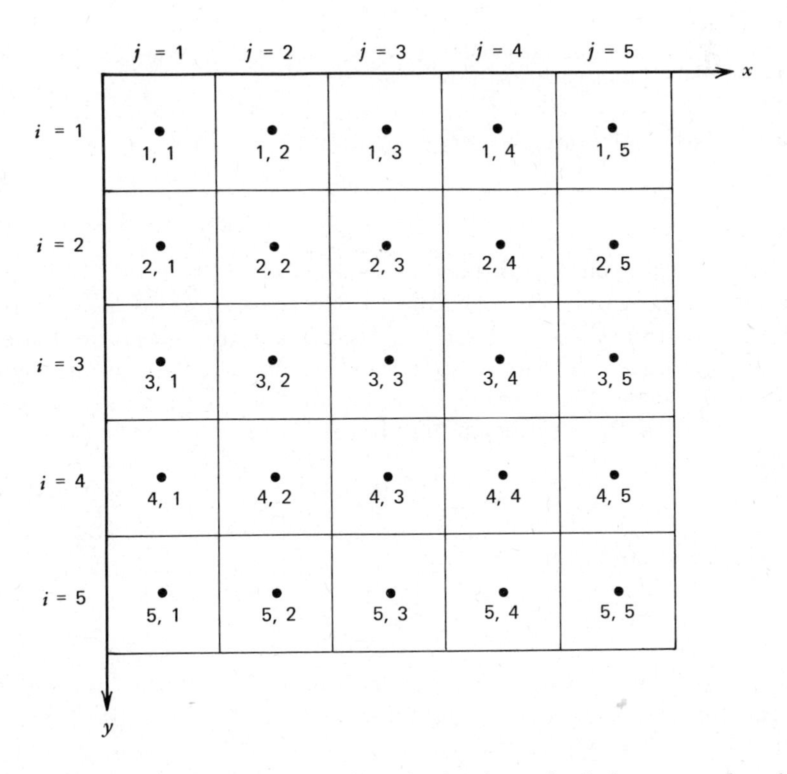

Figure 5-6 Representation of $x-y$ plane that has been divided into a meshwork of square cells, each Δx by Δy in dimension.

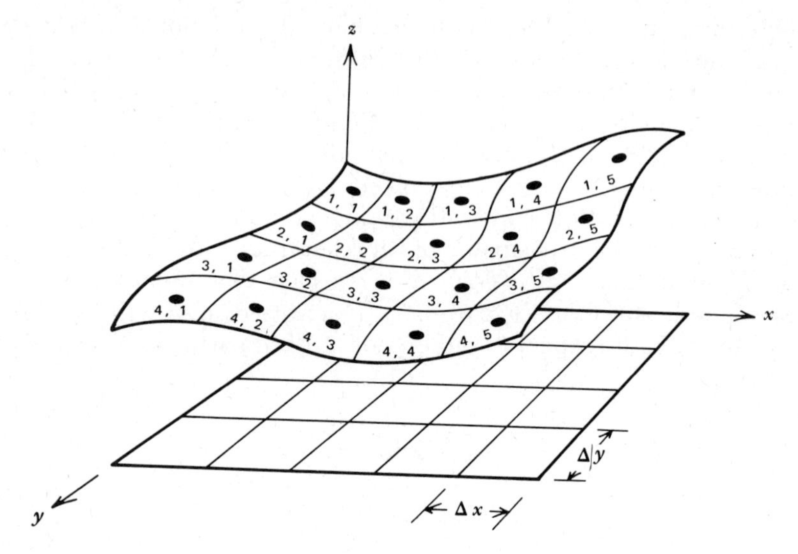

Figure 5-7 Graphic representation of z as function of x and y, in which x and y have been segmented into square cells Δx by Δy in size.

and second partial derivatives are written

$$\frac{\partial^2 z}{\partial x^2} \quad \text{and} \quad \frac{\partial^2 z}{\partial y^2}$$

The transformation of partial derivatives into finite-difference quotients is identical in principle to that involving ordinary derivatives (Equations 5-34 to 5-37). Of course only one independent variable is considered at a time. For example, the finite-difference equivalent of the first partial derivative of z with respect to x at point (3,3) of the function represented in Figure 5-7 is

$$\frac{\partial z}{\partial x} \simeq \frac{z_{3,4} - z_{3,2}}{2\Delta x} \tag{5-43}$$

or at any point (i, j) it is

$$\frac{\partial z}{\partial x} \simeq \frac{z_{i,j+1} - z_{i,j-1}}{2\Delta x} \tag{5-44}$$

The derivative $\partial z/\partial x$ may be regarded as the slope of the intersection of the function and a plane parallel to the z-x plane. Similarly the derivative $\partial z/\partial y$ at point (3,3) is

$$\frac{\partial z}{\partial y} \simeq \frac{z_{4,3} - z_{2,3}}{2\Delta y} \tag{5-45}$$

or at the point (i, j) it is

$$\frac{\partial z}{\partial y} \simeq \frac{z_{i+1,j} - z_{i-1,j}}{2\Delta y} \tag{5-46}$$

The finite-difference equivalents to second partial derivatives are likewise similar to ordinary derivatives. For point (i, j) they are

$$\frac{\partial^2 z}{\partial x^2} \simeq \frac{1}{(\Delta x)^2}(z_{i,j+1} - 2z_{i,j} + z_{i,j-1}) \tag{5-47}$$

$$\frac{\partial^2 z}{\partial y^2} \simeq \frac{1}{(\Delta y)^2}(z_{i+1,j} - 2z_{i,j} + z_{i-1,j}) \tag{5-48}$$

We also have the special case of the mixed partial derivative and its finite-difference equivalent, in which both independent variables are involved:

$$\frac{\partial^2 z}{\partial x\, \partial y} \simeq \frac{1}{4\Delta x\, \Delta y}(z_{i-1,j+1} - z_{i-1,j-1} + z_{i+1,j-1} - z_{i+1,j+1}) \tag{5-49}$$

Our work, however, will involve only the conventional partial derivatives.

Now let us consider the numerical solution of the partial-differential equation

$$\frac{\partial^2 z}{\partial x^2} + \frac{\partial^2 z}{\partial y^2} = -500 \tag{5-50}$$

By substituting Equations 5-47 and 5-48, Equation 5-50 may be represented in finite-difference form at point (i, j) as

$$\frac{1}{(\Delta x)^2}(z_{i,j+1} - 2z_{i,j} + z_{i,j-1}) + \frac{1}{(\Delta y)^2}(z_{i+1,j} - 2z_{i,j} + z_{i-1,j}) = -500 \tag{5-51}$$

If $\Delta x = \Delta y$, the equation can be rearranged as

$$z_{i,j+1} + z_{i,j-1} + z_{i-1,j} + z_{i+1,j} - 4z_{i,j} = -500(\Delta x)^2 \tag{5-52}$$

Examination of the left side of this equation reveals that it is simply the sum of the four points closest to (i, j), less four times the value at (i, j), as shown in Figure 5-8.

Obtaining numerical solutions to the equation is thus a matter of systematically moving across a two-dimensional meshwork, carrying out the arithmetic at each point according to Equation 5-52. This system offers few problems except at the boundaries of the meshwork.

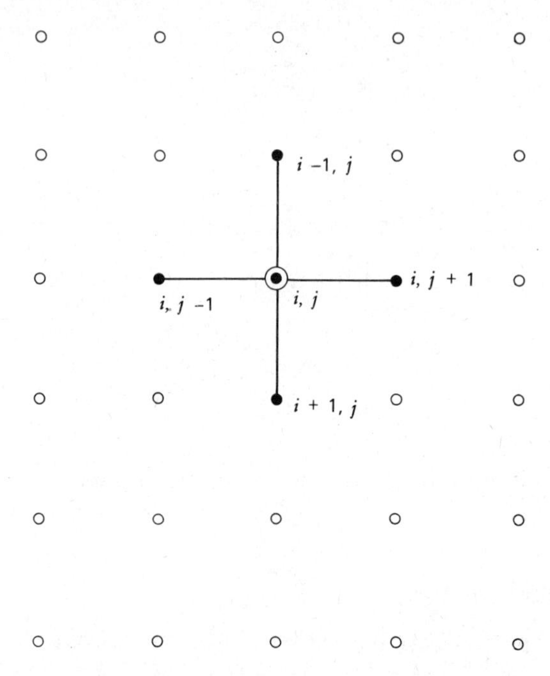

Figure 5-8 Method of indexing points, employing four-point star, in numerical solution of finite-difference equivalent of Equation 5-50. At each interior mesh point defined by i and j, values of z at four immediately adjacent points (black dots) are summed, and four times z value at point (i, j) is subtracted. Star is moved so as to successively occupy all interior mesh points.

There are of course only three immediately adjacent points at an edge boundary and only two at a corner mesh point. Consequently the values at boundary points must be specified in advance. A numerical solution of Equation 5-52 can be obtained in accordance with the boundary values that have been chosen.

For illustrative purposes assume that we have a square meshwork containing five rows and five columns (Figure 5-6). Furthermore let $\Delta x = \Delta y = 1$ and arbitrarily assume that the boundary values are all zero (i.e., the edge and corner values are zero). Thus

$$\begin{aligned} z_{1,1} &= z_{1,2} = z_{1,3} = z_{1,4} = z_{1,5} = z_{2,5} = z_{3,5} = z_{4,5} = z_{5,5} \\ &= z_{5,4} = z_{5,3} = z_{5,2} = z_{5,1} = z_{4,1} = z_{3,1} = z_{2,1} = 0.0 \end{aligned}$$

Our problem is to find the nine unknown values that remain in the interior of the meshwork:

$$z_{2,2}, z_{2,3}, z_{2,4}, z_{3,2}, z_{3,3}, z_{3,4}, z_{4,2}, z_{4,3}, z_{4,4}$$

Nine unknowns, of course, require the solution of nine simultaneous equations. Applying Equation 5-52 to each of the cells containing unknowns in turn, we obtain

$$\begin{aligned}
&z_{2,3}+z_{2,1}+z_{1,2}+z_{3,2}-4z_{2,2}+500=0 \qquad (i=2,j=2)\\
&z_{2,4}+z_{2,2}+z_{1,3}+z_{3,3}-4z_{2,3}+500=0 \qquad (i=2,j=3)\\
&z_{2,5}+z_{2,3}+z_{1,4}+z_{3,4}-4z_{2,4}+500=0 \qquad (i=2,j=4)\\
&z_{3,3}+z_{3,1}+z_{2,2}+z_{4,2}-4z_{3,2}+500=0 \qquad (i=3,j=2)\\
&z_{3,4}+z_{3,2}+z_{2,3}+z_{4,3}-4z_{3,3}+500=0 \qquad (i=3,j=3)\\
&z_{3,5}+z_{3,3}+z_{2,4}+z_{4,4}-4z_{3,4}+500=0 \qquad (i=3,j=4)\\
&z_{4,3}+z_{4,1}+z_{3,2}+z_{5,2}-4z_{4,2}+500=0 \qquad (i=4,j=2)\\
&z_{4,4}+z_{4,2}+z_{3,3}+z_{5,3}-4z_{4,3}+500=0 \qquad (i=4,j=3)\\
&z_{4,5}+z_{4,3}+z_{3,4}+z_{5,4}-4z_{4,4}+500=0 \qquad (i=4,j=4)
\end{aligned} \tag{5-53}$$

These equations can be solved by the Gauss–Seidel method, placing each unknown on the left-hand side of its respective equation. The simultaneous equations then can be rearranged as follows:

$$\begin{aligned}
z_{2,2} &= \frac{(z_{2,3}+z_{2,1}+z_{1,2}+z_{3,2}+500)}{4.0} \qquad (i=2,j=2)\\
z_{2,3} &= \frac{(z_{2,4}+z_{2,2}+z_{1,3}+z_{3,3}+500)}{4.0} \qquad (i=2,j=3)\\
&\vdots\\
z_{4,4} &= \frac{(z_{4,5}+z_{4,3}+z_{3,4}+z_{5,4}+500)}{4.0} \qquad (i=4,j=4)
\end{aligned} \tag{5-54}$$

The Gauss–Seidel method involves successive iterations. Since the interactive sequence must have a beginning, we need to assign initial "guesses" to the nine unknowns. It is convenient to use zero values for the initial guesses, but other arbitrarily chosen values will give the same final results, assuming that enough iterations are carried out. As the iterations progress, the values of the unknowns change, gradually converging on stable values. The boundary values do not change, however. Thus the final stable values for the interior mesh points depend on both the chosen boundary values and the characteristics of the equation whose solution is being sought.

Stepping through the calculations for several of the equations will illustrate the method. In the first equation $z_{1,2}$ and $z_{2,1}$ have values of zero, being boundary values. The other two points, which are interior points $z_{2,3}$ and $z_{3,2}$, will have zero values initially since all interior points were set to zero before the initial iteration. Thus $z_{2,2} = 500/4.0 = 125.0$. The process is repeated for the next equation. But $z_{2,2}$ will have a value of 125.0, and, when this is inserted, a value for $z_{2,3}$ is obtained:

$$z_{2,3} = \frac{125.0 + 500}{4.0} = 156.25$$

The process is repeated until all interior mesh points have been treated during the first pass through the nine simultaneous equations. At each subsequent iteration the interior-mesh-point values are modified, each gradually converging to a stable value as sufficient iterations are carried out.

The computational labor that is involved, if attempted manually, would be overwhelming for even a few iterations and the relatively coarse meshwork employed here. A computer program devised specifically for Equation 5-54 is listed in Table 5-9. In this program

TABLE 5-9 FORTRAN Program for the Iterative Solution of Equation 5-50, Which Has been Transformed into Finite-Difference Form with a 5 × 5 Meshwork (Equation 5-54).

```
C.....GAUSS-SEIDEL SOLUTION OF SIMULTANEOUS EQUATIONS
      DIMENSION Z(5,5)
    1 FORMAT(1H / 4X,'N', 6X,'2,2',7X,'2,3',7X,'2,4',7X,'3,2',7X,'3,3',
     1 7X,'3,4',7X,'4,2',7X,'4,3',7X,'4,4'//)
    2 FORMAT(1H , I5,9F10.3)
C.....SET BOUNDARY CELLS AND INITIAL GUESSES TO ZERO
      DO 90 I=1,5
      DO 90 J=1,5
   90 Z(I,J)=0.0
C.....BEGIN ITERATIVE SOLUTION
      WRITE(6,1)
      DO 110 N=1,30
      DO 100 I=2,4
      DO 100 J=2,4
  100 Z(I,J)=(Z(I-1,J)+Z(I,J+1)+Z(I+1,J)+Z(I,J-1)+500.0)/4.0
  110 WRITE(6,2) N, ((Z(I,J), J=2,4), I=2,4)
      RETURN
      END
```

the string of nine equations has been reduced to one general equation, using I as the row-number index and J as the column-number index. The values of I and J are successively altered as each DO-loop is entered. Table 5-10 lists the output for 30 iterations. Values for the unknowns have converged to a stable result, reaching two-decimal-place accuracy after 17 iterations. The program could be modified to provide an automatic check on accuracy, so that the iterations cease when sufficiently stable results have been obtained.

TABLE 5-10 Output from Program Listed in Table 5-9[a]

Iterations	Cell Indices								
N	2,2	2,3	2,4	3,2	3,3	3,4	4,2	4,3	4,4
1	125.000	156.250	164.063	156.250	203.125	216.797	164.063	216.797	233.398
2	203.125	267.578	246.094	267.578	367.188	336.670	246.094	336.670	293.335
3	258.789	343.018	294.922	343.018	464.844	388.275	294.922	388.275	319.137
4	296.509	389.069	319.336	389.069	513.672	413.036	319.336	413.036	331.518
5	319.534	413.135	331.543	413.135	538.086	425.287	331.543	425.287	337.643
6	331.568	425.299	337.646	425.299	550.293	431.396	337.646	431.396	340.698
7	337.649	431.397	340.698	431.397	556.396	434.448	340.698	434.448	342.224
8	340.698	434.448	342.224	434.448	559.448	435.974	342.224	435.974	342.987
9	342.224	435.974	342.987	435.974	560.974	436.737	342.987	436.737	343.368
10	342.987	436.737	343.368	436.737	561.737	437.118	343.368	437.118	343.559
11	343.368	437.118	343.559	437.118	562.118	437.309	343.559	437.309	343.655
12	343.559	437.309	343.655	437.309	562.309	437.405	343.655	437.405	343.702
13	343.655	437.405	343.702	437.405	562.405	437.452	343.702	437.452	343.726
14	343.702	437.452	343.726	437.452	562.452	437.476	343.726	437.476	343.738
15	343.726	437.476	343.738	437.476	562.476	437.488	343.738	437.488	343.744
16	343.738	437.488	343.744	437.488	562.488	437.494	343.744	437.494	343.747
17	343.744	437.494	343.747	437.494	562.494	437.497	343.747	437.497	343.748
18	343.747	437.497	343.748	437.497	562.497	437.498	343.748	437.498	343.749
19	343.748	437.498	343.749	437.498	562.498	437.499	343.749	437.499	343.750
20	343.749	437.499	343.750	437.499	562.499	437.500	343.750	437.500	343.750
21	343.750	437.500	343.750	437.500	562.500	437.500	343.750	437.500	343.750
22	343.750	437.500	343.750	437.500	562.500	437.500	343.750	437.500	343.750
23	343.750	437.500	343.750	437.500	562.500	437.500	343.750	437.500	343.750

[a]At each iteration values for unknowns $z_{2,2}$ to $z_{4,4}$ are printed out, one row of figures per iteration. Values converge toward true values, reaching two-decimal-place accuracy after 17 iterations.

The numerical solution of the finite-difference equivalent of an equation involving two independent variables can be effectively displayed with a two-dimensional tabulation of values within the cellular meshwork. If the values are contoured, we obtain a convenient representation of the continuous function. Figure 5-9 shows the meshwork of cells pertaining to the numerical solution of Equation 5-50. Figure 5-9*a* shows the values prior to the initial iteration in which the initial guesses have been set to zero; Figure 5-9*b* shows the values, which have been contoured, after a stable solution has been obtained.

If the boundary values are changed, the resulting numerical solution of the equation is different. To illustrate, if a boundary value of 300 (Figure 5-10*a*) is supplied to cell (5, 3) and the other boundary values are kept at zero, the numerical solution (Figure 5-10*b*) strongly reflects the presence of the new boundary value. In Chapter 6 we adjust the boundary values to represent different flow conditions.

Three Independent Variables

Partial-differential equations representing potential flow and involving three spatial coordinates or independent variables x, y, and z, and in which ϕ is the dependent variable, take the general form

$$\frac{\partial^2\phi}{\partial x^2}+\frac{\partial^2\phi}{\partial y^2}+\frac{\partial^2\phi}{\partial z^2}=0 \tag{5-55}$$

which is commonly known as the Laplace equation. Translation of the Laplace equation into its finite-difference equivalent involves steps similar to those involved in translating differential equations containing one or two independent variables.

Figure 5-11 illustrates how three-dimensional space can be divided into a meshwork of rectangular or cubic cells. The locations of the cells can be readily indexed by sequences of three numbers, index i representing the y-direction, j the x-direction, and k the z-direction. Each of the interior cells is surrounded by six immediately adjacent cells that can be represented by a meshwork of points taking the form of a six-pointed star (Figure 5-12). The finite-difference quotients, which approximate the three partial derivatives of Equation 5-55, make use of the approximation to the second derivative (Equation 5-35). In the x-direction

$$\frac{\partial^2\phi}{\partial x^2}\simeq\frac{1}{(\Delta x)^2}\,(\phi_{i,j-1,k}-2\phi_{i,j,k}+\phi_{i,j+1,k}) \tag{5-56}$$

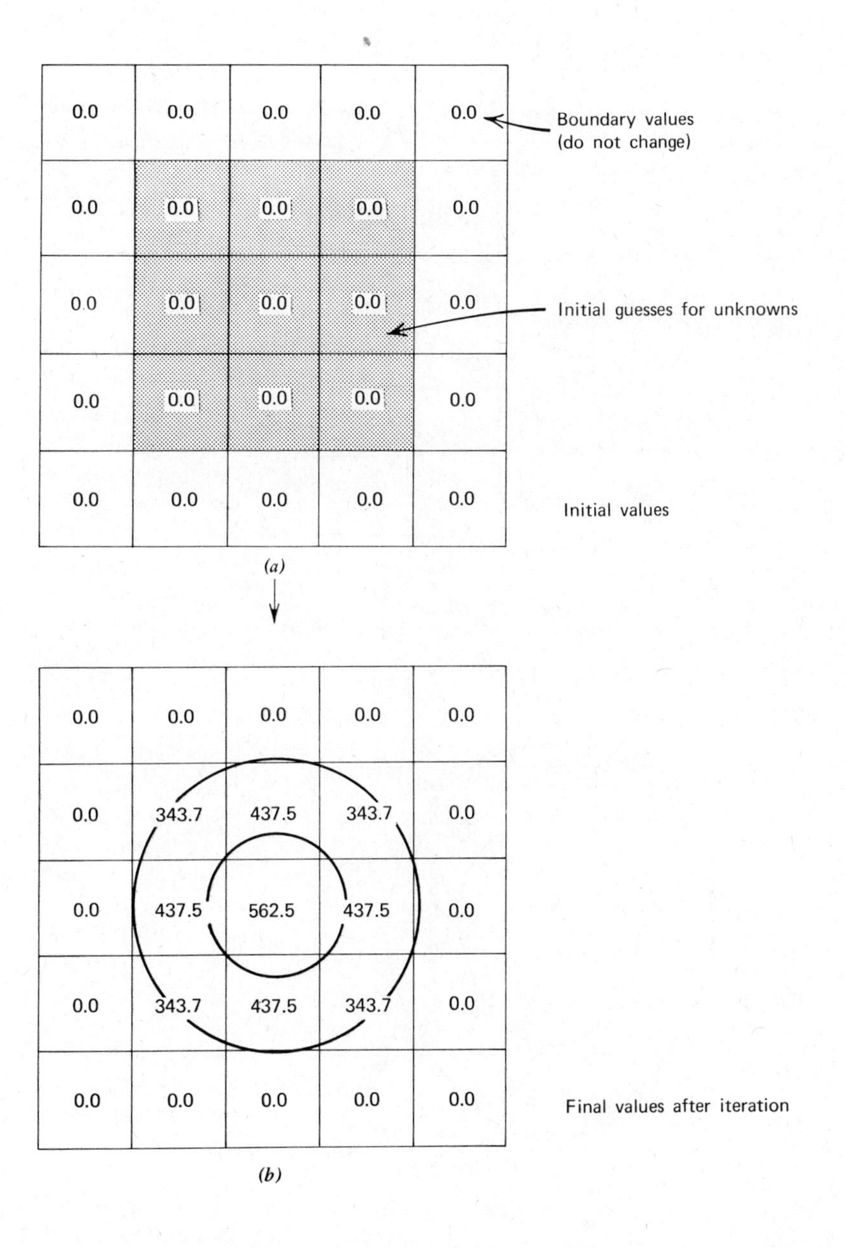

Figure 5-9 Display of numerical solution to Equation 5-50, assuming all boundary values to be zero: (*a*) cell values before iterations have begun; (*b*) solution after sufficient iterations have taken place so that interior cell values are stable. Contour interval is 250.

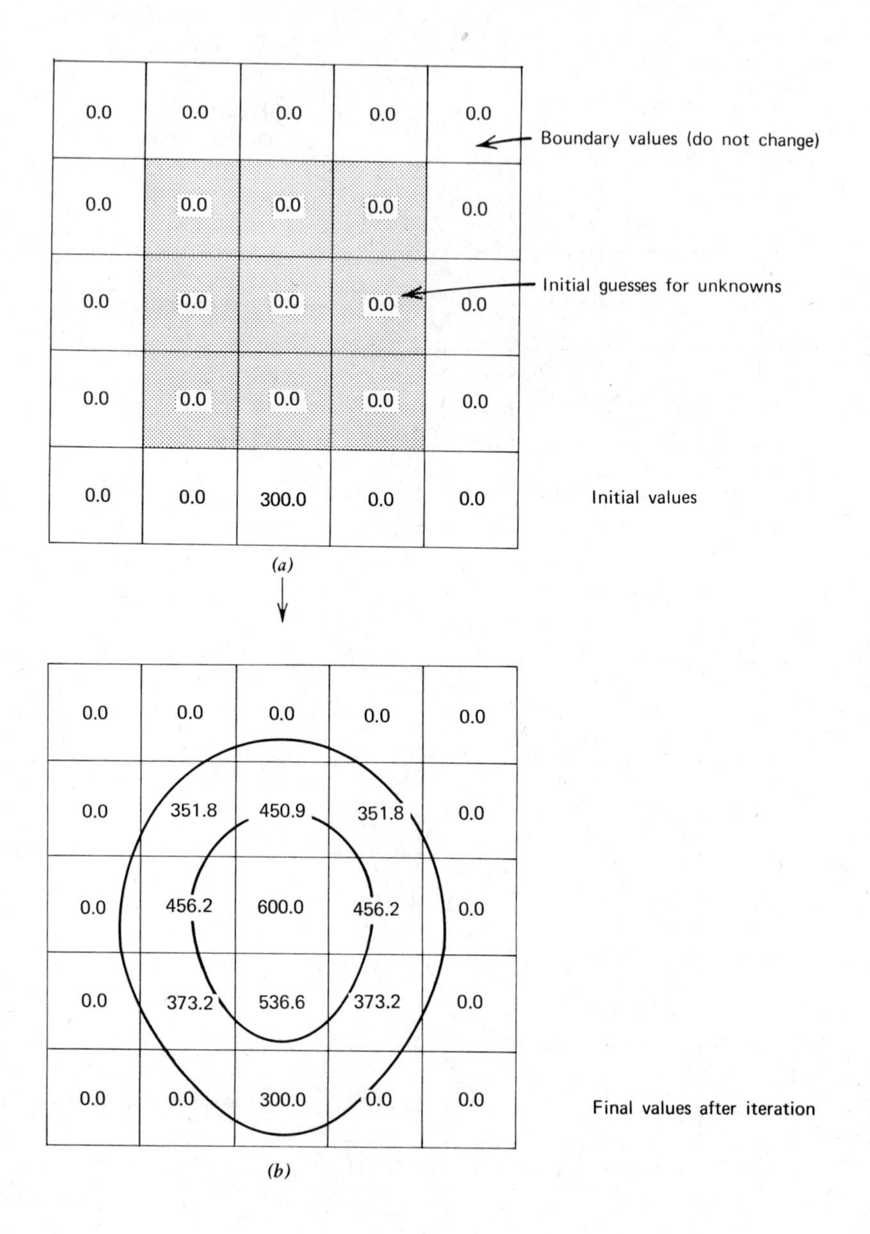

Figure 5-10 Display of numerical solution to Equation 5-50 in which one boundary cell has a value of 300 and remaining boundary cells are zero: (*a*) cell values before iterations have begun; (*b*) stable cell values after sufficient number of iterations have been carried out. Contour interval is 250.

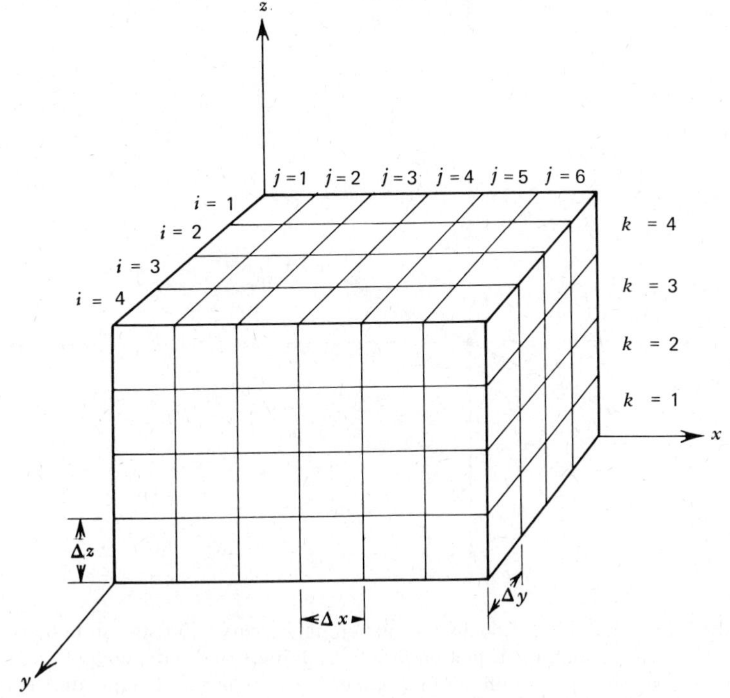

Figure 5-11 Three-dimensional meshwork of cells. Cells, whose dimensions are Δx by Δy by Δz, are indexed by i, j, and k.

in the y-direction

$$\frac{\partial^2 \phi}{\partial y^2} \simeq \frac{1}{(\Delta y)^2}(\phi_{i-1,j,k} - 2\phi_{i,j,k} + \phi_{i+1,j,k}) \tag{5-57}$$

and in the z-direction

$$\frac{\partial^2 \phi}{\partial z^2} \simeq \frac{1}{(\Delta z)^2}(\phi_{i,j,k-1} - 2\phi_{i,j,k} + \phi_{i,j,k+1}) \tag{5-58}$$

Substituting Equations 5-56, 5-57, and 5-58 into Equation 5-55 and assuming that $\Delta x = \Delta y = \Delta z$, we obtain a general finite-difference representation of the Laplace equation for three dimensions:

$$\frac{1}{(\Delta x)^2}(\phi_{i,j-1,k} + \phi_{i,j+1,k} + \phi_{i-1,j,k} + \phi_{i+1,j,k} + \phi_{i,j,k-1,} + \phi_{i,j,k+1} \\ - 6\phi_{i,j,k}) = 0 \tag{5-59}$$

The numerical solution of Equation 5-59 is parallel in principle to the solution of equations containing only two independent vari-

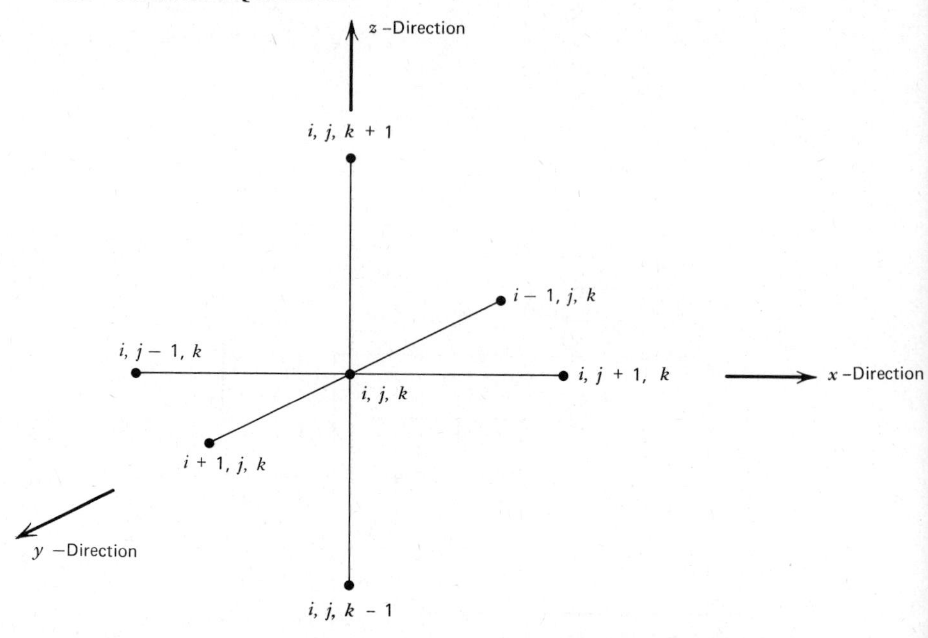

Figure 5-12 Method of indexing points, employing six-point star, in numerical solution of Laplace equation (Equation 5-55) involving three independent variables. At each interior mesh point defined by i, j, and k, values of z at six immediately adjacent points are summed, and six times value of z at i, j, k is subtracted.

ables. Each cell in the rectangular volume under consideration is considered. The cells on the outer surface of the volume are supplied with boundary values. Given guesses for initial values of all the interior cells, stable values representing a numerical solution to the equation are obtained by repeated iterations with the Gauss–Seidel method or other iterative methods. We have selected only one of several iterative methods that can be used for solving simultaneous linear equations. Other methods are discussed by McCalla (1967), Forsythe and Moler (1967), and in other books on numerical analysis.

Problems

Problem 5-1

Write a FORTRAN program to solve the equation

$$y = \frac{c}{\log_e (y/d) - 1.0}$$

by successive approximation. If $d = 10$, $c = -5$, and using an initial value of $y = 1.5$, find the value of y to three decimal places. How many additional iterations are needed to achieve accuracy to four decimal places?

Problem 5-2

Write a FORTRAN program to obtain a numerical solution to the equation

$$\frac{d^2y}{dx^2} - 5\frac{dy}{dx} + 6y = 10x$$

using 11 mesh points (10 increments) in the range $x = 0.0$ to 1.0. Assume boundary values such that $y_0 = 0.0$ and $y_{10} = 10.0$ when $x = 0.0$ and 1.0, respectively. Provide for 50 iterations.

Problem 5-3

a. Prepare a general-purpose FORTRAN program for the numerical solution of the partial-differential equation

$$\frac{\partial^2 z}{\partial x^2} + \frac{\partial^2 z}{\partial y^2} = C$$

Let $\Delta x = \Delta y = 1$. Provide for varying numbers of rows and columns, making provision for up to 20 rows and 20 columns in the meshwork employed for the solution. Provide for input of all boundary values and allow for any prescribed number of iterations. Print the output so that the mesh-point values are arranged in a rectangular grid with square cells.

b. Use the FORTRAN program previously developed to obtain numerical solutions to the equation

$$\frac{\partial^2 z}{\partial x^2} + \frac{\partial^2 z}{\partial y^2} = 100$$

Provide for a 10×10 mesh of cells. Let the boundary values be zero in all boundary cells, except the center-bottom cell, where a value of 50 is assumed, and the center-top cell, where a value of 100 is assumed.

Annotated Bibliography

Franks, R. G. E., 1967, *Mathematical Modeling in Chemical Engineering*, John Wiley and Sons, New York, 285 pp.

Chapters 1 and 2 contain a useful introduction to the various types of equations used in chemical engineering simulation, plus a general overview on mechanical methods of solution.

Forsythe, G., and Moler, C. B., 1967, *Computer Solution of Linear Algebraic Systems*, Prentice-Hall, Englewood Cliffs, N.J., 148 pp.

This book is principally concerned with matrix solutions to equations, but Chapter 24 provides an instructive description of some iterative methods.

Hamming, R. W., 1962, *Numerical Methods for Scientists and Engineers*, McGraw-Hill, New York, 411 pp.

An excellent general text on numerical analysis.

McCalla, T. R., 1967, *Introduction to Numerical Methods and FORTRAN Programming*, John Wiley and Sons, New York, 359 pp.

A medium-level text on numerical analysis that is particularly valuable because of the numerous FORTRAN algorithms included in the text.

Ralston, A., 1965, *A First Course in Numerical Analysis*, McGraw-Hill, New York, 578 pp.

An excellent textbook on numerical analysis, which includes a description of the predictor-corrector method in a chapter on numerical solution of differential equations.

Shaw, F. S., 1953, *An Introduction to Relaxation Methods*, Dover, New York, 396 pp.

A paperback giving a thorough, but readable, account of finite-difference methods for solving differential equations, sometimes known as *relaxation* methods because of their approximating nature.

Stanton, R. G., 1961, *Numerical Methods for Science and Engineering*, Prentice-Hall, Englewood Cliffs, N.J., 266 pp.

A useful source of information on solving differential equations by finite-difference methods (particularly Chapter 10).

Taylor, A. E., 1959, *Calculus with Analytic Geometry*, Prentice-Hall, Englewood Cliffs, N.J., 762 pp.

A standard textbook covering elementary calculus and an introduction to differential equations.

CHAPTER 6

Flow and Transportation

Flow and transport are features of all dynamic systems. In a geological dynamic system, for example, transport of material may be regarded as an essential aspect of all dynamic components of the system. Thus diffusion of ions within a crystal lattice, transportation of silt in rivers, and uplift of continents are all aspects of the flow and transport of materials. Flow and transport processes are not confined to materials, however. The concepts of flow and transport may be extended to include the flow of energy, and in man-made systems the flow of information, money, goods, and people. Clearly, effective mathematical representation of flow processes is vital in many dynamic simulation models.

This chapter is concerned with the representation of flow and transport processes via two methods—namely, *steady-state potential flow* and *time-dependent potential flow*, which is also known as *deterministic diffusion*. In steady-state potential flow the flow velocities are constant with time, but in time-dependent potential flow the flow velocities change with time. There are other methods of mathematically representing flow, including stochastic diffusion, but they are not treated in this chapter. Papers by Hagerstrand (1968), Howard (1968), and Welch and others (1966) provide examples of some of the alternative methods for modeling flow processes. The reader is also referred to well-known textbooks that treat flow in various forms, including those of Rouse (1931), Jost (1952), and Crank (1956).

An overview of examples of major geological applications of flow is provided in Table 6-1. Category I of Table 6-1 lists the materials, water, oil, and ice in order of increasing viscosity. Tangential shear stresses applied to these materials result in permanent deformation or flow. In category II massive rock can deform viscously (e.g., salt) or it can possess elastic properties. Solids will deform permanently only after an elastic limit has been reached. Category III includes phenomena related to the movement of rock particles, such as mass movement on slopes, landslides, turbidity currents, and transport of particles by suspension or by traction in air or water. Category IV pertains to the transport of materials in solution, including solutions at elevated and atmospheric temperatures and pressures. These trans-

TABLE 6-1 Examples of Various Types of Material Flow and Transport Important in Geology

Category	*Type of Material*	*Examples of Flow or Transport in Other Than Porous Media*	*Examples of Flow in Porous Media*
I	Water	Ocean, lake, and river flow	Groundwater flow
	Oil		Oil reservoirs
	Ice	Glacial flow	
II	Massive rock	Viscous flow in salt domes; plastic flowage in folding and deformation of strata	
III	Rock particles	Landslides Turbidity flows Turbulent suspension and traction	
IV	Solutions at high temperatures and pressures		Solutions causing metamorphic alteration
	Solutions at low temperatures and pressures	Flow of seawater in evaporite basins	Dolomitization
V	Passive organisms	Plankton dispersal	
	Active organisms	Animal migration	

port processes are important in the mobilization of materials freed by, for example, chemical weathering. At the other end of the spectrum this category also pertains to the movement of solutions at high pressures and temperatures relevant to the metamorphic alteration of rocks and also to the formation of hydrothermal ore deposits. Finally, category V includes organic transport. It seems useful to distinguish *passive* from *active* transport. Passive transport would include, for example, such phenomena as plankton dispersal, migration of land plants, and the spread of coral reefs. Active transport, by contrast, pertains to the migration of animals, and, if we extend the concept, it can pertain to man's transport activities.

METHODS OF MODELING FLOW AND TRANSPORTATION

Clearly it would be impossible to discuss methods of modeling all the various types of flow phenomena outlined in Table 6-1 in the space of a single chapter. Furthermore many of the flow processes are poorly understood. Under these circumstances the investigator must develop his model heuristically, combining both empirical and theoretical relationships where they are available. For example, the theory pertaining to sediment transport by running water is still poorly understood. There are, however, various empirical relationships based on observations by hydraulic engineers that are useful in describing and predicting the sediment-transporting behavior of streams. For example, several formulas have been developed for estimating sediment discharge from alluvial channels, relating discharge volume of sediment of particular particle size, channel geometry, and other variables. Vanoni, Brooks, and Kennedy (1960) list eight different sediment-discharge formulas. Many of these were derived empirically by using laboratory data or data from a single stream. When applied to a wide range of observed discharge data from real rivers, none of these formulas provides a satisfactory "fit" to all the data points. The dilemma posed for hydraulic engineers by lack of an adequate theoretical foundation is multiplied many times in geology. Most geological processes lack an adequate theoretical underpinning, and for that matter many processes are not even backed by analytical empirical relationships. Hence a major goal of mathematical simulation is to interpret geologic processes as a series of quantitative analytical relationships, whether the relationships be theoretical or empirical.

In spite of inadequate theory for most geologic processes that involve flow and transport, there is, at the other end of the spectrum, considerable fundamental mathematical theory that can be used for modeling fluid-flow and diffusion processes. In the first part of this chapter we are concerned with fluid flow. Fluids may be treated as viscous fluids—or, as a simplification, viscosity may be ignored, in which case we may treat fluid flow as *potential flow*, involving the concept of *velocity potential.* Potential flow implies the displacement of fluid particles without rotation and without friction. If viscosity is introduced, rotation is produced by internal friction, and the concept of a velocity potential is invalid. Real fluids, of course, possess viscosity, and potential flow must necessarily represent an artificial simplification. For some geological applications, however, the loss of realism is not serious and is offset by greater simplicity.

Partial-differential equations describing potential flow may be written in several general forms. One form, known as the Laplace equation, which pertains to steady-state flow, is written

$$\frac{\partial^2 \phi}{\partial x^2}+\frac{\partial^2 \phi}{\partial y^2}+\frac{\partial^2 \phi}{\partial z^2}=0 \tag{6-1}$$

where ϕ is the velocity potential and x, y, and z are rectilinear coordinates in three-dimensional space. The Laplace equation is a linear, homogeneous, partial differential equation. A second form, known as Poisson's equation, is written

$$\frac{\partial^2 \phi}{\partial x^2}+\frac{\partial^2 \phi}{\partial y^2}+\frac{\partial^2 \phi}{\partial z^2}=k \tag{6-2}$$

where k is a constant. Poisson's equation is nonhomogeneous because of the presence of the term on the right side. The Laplace and the Poisson equations are widely applied, being used to represent the steady-state conduction of electricity and heat, and, where viscosity is ignored, the flow of fluids. Both may be translated into finite-difference form for numerical solutions. The flow of viscous fluids can be represented by more complicated differential equations, known as the Navier–Stokes equations, which are not treated in this book. Finite-difference solutions of the Navier–Stokes equations are developed by Welch *et al.* (1966), as well as other authors.

If we deal with non-steady-state, or time-dependent, potential flow, the basic equation for flow is the diffusion equation

$$\frac{\partial c}{\partial t}=k\left(\frac{\partial^2 c}{\partial x^2}+\frac{\partial^2 c}{\partial y^2}+\frac{\partial^2 c}{\partial z^2}\right) \tag{6-3}$$

This equation is of course closely related to the Laplace and Poisson equations. The flow of groundwater and other fluids through porous media may be regarded as a form of time-dependent potential flow. Although water is a viscous fluid, its movement through porous media can be regarded as irrotational, permitting it to be represented by potential-flow equations, as opposed to more complex flow equations that provide for viscosity.

STEADY-STATE POTENTIAL FLOW

We shall begin by considering steady-state potential flow, as defined by the Laplace and Poisson equations, referring to steady-state potential flow simply as potential flow. In this form of representation a velocity vector is expressed as a gradient of the potential, which is a scalar function. It is clear that velocity is a vector term because it possesses both direction and magnitude. In a system of x, y, z Cartesian coordinates, v_x, v_y, and v_z can be defined as quantities denoting the magnitude of velocity in the x-, y-, and z-directions, respectively. In the flow of ideal fluids in which the viscosity is zero the velocity component in each direction can be defined as the gradient of a scalar quantity ϕ known as the velocity potential:

$$v_x = \frac{\partial \phi}{\partial x}; \qquad v_y = \frac{\partial \phi}{\partial y}; \qquad v_z = \frac{\partial \phi}{\partial z} \tag{6-4}$$

This means that each velocity component is independent of the velocity components in the other two directions. For instance, if we wish to calculate the velocity in the x-direction (Figure 6-1), knowledge of the gradient of the potential is not needed for the y- or z-directions.

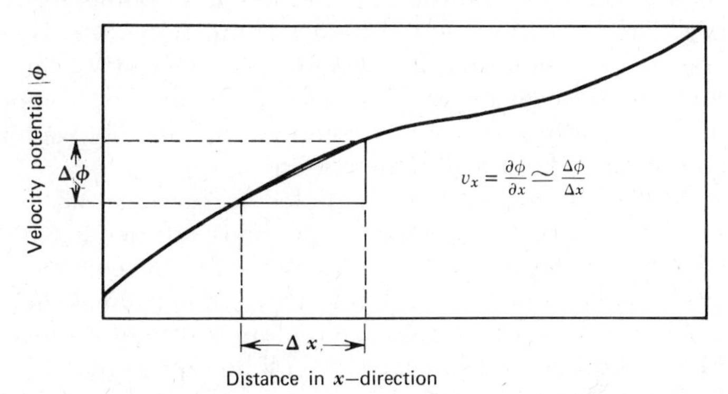

Figure 6-1 Plot of velocity potential ϕ versus distance x. Velocity component in the x-direction is given by slope $\partial\phi/\partial x$, which is approximately equal to $\Delta\phi/\Delta x$. Positive slope in the positive x-direction, thus defined, implies flow from left to right, from points of low velocity potential towards points of high velocity potential.

We realize, of course, that all real fluids possess viscosity. The velocity component in any one direction in a viscous fluid cannot be determined in isolation from velocity components in the other two directions. The presence of internal friction and shear stress, which are aspects of viscosity, cause the fluid to exhibit rotational effects in which velocities are translated from one direction to another. Representation of fluid flow by potential flow is therefore unrealistic for many fluid-flow applications. The solution of problems employing potential flow, however, is considerably easier than dealing with the full equations of fluid motion, which incorporate the rotational effects that result from viscosity. Moreover the geometry of streamlines in potential flowfields is often a reasonable approximation to those of real flowfields. Although internal shear accompanying viscosity tends to reduce velocity magnitudes obtained when viscous flow theory is employed, the dominant directions of flow are, for many cases, approximately the same. Thus models that employ potential flow are sufficiently realistic to have many useful applications in geology and yet are mathematically simple enough to be easily developed.

Calculating Velocities from Potentials

Our first task is to see how velocity vectors are obtained from velocity potentials. Velocity vectors have both direction and magnitude of course, whereas velocity potentials are scalar quantities, which possess magnitude alone. Our objective is to calculate a *velocity-potential field*, which may be defined as an area or volume within which velocities can be represented at all points by velocity potentials. The velocity-potential field is a function of the boundary conditions that surround the flow region as defined. If we are given the velocities at various inlets and outlets to the flow region (i.e., the boundary conditions), we can compute the velocity vectors that define the potential field within the region.

Computational work in our examples is carried out in finite-difference form, employing numerical methods outlined in Chapter 5. In considering potential flow we may begin by considering an extremely simple example involving a two-dimensional meshwork of cells (Figure 6-2a). The rows and columns of the meshwork are numbered. Let the dimensions of each cell be represented by Δx and Δy and for convenience's sake let $\Delta x = \Delta y = 1$. The number within each cell represents the velocity potential at a point in the center of each cell. Since the velocity potentials are scalar quantities, they represent magnitudes only. Our next step is to calculate the velocity

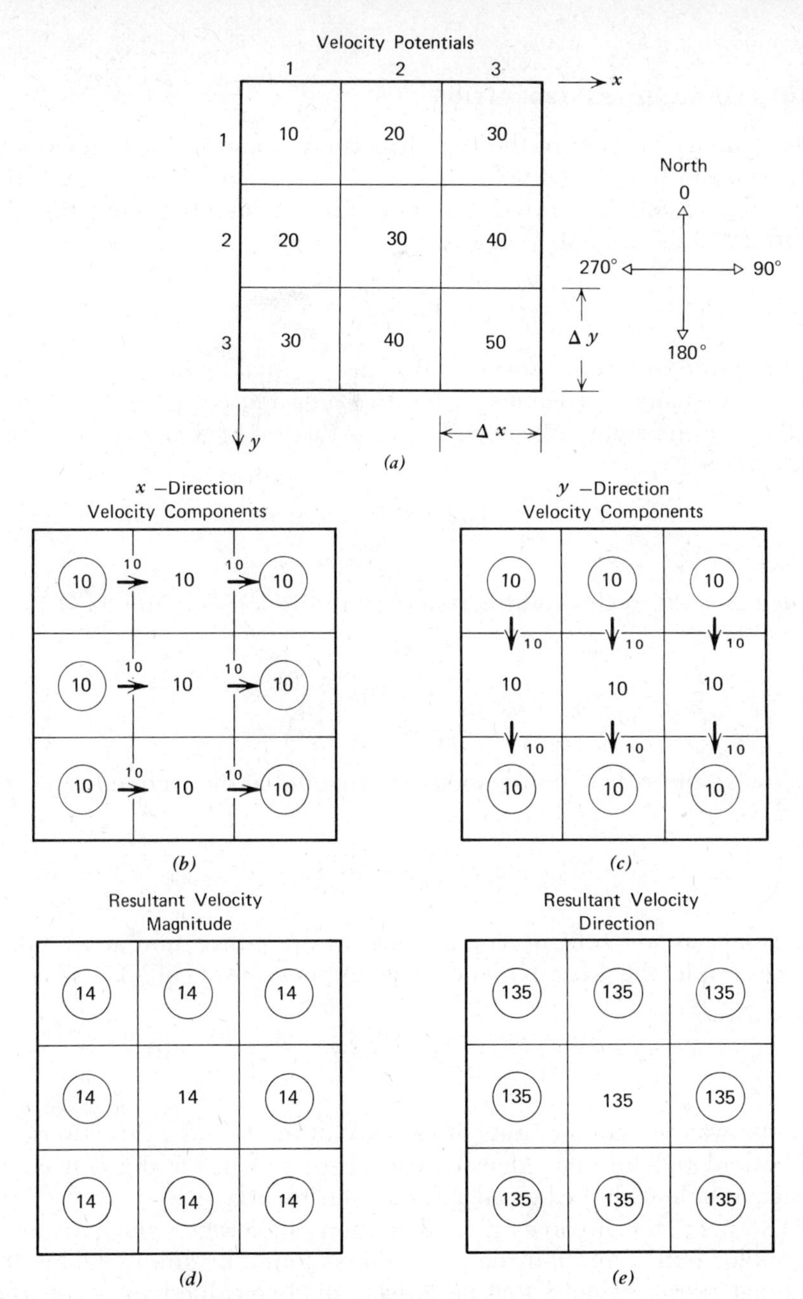

Figure 6-2 Series of diagrams illustrating computation of velocity vectors from velocity potentials: (*a*) meshwork of cells containing three rows and three columns, listing velocity potential at center of each cell; (*b*) velocity components in the *x*-direction (arrows and associated small numbers denote magnitude and east or west direction of flow; large numbers represent average *x*-direction velocity component in each cell; values in boundary cells are encircled); (*c*) velocity components in the *y*-direction; (*d*) vector resultant velocity magnitudes; (*e*) vector resultant velocity azimuth direction measured 0 to 360 degrees clockwise from north, with north toward top of page.

components in each of the two directions, x and y. Consider the x-direction initially. The velocity component in the x-direction at the left edge of cell 1, 2 (row 1, column 2) is obtained by calculating the gradient of the velocity potential

$$v_x = \frac{\partial \phi}{\partial x} \simeq \frac{\Delta \phi}{\Delta x} \tag{6-5}$$

Using finite-difference form, so that $\Delta\phi$ is equal to the difference between the potential values in the two adjacent cells, cell (1, 1) and cell (1, 2) [the right edge of cell (1, 1) is the left edge of cell (1, 2)] we may write

$$v_x = \frac{\text{right cell} - \text{left cell}}{\Delta x} \tag{6-6}$$

Since ϕ is 20 in the right cell and 10 in the left cell and $\Delta x = 1$, we can obtain

$$v_x = \frac{20 - 10}{1} = 10$$

Likewise the x-direction velocity component at the right edge of cell (1, 2) is

$$v_x = \frac{30 - 20}{1} = 10$$

Calculations for velocity components in the y-direction are similar. For example, the y-component for the upper edge of cell (3, 3) is

$$v_y = \frac{\text{lower cell} - \text{upper cell}}{\Delta x} = \frac{50 - 40}{1} = 10$$

In this way the components of velocity in the x- and y-directions can be calculated for the edges of each cell, except for the cell edges that coincide with the boundary of the meshwork.

The next step involves the calculation of average velocity components for both x and y-directions. This is found by simply taking the average of the velocity that pertains to the two cell edges perpendicular to the direction under consideration. For example, the velocity component in the x-direction for cell (2, 2) is

$$v_x = \frac{(10 + 10)}{2} = 10$$

Average velocities, of course, cannot be calculated for cells adjacent to boundaries. For the purposes of this example they have been assumed, however, and in Figure 6-2 they are regarded as equal to the velocity across the inside cell edge. For example, in cell (1, 1) the average velocity $v_x = 10$. Velocity components for boundary cells are encircled in Figure 6-2*b* and *c*.

Once the average velocity components in the two directions have been obtained (Figure 6-2*b* and *c*), the resultant velocity vectors can be calculated readily. The magnitude of the resultant vectors is given by

$$\sqrt{v_x^2 + v_y^2}$$

and the resultant vector direction by

$$\arctan\left(\frac{v_y}{v_x}\right)$$

This gives an azimuth direction of zero in the negative x-direction. Thus in order to bring zero azimuth (north) toward the top of the page, thereby corresponding with conventional map orientation, 90 has been added to each velocity-direction value expressed in degrees of arc. The values may be plotted numerically (Figure 6-2*d* and *e*), or they may be displayed graphically (see Figure 6-3, in which the directions of the arrows denote the vector resultants, and the length of the arrows is proportional to the resultant magnitudes). In Figure 6-3 the point of each arrow is located at the center of its grid cell, and

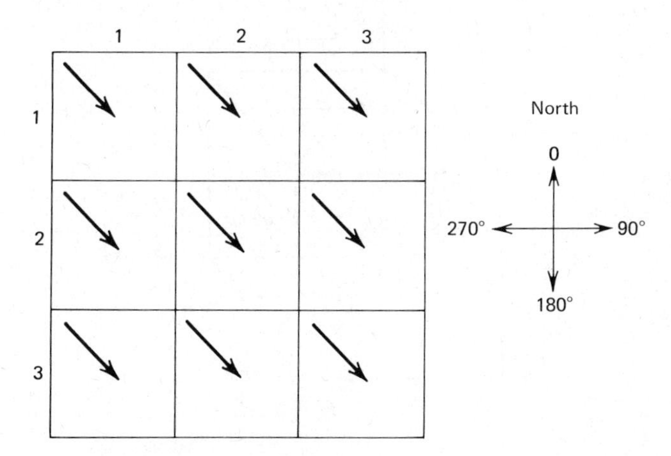

Figure 6-3 Arrows denoting resultant velocity vectors in potential-flow representation of velocity potentials of Figure 6-2*a*.

the vector that it represents pertains to the center of the grid cell. The geographic distribution of velocity arrows provides an excellent method of displaying the flowfield.

Continuity Equation

The conservation of mass is a basic principle of nature. Dynamic simulation models that deal with the transport of materials must necessarily account for materials that enter and leave divisions of the system. The continuity-of-mass equation forms a basic relationship in deterministic models of fluid flow and may be readily derived for flow in three dimensions. The following algebraic development of the continuity equation and subsequently of the Laplace equation closely follows the development by Pollack (1967). Initially let us consider a rectangular volume element of fluid whose dimensions are Δx, Δy, and Δz (Figure 6-4) and whose volume is therefore $\Delta x \Delta y \Delta z$. For reference purposes the sides of the volume element may be identified as follows:

l,r = left and right in the x-direction
f,a = fore and aft in the y-direction
t,b = top and bottom in the z-direction

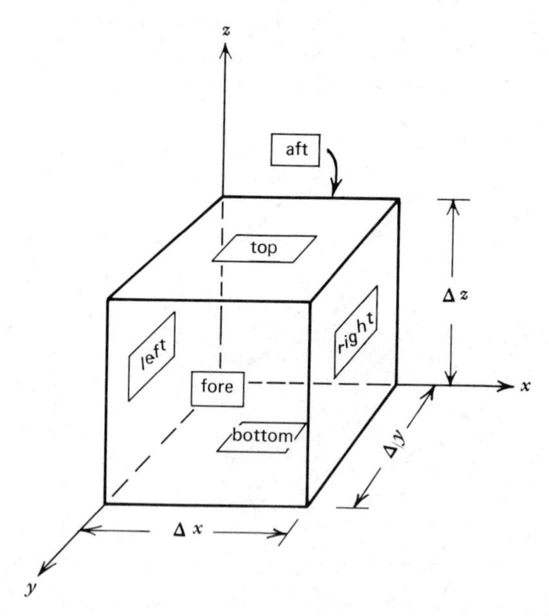

Figure 6-4 Volume element of fluid in x, y, z Cartesian coordinate system.

If fluid enters and leaves the volume element, the basic continuity-of-mass relationship states that

$$\text{input} - \text{output} = \text{accumulation} \tag{6-7}$$

For accounting purposes we need to calculate the rate of mass flux, or quantity of fluid that crosses each face of the volume element in a unit of time. Remembering that density = mass/unit volume, we know that the amount of fluid crossing any face of the volume element equals the cross-sectional area of that face times the velocity times the density. Thus the mass of fluid M_l crossing the left face in a unit of time is

$$M_l = \Delta y \Delta z \rho_l v_l \tag{6-8}$$

where ρ_l = density at the center of the left face
v_l = velocity at the center of the left face

Similarly the mass flux through the right face is

$$M_r = \Delta y \Delta z \rho_r v_r \tag{6-9}$$

where ρ_r is the density and v_r is the velocity at the right face of the volume element. The net difference in the mass flux from the right to the left faces over the interval Δx is ΔM_x and is defined as

$$\begin{aligned} \Delta M_x &= M_r - M_l \\ &= \Delta y \Delta z \rho_r v_r - \Delta y \Delta z \rho_l v_l \\ &= (\rho_r v_r - \rho_l v_l) \Delta y \Delta z \\ &= \Delta(\rho v)_x \Delta y \Delta z \end{aligned} \tag{6-10}$$

where $\Delta(\rho v)_x$ is the net difference in the product of density times velocity in the x-direction across the volume element. Similarly for the y- and z-directions

$$\begin{aligned} \Delta M_y &= M_f - M_a \\ &= (\rho_f v_f - \rho_a v_a) \Delta x \Delta z \\ &= \Delta(\rho v)_y \Delta x \Delta z \end{aligned} \tag{6-11}$$

and

$$\begin{aligned} \Delta M_z &= M_t - M_b \\ &= (\rho_t v_t - \rho_b v_b) \Delta x \Delta y \\ &= \Delta(\rho v)_z \Delta x \Delta y \end{aligned} \tag{6-12}$$

The left-hand side of the basic input–output relationship (input − output = accumulation) can be defined as the algebraic sum of the net differences in mass flux in the three directions, multiplied by the unit time interval Δt. Thus

$$\text{input} - \text{output} = (\Delta M_x + \Delta M_y + \Delta M_z)\,\Delta t \tag{6-13}$$

Accumulation, on the other hand, can be regarded as the change in the mass of the fluid volume element, which is the change in density multiplied by the volume. Thus

$$\text{accumulation} = (\rho_1 - \rho_2)\,\Delta x\,\Delta y\,\Delta z \tag{6-14}$$

where ρ_1 and ρ_2 are the average densities over the whole volume element at the beginning and end, respectively, of time interval Δt. Since input − output = accumulation, we may equate Equations 6-13 and 6-14, obtaining

$$(\Delta M_x + \Delta M_y + \Delta M_z)\,\Delta t = (\rho_1 - \rho_2)\,\Delta x\,\Delta y\,\Delta z \tag{6-15}$$

If we substitute the equivalents of ΔM_x, ΔM_y, and ΔM_z, and let $\Delta\rho_t = \rho_1 - \rho_2$, we obtain

$$[\Delta(\rho v)_x\,\Delta y\,\Delta z + \Delta(\rho v)_y\,\Delta x\,\Delta z + \Delta(\rho v)_z\,\Delta x\,\Delta y]\,\Delta t = \Delta\rho_t\,\Delta x\,\Delta y\,\Delta z \tag{6-16}$$

On dividing through by $\Delta x\,\Delta y\,\Delta z\,\Delta t$, we in turn obtain

$$\frac{\Delta(\rho v)_x}{\Delta x} + \frac{\Delta(\rho v)_y}{\Delta y} + \frac{\Delta(\rho v)_z}{\Delta z} = \frac{\Delta\rho_t}{\Delta t} \tag{6-17}$$

If the volume element is infinitesimally small, the expression becomes

$$\frac{\partial(\rho v)_x}{\partial x} + \frac{\partial(\rho v)_y}{\partial y} + \frac{\partial(\rho v)_z}{\partial z} = \frac{\partial\rho_t}{\partial t} \tag{6-18}$$

If we deal with incompressible and constant density fluids under steady-state conditions in which no net accumulation occurs, this relationship simplifies to

$$\frac{\partial v_x}{\partial x} + \frac{\partial v_y}{\partial y} + \frac{\partial v_z}{\partial z} = 0 \tag{6-19}$$

In dealing with compressible fluids such as gases, however, this simplification cannot be made because density is not constant and must be related to pressure and temperature by the *equation of state*:

$$\rho = k\frac{P}{T} \tag{6-20}$$

where P = pressure
T = temperature
k = constant, depending on the fluid

Laplace Equation for Potential Flow

If we are to deal with incompressible fluids under steady-state conditions, we may adapt the no-net-accumulation relationship, as defined above, to yield the Laplace equation. For potential flow the velocity component in each direction is given by the gradient in the velocity potential in that direction. Then the three velocity components are

$$v_x = \frac{\partial \phi}{\partial x}; \qquad v_y = \frac{\partial \phi}{\partial y}; \qquad v_z = \frac{\partial \phi}{\partial z}$$

The three partial derivatives, or gradients in the velocity potential, represent the directional rate of change in the velocity potential. We may substitute these partial derivatives into the continuity equation (Equation 6-19), yielding

$$\frac{\partial(\partial\phi/\partial x)}{\partial x} + \frac{\partial(\partial\phi/\partial y)}{\partial y} + \frac{\partial(\partial\phi/\partial z)}{\partial z} = 0 \tag{6-21a}$$

or

$$\frac{\partial^2\phi}{\partial x^2} + \frac{\partial^2\phi}{\partial y^2} + \frac{\partial^2\phi}{\partial z^2} = 0 \tag{6-21b}$$

which is the Laplace equation. In the literature we often see a shorthand form of Equation 6-21b, namely,

$$\nabla^2\phi = 0 \tag{6-22}$$

where

$$\nabla^2 \equiv \frac{\partial^2}{\partial x^2} + \frac{\partial^2}{\partial y^2} + \frac{\partial^2}{\partial z^2}$$

∇^2 is a scalar operator, known as the "Laplacian," which permits second partial derivatives with respect to the three spatial variables to be represented by a single symbol. Various other shorthand symbols (such as the "del," or differential operator, ∇) and words (such as grad, div, curl) are commonly found in literature pertaining to fluid mechanics. For a full understanding of these terms interested readers are referred to books that deal with vector calculus, such as that by Miller (1962).

Numerical solutions of partial-differential equations of the Laplace type are discussed in Chapter 5. The numerical solution demands, of course, that they be transformed into finite-difference form. Let us now consider the solution of the Laplace equation involving only two spatial dimensions:

$$\frac{\partial^2 \phi}{\partial x^2} + \frac{\partial^2 \phi}{\partial y^2} = 0 \tag{6-23}$$

For point (i,j) in the (x, y) plane the finite-difference form of Equation 6-23 is

$$\frac{(\phi_{i,j+1} - 2\phi_{i,j} + \phi_{i,j-1})}{(\Delta x)^2} + \frac{(\phi_{i-1,j} - 2\phi_{i,j} + \phi_{i+1,j})}{(\Delta y)^2} = 0 \tag{6-24}$$

If we let $\Delta x = \Delta y = 1.0$, this simplifies to

$$\phi_{i,j+1} + \phi_{i,j-1} + \phi_{i-1,j} + \phi_{i+1,j} - 4\phi_{i,j} = 0 \tag{6-25}$$

and in turn to

$$\phi_{i,j} = \frac{\phi_{i,j+1} + \phi_{i,j-1} + \phi_{i-1,j} + \phi_{i+1,j}}{4} \tag{6-26}$$

Following the procedures outlined in Chapter 5, the iterative Gauss–Seidel technique can be used to obtain values of ϕ for each cell of a two-dimensional grid, provided that the boundary values are given (i.e., that the velocity potentials are known for each cell at the edges of the meshwork over which the numerical solution is to be obtained).

Tables 6-2, 6-3, and 6-4 list a FORTRAN program that accepts boundary potentials as input and calculates the total velocity-potential field. Then, employing the steps outlined on pp. 210–214, a final velocity-vector map is produced, using a subroutine listed in Table 6-4. Figure 6-5*a* shows the input of boundary-potential values. Their transformation into the full field of velocity potentials by this program is shown in Figure 6-5*b*. Figure 6-6 shows the resulting velocity-vector map plotted by CALCOMP digital plotter.

Boundary Conditions

We can summarize the computational steps so far as follows:

1. Assign potential values to each boundary cell.
2. Iteratively calculate the velocity potential for every cell in the grid by solving the Laplace equation.

3. Obtain velocity components in each of the principal coordinate directions.
4. Calculate and plot the resultant velocity vectors.

Once the boundary-potential values have been obtained, steps 2 through 4 are quite straightforward. But how do we obtain these boundary-potential values? What we often wish to use as input are the velocity values of fluid entering or leaving the flow region across each of the outer margins of each boundary cell. For the simplest case we can make the cells square and equal in size, and mass flux is directly proportional to the average velocity of material entering or leaving normal to each boundary cell, as shown in Equation 6-8. Thus in the x-direction

$$\text{mass flux} = \text{velocity} \times \text{cross-sectional area} \times \text{density}$$

or

$$M_x = v_x \Delta x \Delta h \rho \tag{6-27}$$

where M_x = mass flux crossing the margin of a boundary cell in the x-direction
v_x = velocity component at the same cell margin, normal to the boundary
Δx = length of cell side
Δh = depth of flow, which is constant throughout the two-dimensional flow meshwork
ρ = density, which is constant

Thus mass flux and velocity are directly proportional to one another:

$$M_x = k v_x \tag{6-28}$$

where $k = \Delta x \Delta h \rho$. In subsequent discussions we assume that $k = 1$, and we refer to velocity components instead of mass-flux terms. Mass flux can always be recalculated, however, given velocities and the value of k.

Boundary Velocities at Side of Grid

In order to see how we can use the boundary velocities as input let us return to the finite-difference form of the Laplace equation for a two-dimensional grid in which $\Delta x = \Delta y = 1.0$:

$$\phi_{i+1,j} + \phi_{i-1,j} + \phi_{i,j+1} + \phi_{i,j-1} - 4\phi_{i,j} = 0 \tag{6-29}$$

or

$$\phi_{i,j} = \frac{\phi_{i+1,j} + \phi_{i-1,j} + \phi_{i,j+1} + \phi_{i,j-1}}{4} \tag{6-30}$$

TABLE 6-2 Program for Calculating Velocity Potentials and Resultant Velocity Vectors, Given Input Boundary-Potential Values

```
C.....ALGORITHM FOR POTENTIAL FLOW IN TWO DIMENSIONS
C.....  VELPOT       VELOCITY POTENTIAL ARRAY
C.....  RESVEL       RESULTANT VELOCITY DIRECTIONS
C.....  VELDIR       RESULTANT VELOCITY MAGNITUDES
C.....  ITMAX        MAX ITERATIONS IN GAUSS-SEIDEL SOLUTION
C.....  TOL          TOLERANCE LIMIT BELOW WHICH ITERATIONS TERMINATED
      DIMENSION VELPOT(20,20),NFXPOT(20,20),RESVEL(20,20),VELDIR(20,20)
    1 FORMAT(3I5, 2F5.0)
    2 FORMAT(16F5.0)
    3 FORMAT(1H1, 'ERROR', F10.4, 5X, 'ITERATIONS', I5//
     1 5X, 'VELOCITY POTENTIAL ARRAY'//)
    5 FORMAT(1H1, 'RESULTANT VELOCITY MAGNITUDES'//)
    6 FORMAT(1H1, 'RESULTANT VELOCITY DIRECTIONS'//)
      READ(5,1) NROWS, NCOLS, ITMAX, TOL, DX
C.....READ INITIAL VELOCITY POTENTIAL VALUES (AT BOUNDARIES ETC.)
      DO 20 I=1,NROWS
      READ(5,2) (VELPOT(I,J), J=1,NCOLS)
      DO 20 J=1,NCOLS
      NFXPOT(I,J)=0
   20 IF (VELPOT(I,J).NE.0.000) NFXPOT(I,J)=1
      NRL1=NROWS-1
      NCL1=NCOLS-1
C.....USE GAUSS-SEIDEL METHOD TO SOLVE SIMULTANEOUS EQUATIONS
      DO 50 IT=1,ITMAX
      ERROR=0.0
      DO 40 I=2,NRL1
      DO 40 J=2,NCL1
      IF (NFXPOT(I,J).EQ.1) GO TO 40
      DUMMY=(VELPOT(I,J+1)+VELPOT(I,J-1)+VELPOT(I+1,J)+VELPOT(I-1,J))/
     1  4.0
      ERROR=ERROR+ ABS(DUMMY-VELPOT(I,J))
      VELPOT(I,J)=DUMMY
   40 CONTINUE
      IF (ERROR.LT.TOL) GO TO 60
   50 CONTINUE
C.....WRITE OUT VELOCITY POTENTIAL ARRAY
   60 WRITE(6,3) ERROR, IT
      CALL PRINT (NROWS, NCOLS, VELPOT,20,20,10.)
C.....CALCULATE VELOCITY VECTORS
      DO 150 I=1,NROWS
      DO 150 J=1,NCOLS
      IF (I.EQ.1) GO TO 90
      IF (I.EQ.NROWS) GO TO 100
      VELOCY=(VELPOT(I+1,J)-VELPOT(I-1,J))/(2.0*DX)
      GO TO 110
   90 VELOCY=(VELPOT(I+1,J)-VELPOT(I,J))/DX
      GO TO 110
  100 VELOCY=(VELPOT(I,J)-VELPOT(I-1,J))/DX
  110 IF (J.EQ.1) GO TO 120
      IF (J.EQ.NCOLS) GO TO 130
      VELOCX=(VELPOT(I,J+1)-VELPOT(I,J-1))/(2.0*DX)
      GO TO 140
  120 VELOCX=(VELPOT(I,J+1)-VELPOT(I,J))/DX
      GO TO 140
  130 VELOCX=(VELPOT(I,J)-VELPOT(I,J-1))/DX
```

TABLE 6-2 *contd.*

```
C.....CALCULATE RESULTANT MAGNITUDE AND DIRECTION
  140 RESVEL(I,J)=SQRT(VELOCX**2+VELOCY**2)
      IF (RESVEL(I,J).LT.0.0001) GO TO 145
      VELDIR(I,J)=ATAN2(VELOCY,VELOCX)*57.324+90.0
      GO TO 150
  145 VELDIR(I,J)=0.0
  150 CONTINUE
C.....PRINT OUT VELOCITY ARRAYS
      WRITE(6,5)
      CALL PRINT(NROWS, NCOLS,RESVEL,20,20,1.)
      WRITE(6,6)
      CALL PRINT(NROWS,NCOLS,VELDIR,20,20,1.)
C.....PLOT VELOCITY VECTOR MAP
      CALL VECTOR(RESVEL,VELDIR,NROWS,NCOLS,20,20)
      RETURN
      END
```

We can see from Figure 6-7*a* that arithmetic operations with the Laplace equation in this form simply involve summing the potential values for the four cells closest to cell (i,j) and dividing by 4. At a boundary of the grid meshwork, however, one of the adjacent cells would be outside the boundary—for example, cell $(i,j-1)$ at the left boundary (Figure 6-7*b*). Since we have no direct knowledge of the velocity potential in an outside cell, we must replace $\phi_{i,j-1}$ with a value so that we may calculate $\phi_{i,j}$. A value for $\phi_{i,j-1}$ may be obtained from the velocity v_x of fluid flowing normal to the boundary (whether it is inflow or outflow depends on the algebraic sign and position of the coordinate origin). The velocity v_x is therefore a type of boundary condition. We know that by definition velocity v_x is related to the difference in potential between the two adjacent cells $\phi_{i,j}$ and $\phi_{i,j-1}$:

$$v_x = \frac{\partial \phi}{\partial x} \simeq \frac{\phi_{i,j} - \phi_{i,j-1}}{\Delta x} \tag{6-31}$$

TABLE 6-3 Subroutine for Printing Maps from Two-Dimensional Arrays of Data, Employed by Program Listed in Table 6-2

```
      SUBROUTINE PRINT(NROWS, NCOLS,VAL,NR,NC,FACT)
      DIMENSION VAL(NR,NC), NUM(21)
    1 FORMAT(1H ,'ALL VALUES MULTIPLIED BY ', F12.3/5X, 21I5)
    2 FORMAT(1H // 22I5)
C.....PRINT OUT MAPS OF VAL
      WRITE(6,1) FACT, (J,J=1,NCOLS)
      DO 20 I=1,NROWS
      DO 10 J=1,NCOLS
      A=VAL(I,J)*FACT
      FACTOR=0.5
      IF (A.LE.0.00000) FACTOR=-0.5
   10 NUM(J)=A+FACTOR
   20 WRITE(6,2) I, (NUM(J), J=1,NCOLS)
      RETURN
      END
```

TABLE 6-4 Subroutine VECTOR Used for Plotting Vector Diagrams with On-Line CALCOMP Plotter[a]

```
      SUBROUTINE VECTOR(RESVEL,VELDIR,NROWS,NCOLS,NR,NC)
C.....PLOTS VELOCITY VECTOR MAP ON CALCOMP
      DIMENSION  X(4), Y(4),AMODES(200), RESVEL(NR,NC), VELDIR(NR,NC)
C.....  NROWS    NO OF ROWS IN GRID
C.....  NR       MAX NO OF ROWS PERMISSIBLE
C.....  NCOLS    NO OF COLS IN GRID
C.....  NC       MAX NO OF COLS IN GRID
      CALL MODESG(AMODES,'BONHAM-CARTER, BIN 203, VECTOR MAP', 34)
      AMODES(97)=1200.
      CALL PICTRG(AMODES)
      AMODES(97)=1200.
      RAD=0.0174533
      DX=80.0
      IF (NCOLS.GT.8)  DX=800.0/FLOAT(NCOLS)
C.....FIND LARGEST VELOCITY VALUE
      BIG=0.0
      DO 10 I=1,NROWS
      DO 10 J=1,NCOLS
   10 IF (BIG.LT.RESVEL(I,J)) BIG=RESVEL(I,J)
C.....PLOT VELOCITY VECTORS
      HEAD=DX/5.0
      SCALE=DX/BIG
      DO 40 I=1,NROWS
      X(1)=FLOAT(I)*DX
      DO 40 J=1,NCOLS
      Y(1)=FLOAT(J)*DX
      IF (RESVEL(I,J).LT.0.00001) GO TO 40
      THETA=(180.0+VELDIR(I,J))*RAD
      SHAFT=RESVEL(I,J)*SCALE
C.....PLOT SHAFT
      COSA=COS(THETA)
      SINA=SIN(THETA)
      X(2)=X(1)-SHAFT*COSA
      Y(2)=Y(1)+SHAFT*SINA
      CALL LINESG(AMODES,2,X,Y)
C.....PLOT HEAD OF ARROW (20 DEGREES EITHER SIDE OF SHAFT)
      DO 20 N=2,3
      ALPHA=THETA-0.174533
      IF (N.EQ.3) ALPHA=THETA+0.174533
      SINA=SIN(ALPHA)
      COSA=COS(ALPHA)
      X(N)=X(1)-HEAD*COSA
      Y(N)=Y(1)+HEAD*SINA
   20 CONTINUE
      X(4)=X(1)
      Y(4)=Y(1)
      CALL LINESG(AMODES,4,X,Y)
   40 CONTINUE
      CALL EXITG(AMODES)
      RETURN
      END
```

[a]All the call statements in the program are specific to plotting routines in use at Stanford University, but can probably be modified for the requirements of other computation centers with only minor difficulty.

--1.0	--2.0	--5.0	--2.0	--1.0
0.0	0.0	0.0	0.0	0.0
2.0	0.0	0.0	0.0	2.0
1.0	0.0	0.0	0.0	1.0
1.0	1.0	1.0	1.0	1.0

(a)

--1.0	--2.0	--5.0	--2.0	--1.0
0.0	--0.8	--1.7	--0.8	0.0
2.0	0.5	0.0	0.5	2.0
1.0	0.8	0.6	0.8	1.0
1.0	1.0	1.0	1.0	1.0

(b)

Figure 6-5 Two-dimensional velocity-potential field: (*a*) meshwork prior to computation in which values in boundary cells have been supplied, but values for inner cells (shaded) have been arbitrarily set to zero as initial guesses; (*b*) values in meshwork cells when interior values have stabilized after a sufficient number of iterations with the Gauss–Seidel method, rounded to one decimal place.

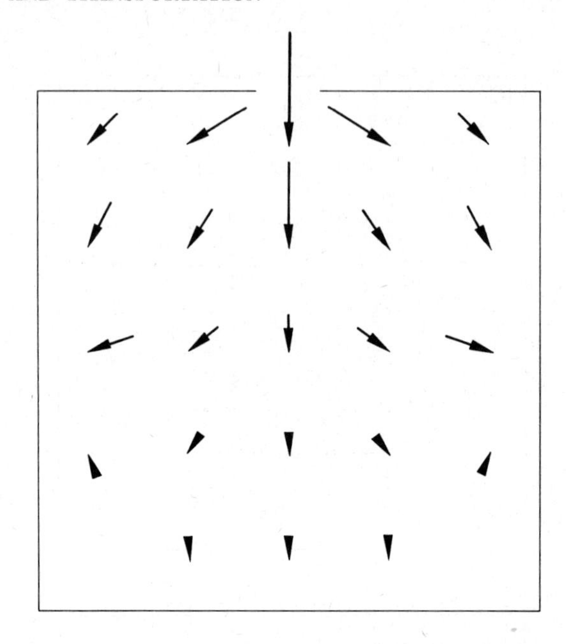

Figure 6-6 Velocity-vector map produced from velocity potentials shown in Figure 6-5. An on-line CALCOMP plotter drew diagram, controlled by statements in subroutine VECTOR listed in Table 6-4.

which in turn permits us to obtain an expression for $\phi_{i,j-1}$

$$\phi_{i,j-1} = \phi_{i,j} - v_x \Delta x \tag{6-32}$$

Substituting back into Equation 6-30, we find that

$$4\phi_{i,j} = \phi_{i+1,j} + \phi_{i-1,j} + \phi_{i,j+1} + \phi_{i,j} - v_x \Delta x \tag{6-33}$$

Subtracting $\phi_{i,j}$ from both sides and dividing through by 3, we obtain

$$\phi_{i,j} = \frac{\phi_{i+1,j} + \phi_{i-1,j} + \phi_{i,j+1} - v_x \Delta_x}{3} \tag{6-34}$$

This equation is now used instead of the basic equation (Equation 6-29) for all cells on the left boundary (except at the corners) given the values of v_x as boundary values in each case. Similar expressions can be derived for lower, upper, and right boundary cells.

Boundary Velocities at Corners

The cells at the corners of the grid mesh must be treated separately from other boundary cells because there are two unknown potential

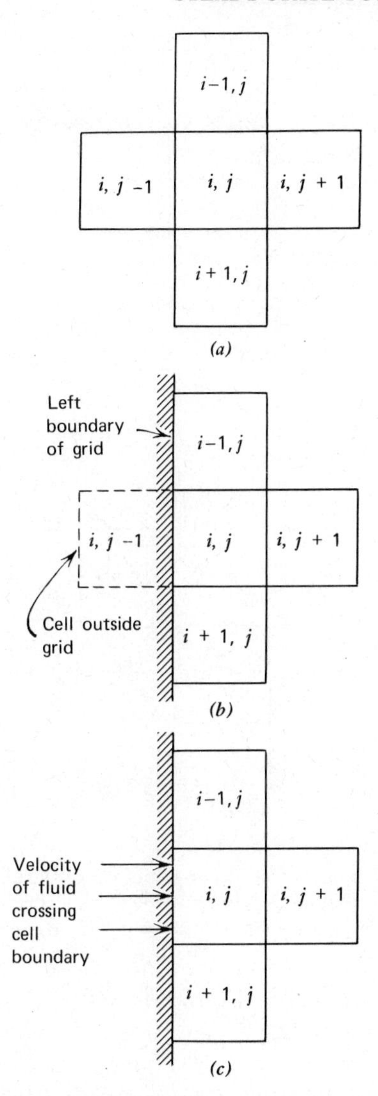

Figure 6-7 Series of diagrams illustrating treatment of boundary conditions: (*a*) cell (i, j) surrounded by four closest neighbors, all of which are inside grid meshwork; (*b*) left boundary with cell $(i, j-1)$ which, if present, would lie outside boundary; (*c*) same as (*b*) but showing velocity normal to boundary as boundary condition.

values. For example, consider the upper-left grid corner (Figure 6-8*a*), for which the values of ϕ in cells $(i, j-1)$ and $(i-1, j)$ are unknown. Given the velocities normal to the two boundaries v_x and v_y, however, the two potential values may be obtained in a manner similar to that

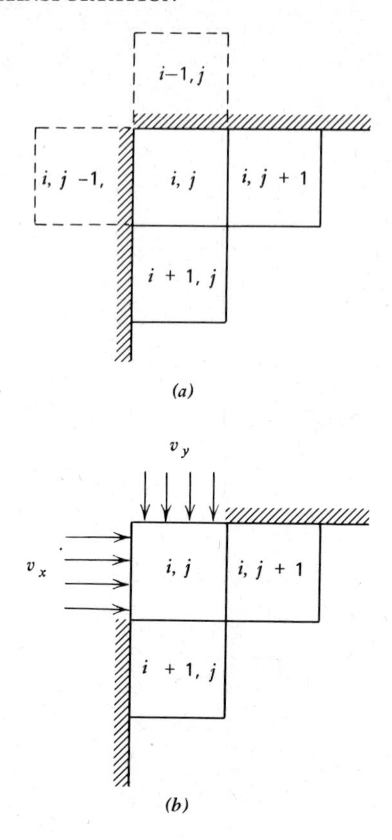

Figure 6-8 Boundary conditions at upper-left corner of grid: (*a*) cell (i, j) has two of its immediately adjacent neighbors lying outside grid; (*b*) velocities of flow v_x and v_y, perpendicular to boundaries, permit potential values in missing cells to be calculated.

employed for the noncorner boundary cells, described previously. As defined in Equation 6-32,

$$\phi_{i,j-1} = \phi_{i,j} - v_x \Delta x$$

Similarly

$$v_y = \frac{\partial \phi}{\partial y} \simeq \frac{\phi_{i,j} - \phi_{i-1,j}}{\Delta y} \tag{6-35}$$

leading to

$$\phi_{i-1,j} = \phi_{i,j} - v_y \Delta x \tag{6-36}$$

where $\Delta y = \Delta x$. Substituting for $\phi_{i-1,j}$ and $\phi_{i,j-1}$ in the basic equation (Equation 6-29) gives

$$4\phi_{i,j} = \phi_{i+1,j} + \phi_{i,j+1} + \phi_{i,j} - v_x \Delta x + \phi_{i,j} - v_y \Delta x \tag{6-37}$$

Subtracting $2\phi_{i,j}$ from both sides and dividing through by 2 yields

$$\phi_{i,j} = \frac{\phi_{i+1,j} + \phi_{i,j+1} - \Delta x(v_x + v_y)}{2} \tag{6-38}$$

which is the equation for the upper-left-corner cell. Similar expressions can be derived for the other corners of the grid.

Although not discussed here, some applications of potential flow involve specifying velocity potentials at the boundaries instead of velocity components. Mixed boundary conditions employing both velocities and potentials in different parts of the grid are also possible. A zero flux boundary—that is, a boundary across which flow is zero—is readily represented by assigning a velocity component of zero normal to the boundary.

FORTRAN Program for Two-Dimensional Potential Flow

Table 6-5 lists a FORTRAN program for solving the Laplace equation for potential flow, using velocities normal to all boundary cells as the main input requirement. The two subroutines PRINT and VECTOR employed by the program are the same as those listed in Tables 6-3 and 6-4, respectively. The program is also useful in modeling the movement of fluids through flowfields in which obstacles occur. In addition to the boundary velocities that must be read in as input, it is also necessary to read in an integer array that assigns a label to each cell in the grid. The code for the labeling method is as follows:

1 = top left corner of grid
2 = top margin
3 = top right corner
4 = right margin
5 = bottom right corner
6 = bottom margin
7 = bottom left corner
8 = left margin
9 = obstacle cell, with "no flow" condition across all four faces
10 = internal cell in which there is no restriction on flow

By using these labels, the flow region can be outlined with a rectangular grid (the boundaries of the flow area need not necessarily coincide with the boundaries of the grid), and internal "boundaries" within the flow area can be included by inserting obstacle cells. By

TABLE 6-5 FORTRAN Program for Obtaining Flowfield for Two-Dimensional Potential Flow Using Boundary Velocities as Input[a]

```
C.....POTENTIAL FLOW - BOUNDARY VELOCITIES SUPPLIED AS INPUT
C....ARRAY NAMES MOSTLY SIMILAR TO PROGRAM IN TABLE 6-2
C.....  IOX        NO. OF I/O CELL BOUNDARIES IN X DIRECTION
C.....  IOY        NO. OF I/O CELL BOUNDARIES IN Y DIRECTION
C.....  LABEL(I,J)ARRAY DENOTING CELL TYPE (BOUNDARY, OBSTACLE, ETC.)
      DIMENSION VELPOT(20,20), LABEL(20,20), VELX(20,21), VELY(21,20),
     1 RESVEL(20,20), VELDIR(20,20)
    1 FORMAT(5I5, 2F5.0)
    2 FORMAT(16I5)
    3 FORMAT(2I5, F5.0)
    4 FORMAT(1H1, 'ERROR', F10.4, 5X, 'ITERATIONS', I10//1X,
     1 'VELOCITY POTENTIAL'//)
    5 FORMAT(1H1, 'VELOCITY COMPONENT IN X-DIRECTION'//)
    6 FORMAT(1H1, 'VELOCITY COMPONENT IN Y-DIRECTION'//)
    7 FORMAT(1H1, 'RESULTANT VELOCITY MAGNITUDE'//)
    8 FORMAT(1H1, 'RESULTANT VELOCITY DIRECTION'//)
C.....READ INPUT PARAMETERS
      READ(5,1) NROWS, NCOLS, ITMAX, IOX, IOY, TOL, DX
      NRP1=NROWS+1
      NCP1=NCOLS+1
C.....READ CELL LABELS
      DO 20 I=1,NROWS
   20 READ(5,2) (LABEL(I,J), J=1,NCOLS)
C.....INITIALIZE VELOCITY ARRAYS
      DO 30 I=1,NROWS
      DO 30 J=1,NCP1
   30 VELX(I,J)=0.0
      DO 40 I=1,NRP1
      DO 40 J=1,NCOLS
   40 VELY(I,J)=0.0
C.....READ VELOCITIES AT INLETS AND OUTLETS
      IF (IOX.LT.1) GO TO 60
      DO 50 IO=1,IOX
   50 READ(5,3) I, J, VELX(I,J)
   60 IF (IOY.LT.1) GO TO 80
      DO 70 IO=1,IOY
   70 READ(5,3) I, J, VELY(I,J)
C.....INITIALIZE VELOCITY POTENTIAL ARRAY
   80 DO 90 I=1,NROWS
      DO 90 J=1,NCOLS
   90 VELPOT(I,J)=0.0
C.....USE GAUSS-SEIDEL ITERATIVE METHOD TO SOLVE POTENTIAL FIELD
      DO 215 IT=1,ITMAX
      ERROR=0.0
      DO 210 I=1,NROWS
      DO 210 J=1,NCOLS
C.....CALCULATION DEPENDS ON TYPE OF CELL LABEL
      L=LABEL(I,J)
      GO TO (110,120,130,140,150,160,170,180,190,100), L
C.....NORMAL CELL (L=10)
  100 DUMMY=(VELPOT(I+1,J)+VELPOT(I-1,J)+VELPOT(I,J+1)+VELPOT(I,J-1))/
     1  4.0
      GO TO 200
C.....UPPER LEFT CORNER CELL (L=1)
  110 DUMMY=(VELPOT(I+1,J)+VELPOT(I,J+1)-(VELY(I,J)+VELX(I,J))*DX)/2.0
      GO TO 200
C.....UPPER BOUNDARY CELL (L=2)
  120 DUMMY=(VELPOT(I+1,J)+VELPOT(I,J-1)+VELPOT(I,J+1)-VELY(I,J)*DX)/3.0
      GO TO 200
```

TABLE 6-5 *contd.*

```
C.....UPPER RIGHT CORNER (L=3)
  130 DUMMY=(VELPOT(I+1,J)+VELPOT(I,J-1)-(VELY(I,J)-VELX(I,J+1))*DX)/2.0
      GO TO 200
C.....RIGHT BOUNDARY (L=4)
  140 DUMMY=(VELPOT(I+1,J)+VELPOT(I,J-1)+VELPOT(I-1,J)+(VELX(I,J+1)*DX))
     1  /3.0
      GO TO 200
C.....LOWER RIGHT CORNER (L=5)
  150 DUMMY=(VELPOT(I,J-1)+VELPOT(I-1,J)+(VELX(I,J+1)+VELY(I+1,J))*DX)/
     1  2.0
      GO TO 200
C.....LOWER BOUNDARY (L=6)
  160 DUMMY=(VELPOT(I,J-1)+VELPOT(I-1,J)+VELPOT(I,J+1)+VELY(I+1,J)*DX)/
     1  3.0
      GO TO 200
C.....LOWER LEFT CORNER (L=7)
  170 DUMMY=(VELPOT(I-1,J)+VELPOT(I,J+1)-(VELX(I,J)-VELY(I+1,J))*DX)/2.0
      GO TO 200
C.....LEFT BOUNDARY (L=8)
  180 DUMMY=(VELPOT(I-1,J)+VELPOT(I,J+1)+VELPOT(I+1,J)-(VELX(I,J)*DX))/
     1  3.0
      GO TO 200
C.....INTERIOR OBSTACLE CELL (L=9)
  190 DUMMY=VELPOT(I,J)
C.....TEST FOR CONVERGENCE
  200 ERROR=ERROR+ABS(DUMMY-VELPOT(I,J))
      VELPOT(I,J)=DUMMY
  210 CONTINUE
      IF (ERROR.LT.TOL) GO TO 220
  215 CONTINUE
C.....WRITE OUT VELOCITY POTENTIAL ARRAY
  220 WRITE(6,4) ERROR, IT
      CALL PRINT(NROWS,NCOLS,VELPOT,20,20,10.)
C.....CALCULATE VELOCITY COMPONENTS ACROSS EACH CELL BOUNDARY
      DO 260 I=1, NROWS
      DO 260 J=1,NCOLS
      LL=LABEL(I,J)
C.....FIRST IN X-DIRECTION
      GO TO (240,230,230,230,230,230,240,240,240,230), LL
C.....CALCULATE VELOCITY ON LEFT SIDE OF CELL
  230 VELX(I,J)=(VELPOT(I,J)-VELPOT(I,J-1))/DX
C.....NOW IN Y-DIRECTION
  240 GO TO (260,260,260,250,250,250,250,250,260,250), LL
C.....CALCULATE VELOCITY ON UPPER SIDE OF CELL
  250 VELY(I,J)=(VELPOT(I,J)-VELPOT(I-1,J))/DX
  260 CONTINUE
C.....CALCULATE AVERAGE RESULTANT VELOCITY MAGNITUDE AND DIRECTION FOR
C.....EACH CELL CENTER
      DO 280 I=1,NROWS
      DO 280 J=1,NCOLS
      VX=(VELX(I,J+1)+VELX(I,J))/2.0
      VY=(VELY(I+1,J)+VELY(I,J))/2.0
      RESVEL(I,J)=SQRT(VX*VX+VY*VY)
      IF (ABS(RESVEL(I,J)).LT.0.001) GO TO 270
      VELDIR(I,J)=ATAN2(VY,VX)*180.0/3.14159 +90.0
      GO TO 280
  270 VELDIR(I,J)=0.0
  280 CONTINUE
```

TABLE 6-5 *contd.*

```
C.....WRITE OUT VELOCITY COMPONENTS ARRAYS
      WRITE(6,5)
      CALL PRINT(NROWS, NCP1, VELX,20,21,1.)
      WRITE(6,6)
      CALL PRINT(NRP1,NCOLS,VELY,21,20,1.)
C.....WRITE OUT VELOCITY RESULTANT ARRAYS
      WRITE(6,7)
      CALL PRINT(NROWS,NCOLS,RESVEL,20,20,1.)
      WRITE(6,8)
      CALL PRINT(NROWS,NCOLS,VELDIR,20,20,1.)
C.....PLOT VELOCITY VECTOR MAP
      CALL VECTOR(RESVEL,VELDIR,NROWS,NCOLS,20,20)
      RETURN
      END
```

[a]Subroutines PRINT and VECTOR are the same as those listed in Tables 6-3 and 6-4, respectively.

definition obstacle cells have zero velocity components across each face. Obstacle cells must be surrounded by appropriate boundary cells, as an "internal" obstacle is equivalent to an "external" boundary at the margins of the grid. These conditions will be more easily understood after working through some actual examples.

Figure 6-9 shows the input for a sample run in diagrammatic form. The integer values in each cell are the cell labels, read in as an array called LABEL(I,J). Boundary velocities are assumed to be zero at all margins and corners except where input values have been assigned, as shown in the diagram. Velocity components in the positive x-direction (across the page from left to right) have been specified at an inlet at the left boundary, two cells wide, and at an outlet eight cells wide on the right boundary. Assume that the velocities are expressed in consistent units (e.g., centimeters per second). The boundary velocities in the y-direction (positive downward) are all set to zero, indicating no flow. Because the fluid is assumed to be ideal, there is no frictional resistance, and the sum of the input velocities (equivalent to mass flux times a constant) must equal the sum of the output velocities. This requirement has been satisfied here, inasmuch as the sum of the input velocities is equal to the sum of the output velocities, except for the difference in algebraic sign.

The actual input data that correspond to the diagram of Figure 6-9 have been listed in Table 6-6. The first line in the table contains integer values for number of rows, number of columns, maximum number of iterations used in solving the equations, number of cell margins with nonzero input velocities in the x-direction, and number of cell margins with nonzero input velocities in the y-direction. The first of the two real numbers in the upper line pertains to

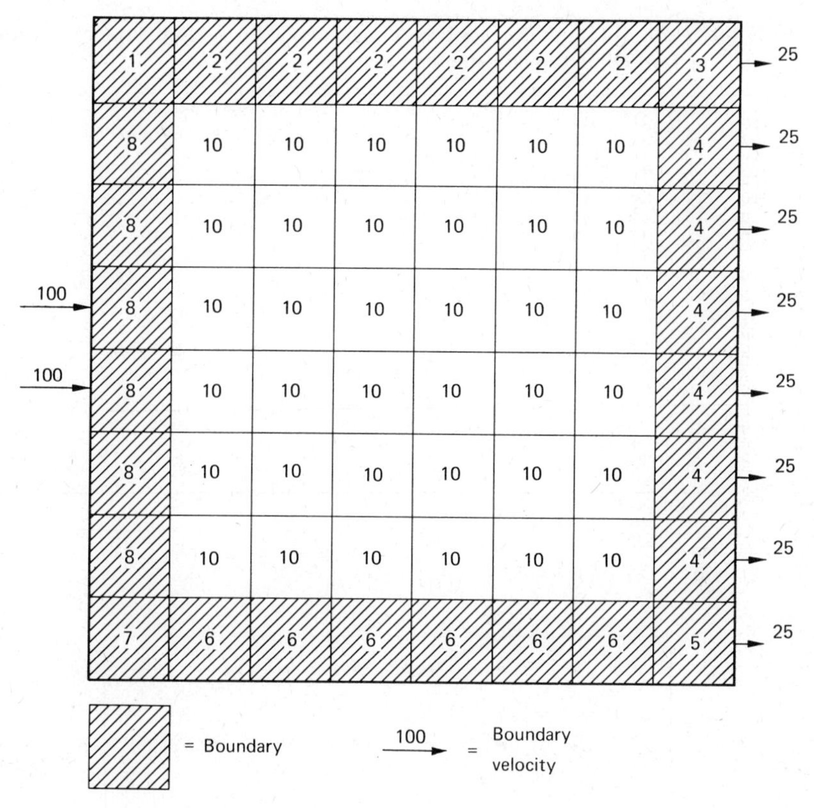

Figure 6-9 Diagram illustrating initial conditions used as input to the FORTRAN program listed in Table 6-5. Integers within cells identify cells according to method outlined above. Boundary velocities that have been assumed are shown with arrows. Inlet velocity is 100 (arbitrary units), whereas outlet velocity is 25. No-flow conditions pertain across all boundaries not labeled with vector arrow.

the tolerance level for stopping the iterative solution of equations if coefficient values have converged before the maximum number of iterations has been reached; the second pertains to the length of the cell side in any unit of length that is consistent with the velocity units. Lines 2 through 9 hold the values of LABEL(I,J). Lines 10 through 19 hold the nonzero boundary-velocity values for the inlets and outlets. In each of these lines the first two numbers indicate the row and column number respectively, of the cell margin under consideration, and the third number is the velocity value.

Output from the program employing the input data described (Figure 6-9 and Table 6-6) is shown in Figure 6-10. The vectors represent the velocity and direction of flow in each cell, effectively

TABLE 6-6 Listing of Input Data for First Experimental Run of FORTRAN Program (Table 6-5) for Two-Dimensional Potential Flow

8	8	50	10	0	.1	1.	
1	2	2	2	2	2	2	3
8	10	10	10	10	10	10	4
8	10	10	10	10	10	10	4
8	10	10	10	10	10	10	4
8	10	10	10	10	10	10	4
8	10	10	10	10	10	10	4
8	10	10	10	10	10	10	4
7	6	6	6	6	6	6	5
4	1	100					
5	1	100					
1	9	25					
2	9	25					
3	9	25					
4	9	25					
5	9	25					
6	9	25					
7	9	25					
8	9	25					

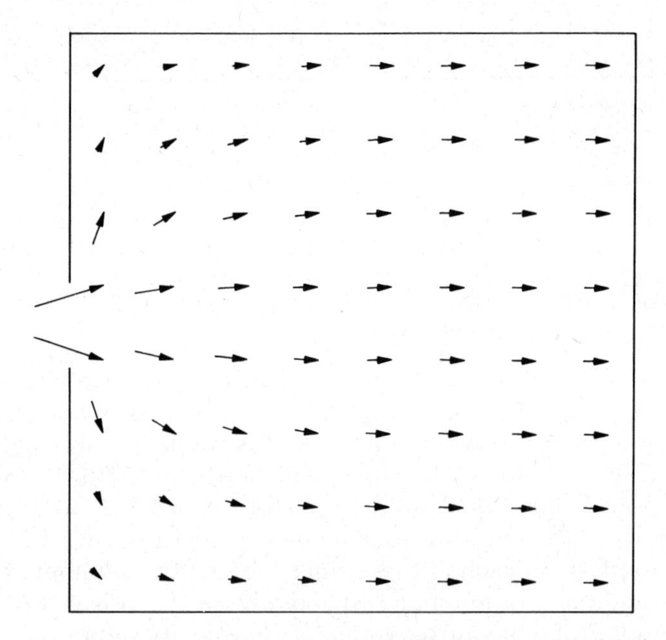

Figure 6-10 Velocity-vector map representing flowfield produced by FORTRAN program listed in Table 6-5, using input data of Table 6-6. Flowfield, as defined by boundary velocities, has inlet that is two cells wide in the center of the left boundary, and outlet occupies the entire right boundary.

portraying the flowfield within the grid meshwork. Other types of output are also produced by the program. Although not shown here, these include line-printed maps of velocity potentials, velocity components of fluid crossing each cell boundary, and resultant velocity-vector magnitudes and directions.

The power and flexibility of the program are shown by subsequent examples in which the input data have been varied. Because of the ability to define the types of cells within the grid meshwork, it is feasible to represent obstacles in the flowfield. Cells labeled with 9's (Figure 6-11*a*) are defined as being incapable of transmitting fluid flow. When obstacle cells are inserted, the calculated flowfield faithfully reflects their presence, the flow paths bending around the obstacles much as they would in the flow of an ideal fluid.

The program may be used to calculate flowfields in which there is more than one inflow direction. Suppose that we have an inlet in the left boundary that admits fluid flowing toward the right, as in the preceding two examples. In addition, however, fluid is also admitted along the upper boundary (Figure 6-12*a*). Outflow is specified along the right and lower boundaries, taking care to balance total inflow with total outflow. The velocity vectors representing the resulting flowfield (Figure 6-12*b*) faithfully portray the overall drift in the positive y-direction.

Relatively complex flowfields can be modeled with the two-dimensional potential-flow program. Figure 6-13 portrays a flowfield containing a single inlet and a single outlet, both located in the same boundary edge. Figure 6-14 represents a flowfield with a nonrectangular shape that has three inlets and one outlet. These examples demonstrate the ability of potential-flow models to deal, in principle, with flowfields of virtually any shape and any configuration of inlets and outlets. Of course the fineness of the cellular meshwork must be appropriate for the degree of detail to be represented. In theory a rectangular grid containing any number of cells in length and width may be employed. Thus greater detail is possible by reducing the cell dimensions and increasing the number of cells. In practice, however, the computing effort that is involved in grids with many cells exerts a restraining influence on the number of cells that may be dealt with effectively.

The accuracy of the results obtained by this method depends not only on the mesh size used but also on the number of iterations employed in the Gauss–Seidel solution. With a small mesh size and with uncomplicated boundary conditions, relatively few iterations are required. All the examples illustrated here employed 50 itera-

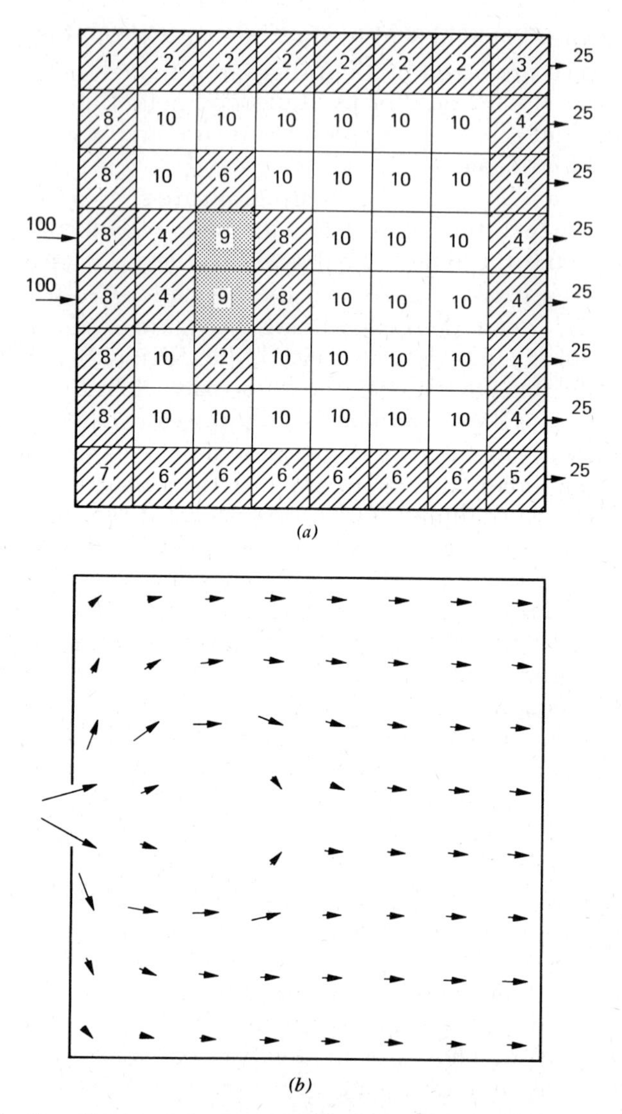

Figure 6-11 Two-dimensional potential flow in which flowfield has same boundary velocities as shown in Figures 6-9 and 6-10, but two interior obstacle cells have been added. (*a*) Mesh diagram showing how cells are labeled to achieve a particular configuration. Obstacle cells (shaded) labeled with 9's must be flanked by boundary cells which have appropriate labels. All boundary cells are diagonally ruled. Input and output velocities are shown as vectors perpendicular to boundary. Note that the sum of input velocity values is equal to the sum of output values. (*b*) Resulting velocity-vector map. Point of each vector arrow touches center of cell, and vector pertains to velocity at that point.

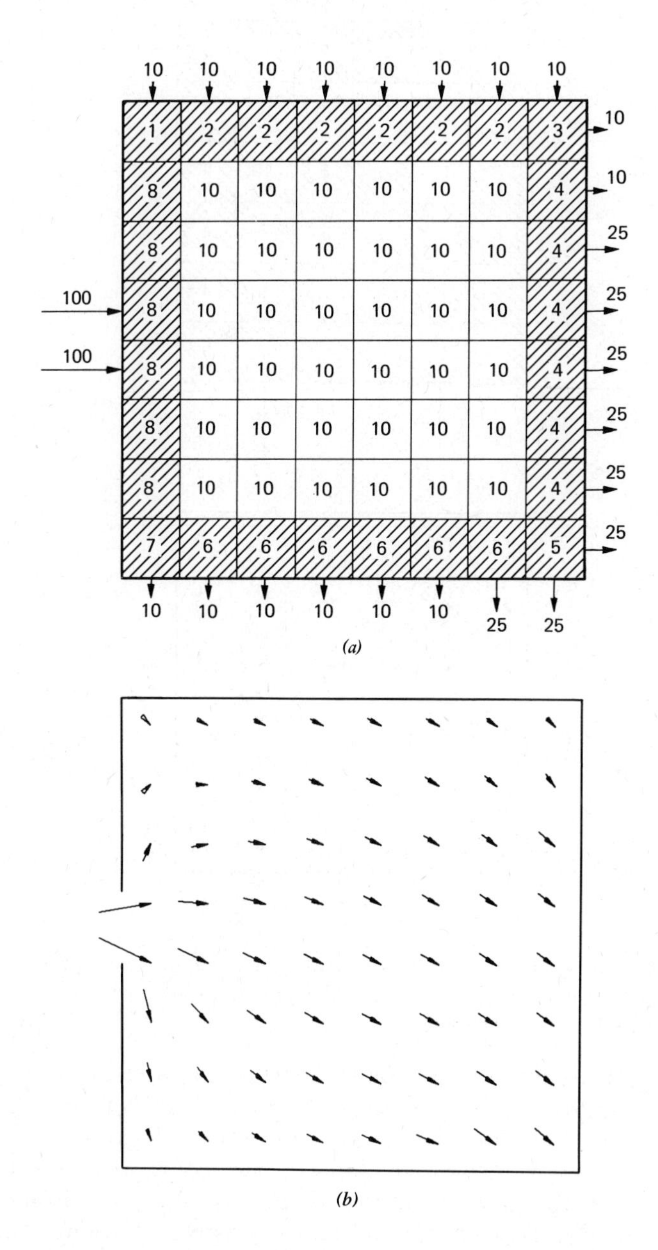

Figure 6-12 Flowfield similar to that of Figure 6-10, but with overall drift in the positive y-direction. (a) Input diagram in which sum of inflow-velocity values is equal to sum of outflow-velocity values; (b) velocity-vector diagram obtained as output.

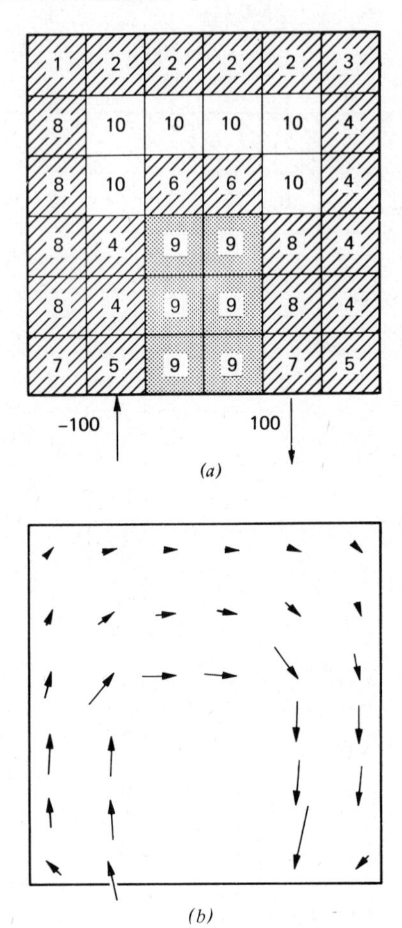

Figure 6-13 Flowfield containing single inlet and single outlet in lower boundary. Fluid must flow around elongate obstacle before emerging at outlet. (*a*) Input diagram; (*b*) output velocity-vector diagram.

tions only. If the boundary configuration is complex, however, being highly indented or with peculiarly shaped obstacles, or if the grid contains a large number of rows and columns, several hundred iterations may be required. A rough check on the accuracy can be obtained by tabulating the overall improvement in the velocity-potential field as the iterations progress. One measure of improvement is simply the difference in potential from one iteration to the next, summed algebraically over all the cells in the grid. This measure is included in the program in Table 6-5.

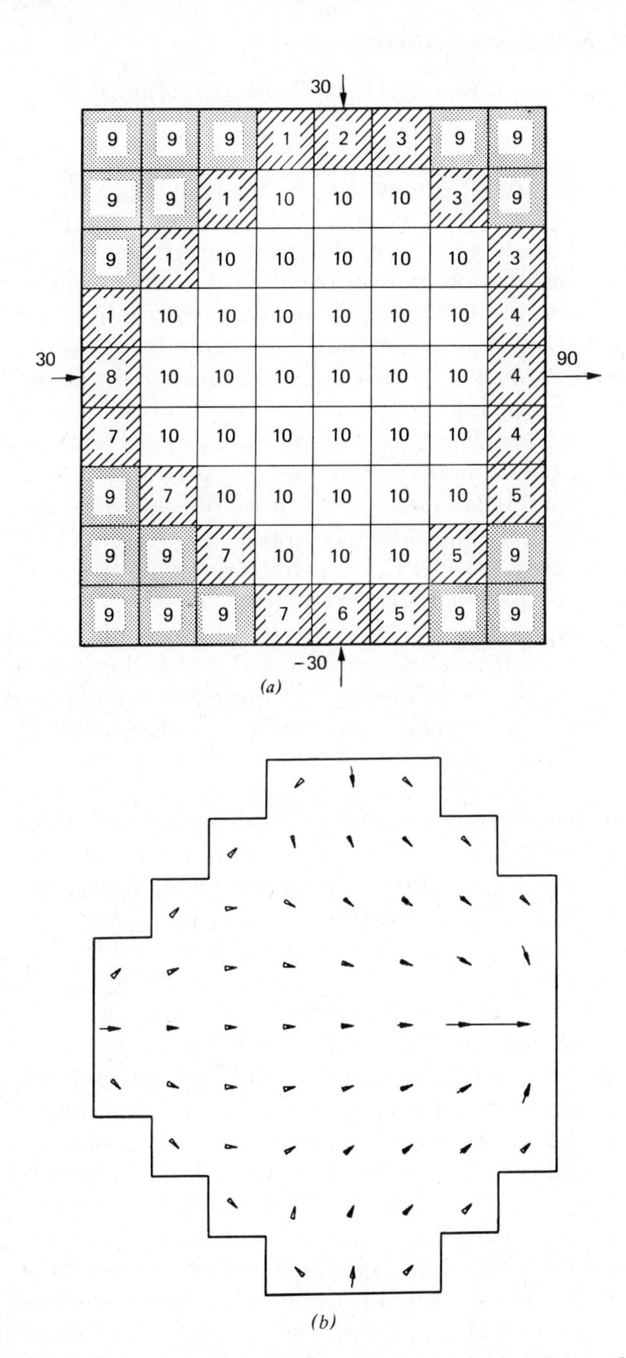

Figure 6-14 Flowfield of irregular shape containing three inlets and one outlet. (*a*) Input diagram; (*b*) output velocity-vector diagram.

TIME-DEPENDENT POTENTIAL FLOW (DETERMINISTIC DIFFUSION)

As we have already pointed out, deterministic diffusion can be regarded as time-dependent potential flow. Diffusion processes also can be modeled stochastically (Hagerstrand, 1968), although we do not treat stochastic models in this chapter. Diffusion processes may be regarded as tending to bring about an equalization of concentration of material in space. For example, if we drop a crystal of potassium permanganate into a beaker of water, solution and diffusion processes will eventually result in a uniform concentration of potassium permanganate throughout the water. This process will occur in the absence of any fluid motion produced by stirring or agitation. Diffusion in this example results from the random motion of molecules. In a gas, where molecules move with relatively high velocities, diffusion tends to be rapid. In liquids diffusion is slower, although it is still much more rapid than in solids.

It is possible to make effective use of diffusion methods in simulation, even though the actual processes to which they are applied do not pertain strictly to "diffusion" in the conventional sense of the word. For example, transportation of suspended material in water can be treated as a diffusion process. If sand is dyed so that it is readily recognizable and then released at a point in the breaker zone on a beach, it will be rapidly dispersed by the turbulent action of waves. Much of this dispersion will result simply from the random motion of sediment particles due to the orbital motion and turbulence associated with shoaling waves (Murray, 1967). If diffusion were the sole transporting factor, a circular dispersion pattern would result. In reality, however, the convective motion of currents imposes a directional drift, creating a noncircular dispersion pattern.

Diffusion methods can be employed in very diverse applications. The dispersal of marine organisms could be regarded as a diffusion process on which the effects of ocean currents are superimposed. As an extreme simplification, even erosional processes can be regarded as diffusion processes. For example, hills can be regarded as areas of "high concentration," whereas valleys are of "low concentration." Diffusion processes can be thought of as leading to a gradual leveling of topography, removing material from the high places and filling in the low places. Finally, diffusion methods may be effectively applied to nonmaterial properties, such as a sociologist's treatment of the spatial diffusion of ideas or a statistician's analysis of the spread of disease epidemics.

Deterministic Diffusion Model

Our next objective is to outline the mathematical theory of deterministic diffusion. We shall make the assumption that the material or property that undergoes diffusion can be treated as if it were capable of flow, regardless of whether the actual material is capable of flowing. The basic laws of diffusion, originally formulated by Fick, relate the rate of diffusion to the concentration gradient. The diffusion flow, or mass flux J_x of a material, can be defined as the mass of this material passing through a reference surface of unit area per unit time. Fick's first law of diffusion states that

$$J_x = -k\frac{\partial c}{\partial x} \tag{6-39}$$

where k = coefficient of diffusion
c = concentration of material
x = coordinate direction parallel to flow and perpendicular to reference surface

In the y- and z-directions J_y and J_z are defined similarly as

$$J_y = -k\frac{\partial c}{\partial y}; \qquad J_z = -k\frac{\partial c}{\partial z} \tag{6-40}$$

Fick's first law of diffusion is mathematically similar to Ohm's law describing the flow of an electric current through a conducting medium and to Darcy's law, which describes the flow of fluid through porous media, even though actual diffusion processes are quite different from the flow of electricity or of fluids.

We can combine Fick's first law with an equation of continuity to obtain Fick's second law of diffusion. Using a similar line of reasoning to that employed in deriving the continuity equation (Equation 6-18), we can write a continuity equation employing mass flux J and concentration c:

$$\frac{\partial J_x}{\partial x} + \frac{\partial J_y}{\partial y} + \frac{\partial J_z}{\partial z} = \frac{\partial c}{\partial t} \tag{6-41}$$

The expression on the left side of Equation 6-41 represents input minus output, and $\partial c/\partial t$ is an accumulation term representing the rate of change in concentration with time (Crank, 1956). By substituting Equations 6-39 and 6-40 into Equation 6-41 and assuming the diffusion coefficient k to be the same in all directions (isotropic), we arrive at Fick's second law:

$$\frac{\partial c}{\partial t} = -k\left(\frac{\partial^2 c}{\partial x^2} + \frac{\partial^2 c}{\partial y^2} + \frac{\partial^2 c}{\partial z^2}\right) \tag{6-42}$$

Using the Laplacian operator ∇^2, we can write

$$\frac{\partial c}{\partial t} = -k\nabla^2 c \tag{6-43}$$

If the rate of diffusion flow varies according to direction (i.e., the medium through which diffusion takes place is anisotropic), the diffusion equation is modified so that a separate coefficient pertains to each direction. This assumes that the coordinate axes coincide with the anisotropic axes, that the three diffusion coefficients k_x, k_y, and k_z are constant from point to point in the medium (i.e., it is homogeneous), and that the diffusion coefficients are independent of concentration. If we drop the minus sign by redefining k, the equation becomes

$$\frac{\partial c}{\partial t} = k_x \frac{\partial^2 c}{\partial x^2} + k_y \frac{\partial^2 c}{\partial y^2} + k_z \frac{\partial^2 c}{\partial z^2} \tag{6-44}$$

where k_x = diffusion coefficient in the x-direction
k_y = diffusion coefficient in the y-direction
k_z = diffusion coefficient in the z-direction.

Finite-Difference Representation of the One-Dimensional Diffusion Equation

For computing purposes the diffusion equations must be translated into finite-difference form. The translation is very similar to that for the Laplace equation, discussed earlier.

To illustrate the principles involved let us deal with the equation for diffusion in one dimension and employ it in a rather artificial geological application dealing with the movement of material from places of high topographic elevation to those of low elevation. The equation may be written

$$\frac{\partial c}{\partial t} = k \frac{\partial^2 c}{\partial x^2} \tag{6-45}$$

where c = topographic elevation
x = distance from an arbitrary coordinate origin (Figure 6-15)

This equation could of course be solved analytically. We shall obtain a numerical solution for illustrative purposes. We shall regard the solution as pertaining to diffusion in only one spatial dimension. This seems to contradict the graph of Figure 6-15, which involves two dimensions, height and horizontal distance along the x-axis. From the point of view of applying the diffusion equation, however, we are

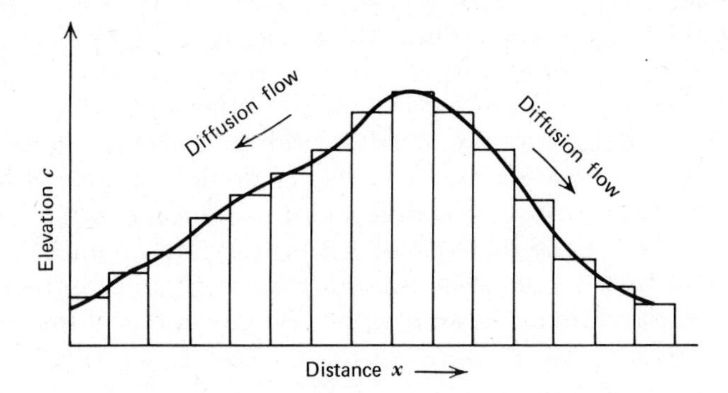

Figure 6-15 Diagram to illustrate finite-difference approximation of topographic profile (continuous curved line) by series of columns. Changes in elevation c with time t result from progressive diffusion of "elevation" along the x-axis. Result is analogous to transportation of material from higher to lower places.

considering elevation c simply as if it represented concentration at points along the x-axis, and not as a second spatial dimension. For this reason only the term $\partial^2 c/\partial x^2$ is used, and we do not consider $\partial^2 c/\partial y^2$ or $\partial^2 c/\partial z^2$.

Solution of the equation in numerical form involves division of the x-axis into columns each Δx wide and division of time into uniform finite increments, Δt. We can write the finite-difference form of Equation 6-45 as

$$\frac{\partial c}{\partial t} \simeq \frac{c_{t+1,j} - c_{t,j}}{\Delta t} \tag{6-46}$$

and

$$\frac{\partial^2 c}{\partial x^2} \simeq \frac{c_{t,j+1} + c_{t,j-1} - 2c_{t,j}}{(\Delta x)^2} \tag{6-47}$$

where $c_{t,j}$ = topographic elevation at column j at time t
$c_{t+1,j-1}$ = topographic elevation at column $(j-1)$ at time $(t+1)$ (and so on for other combinations of subscripts)
Δt = time increment
Δx = distance increment

Substituting Equations 6-46 and 6-47 into Equation 6-45, we obtain

$$\frac{c_{t+1,j} - c_{t,j}}{\Delta t} = k\left[\frac{c_{t,j+1} + c_{t,j-1} - 2c_{t,j}}{(\Delta x)^2}\right] \tag{6-48}$$

which simplifies to

$$c_{t+1,j} = c_{t,j} + k\frac{\Delta t}{(\Delta x)^2}\,(c_{t,j+1} + c_{t,j-1} - 2c_{t,j}) \tag{6-49}$$

The significance of Equation 6-49 can be better appreciated if we think of the change of elevation c with respect to distance x, through time t. Since we are dealing with a total of three variables, a single two-dimensional diagram cannot adequately portray the interrelationships. Instead, we can employ a series of profiles (Figure 6-16) that relate elevation and distance at different time increments.

Equation 6-49 states that the elevation $c_{t+1,j}$ in column j at time $t+1$ is equal to the sum of its previous value $c_{t,j}$ plus an increment equal to a constant times the sum of the two elevations of the immedi-

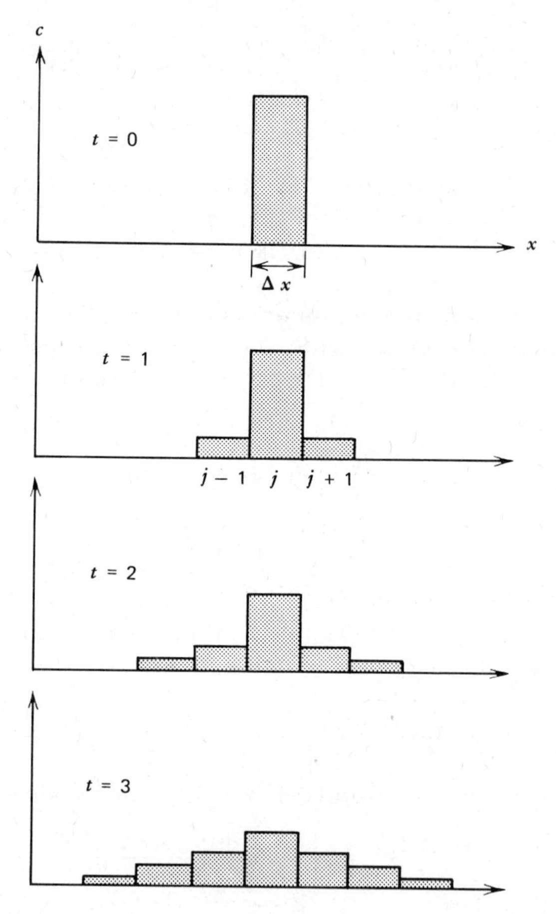

Figure 6-16 Diagrams to illustrate changes in elevation c that result from lateral diffusion of material along the x-axis during successive time increments, each separated by Δt. Diagrams could be regarded as portraying changes in topographic profiles through time.

ately adjacent columns during the previous moment in time ($c_{t,j-1}+c_{t,j+1}$), minus twice the elevation of column j during the previous moment in time ($2c_{t,j}$). The constant $k\Delta t/(\Delta x)^2$ combines the values of Δt, Δx, and the diffusion coefficient k. The equation is thus in complete accord with the principles of conservation of mass because the amount credited to, or debited from, a particular column j during a time step is exactly equal to the amount obtained from, or supplied to, each of the two immediately adjacent columns. These "bookkeeping" operations are illustrated graphically in Figure 6-17.

As we have already emphasized, the solution of an equation in finite-difference form by numerical methods depends in large part on the boundary values that are assumed. For the one-dimensional equation at hand this involves a series of initial values of c for each column at $t=0$ (this could be considered to be a "time boundary"). Second, the values of c in the columns at each end of the x-axis must be specified. The solution that is then obtained will consist of a series of values of c for the interior cells for different values of t. In our example solution we first assume that c remains constant at zero in both boundary columns. If we make the number of divisions on the x-axis sufficiently large and place the initial high concentration toward the center of the x-range, the influence of the boundary condition will not be important. We shall discuss the problems associated with different choices of boundary conditions later.

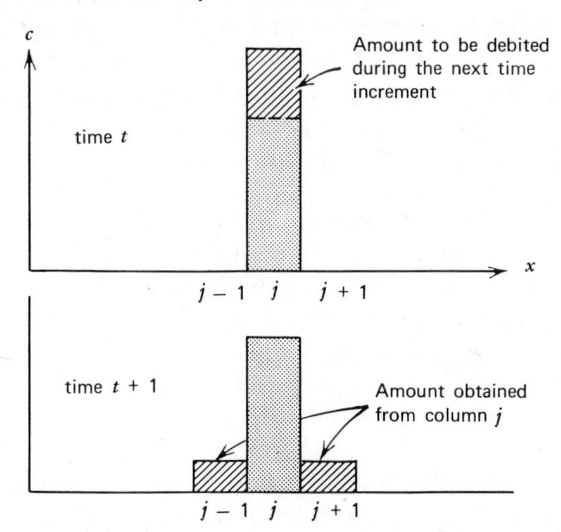

Figure 6-17 Schematic diagram to illustrate bookkeeping concept in diffusion equation. Value of c at time $t+1$ at column j depends on previous value of c in that column less the amount lost to (or alternatively supplied by) immediately adjacent columns.

In the solution we must also specify the value of the factor that combines $k\Delta t/(\Delta x)^2$ so it falls in the range 0.0 to 0.5. If it is zero, the diffusion rate is zero. If it is greater than 0.5 in the one-dimensional case, concentration highs are immediately changed into concentration lows, and the results are contrary to physical laws in a diffusion process.

Adhering to the conditions specified above, the steps in the calculation are as follows:

1. Assign initial concentration values for $t = 0$,
2. Advance time by Δt to the next time value,
3. Omitting the boundary columns, calculate c for each column by using Equation 6-49.
4. Return to step 2 for each successive time increment. Repeat steps 2, 3, and 4 in succession for each time increment.

The method that we have illustrated above is an *explicit* method that yields a solution directly at each time step. This method contrasts with the *implicit* method, which we used to solve the Laplace equation and which converges iteratively toward a stable solution. For background the reader is referred to McCalla (1967, Chapter 9) and Crank (1956, Chapter 10). Both these books describe a variety of numerical methods for solving diffusion-type equations.

FORTRAN Program for One-Dimensional Diffusion

Table 6-7 lists a FORTRAN program for diffusion in one spatial dimension, employing the finite-difference method described above. For the example run, whose output is shown in Figure 6-18, the following parameters were read in as input:

N = 20 = number of columns on the x-axis
NTIM = 20 = number of time increments
DX = 1.0 = incremental step on the x-axis
COEF = 0.2 = value of diffusion coefficient
DT = 0.5 = length of time increment
VAL(10) = 100.0 = elevation at column 10 (all other values of VAL(J) were set to zero)

In this run the central "spike" representing the initial distribution at time 0 is gradually spread out laterally. The progressive variations in height of the columns approximate (as histograms) a succession of Gaussian curves. The similarity is more than incidental; the analytical solution to the diffusion equation is Gaussian in form.

TABLE 6-7 FORTRAN Program for Diffusion in One Spatial Dimension, Assuming a Constant Concentration of Zero at Boundaries

```
C.....ONE DIMENSIONAL DIFFUSION
      REAL VAL(20), NEWVAL(20), GRAPH(100)
      DATA GRAPH/100*'*'/
    1 FORMAT(2I5, 3F5.0)
    2 FORMAT(16F5.0/4F5.0)
    3 FORMAT(1H1, 'TIME INCREMENT', I5//)
    4 FORMAT(1H , I5, 100A1)
C.....READ INPUT PARAMETERS
      READ(5,1) N, NTIM, DX, COEF,DT
      READ(5,2) (VAL(I), I=1,N)
      NL1=N-1
      FACTOR=COEF*DT/(DX*DX)
C.....BEGIN TIME INCREMENT LOOP
      DO 40 NT=1,NTIM
      WRITE(6,3) NT
      DO 10 I=2,NL1
C.....CALCULATE NEWVAL FROM OLD VALUES
   10 NEWVAL(I)=VAL(I)+FACTOR*(VAL(I-1)-2.*VAL(I)+VAL(I+1))
C.....TRANSFER NEWVAL TO VAL ARRAY
      DO 20 I=2,NL1
   20 VAL(I)=NEWVAL(I)
C.....GRAPH THE RESULTS
      DO 40 I=1,N
      INDEX=VAL(I)+0.5
      IF (INDEX.LT.1) GO TO 30
      WRITE(6,4) I, (GRAPH(IN), IN=1,INDEX)
      GO TO 40
   30 WRITE(6,4) I
   40 CONTINUE
      RETURN
      END
```

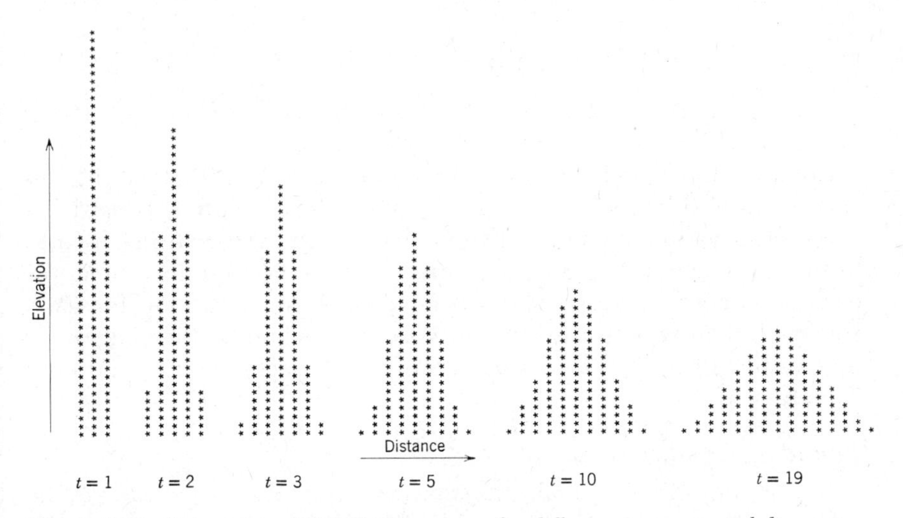

Figure 6-18 Output from FORTRAN program for diffusion in one-spatial dimension (Table 6-7). Output consists of succession of profiles for time steps as numbered.

Finite-Difference Representation of the Two-Dimensional Diffusion Equation

The diffusion equation for two spatial dimensions is

$$\frac{\partial c}{\partial t} = k_x \frac{\partial^2 c}{\partial x^2} + k_y \frac{\partial^2 c}{\partial y^2} \tag{6-50}$$

where k_x and k_y are the diffusion coefficients for the x- and y-directions, respectively, and the coordinate axes are assumed to correspond to the axes within a homogeneous anisotropic medium. Finite-difference representation of this equation requires that it be considered as a function extending over a rectangular meshwork of cells in the x- and y-directions. The rows and columns of cells are conveniently indexed with i and j, respectively. Translation of the derivatives to finite-difference quotients is similar to the procedure of Equations 6-46 and 6-47, except, of course, that two spatial dimensions are involved instead of one. The numerical value at cell (i,j) at time $t+1$ may be obtained as follows:

$$\frac{(c_{t+1,i,j} - c_{t,i,j})}{\Delta t} = k_x \left[\frac{c_{t,i,j+1} + c_{t,i,j-1} - 2c_{t,i,j}}{(\Delta x)^2} \right] + k_y \left[\frac{c_{t,i+1,j} + c_{t,i-1,j} - 2c_{t,i,j}}{(\Delta y)^2} \right] \tag{6-51}$$

If $\Delta x = \Delta y$, the equation simplifies to

$$c_{t+1,i,j} = c_{t,i,j} + \frac{\Delta t}{(\Delta x)^2} [k_x(c_{t,i,j-1} + c_{t,i,j+1} - 2c_{t,i,j}) + k_y(c_{t,i+1,j} + c_{t,i-1,j} - 2c_{t,i,j})] \tag{6-52}$$

Equation 6-52 can be solved in the same manner as the one-dimensional finite-difference diffusion equation. The solution consists of a series of values for each of the cells forming the grid meshwork, through an interval of time containing a specified number of steps, each separated by increment Δt. Given the concentration for each cell at $t = 0$, as well as the constants Δx, Δt, k_x, and k_y, the concentration values in each cell can be obtained by successive evaluation of Equation 6-52.

Boundary Conditions

In the one-dimensional diffusion example described above the concentration in the "columns" at the boundary were assumed to remain fixed at zero for each time increment. This was handled in the FORTRAN program (Table 6-7) by omitting the first and last columns from the main DO-loop, so that the concentration at these columns was never recalculated. If, instead of zero concentration, we had assigned positive concentration values to these boundary columns, the results would have been different. Clearly, the assumed boundary conditions strongly influence the numerical solution of the diffusion equation.

The finite-difference representation of the two-dimensional diffusion equation involves arithmetic operations that have the effect of smoothing concentration values. This accords of course with the process of diffusion itself in which the tendency is to equalize differences in concentration from place to place. The arithmetic operations involve summing the values in the four immediately adjoining cells, two in the x-direction and two in the y-direction. At boundaries, however, the summation values must be altered because the concentration values in cells outside the grid are unknown. At each of the four cells at the corners of the grid two adjacent cells lie outside, and elsewhere along the edges of the grid one cell lies outside each boundary cell. The geometrical relationship of boundary cells to outside cells is identical to that in the finite-difference representation of steady-state potential flow.

Although there are a number of methods of treating boundary conditions with the diffusion equations (Crank, 1956, p. 198), we touch on only two. One of the methods involves the assumption of a constant concentration *gradient* across the boundary through time, and the other involves a constant concentration *value* across the boundary through time. The two methods are illustrated in Figure 6-19.

The use of either of these two methods involves adapting the finite-difference form (Equation 6-52) of the diffusion equation. For the two-dimensional case this equation involves use of the previous concentration values in the four cells that lie immediately adjacent to the particular cell for which the calculations are being performed. The problem is that at the grid boundary one of the adjacent cells is missing (two are missing at corners). We modify the finite-difference equation to make use of information at the boundary itself, in place of information for a cell lying outside the boundary. As pointed out above, one way of doing this is to specify the slope $\partial c/\partial x$ of the concentration gradient with distance at the outer margin of the particular boundary cell. Alternatively we can specify a constant concentration value at the outer margin of the particular boundary cell. Depending on the method, the equations need to be modified

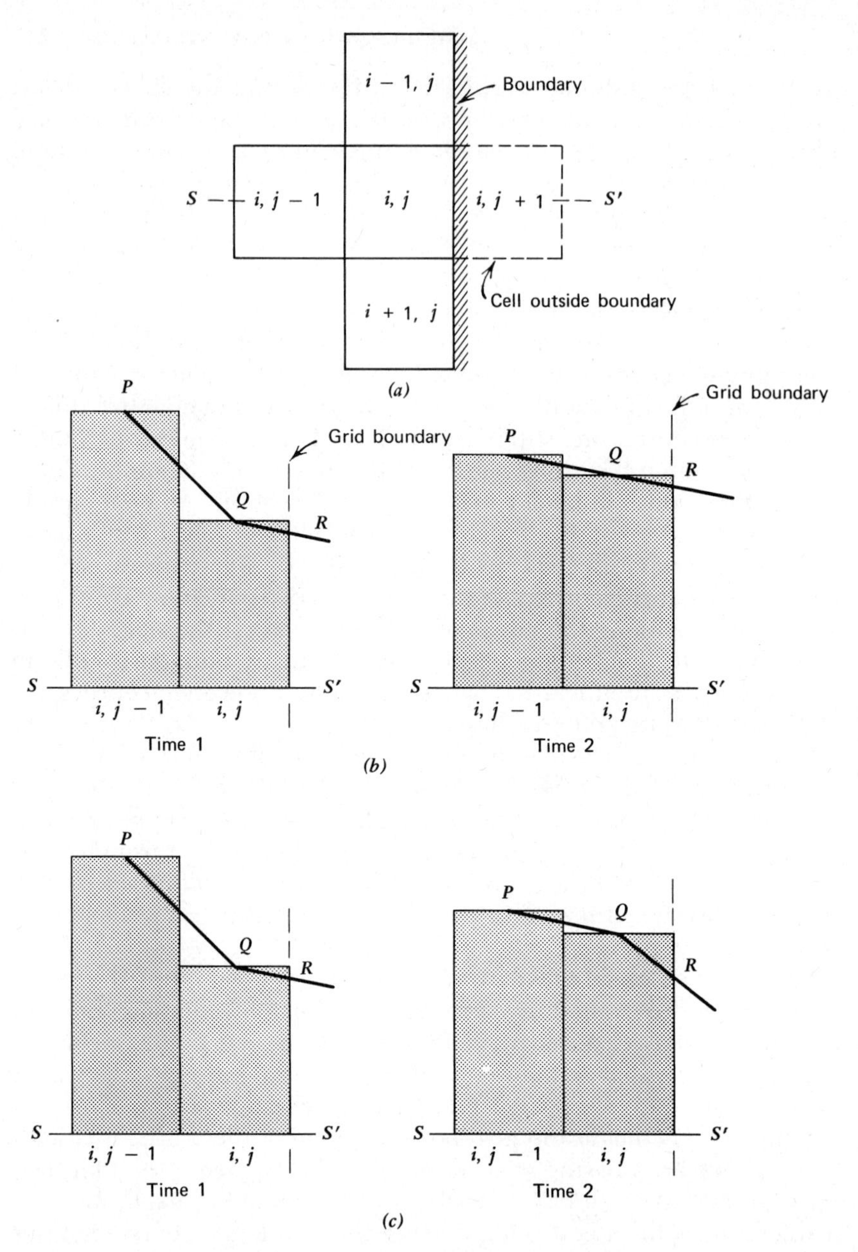

Figure 6-19 Diagrams illustrating constant-gradient method versus constant-concentration method for treating boundary conditions in diffusion equation: (*a*) plan view showing boundary cell (*i*, *j*) and immediately adjacent cells, including cell outside boundary; (*b* section *S*-*S*′ at time 1 and time 2, showing concentration–distance profile in constant-gradient method (slope of line *QR* remains constant, although concentration values at *Q* and *R* vary through time); (*c*) constant-concentration method in which value of *R* remains constant, although slope of line *QR* varies with time.

at each type of boundary (top, bottom, left and right edges, and each of the four corners). The equations developed for the boundary conditions in steady-state potential flow are similar.

FORTRAN Program for Two-Dimensional Diffusion

Table 6-8 lists a FORTRAN program for diffusion in two spatial dimensions. Besides being expanded from one to two spatial dimensions, this program is more complex than the program listed in Table 6-7 for the following reasons: (a) the diffusion coefficient can be assigned different values in the x- and y-directions, and (b) the boundary conditions involve a constant gradient in which $\partial c/\partial x = 0$. In the program boundary cells are labeled according to the method adopted for the two-dimensional potential-flow program (Table 6-5), with the exception that obstacle cells are omitted and there are no internal boundaries. Cell label 9 refers to a "normal" interior cell in the grid, and the label 10 is not used.

Output from an experimental run made with the program, involving a 10×10 meshwork of cells, is shown in Figures 6-20 and 6-21. Input data for the run are listed in Table 6-9. During the run an initial concentration value of 1000 (at time 0) in a single cell was allowed to diffuse outward into surrounding cells with initial concentrations of zero. The diffusion coefficients were set so that the diffusion rate in the y-direction was faster than that in the x-direction. The resulting spatial distribution of concentration values forms a series of elliptical patterns that are elongate in the y-direction.

The results of a second experimental run are shown in Figure 6-22. In this run two centers of high initial concentration (time 0) were used, a value of 500 in a cell in the upper left and 1000 in a cell in the lower right. The remaining cells in the grid were assigned initial concentrations of zero. As in the preceding run, the diffusion coefficients for the two directions were set so as to cause diffusion in the y-direction to proceed faster than that in the x-direction.

Heterogeneous Diffusion

Many materials through which diffusion takes place are not uniform in their capability of transmitting the material that diffuses. In other words, they are anisotropic. We may distinguish between (a) anisotropism in which the diffusion coefficients differ with direction in the materials and (b) local variations of diffusion coefficients due to heterogeneity of the materials. Diffusion in heterogeneous materials can be simulated in one, two, or three dimensions. Table 6-10 lists a FORTRAN program for two-dimensional applications. The program

TABLE 6-8 FORTRAN Program for Finite-Difference Representation of Diffusion in Two Spatial Dimensions[a]

```
C.....TWO DIMENSIONAL DIFFUSION. DIFFUSION COEFFICIENT DIFFERENT IN X
C.....AND Y DIRECTIONS.
      DIMENSION VAL(20,20), VALNEW(20,20), LABEL(20,20)
    1 FORMAT(3I5, 4F5.0)
    2 FORMAT(16F5.0)
    3 FORMAT(16I5)
    4 FORMAT(1H1, 'TIME INCREMENT', I5// 5X, 'CONCENTRATION VALUES'//)
C.....READ INPUT PARAMETERS
      READ(5,1) NROWS, NCOLS, NTIM, DXY, COEFX, COEFY, DT
C.....READ INITIAL CONCENTRATION VALUES
      DO 10 I=1, NROWS
   10 READ(5,2) (VAL(I,J), J=1,NCOLS)
C.....READ CELL LABELS, DENOTING BOUNDARY CELL TYPES
      DO 20 I=1,NROWS
   20 READ(5,3) (LABEL(I,J), J=1,NCOLS)
C.....WRITE OUT INITIAL CONCENTRATION VALUES
      NT=0
      WRITE(6,4) NT
      CALL PRINT(NROWS, NCOLS, VAL,20,20,1.)
C.....SET MULTIPLYING FACTORS
      FACTX=COEFX*DT/(DXY*DXY)
      FACTY=COEFY*DT/(DXY*DXY)
C.....BEGIN TIME ITERATIONS
      DO 220 NT=1,NTIM
      DO 200 I=1,NROWS
      DO 200 J=1,NCOLS
      L=LABEL(I,J)
C.....GO TO APPROPRIATE SECTION DEPENDING ON CELL TYPE
      GO TO (110,120,130,140,150,160,170,180,190), L
C.....TOP LEFT CORNER CELL
  110 VALNEW(I,J)=VAL(I,J)+FACTX*(VAL(I,J+1)-VAL(I,J))+
     1 FACTY*(VAL(I+1,J)-VAL(I,J))
      GO TO 200
C.....TOP MARGIN OF GRID
  120 VALNEW(I,J)=VAL(I,J)+FACTX*(VAL(I,J-1)+VAL(I,J+1)-2.*VAL(I,J))+
     1 FACTY*(VAL(I+1,J)-VAL(I,J))
      GO TO 200
C.....UPPER RIGHT CORNER
  130 VALNEW(I,J)=VAL(I,J)+FACTX*(VAL(I,J-1)-VAL(I,J))+
     1 FACTY*(VAL(I+1,J)-VAL(I,J))
      GO TO 200
C.....RIGHT MARGIN
  140 VALNEW(I,J)=VAL(I,J)+FACTX*(VAL(I,J-1)-VAL(I,J))+
     1 FACTY*(VAL(I-1,J)+VAL(I+1,J)-2.*VAL(I,J))
      GO TO 200
C.....LOWER RIGHT CORNER
  150 VALNEW(I,J)=VAL(I,J)+FACTX*(VAL(I,J-1)-VAL(I,J))+
     1 FACTY*(VAL(I-1,J)-VAL(I,J))
      GO TO 200
C.....LOWER MARGIN
  160 VALNEW(I,J)=VAL(I,J)+FACTX*(VAL(I,J-1)+VAL(I,J+1)-2.*VAL(I,J))+
     1 FACTY*(VAL(I-1,J)-VAL(I,J))
      GO TO 200
C.....LOWER LEFT CORNER
  170 VALNEW(I,J)=VAL(I,J)+FACTX*(VAL(I,J+1)-VAL(I,J))+
     1 FACTY*(VAL(I-1,J)-VAL(I,J))
      GO TO 200
```

TABLE 6-8 *contd*

```
C.....LEFT MARGIN
  180 VALNEW(I,J)=VAL(I,J)+FACTX*(VAL(I,J+1)-VAL(I,J))+
     1 FACTY*(VAL(I+1,J)+VAL(I-1,J)-2.*VAL(I,J))
      GO TO 200
C.....NORMAL (NON-BOUNDARY) CELL
  190 VALNEW(I,J)=VAL(I,J)+FACTX*(VAL(I,J-1)+VAL(I,J+1)-2.*VAL(I,J))+
     1 FACTY*(VAL(I+1,J)+VAL(I-1,J)-2.*VAL(I,J))
  200 CONTINUE
C.....TRANSFER VALNEW TO VAL
      DO 210 I=1,NROWS
      DO 210 J=1,NCOLS
  210 VAL(I,J)=VALNEW(I,J)
C.....WRITE OUT CONCENTRATION VALUES
      WRITE(6,4) NT
      CALL PRINT(NROWS,NCOLS,VAL,20,20,1.)
  220 CONTINUE
      RETURN
      END
```

[a]Subroutine PRINT, called within program, is listed in Table 6-3.

(which assumes that the axes of anisotropism are parallel to the coordinate axes) provides for input of diffusion coefficients on a cell-by-cell basis, employing two arrays (Table 6-11), one containing the coefficients for the x-direction, and the other for the y-direction. The two arrays containing the diffusion coefficients have the same number of rows and columns as the main grid meshwork.

TABLE 6-9 Input Data for Experimental Run of Two-Dimensional-Diffusion Program Listed in Table 6-8[a]

10	10	10	1.0	0.3	0.8	0.25			
0	0	0	0	0	0	0	0	0	0
0	0	0	0	0	0	0	0	0	0
0	0	0	0	0	0	0	0	0	0
0	0	0	0	0	0	0	0	0	0
0	0	0	1000	0	0	0	0	0	0
0	0	0	0	0	0	0	0	0	0
0	0	0	0	0	0	0	0	0	0
0	0	0	0	0	0	0	0	0	0
0	0	0	0	0	0	0	0	0	0
0	0	0	0	0	0	0	0	0	0
1	2	2	2	2	2	2	2	2	3
8	9	9	9	9	9	9	9	9	4
8	9	9	9	9	9	9	9	9	4
8	9	9	9	9	9	9	9	9	4
8	9	9	9	9	9	9	9	9	4
8	9	9	9	9	9	9	9	9	4
8	9	9	9	9	9	9	9	9	4
8	9	9	9	9	9	9	9	9	4
7	6	6	6	6	6	6	6	6	5

[a]Lines 2 to 11 contain initial concentration values. Lines 12 to 21 are code labels for cell types that describe role of each according to position in grid.

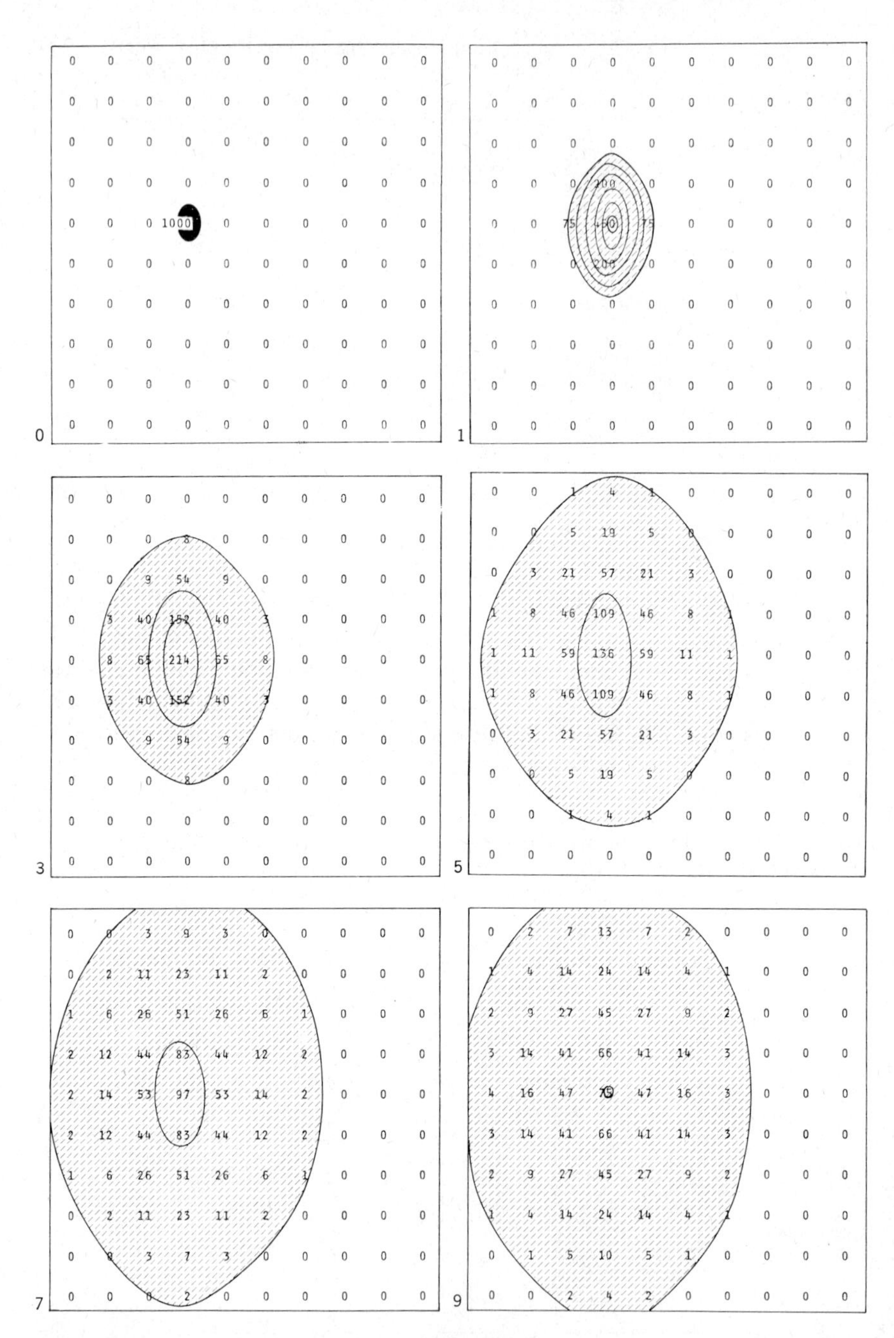

Figure 6-20 Output from two-dimensional diffusion program listed in Table 6-8. Initial concentration values (lines 2 to 11 in Table 6-9) were all zero except for value of 1000 in cell in fifth row and fourth column. Maps of grid mesh show concentration values at times 0, 1, 3, 5, 7, and 9. Contour interval is 75.

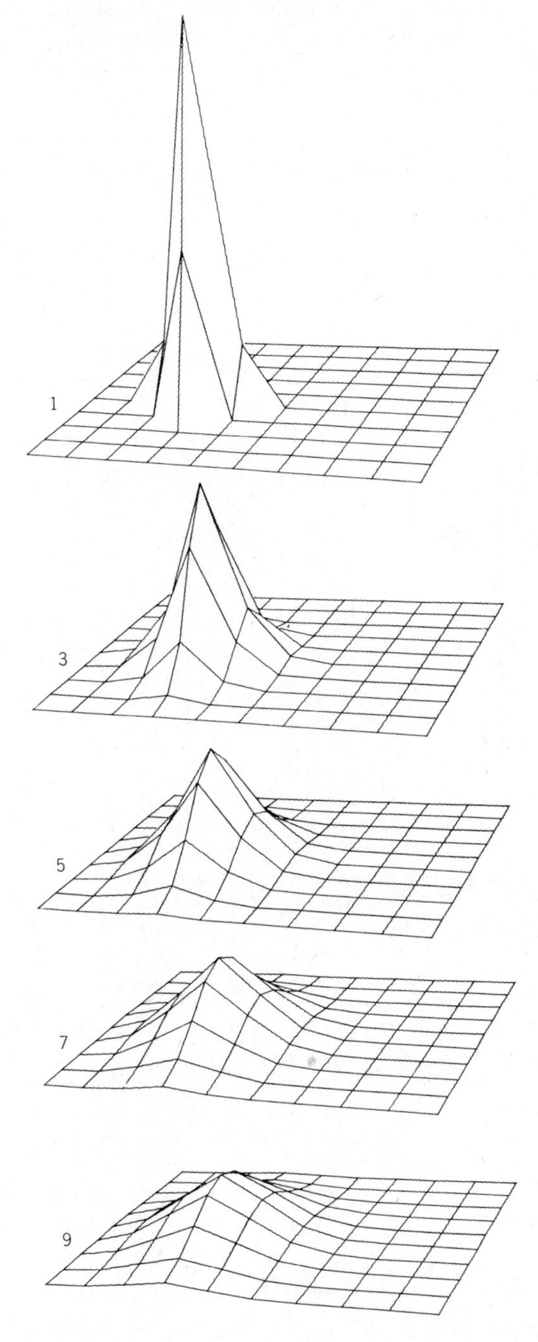

Figure 6-21 Series of computer-drawn perspective diagrams corresponding to Figure 6-20, showing concentration surfaces produced by two-dimensional-diffusion program at time increments 1, 3, 5, 7, and 9.

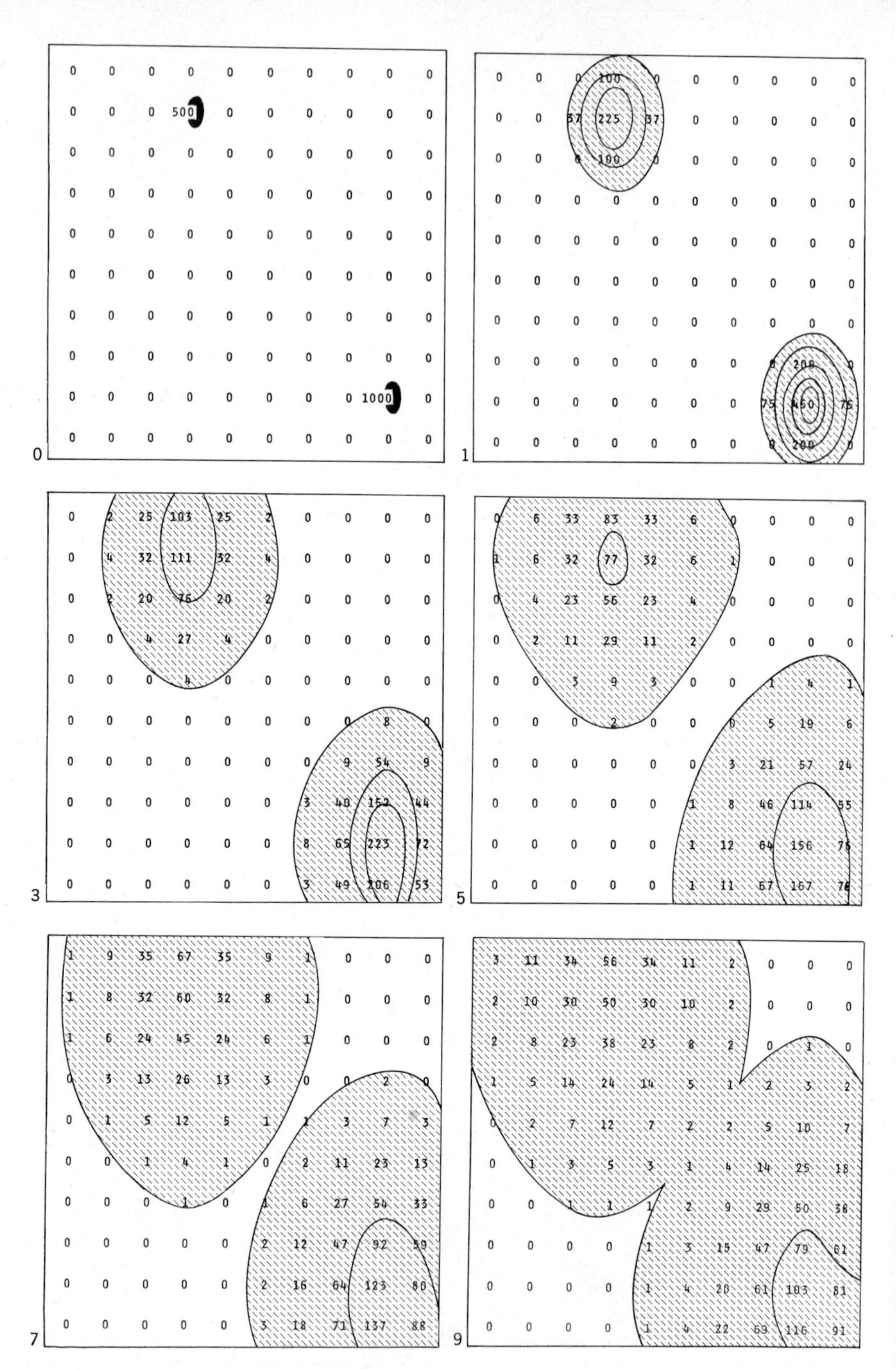

Figure 6-22 Output from second experimental run of two-dimensional-diffusion program, portraying initial concentration values at (time 0) and values at subsequent times.

TABLE 6-10 FORTRAN Program for Two-Dimensional Diffusion in Heterogeneous Materials

```
C.....TWO DIMENSIONAL DIFFUSION WITH VARIABLE DIFFUSION COEFFICIENTS
      REAL VAL(12,12), NEWVAL(12,12), COEFX(12,12), COEFY(12,12)
      DIMENSION NUM(12)
    1 FORMAT(3I5, 2F5.0)
    2 FORMAT(12F5.0)
    3 FORMAT(1H1, 'TIME INCREMENT', I5// 6X, 12I5//)
    4 FORMAT(1H , I5, 12I5//)
C.....READ INPUT PARAMETERS
      READ(5,1) NROWS, NCOLS, NTIM, DXY, DT
      NT=0
      WRITE(6,3) NT, (J,J=1,NCOLS)
      DO 15 I=1,NROWS
      READ(5,2) (VAL(I,J), J=1,NCOLS)
      DO 10 J=1,NCOLS
   10 NUM(J)=VAL(I,J)+0.5
   15 WRITE(6,4) I, (NUM(J), J=1,NCOLS)
      DO 20 I=1,NROWS
   20 READ(5,2) (COEFX(I,J), J=1,NCOLS)
      DO 30 I=1, NROWS
   30 READ(5,2) (COEFY(I,J), J=1,NCOLS)
      NCL1=NCOLS-1
      NRL1=NROWS-1
      FACTOR=DT/(DXY*DXY)
C.....BEGIN TIME STEP LOOP
      DO 60 NT=1, NTIM
      WRITE(6,3) NT, (J, J=1,NCOLS)
C.....FOR EACH CELL IN GRID
      DO 40 I=2,NRL1
      DO 40 J=2,NCL1
C.....FIND AVERAGE DIFFUSION COEFFICIENT ON CELL MARGINS
      CXPHLF=(COEFX(I,J)+COEFX(I+1,J))/2.
      CXLHLF=(COEFX(I-1,J)+COEFX(I,J))/2.
      AVCX=(CXPHLF+CXLHLF)/2.
      CYPHLF=(COEFY(I,J)+COEFY(I,J+1))/2.
      CYLHLF=(COEFY(I,J-1)+COEFY(I,J))/2.
      AVCY=(CYPHLF+CYLHLF)/2.
C.....CALCULATE NEWVAL FROM OLD VALUES, USING AVERAGE DIFFUSION VALUES
   40 NEWVAL(I,J)=VAL(I,J)+FACTOR*(VAL(I-1,J)*CXLHLF+VAL(I+1,J)*CXPHLF-
     1 2.*VAL(I,J)*AVCX+VAL(I,J-1)*CYLHLF+VAL(I,J+1)*CYPHLF-2.*VAL(I,J)*
     2 AVCY)
      DO 45 I=2,NRL1
      DO 45 J=2,NCL1
   45 VAL(I,J)=NEWVAL(I,J)
C.....WRITE OUT NEW VALUES
      DO 60 I=1,NROWS
      DO 50 J=1,NCOLS
   50 NUM(J)=VAL(I,J)+0.5
   60 WRITE(6,4) I, (NUM(J), J=1,NCOLS)
      RETURN
      END
```

Output from an experimental run of the two-dimensional heterogeneous diffusion program is shown in Figure 6-23. Input to the program consists of various controlling parameters and other data as used with the previously described two-dimensional-diffusion program (Table 6-8), and in addition the two diffusion-coefficient arrays (Table 6-11). This particular input has the effect of creating a one-cell-wide permeable channel that extends from the left grid

TABLE 6-11 Diffusion-Coefficient Arrays for the *x*- and *y*-Directions[a] Employed with Heterogeneous Two-Dimensional-Diffusion Program (Table 6-10), Producing Output Shown in Figure 6-23

Diffusion Coefficients for x*-Direction*

.0	.0	.2	.2	.2	.2	.2	.2	.2	.2
.0	.0	.2	.2	.2	.2	.2	.2	.2	.2
.0	.0	.2	.2	.0	.0	.2	.2	.2	.2
.0	.0	.2	.2	.0	.0	.2	.2	.2	.2
.2	.2	.2	.2	.0	.0	.2	.2	.2	.2
.0	.0	.2	.2	.0	.0	.2	.2	.2	.2
.0	.0	.2	.2	.0	.0	.2	.2	.2	.2
.0	.0	2	.2	.2	.2	.2	.2	.2	.2
.0	.0	.2	.2	.2	.2	.2	.2	.2	.2
.0	.0	.2	.2	.2	.2	.2	.2	.2	.2

Diffusion Coefficients for y*-Direction*

.0	.0	.2	.2	.2	.2	.2	.2	.2	.2
.0	.0	.2	.2	.2	.2	.2	.2	.2	.2
.0	.0	.2	.2	.2	.2	.2	.2	.2	.2
.0	.0	.2	.2	.2	.2	.2	.2	.2	.2
.0	.0	.2	.2	.2	.2	.2	.2	.2	.2
.0	.0	.2	.2	.2	.2	.2	.2	.2	.2
.0	.0	.2	.2	.2	.2	.2	.2	.2	.2
.0	.0	.2	.2	.2	.2	.2	.2	.2	.2
.0	.0	.2	.2	.2	.2	.2	.2	.2	.2
.0	.0	.2	.2	.2	.2	.2	.2	.2	.2

[a]In the x-direction material can diffuse along a narrow channel. Movement is slowed by zone of zero x-direction diffusion coefficients beyond the mouth of the channel. In the y-direction diffusion coefficients are uniform except in the first two columns.

boundary toward the interior of the grid. The channel consists of a row of diffusion coefficients in the x-direction with a value of 0.2, surrounded by cells with values of 0.0 in both x- and y-directions. At the "mouth" of the channel diffusion is permitted in both x- and y-directions, for simplicity being set everywhere to 0.2. To complicate the diffusion pattern, however, an "obstruction" consisting of a group of 10 cells with zero diffusion coefficients in the x-direction has been placed in the center of the area. In the run an initial concentration value of 5000 units was assumed in a single cell, and all other cells contained initial values of zero. During the run the concentra-

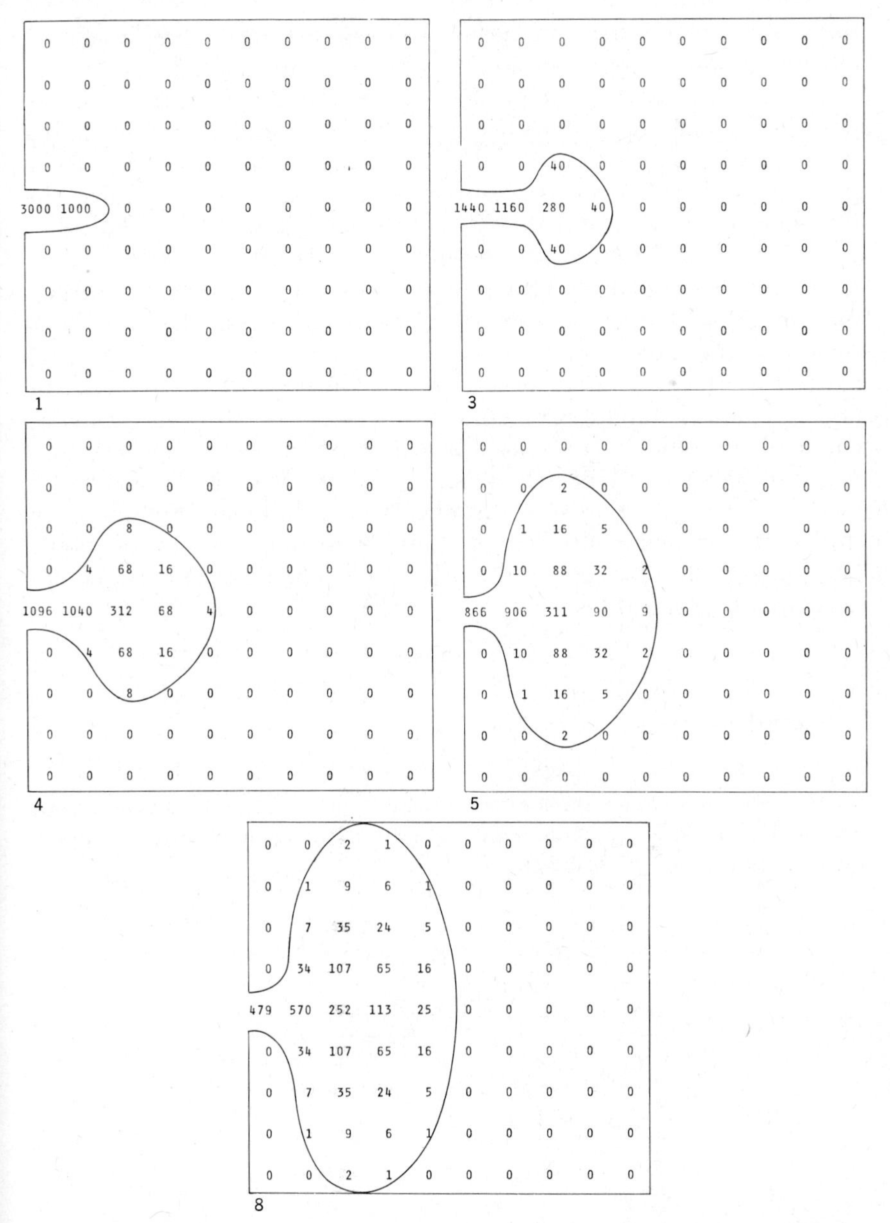

Figure 6-23 Output from two-dimensional heterogeneous diffusion model at different time steps. Diffusion coefficients used as input are listed in Table 6-11. Initial concentration (time 0) consists of 5000 in cell in center of left column (cell 5, 1). Continual loss of material at left boundary is responsible for progressively smaller totals of material inside boundaries.

tions are affected by the permeabilities. Material also "escapes" outward from the boundary and is "lost" from the system.

The results of the computer run suggest the capability of the program in modeling various geological situations. For example, the channel might be regarded as a permeable fracture from which ions in a hydrothermal solution diffuse. Beyond the fracture the diffusion pattern is sensitive to variations in lithology. An alternative geological representation might be the diffusion of sediment from a river mouth during the low river stage. Under these conditions the rate of discharge and velocity of flow in the river channel would be small, and a diffusion model might present a reasonable approximation for the dispersal of sediment. The barrier in front of the channel mouth might represent a river-mouth bar. The diffusion coefficients could be used to approximate the geographic arrangemand of water and land (river banks, barrier bars) and conceivably could be used to represent the effects of turbulence. For example, the sediment-dispersal effect of a highly turbulent surf zone could be approximated by setting the diffusion coefficients in certain cells at relatively high values.

Problems

Problem 6-1

Given the following values of the velocity potential in a two-dimensional grid, calculate velocity components and resultant velocity vectors, similar to those shown in Figures 6-2 and 6-3.

−27	−69	−159	−204	−223
16	−22	−103	−129	−143
96	68	0	−66	−77
205	130	77	7	−21
288	171	95	36	8

Problem 6-2

Write out the Laplace equation for three-dimensional flow in finite-difference form. Use i, j, k as the indices for the y-, x-, and z-directions, respectively. Assume that $\Delta x \neq \Delta y \neq \Delta z$.

Problem 6-3

Write a FORTRAN program for obtaining velocity potentials over a three-dimensional grid, assuming that the boundary potentials are known. This should include some means of representing "slices" through the grid to show the distribution of the velocity potential in any plane. It is suggested that a separate subroutine be used to display the velocity potentials.

Problem 6-4

Obtain finite-difference equations for calculating the velocity potential ϕ at a cell (i, j) in a two-dimensional grid where (a) cell (i, j) lies at the lower boundary of the grid (not in a corner) and (b) where cell (i, j) lies at the upper left corner of the grid. Assume that velocity components normal to the outer side of each boundary cell are assigned as boundary values.

Problem 6-5

Using either the FORTRAN program listed in Table 6-5 or a modification of it, conduct several experiments on two-dimensional potential flow. Obtain solutions for various boundary conditions.

Problem 6-6

Using the diffusion equation for one spatial dimension in finite-difference form, determine the effect of variation in the factor $k\Delta t/(\Delta x)^2$.

Problem 6-7

Write a FORTRAN program for diffusion in one spatial dimension similar to that of Table 6-7, but employ the boundary condition $\partial c/\partial x = 0$.

Problem 6-8

Write out the diffusion equation for three spatial dimensions in finite-difference form. How may this be simplified if the diffusion coefficient is the same in all three directions (isotropic)?

Annotated Bibliography

Crank, J., 1956, *The Mathematics of Diffusion*, Oxford University Press, Oxford, 347 pp.

Authoritative treatment, with a particularly valuable chapter on finite-difference methods.

Fayers, F. J., and Sheldon, J. W., 1962, "The Use of a High-Speed Digital Computer in the Study of the Hydrodynamics of Geologic Basins", *Journal of Geophysical Research*, v. 67, pp. 2421–2431.

Discusses the derivation of the basic equations for flow through porous media. These include forms of the Laplace equation, the Poisson equation, and the diffusion equation. Methods of solving these equations numerically are introduced, and a three-dimensional-flow problem is illustrated. This paper is quite technical but should be understood without too much difficulty after reading Chapter 6 of this volume.

Hagerstrand, T., 1968, "A Monte Carlo Approach to Diffusion," in *Spatial Analysis: a Reader in Statistical Geography*, B. J. Berry and D. F. Marble, eds., Prentice-Hall, Englewood Cliffs, N.J., pp. 368–384.

Develops a stochastic model for spatial diffusion of ideas through geographic regions. The materials accounting and subdivision of space using a meshwork of cells is similar to the deterministic models discussed in this chapter. The diffusion mechanism is stochastic, however, employing frequency distributions for assigning probabilities of transport from cell to cell.

Howard, J. C., 1968, "Monte Carlo Simulation Model for Piercement Salt Domes," Computer Contribution 22, Kansas Geological Survey, pp. 22–34.

This model employs a stochastic mechanism for simulating the rise of salt in a piercement dome, and is discussed in some detail in Chapter 10.

Jost, W., 1952, *Diffusion in Solids, Liquids, and Gases*, Academic Press, New York, 558 pp.

A classic book on diffusion. The first chapter contains a thorough treatment of the basic laws of diffusion. The inclusion of terms for the influence of external force is also introduced.

McCalla, T. R., 1967, *Introduction to Numerical Methods and FORTRAN Programming*, John Wiley and Sons, New York, 359 pp.

A valuable book on numerical analysis, containing numerous FORTRAN programs. Chapter 9 presents an authoritative treatment of the numerical solution of differential equations.

Miller, K. S., 1962, *A Short Course in Vector Analysis*, Charles E. Merrill Company, Columbus, Ohio, 104 pp.

A basic text on elements of vector analysis. Vector calculus is introduced in Chapter 2, and the shorthand notation using the "del" operator ∇, involving the terms "gradient," "divergence," and "curl," is clearly explained. These terms are frequently encountered in literature dealing with fluid mechanics.

Murray, S. P., 1967, "Control of Grain Dispersion by Particle Size and Wave State," *Journal of Geology*, v. 75, pp. 612–634.

A study of sand dispersal under the action of shoaling waves, which demonstrates the applicability of a diffusion model to sediment transport.

Pollack, H. N., 1967, *Deterministic Modeling in Geology and Geophysics*, (mimeographed notes for an engineering summer conference course given at the University of Michigan, Ann Arbor, May 1967), 21 pp.

These notes may be difficult to obtain, but they contain an excellent introduction to the Laplace and Poisson equations for representing the potential flow of electricity and heat as well as potential fluid flow. The numerical solution of these equations is briefly described, and a model for the transport of solutes via fluid flow is developed.

Pollack, H. N., 1968, "On the Interpretation of State Vectors and Local Transformation Operators," Computer Contribution 22, Kansas Geological Survey, pp. 47–51.

Beneath this obscure-sounding title is a short and interesting discussion of the finite-difference solution of the diffusion equation and some other differential equations. An application to the erosional cutting of canyons is developed.

Pollack, H. N., 1969, "A Numerical Model of the Grand Canyon," in *Geology and Natural History of the Grand Canyon Region*, Four Corners Geological Society Guidebook to the Fifth Field Conference, D. C. Baars, ed., pp. 61–62.

Describes a mathematical model for simulating the processes of fluvial downcutting and valley widening in strata with varying resistance to erosion. The basic equation that is employed is a modified form of the diffusion equation in one dimension.

Remson, I., Molz, F. J., and Hornberger, G. M., *Numerical Methods in Subsurface Hydrology*, John Wiley and Sons, New York (in press).

Treats a wide variety of numerical methods useful for solving problems in groundwater flow.

Rouse, H., 1931, *Fluid Mechanics for Hydraulic Engineers*, Dover Publications, New York, 422 pp.

The paperback edition of this engineering classic contains excellent sections on the elementary principles of fluid flow and the derivation of the equations of motion for ideal and viscous fluids.

Stagg, K. G., and Cheung, Y. K., 1965, "Finite Elements in the Study of Field Problems," *The Engineer*, pp. 507–510.

Potential-flow problems can be solved by finite-element methods as well as by finite-difference methods (see next reference).

Stagg, K. G., and Zienkiewicz, O. C., (editors), 1968, *Rock Mechanics in Engineering Practice*, John Wiley and Sons, New York, 442 pp.

Chapter 8 deals with the application of finite-element methods (see reference to Zienkiewicz and Cheung, 1967) to problems of stress-strain analysis. Elastic and plastic behavior of rocks can be numerically simulated by this method. The finite-element method is similar to the finite-difference method for solving flow equations in that the spatial dimensions are subdivided into discrete elements. The finite-element method is more flexible in some ways, however, for each element can be different in shape and size (Figure 6-24). Finite-element methods have been applied by engineers for studying stress-strain relationships in rock masses, and by some geologists for studying folding.

Tobler, W., 1967, "Of Maps and Matrices," *Journal of Regional Science*, v. 7, pp. 275–280.

A stimulating discussion which shows a link between stochastic and deterministic methods of modeling diffusion. Matrices obtained by finite-difference solution of diffusion equations are similar to Markov transition matrices.

Vanoni, V. A., Brooks, N. H., and Kennedy, J. F., 1960, *Lecture Notes on Sediment Transportation and Channel Stability*, W. M. Keck Laboratory for Hydraulics and Water Research, California Institute of Technology Report No. KH-R-1.

Weinaug, C. F., 1968, "Mathematical Modeling of Reservoir Behavior," Computer Contribution 22, Kansas Geological Survey, pp. 52–58.

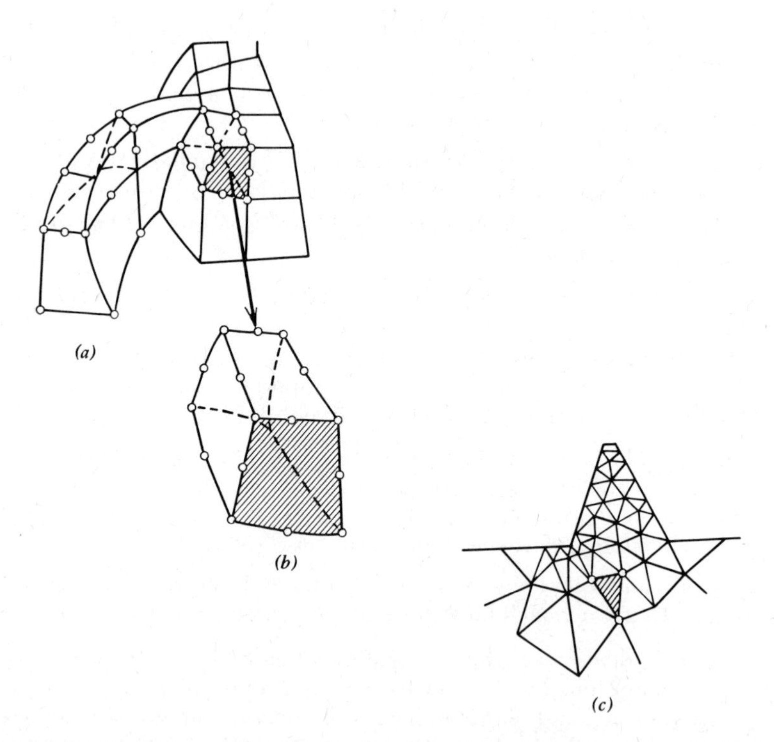

Figure 6-24 Diagrams to show body of rock divided into finite elements: (*a*) three-dimensional body composed of trapezoids of various shapes and sizes; (*b*) single trapezoid enlarged from (*a*); (*c*) two-dimensional section of another body, showing triangle-shaped finite elements of different sizes. From Stagg and Zienkiewicz (1968).

Written by a petroleum engineer, this paper shows how the finite-difference equations for the flow of gas through a porous medium are derived and solved numerically. Equations for various boundary conditions are also introduced.

Welch, J. E., Harlow, F. H., Shannon, J. P., and Daly, B. J., 1966, *The MAC Method: A Computing Technique for Solving Viscous, Incompressible, Transient Fluid Flow Problems Involving Free Surfaces*, Report LA-3425, Los Alamos Scientific Laboratory, Los Alamos, New Mexico, 146 pp.

Discusses in detail how the finite-difference form of the Navier-Stokes equations is obtained and solved numerically. This is a valuable reference for those who wish to use finite-difference methods for simulating viscous-flow phenomena.

Winslow, J. D., and Nuzman, C. E., 1966, *Electronic Simulation of Groundwater Hydrology in the Kansas River Valley Near Topeka, Kansas*, Special Distribution Publication 29, Kansas Geological Survey, 24 pp.

Shows how both the flow of an electric current and the flow of water through a porous medium may be represented by the Laplace equation. Description of an electric analog simulation model for groundwater movement comprises most of the paper.

Witherspoon, P. A., Javandel, I., and Neuman, S. P., 1968, "Use of Finite-Element Method in Solving Transient Flow Problems in Aquifer Systems," in *Use of Analog and Digital Computers in Hydrology*, International Association of Scientific Hydrology, UNESCO, Publication No. 81, v. 2, pp. 687–698.

Shows how groundwater-flow problems may be modeled by using the finite-element method. Results obtained numerically compare favorably with those obtained by exact analytical means.

Zienkiewicz, O. C., and Cheung, Y. K., 1967, *The Finite-Element Method in Structural and Continuum Mechanics*, McGraw-Hill, New York, 274 pp.

Detailed treatment of finite-element methods.

CHAPTER 7

System Control

This chapter deals with the application of some elementary principles of system control. The technology of control systems is highly advanced in industry. Most industrial or technological control-system concepts deal with the short-term response of electrical, thermal, hydraulic, or mechanical systems. The control elements in these systems can be treated theoretically and experimentally. Geological dynamic systems also possess control elements, but the control elements are difficult to define and analyze in terms of formal control theory. Consider, for example, the conceptual model of shallow-water marine sedimentation shown in Figure 7-1. Accumulation of sediment affects isostatic adjustment, which in turn affects basin configuration, affecting accumulation of sediment, and so on. The components of this loop tend to regulate each other. In addition, there are other loops that provide a complex network of interdependent, overlapping control systems. Because of the continuously linked nature of the model, most of its components may be regarded as control elements.

The distinction between an endogenous and an exogenous component in a dynamic system depends on the presence of an interdependent relationship—that is, on the presence of a feedback loop or loops. Exogenous components, by definition, are not linked by loops, whereas endogenous components are. Depending on the

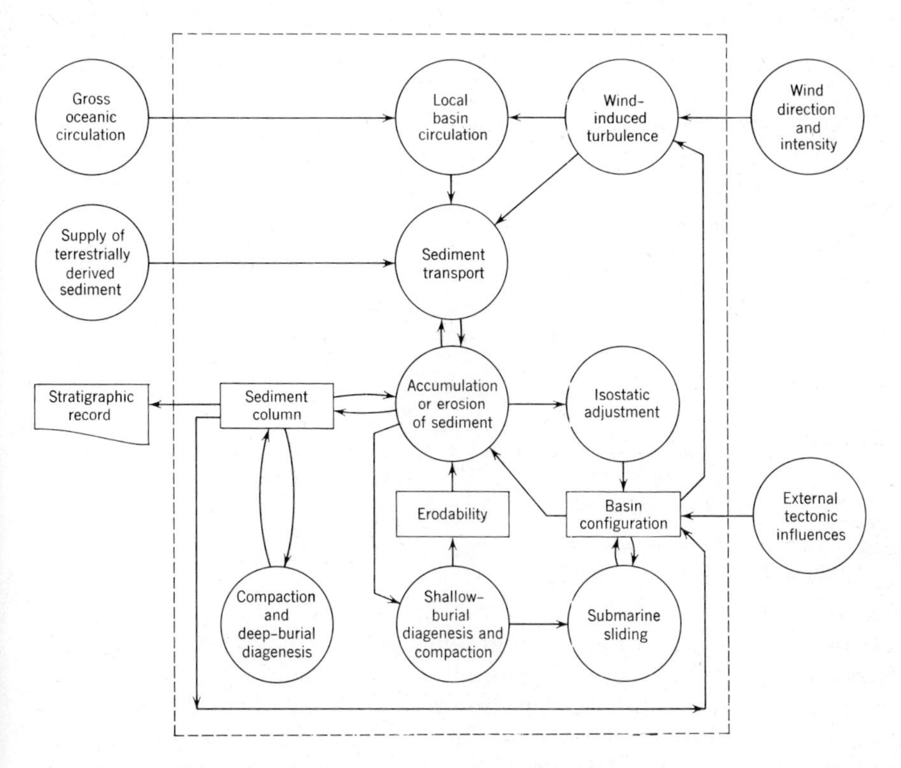

Figure 7-1 Theoretical system diagram showing principal processes that affect shallow-water marine sedimentation. Dashed outline separates exogenous processes that supply inputs to system but do not receive feedback from system. Endogenous processes are inside dashed outline. From Harbaugh and Merriam (1968).

degree of complexity, the loops may be direct and immediately obvious, or they may be highly indirect. A loop linking reactions in a stream system is shown in Figure 7-2. Overall, the loop may be regarded as a negative-feedback loop because it represents a set of relationships that tend to negate each other. Decreased velocity in a sediment-carrying stream results in the deposition of sediment, which in turn causes steepening of the stream gradient, which in turn causes an increase in velocity, which causes an increase in sediment-carrying capacity, and so on. The system is continuously self-regulating.

Positive feedback is also present in dynamic geologic systems. An example is shown in Figure 7-3, which represents a reef or marine bank deposit of limestone forming in a shallow sea. Many organisms that secrete calcium carbonate are depth dependent. For example,

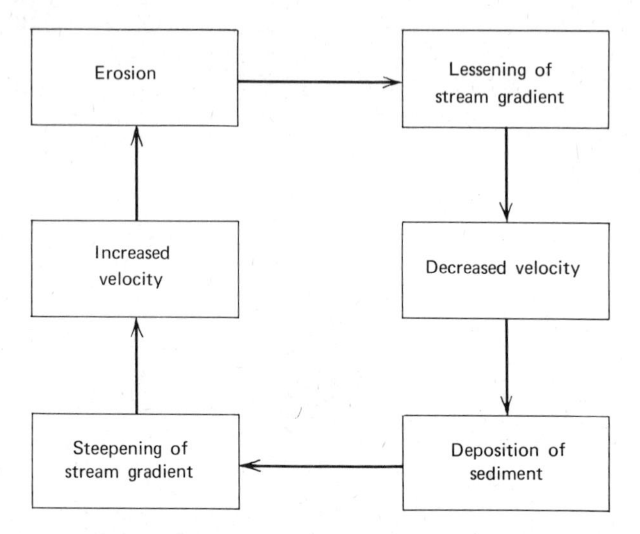

Figure 7-2 Negative feedback loop linking velocity changes, stream gradient, erosion, and deposition within hypothetical stream system.

calcareous algae require light and therefore tend to grow most vigorously and secrete calcium carbonate at maximum rates at relatively shallow depths. Accordingly calcareous algae growing on a shallowly submerged bank produce carbonate at a greater rate than algae growing at greater depths, creating a positive-feedback relationship. Increased algal productivity brings about a decrease in depth, which in turn causes increased productivity, and so on, until some limiting depth is reached where growth ceases. Wave erosion close to sea level, coupled with the inability of carbonate-secreting organisms to grow above sea level, provides negative feedback. The elevation of the top of the bank with respect to sea level determines whether positive or negative feedback prevails (Figure 7-4).

CONTROL IN A SIMPLE SEDIMENTATION MODEL

At this point it is convenient to discuss a relatively simple conceptual model of shallow-water sedimentation processes. As a first step we can envision the model as a series of simple, abstract system diagrams that distinguish between exogenous and endogenous components.

The simplest model (Figure 7-5*a*) is one in which there is neither feedback nor interaction between exogenous inputs. A slightly more complex model (Figure 7-5*b*) lacks feedback but incorporates interac-

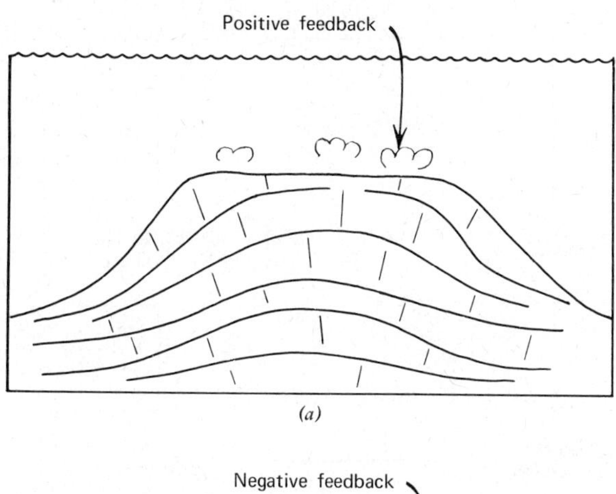

(a)

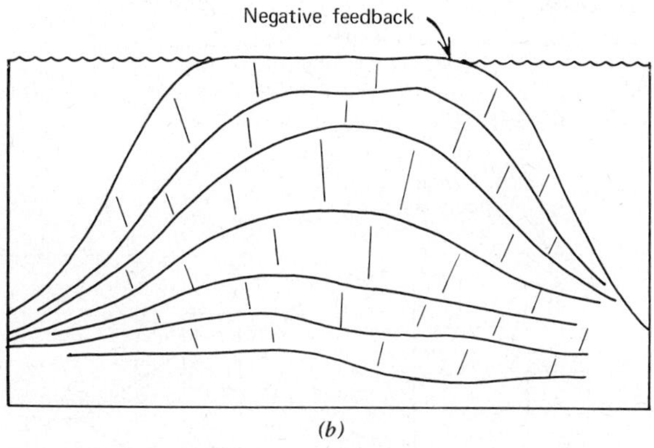

(b)

Figure 7-3 Feedback relationships linking depth with the productivity of calcareous algae growing on submerged marine bank: (*a*) vigorous upward growth of calcareous algae, providing increasingly favorable environment; (*b*) growth checked on reaching sea level.

tion between multiple exogenous inputs. In Figure 7-5*a* and *b* the exogenous inputs exert control over the system since there is no feedback and therefore no other controlling influence. Figure 7-5*c*, on the other hand, includes feedback. If the feedback loop symbolizes negative feedback, then we can regard one of the components as exerting control over the other, and the exogenous inputs provide "disturbances" to the system. Although the conceptual models represented by the diagrams are very simple, they illustrate two important aspects of control—namely, *open-loop control* (Figure 7-5*a* and *b*) and *closed-loop control* (Figure 7-5*c*). We shall return to these aspects later.

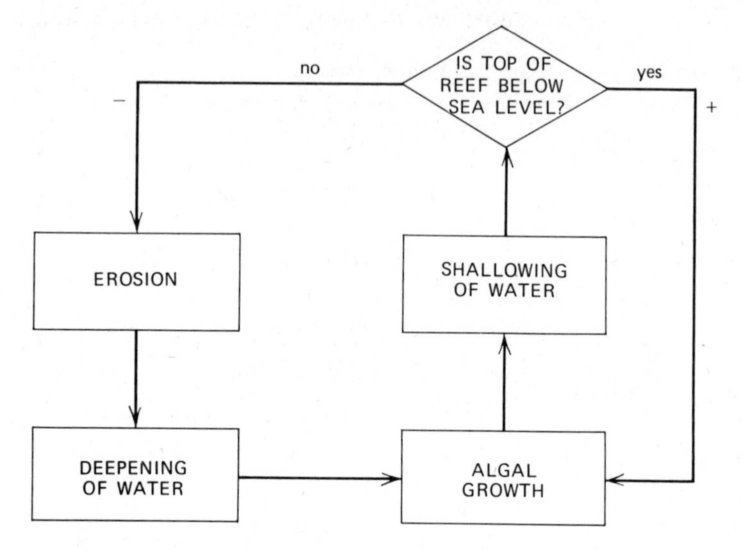

Figure 7-4 Flow chart representing the role of sea level as controller in regulating algal growth on hypothetical calcareous reef.

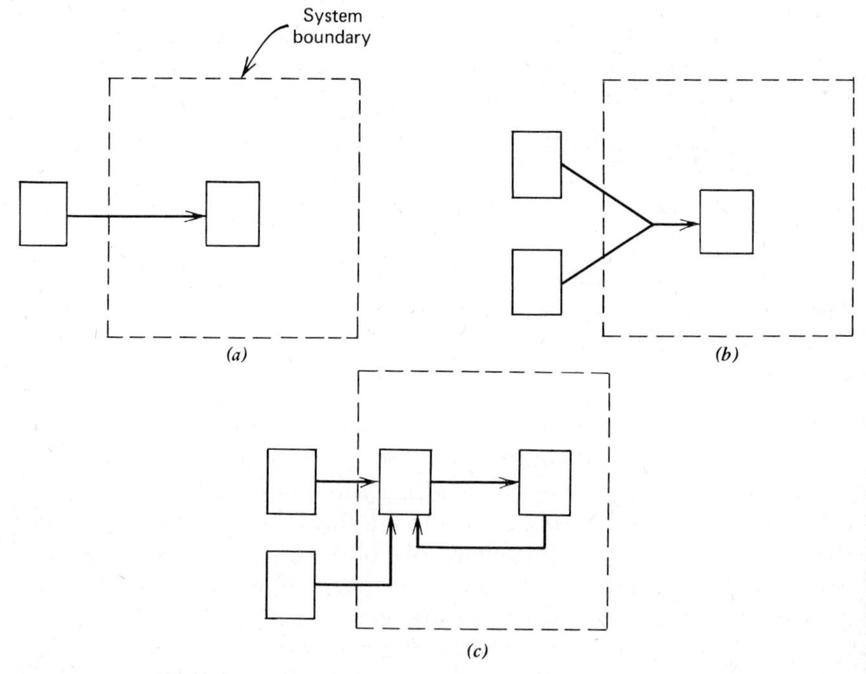

Figure 7-5 Diagram to illustrate effects of (*a*) single exogenous input without feedback within system (open-loop control); (*b*) multiple exogenous inputs which interact but in which there is no feedback; (*c*) internal feedback (closed-loop control) in addition to exogenous inputs.

Now let us adapt these elementary systems concepts to a simple conceptual model involving sedimentation in a shallow marine basin (Figure 7-6). Sediment entering the basin is an exogenous input that directly controls the amount of sediment transported within the basin during each interval of time. This represents an open-loop-control component of the model. In turn the transported-sediment component forms an input to the component labeled base-level control. Depth to base level is the second exogenous input, and it forms an open-loop-control link with the base-level-control component. The base-level-control component, however, is linked via a closed loop with sediment deposition and water depth.

This conceptual model has been transformed into a simple computer model, which is explained in detail in Chapter 9 (the FORTRAN program containing the model is listed in Table 9-3). Output from the model is in the form of printed two-dimensional vertical slices through the sedimentary basin (Figures 7-7, 7-9, and 7-10). The basin is represented by a series of individual columns which are assumed to behave independently of each other. Sediment from the exogenous input source is supplied at the right side of the basin and is moved toward the left. All materials entering the system and manipulated within the system are rigidly accounted for. Two types of sediment, silt and sand, are supplied in equal proportions, and the same amount is supplied during each time increment as the model is advanced forward through time in uniform steps. Each sediment type is assigned an equilibrium water depth. If the depth is equal to or less than the equilibrium depth, no sediment is deposited. The

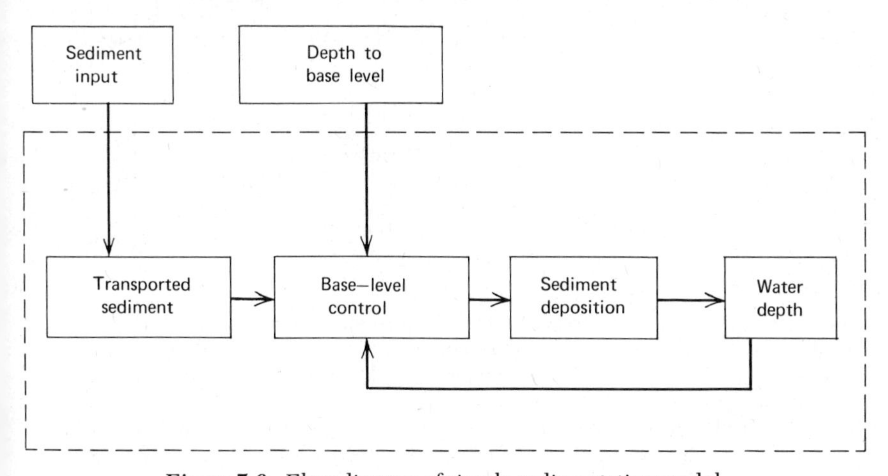

Figure 7-6 Flow diagram of simple sedimentation model.

equilibrium depth for sand is equivalent to the width of three printed characters in a column, and the equilibrium depth for silt is five printed characters. The printed cross sections are turned sideways for viewing. Sand is represented by 0, silt by $, water by I, and the sea floor (on which the sediment is deposited) by <.

The response of the model shown in Figure 7-7 assumes an initial wedge of water into which sediment is progressively deposited. Since there is no provision for subsidence, the form of the deposits reflects the control elements incorporated in the model, plus the initial configuration of the sea floor. The depth to base level provides a minimum depth for sediment accumulation which prohibits deposition of

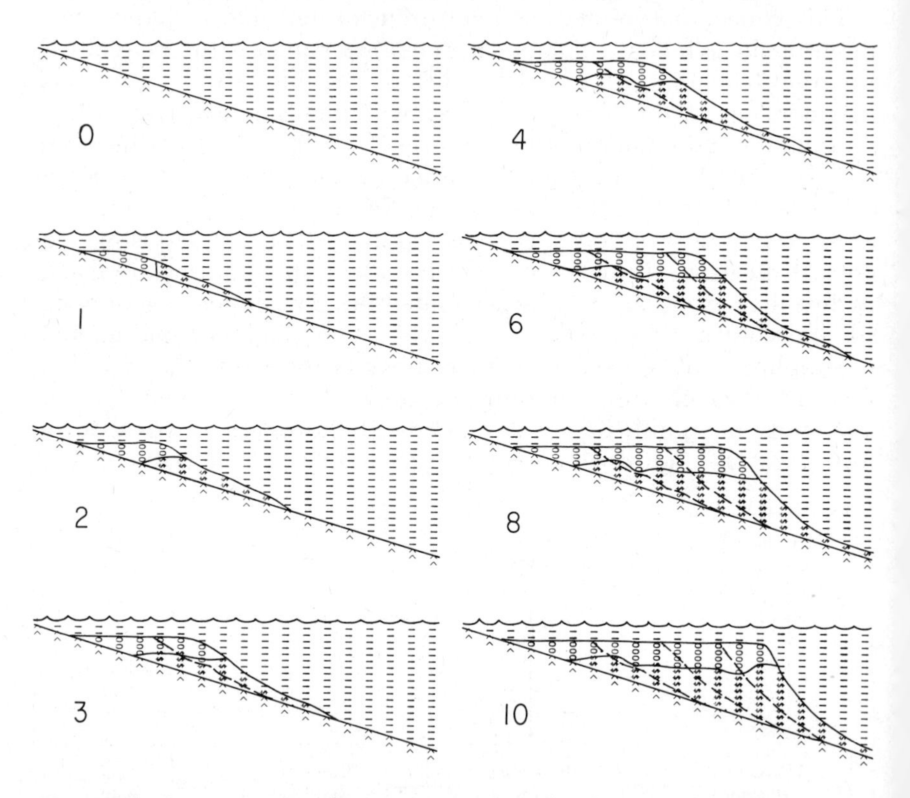

Figure 7-7 Computer-printed cross sections through sedimentary basin that does not subside. Numbers identify time increments. Water is symbolized by prostrate I's, sand by O's, and silt by $'s. Sea floor is presumed to have sloped uniformly to right at start of simulation run. Five "characters worth" of sand and five of silt were supplied from left to right during each time increment. Time lines, which are added every third time increment, are represented by dots. Cross sections, being purely hypothetical, lack scale.

sediment at shallower depths. In addition, the behavior of the base-level-control component is such that silt is transported beyond and deposited at greater depths than sand. The configuration of the deposits is that of a simple, hypothetical regressive sequence, in which sandy facies progressively extend out over silt facies.

Deposition of a large delta deposit unaccompanied by crustal subsidence to preserve isostatic equilibrium might be regarded as extremely unrealistic geologically. The question could be raised, then, as to the behavior of the model if subsidence took place in response to sedimentation. We can revise the model by incorporating a second closed loop that links deposition, subsidence, and water depth (Figure 7-8). If subsidence occurs, the geometrical configuration of the deposits should be strongly influenced. Furthermore the rapidity of response has an important influence. An instantaneous response will cause the depositional facies to assume a geometrical form which differs from that of facies formed under conditions in which there is an appreciable time lag between deposition and subsidence.

The behavior of the sedimentation model when it incorporates subsidence is shown in Figures 7-9 and 7-10. In Figure 7-9 subsidence occurs in each time increment in response to deposition that occurred in each preceding time increment. If the time increments were infinitesimally short, subsidence would respond instantaneously to deposition. Because of the finite nature of each time increment,

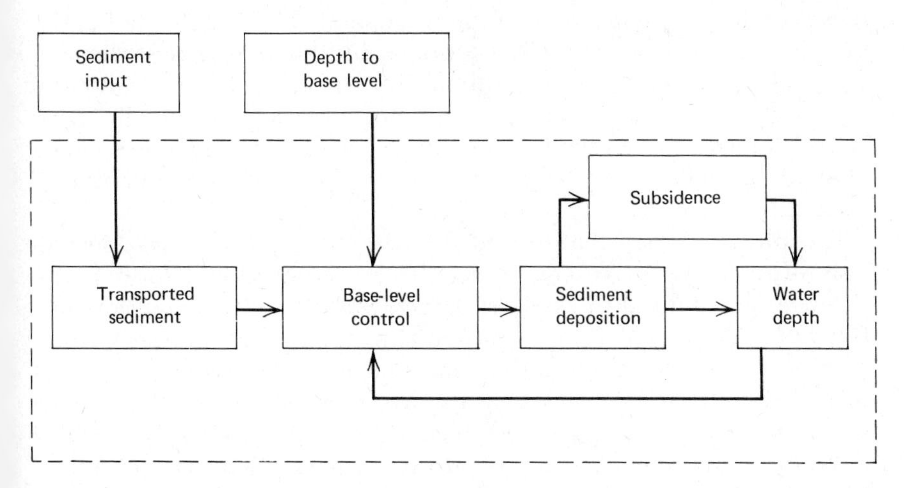

Figure 7-8 Flow diagram of simple sedimentation model that incorporates crustal subsidence.

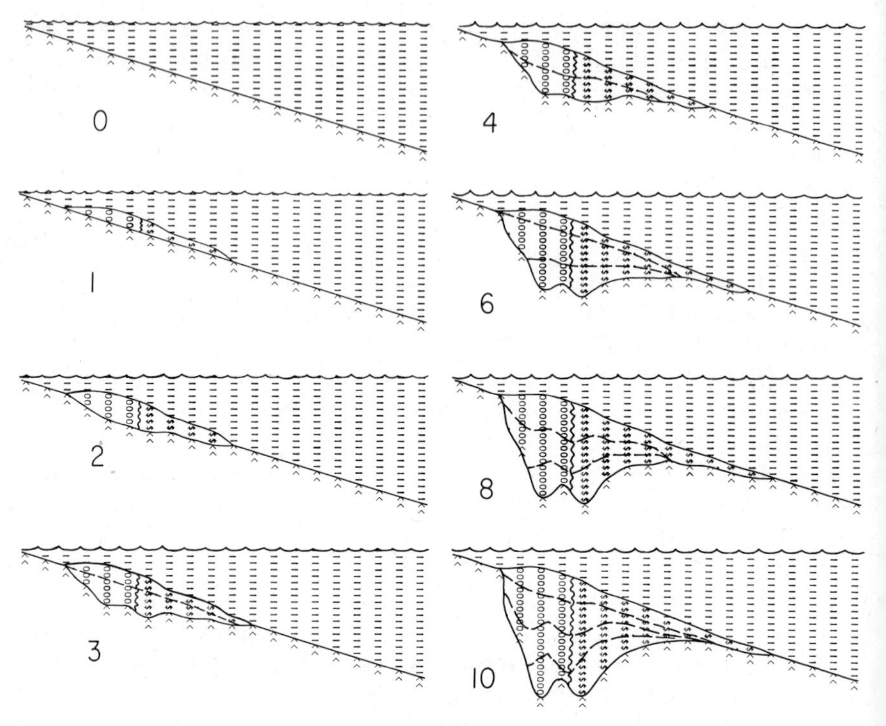

Figure 7-9 Results in which subsidence, following initial deposition, keeps pace with deposition of sediment.

however, there is a lag of one time increment. This is reflected in the modification of the sea-bottom topography during the first time increment (Figure 7-9). In subsequent time increments the sea-bottom topography remains unchanged, although the deposits are progressively deformed as a result of subsidence. The resulting lensing mass of sand and silt has a maximum thickness near the "delta-platform" edge.

It could be argued that the subsidence of the actual crust does lag behind deposition. We can adjust the computer model so that subsidence lags behind by some specified number of time increments. Furthermore the subsidence can be adjusted so that it either occurs continuously (i.e., in each time increment) or periodically (i.e., at intermittent time increments). Figure 7-10 illustrates the results when the model is so adjusted that subsidence occurs only at every third time increment. The resulting sequence consists of interfingering sand and silt deposits, with the locus of maximum thickness farther offshore than that in Figure 7-9.

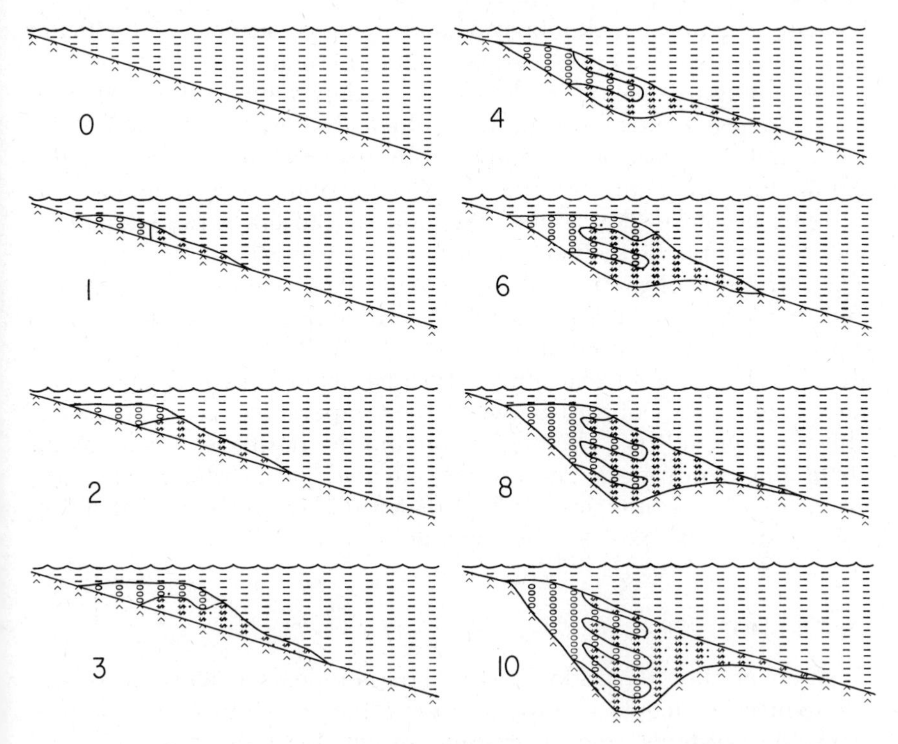

Figure 7-10 Results in which crustal subsidence lags behind deposition. Subsidence compensating for sediment load takes place every third time increment.

This model is described in greater depth in Chapter 9. From this brief illustration, however, it is clear that the control mechanisms of the model exert a strong influence on the behavior of the model. We now discuss some elementary aspects of formal control theory and their relevance to dynamic geological models.

TYPES OF CONTROL APPLICATIONS

Most control theory deals with technological control systems. By a technological control system we refer to control of technological processes employed in heating and air-conditioning systems, flow processes in chemical plants, electrical circuits, compressed-air systems, space vehicles, and so on. The application of control theory to these systems is exceedingly advanced and is supported by extensive mathematical theory, much of which is centered about representation with sets of linear differential equations. Technological control systems embrace on–off control and various types of continuous control. In general, technological control theory is concerned with the short-term response of systems to changes in input.

A challenging task for geologists is to adapt some of the aspects of technological control theory to geological simulation models. Unfortunately application is hampered by the difficulty in defining the parameters of natural systems. The attributes of technological systems—such as flow rates, resistances, capacities, and temperatures—can often be defined and their numerical parameters specified. The attributes of their geological counterparts are difficult to define and to specify, however. One of the problems lies in distinguishing between a controlling variable (or controller) and a controlled variable. In technological systems the distinction is readily apparent—for example, the contrast between the temperature (controlled variable) in a vessel in a chemical plant and the position of a valve in a steam line (controller). The variables in geological systems, however, are normally too interlocked for the distinction between controlling and controlled variables to be meaningful.

OPEN-LOOP SYSTEMS

Much of the remainder of this chapter deals with some of the elementary concepts of formal control theory. Initially we discuss open-loop systems—that is, systems in which feedback is absent, in contrast to closed-loop systems, in which feedback is present. In an open-loop system (Figure 7-11) the control process consists of modifying an input to the system to yield an output, as a function of input. The relationship of the output variable to the input variable depends solely on the value of the input variable and the nature of the control process, or *transfer function*, that takes place within the control system. We may classify control systems according to order, such as zero order, first order, and so on. The classification is closely related to the classification of differential equations according to order. Open-loop-control theory is described in numerous textbooks on control systems (e.g., Grodins, 1963).

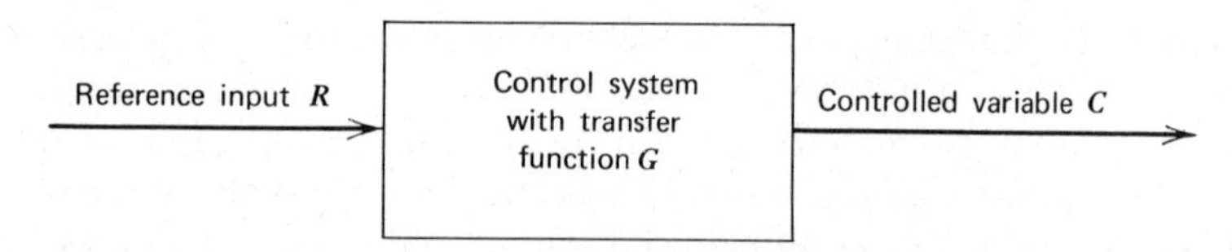

Figure 7-11 Schematic representation of open-loop control system.

A Zero-Order System

A simple physical system is shown in Figure 7-12; it consists of a spring of stiffness K. A force F is applied to the spring, stretching it by an amount y, as measured on the scale. In terms of our schematic open-loop-system diagram (Figure 7-11), force F is the input variable, stiffness is an attribute of the transfer function of the control system, and the amount of stretching is the controlled variable. When no force is applied, the spring is at its resting position, where the recorded value of y is zero. Furthermore we shall understand that force F and response y are functions of time.

The spring can be thought of as a dynamic system. To obtain the equation of motion we employ a basic law of mechanics that states that at each moment of time the force that is applied is equal in magnitude, and opposite in direction, to the opposing force. The system has only one component, which provides the opposing force, the spring itself. Hooke's law relates the stretching of an ideal spring to the force applied, the amount of stretching being directly proportional to the force applied. Thus in our system the opposing force generated within the spring, F_k, is equal to the applied force F and is

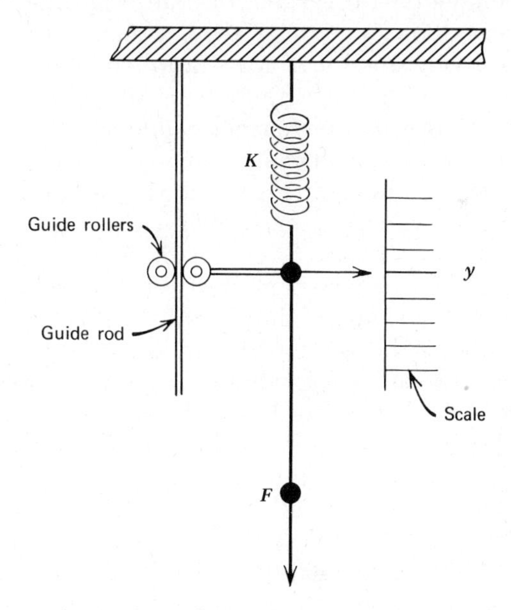

Figure 7-12 Zero-order mechanical system consisting of spring of stiffness K, attached to rigid support above, to which force F is applied. Amount of stretching, y, is recorded on scale. Spring moves only in the vertical direction, being guided by rollers and vertical rod. Friction is assumed to be absent.

proportional to the displacement y. Since $F_k = F$, we need not distinguish F_k from F. Thus we may write

$$F = Ky \tag{7-1}$$

where F = either the opposing or applied force
K = stiffness of spring
y = displacement (amount of stretching) of spring

The equation of motion—that is, the amount of stretching of the spring—is obtained by rearranging the equation:

$$y = \left[\frac{1}{K}\right]F \tag{7-2}$$

We have placed the reciprocal of K in brackets to emphasize that it represents the system's transfer function, which we define here as the quantity by which the system multiplies the input to generate the output or, alternatively, as the ratio of output to input. This direct proportionality defines our spring system as being a zero-order system.

The significance of its zero order will be apparent if we consider the system's behavior with respect to time. Figure 7-13 relates the response of the system (amount of displacement of the spring) that accords with a given amount of force applied. When the input, or forcing, function is varied as a single step instantaneously jumping from zero force F_0 to some prescribed value F_1, a proportional displacement of the spring also occurs instantaneously. Although such simple graphs seem to belabor the obvious, they illustrate concepts that are less simple when higher-order systems are concerned. The point to note in comparing the forcing and response diagrams is that displacement y follows F instantaneously. There is no lag and there is no difference in the shape of the curves. An effect is that the value of y depends only on the value of F and is not at all dependent on time. If we graph Equation 7-2, where y is a function of F, we obtain a plot (Figure 7-14) that is valid for all instants of time. To put it another way, the functional relationship as expressed by Equation 7-2 is such that the "input" (the applied force) is multiplied by a constant to produce the "output," thereby changing its size but not its timing or "shape." We may speak of this relationship as the *gain* of the system, with the understanding that the gain may be greater or less than unity.

We have labored over the description of this exceedingly simple system to develop the principles involved. We encounter zero-order open-loop systems as components in many simulation models where

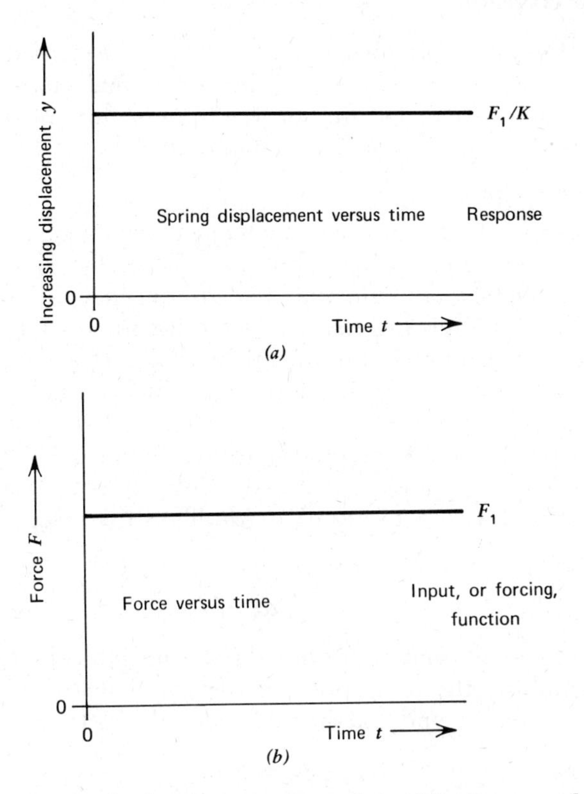

Figure 7-13 Two graphs that illustrate time relationships in zero-order mechanical system consisting of frictionless stretched spring: (*a*) displacement y (stretching) of spring versus time; (*b* force F versus time.

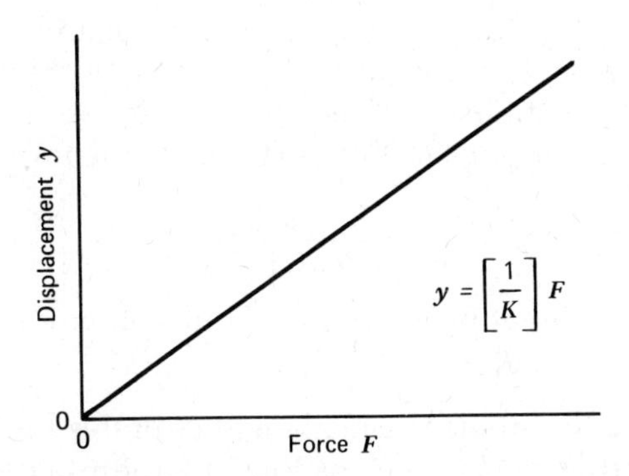

Figure 7-14 Graph of equation relating spring displacement y to force F. Slope of curve is $1/K$, which may be regarded as gain of system.

the value of the output of a particular component (e.g., the amount of sand in the transported sediment of the conceptual model shown in Figure 7-6) is a linear proportion of the input (sediment supplied) to the component and there is no time dependence.

A First-Order System

The analysis of the zero-order mechanical spring system outlined above has the merit of simplicity, but it is not wholly realistic mechically because friction is neglected. The rollers that move the guide rod (Figure 7-12) would necessarily introduce some friction into the system. There is also friction in the spring itself. If we consider friction, there is a second opposing force F_r, which is the viscous frictional resistance of the system. Viscous frictional resistance can be regarded as proportional to velocity (there being no viscous resistance when the components are at rest). Velocity is the derivative of displacement with respect to time, dy/dt. Thus the viscous frictional resistance is

$$F_r = R\frac{dy}{dt} \tag{7-3}$$

where R is a constant relating frictional resistance to velocity.

We can combine the two opposing forces—that due to the resistance of the spring, F_k, and that due to frictional resistance:

$$F_r + F_k = F \tag{7-4}$$

Substituting Equations 7-3 and 7-1 into 7-4, we obtain a new equation of motion

$$R\frac{dy}{dt} + Ky = F \tag{7-5}$$

Equation 7-5 is a first-order differential equation. We shall not dwell on the details, but it is possible to obtain a solution consisting of a functional relationship, free of derivatives, that relates the system's input and output variables to time t in the form

$$y = \left[\frac{1/K}{\tau s + 1}\right]F \tag{7-6}$$

where s = symbol denoting operation of differentiation [$s(y) \equiv dy/dt$]
τ = ratio of R to K

Readers wishing to pursue the mathematics of this topic are referred to books on control theory, such as that of Grodins (1963). As before, the expression in the square brackets is the transfer function, which, when multiplied by the input, yields the output.

At this point we can examine the behavior of the first-order system, employing again an input that changes in an instantaneous step (*step function*), as we did with the zero-order system graphed in Figure 7-13. The step-function forcing graph for a particular value of τ is shown in Figure 7-15. Again, we can specify the same input conditions as before—namely, that $F = 0$ at $t = 0$ and that $F = F_1$ when t is greater than zero. In other words, the value of F changes instantaneously from 0 to F_1. However, because of the effect of viscous resistance due to friction, the displacement y no longer follows F instantaneously. There is a lag before y finally reaches a steady-state value, which is asymptotic to the horizontal line when graphed. Figure 7-15*b* shows the response with a value of τ that is greater than that in Figure 7-15*a*. The final steady-state value is the product of the transfer function and F_1, but the value of y depends on the time when it is measured. The graph of force versus time shows, however, that force F remains at the value of the force applied, F_1, when plotted against time. Thus force F is not a function of time, but the displacement y is a function of time.

The steady-state performance of this first-order mechanical system is the same as that of the zero-order system. If different steady-state values of F are plotted against y, exactly the same straight-line gain relationship will result, as graphed in Figure 7-14. The gain, however, must be distinguished as a steady-state gain, for the relationships of y and F are quite different when non-steady-state conditions exist. The constant term $1/K$ of the transfer function of Equation 7-6 describes the steady-state gain, whereas the denominator $(\tau s + 1)$ pertains to its time-dependent aspects of behavior.

A Second-Order System

The mechanical system described so far is still not wholly realistic. Although it considers the effect of friction, it ignores inertia. Let us revise the system by incorporating a mass that hangs on the spring. The system can be adjusted so that the scale reads zero when only the force of gravity is acting (Figure 7-16), so that the influence of the mass in terms of displacement is perceived only when the system is in motion, and not at rest. When the system is in motion, a third opposing force F_m occurs because of the reaction due to inertia of the mass. This force is equal to mass times acceleration:

$$F_m = MA \tag{7-7}$$

where M = mass
A = acceleration

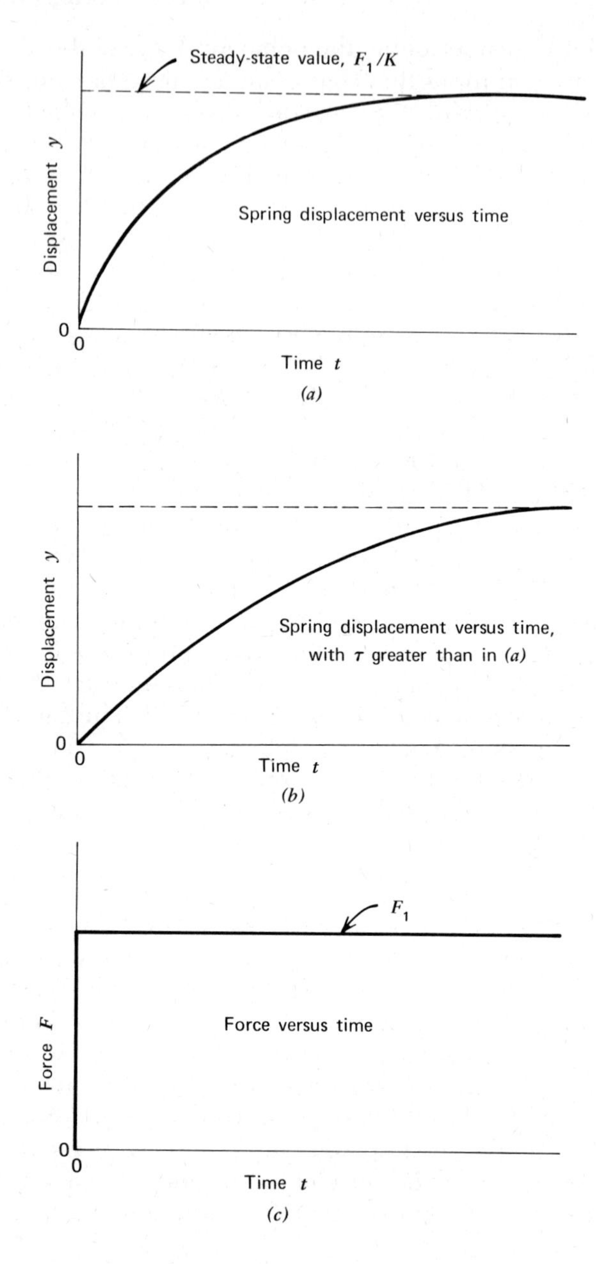

Figure 7-15 Graphs pertaining to first-order mechanical system consisting of stretched spring. Viscous resistance due to friction is present. (*a*) Displacement of spring y versus time t, with particular value of τ; (*b*) displacement of spring with value of τ that is greater than that shown in (*a*); (*c*) graph of force versus time.

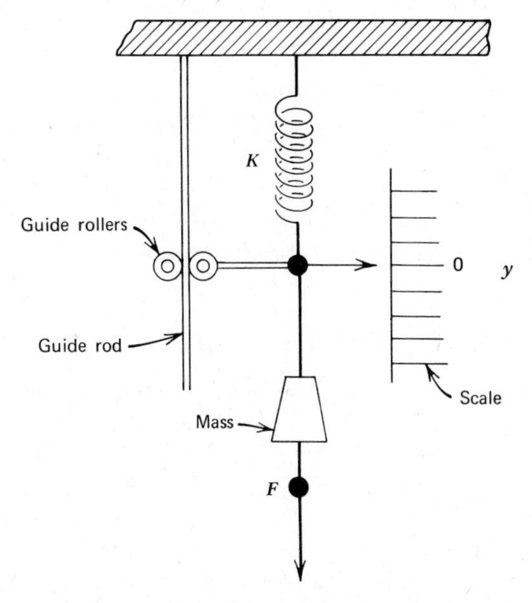

Figure 7-16 Second-order mechanical system consisting of spring to which mass has been attached. Viscous resistance due to friction is assumed to be present.

Acceleration, of course, is the second derivative of displacement with respect to time, or d^2y/dt^2. Thus we may write

$$F_m = M\frac{d^2y}{dt^2} \tag{7-8}$$

and this term may be incorporated into the equation of motion:

$$F_m + F_r + F_k = F \tag{7-9}$$

Substituting Equations 7-8, 7-3, and 7-1 into 7-9, we obtain a second-order differential equation of motion:

$$M\frac{d^2y}{dt^2} + R\frac{dy}{dt} + Ky = F \tag{7-10}$$

where M, R, and K are properties of the system. These can be combined to yield two relational parameters that in turn can be used to define a transfer function. The relational parameters are

$$\omega = \sqrt{\frac{K}{M}} \tag{7-11}$$

$$\zeta = \frac{R}{2KM} \tag{7-12}$$

where ω = natural angular frequency
ζ = damping ratio

After combining into a transfer function with shorthand notation, letting $s(y) \equiv dy/dt$ and $s^2(y) \equiv d^2y/dt^2$ and rearranging so as to solve for y, it is possible to obtain a solution of the form

$$y = \left[\frac{1/K}{(1/\omega^2)s^2 + (2\zeta/\omega)s + 1}\right]F \tag{7-13}$$

where the expression within the brackets is the transfer function. As before, the constant term $1/K$ governs the steady-state gain, which linearly relates F and K (Figure 7-17). The time behavior of the system is another matter, being governed by the denominator of the transfer function.

The behavior of the system can be described by step-function graphs (Figure 7-17). The curve of displacement versus time is strongly affected by the value of the damping ratio ζ. If ζ is unity or greater, the curve approaches the steady-state value asymptotically (Figure 7-17*a*). If, however, ζ is less than unity (but greater than zero), y undergoes a series of progressively damped oscillations around the steady-state value before reaching the steady state (Figure 7-17*b*). The oscillation results from the transfer of energy to and from the spring and the moving mass. Figure 7-18 represents a family of curves illustrating the response of second-order systems to a unit-step impulse to the system. When the unit-step impulse is applied, variations in response y depend on the value of the damping ratio ζ. As ζ approaches zero, the oscillations become extreme.

Formal treatment of oscillation in simple control systems has been applied relatively little in geology. A notable exception is the work by Nye (1965), dealing with the advance and retreat of the lower ends of mountain glaciers as they respond to step variations in the accumulation of snow.

CLOSED-LOOP SYSTEMS

Much of the science of technological control is concerned with closed-loop, or feedback-control, systems. A schematic diagram of an idealized feedback-control system is shown in Figure 7-19. This system has two major components that can be regarded as lesser systems—namely, a controlling system and a controlled system. The

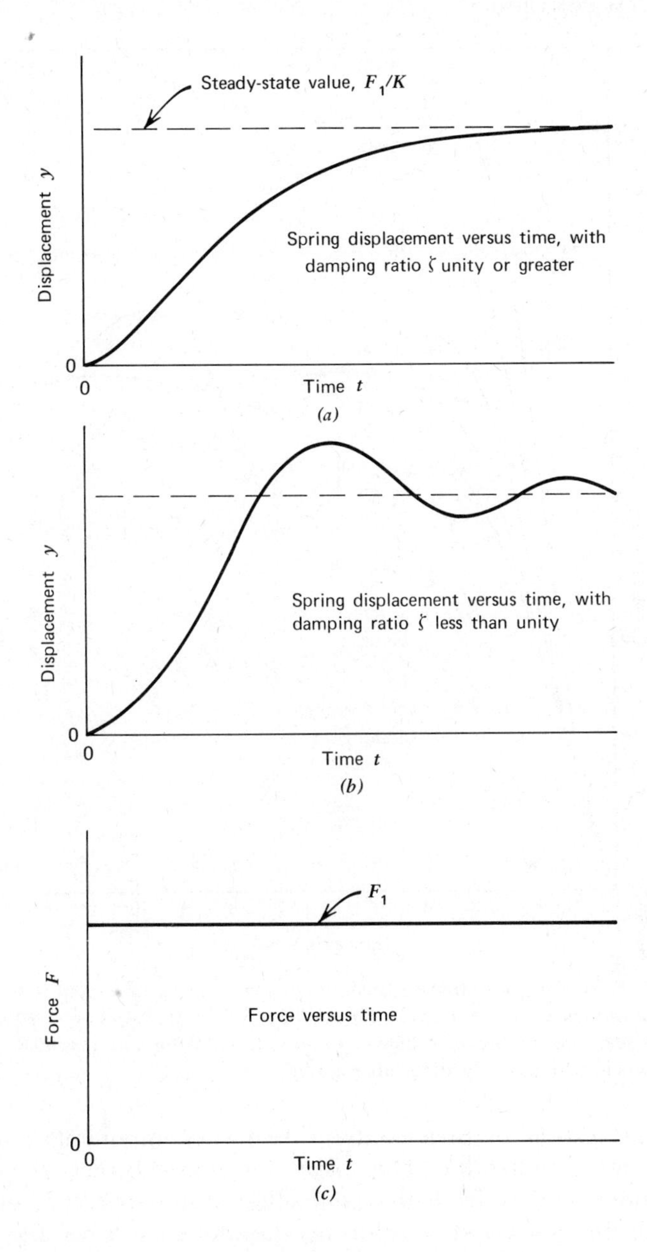

Figure 7-17 Step-function graphs pertaining to second-order mechanical system consisting of stretched spring with attached mass. Viscous resistance due to friction is present. (*a*) Displacement versus time in which damping ratio ζ is unity or greater (*b*) displacement versus time in which damping ratio is less than unity (e.g., 0.5); (*c*) force versus time.

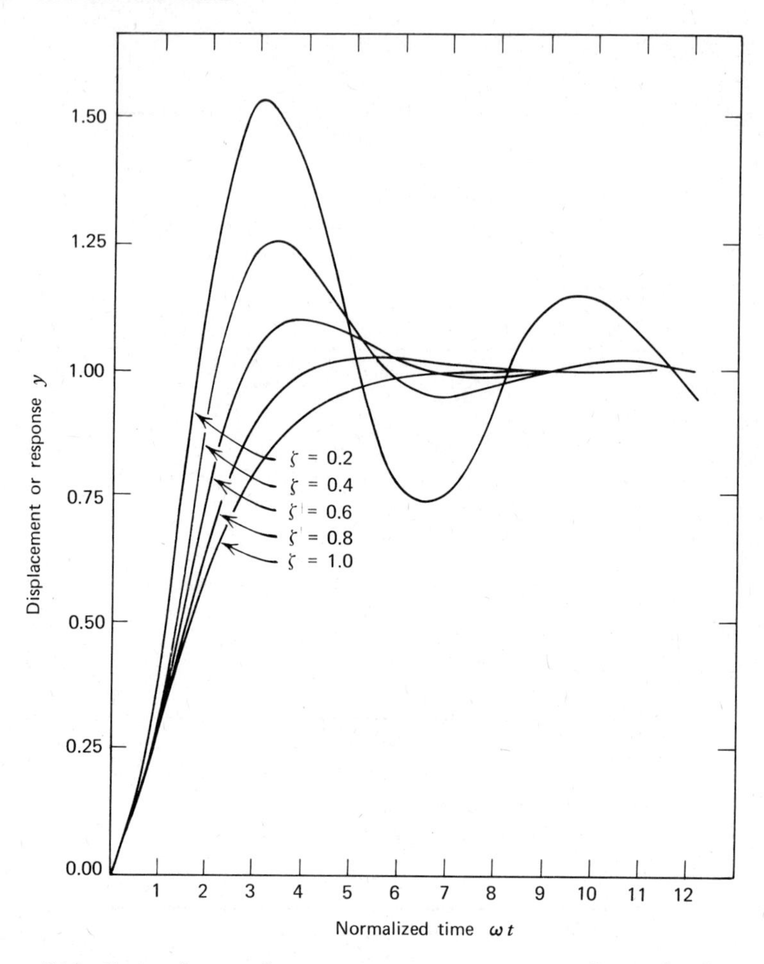

Figure 7-18 Series of curves that portray unit-step response y, of second-order system, for different values of damping ratio ζ ranging from 0.2 to 1.0. Time expressed on normalized scale consisting of radians per unit time. Values of response y are in response units in which steady-state value is 1.0.

controlling system in turn contains two subcomponents, an error detector and a controller. The error detector subtracts the output signal, which comes from the controlled system via the feedback loop, from the command or reference input signal. Since the output signal is subtracted from the input signal, the feedback is negative. The error detector makes the comparison and supplies an error signal to the controller. The controller then responds by sending a "controlling signal" so that the controlled system will respond so as to narrow

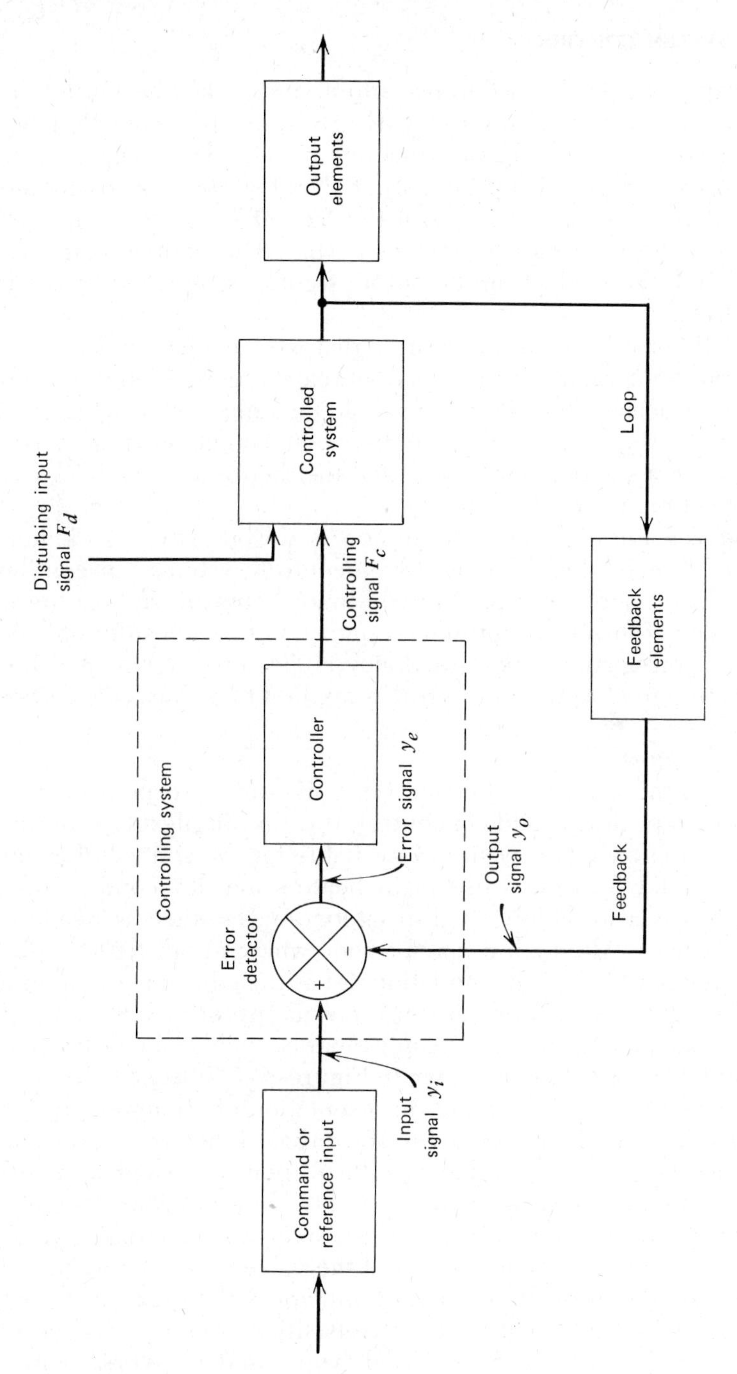

Figure 7-19 Generalized diagram of feedback-control system.

the gap between the reference input signal and the output signal. There may be other input signals acting on the controlled system, which we may regard as disturbing signals. The major job of the control system is to keep the output signal as close as possible to the command or reference input in the face of such disturbances. The only way the controller recognizes the presence of these disturbances is through their effect on the output signal, as reported by the feedback loop.

The theoretical feedback system that we have just outlined can be illustrated in practice by a number of real systems. A familiar example is a thermostat-controlled household heating system. Figure 7-20 illustrates the correspondence between components of a control system for a steam-heated room and the theoretical feedback-control system of Figure 7-19.

The remainder of our discussion is devoted to two elementary forms of technological closed-loop-control systems—discontinuous control (or on–off control) and continuous control. In turn there are a number of forms of continuous control, but we consider only one—namely, proportional control. Lajoy (1954) provides a readable introduction to closed-loop control as applied to technological systems.

On–Off Control

One of the most widely used types of control response is the on–off, or two-position, mode of control. It is the simplest type of control, there being only two positions available for the control element. Its principal technological use is in heating and air conditioning, although it is also widely used in safety devices, such as steam-boiler safety valves. Although it may be somewhat unrealistic to apply on–off control in geological simulation, it has the advantage of simplicity. Furthermore, because of the discrete nature of on–off control, it is readily adapted to digital dynamic models, such as the sedimentation model whose response is shown in Figures 7-7, 7-9, and 7-10.

Figure 7-21 is a schematic representation of an automatic control system for a steam-heated room. If we regard the system as an on–off system, the radiator's valve is either open or closed, causing the room temperature to fluctuate as the valve alternatively opens and closes. Because of the time lag in the response of the radiator's temperature to the opening and closing of the valve, the rise and fall of the room's temperature curve does not coincide with the graph of the valve position (Figure 7-21*b*). The valve-position graph is that of a step function; the transition from closed to open and vice-versa represents a step that can be considered to be instantaneous for our purposes.

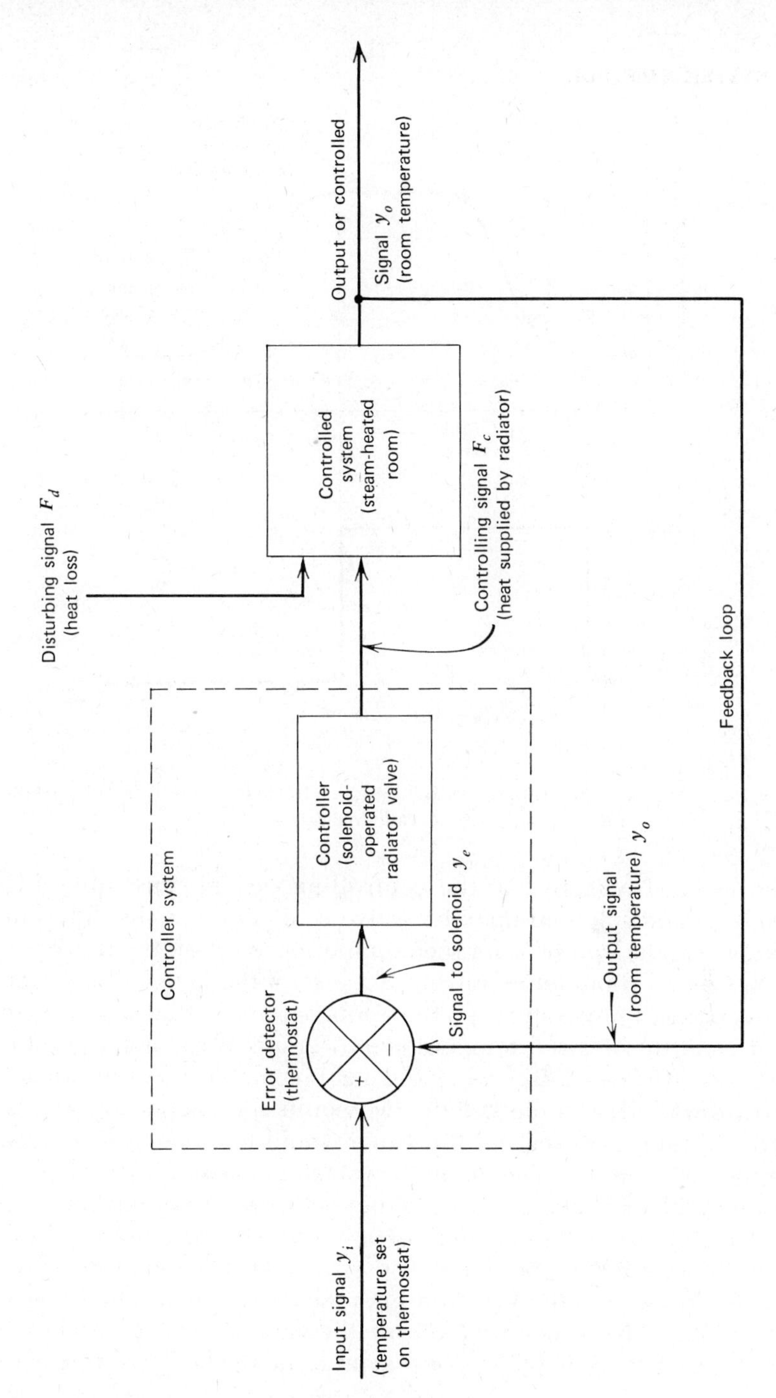

Figure 7-20 Flow diagram of control system for steam-heated room. Arrows pertain to flow of "signals," not to direction of heat flow.

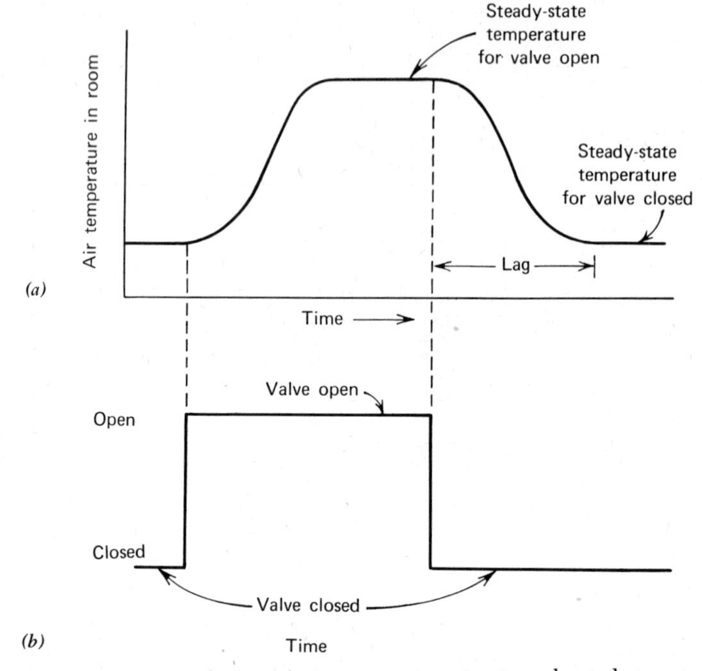

Figure 7-21 Diagrams relating (*a*) air temperature in steam-heated room to (*b*) position of valve controlling access of steam to radiator.

The general objective of the controlling system consisting of the thermostat and solenoid-operated valve is to minimize the difference between the desired or reference temperature set on the thermostat and the actual room temperature. Because of the lags in the system, however, the temperature in the room is alternately warmer and cooler than the desired temperature. A graph of room temperature versus time (Figure 7-22) may yield a curve similar to a sine wave. If there were no lags in the system, we would approach a straight-line control of temperature, and the valve would be opened and closed extremely frequently. The lag in the transfer of heat (transfer lag) and frequency of oscillation have a general inverse relationship. Long transfer lag results in oscillations of relatively long period length and high amplitude. The diagram portrays the valve-on intervals as equal in duration to the valve-off intervals. The actual ratio between the time the valve is on and the time the valve is off is related to the magnitude of the radiator as a heat source and the heat-loss rate of the room.

So far we have assumed that there is a single input-reference

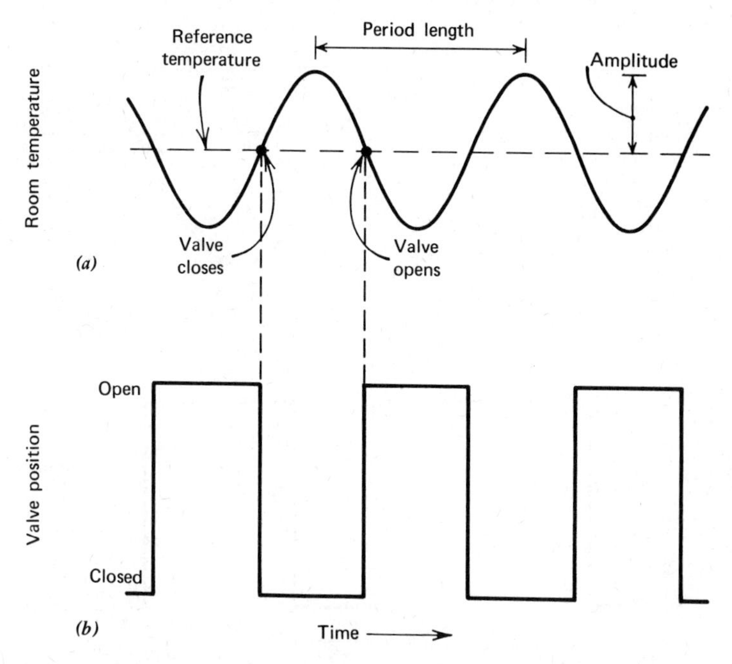

Figure 7-22 Graphs relating valve position and room temperature for on–off controller (thermostat) that has no differential gap. Process has transfer lag.

temperature. Alternatively the thermostat can be constructed so that there is a temperature differential, in which the upper temperature value affects the point at which the valve is closed, and the lower value affects the signal to open the valve. The heat-transfer lag is still present of course, and the effect of the combined differential gap and the lag is to cause both the period length and the amplitude of the temperature oscillations to be greater (Figure 7-23) than those in the system that lacks a differential gap. The magnitude of the differential gap will influence the frequency and amplitude of the oscillations, larger gaps producing oscillations of lower frequency and greater amplitude.

Our assumption of equal on and off times implies that the rise in temperature occurs at the same rate as the fall in temperature. In most actual examples this is unrealistic because a radiator usually heats up faster than it cools. Thus it is more common for a graph of the temperature variations to display a pattern of asymmetric cycles (Figure 7-24).

If the controlled process in an on–off control system has no transfer lag, the response of the controlled system will not exhibit the

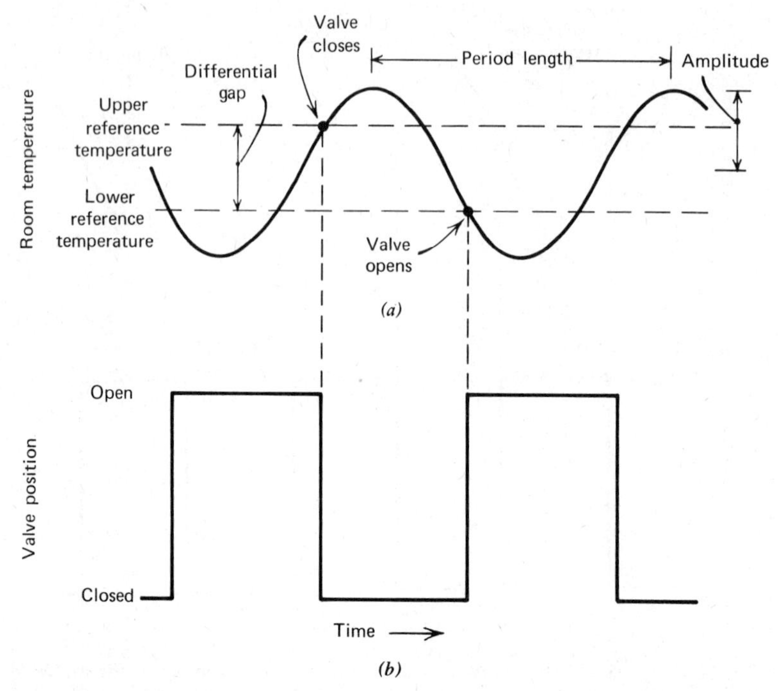

Figure 7-23 Graphs relating valve position and room temperature for on–off thermostat having differential gap. Process has transfer lag.

sinuous curves shown in Figures 7-21 to 7-24. Instead the response curve will have a saw-tooth appearance (Figure 7-25). A differential gap must be present if transfer lag is absent, because otherwise the oscillation period length would be infinitesimally short. Technological systems that lack transfer lag are relatively common, familiar examples including systems that involve changes in the level of a fluid. When a valve is opened, for example, the fluid may flow into or out from a reservoir almost instantaneously, without the lag associated with the flow of heat, for example. In any real system that lacks transfer lag there is overshoot and undershoot beyond the upper and lower reference limits of the differential gap. The reason for this is failure of the controlling devices to respond instantaneously. For example, when the level of the liquid in a tank that is part of a controlled system reaches an upper reference level, a float-operated switch may energize a solenoid that in turn closes a valve that stops the flow of liquid into the tank. Even though the flow stops instantaneously once the valve is closed, some time elapses between the time that the liquid reaches the upper reference level and the closing of the valve.

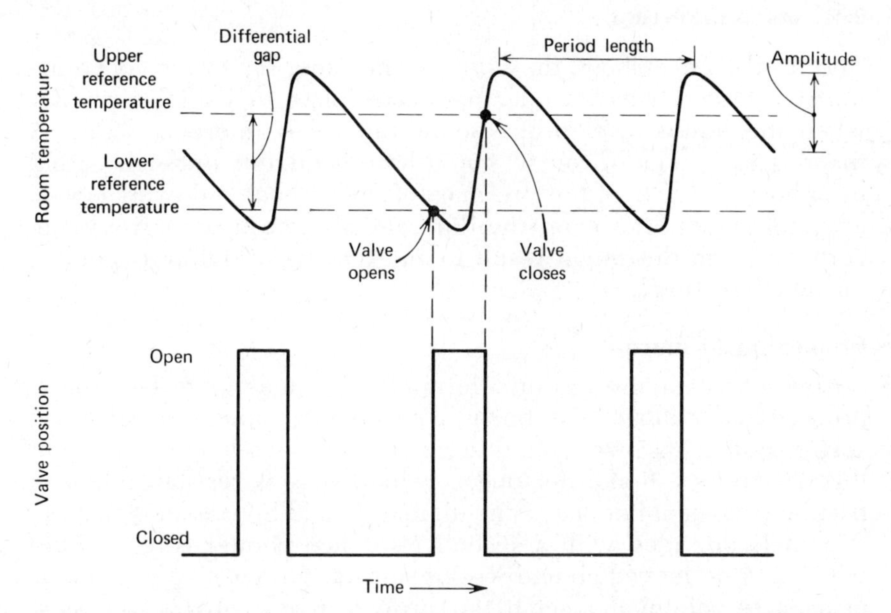

Figure 7-24 Graphs relating valve position and room temperature for on–off thermostat-controlled system which has differential gap and in which heat inflow rate is greater than outflow rate. Process has transfer lag.

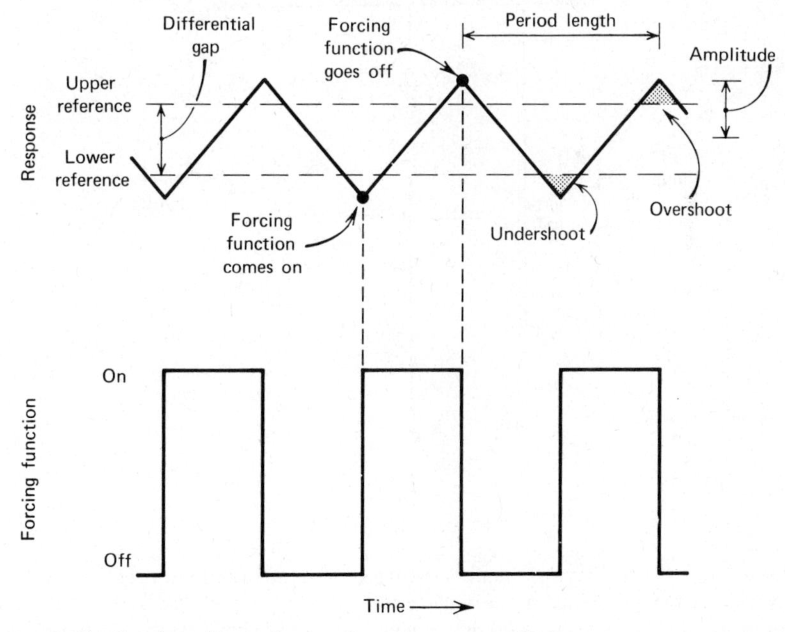

Figure 7-25 Graphs relating forcing function to controlled system response in on–off control system that has no transfer lag.

The result is overshoot, the liquid level rising somewhat above the upper reference level. Undershoot takes place in a similar manner when the liquid level drops below the lower reference value. A theoretical system, of course, could have negligible undershoot and overshoot. Because of the presence of undershoot and overshoot in any real system, however, the differential gap can be narrowed to zero, although this might result in an excessively high frequency of on–off alternations.

Proportional Control

Now let us explore a simple form of continuous control—namely, proportional control. To begin consider the simple mechanical spring system that we have described in an open-loop context. Let the system be a first-order one in which viscous resistance due to friction is present but mass is negligible. In addition, assume that the system is provided with a second reference pointer (Figure 7-26) which can be moved about by some outside source. The problem is to keep the pointer attached to the spring y_0 in line with the reference

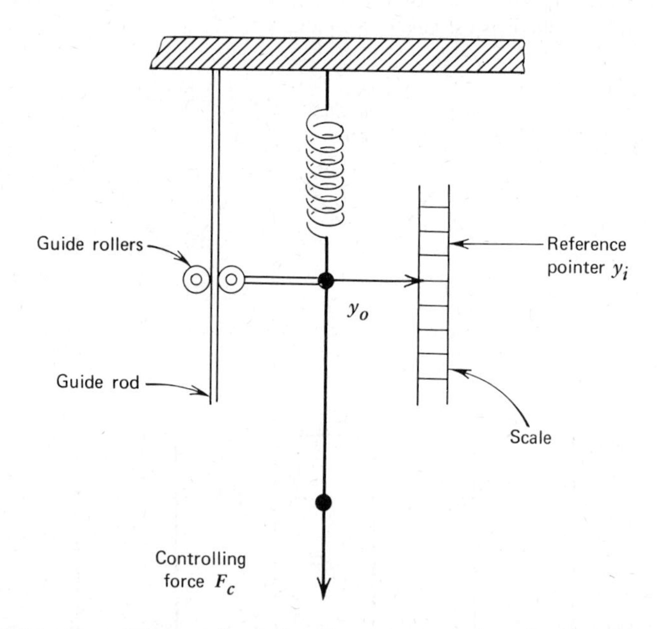

Figure 7-26 First-order mechanical spring system equipped with reference pointer y_i, which is moved about by an outside influence. A human controller could bring pointer y_o into line with y_i by adjusting controlling force F_c. Viscous resistance due to friction is assumed to be present, but mass is negligible.

pointer y_i. An automatic device might be devised to do this, but assume for the moment that we engage a person to do it. His job will be to compare the y_i and y_0, and then adjust the controlling force F_c so that the pointer attached to the spring is brought into line with the reference pointer. The difference between y_i and y_0 is the error y_e. Since force F_c can be continuously varied in an attempt to eliminate the difference between y_0 and y_i, and since the system is linear, we are dealing with a proportional control system. The human operator is in essence both the error detector and the controller because he compares the pointers and translates this into a controlling signal, which consists of varying the force on the spring. The mechanical system combined with the person thus forms a proportional control closed-loop system in which the controlling signal is directly proportional to the error signal. The controlling force F_c and the error y_e are related by the proportionality factor k:

$$F_C = k(y_i - y_0) = ky_e \tag{7-14}$$

When $y_0 = y_i$, $F_c = 0$; in other words, the controlling force is zero when the displacement pointer is opposite the reference pointer. If the controller could be sure that there would be no input other than the controlling force F_c, and that the reference pointer y_i would always be stationary, the feedback loop would be unneccessary, because the behavior of the system could be exactly predicted ahead of time. In practice, however, the inputs to the system are subject to outside disturbances, and the reference position may change with time.

An external disturbing force may be labeled F_d (Figure 7-19). The disturbing signal affects the controlled system, causing its output signal y_0 to change, and this is detected as a difference between y_0 and y_i. In systems where the reference pointer is unchanged with time, the task of the controller is simply to minimize $(y_i - y_0)$. Such control systems are known as *regulators*. A constant-temperature laboratory water bath is a good example of a regulator system. On the other hand, some control systems deal with a continuously varying reference position. Again the error detector is sensitive to differences between y_i and y_0, and the controlling parameter F_c is adjusted in an attempt to minimize the difference. Such systems are known as *servo* systems. An anti-aircraft-fire control system is an example of a servo system, where the reference pointer (target) is continually moving. Of course many technological control systems serve both in a servo and in a regulator role.

ADAPTING TECHNOLOGICAL CONTROL CONCEPTS TO GEOLOGICAL DYNAMIC SYSTEMS

Adapting mathematical concepts of technological control theory to geological dynamic systems is difficult. As we have seen, technological control theory is principally centered about systems that may be described by sets of equations that are relatively simple, generally consisting of low-order linear differential equations. We suspect that many geological systems contain components that serve the same functions as their technological counterparts, but important differences remain. A technological system is a system that man has designed for a specific purpose. A natural system exists, however, as a consequence of the physical and chemical laws that govern the behavior of its constituent parts. Physiological systems in biology, for example, have arisen through the selective processes during millions of years of organic evolution.

The "well-designed" systems have tended to be perpetuated, whereas the poorly designed ones have disappeared. Our problem is not to design a natural system but rather to try to discover the plan of the natural system and attempt to approximate it by applying concepts borrowed from technological control systems.

Our greatest problem in contrasting technological and geological dynamic systems is that we know so little of the nature of the geological systems. One basic difference is that technological systems are usually linear, whereas most geological and biological systems are probably nonlinear. Linear systems are not necessarily better than nonlinear ones, but the mathematical formulation is simpler for linear systems.

The limitations of attempting to apply linear-system concepts to natural systems may seem very great at first glance. A geological control system, such as the behavior of the earth's crust due to loading by deposition of sediment, is characterized by multiple feedback loops. Not only are the relevant variables difficult to identify, they are also exceedingly difficult to measure. Furthermore it is unrealistic to consider most geological systems in isolation. Despite the difficulties, geological systems are undoubtedly governed by complex control mechanisms. Simulation models concerned with geological systems therefore cannot avoid aspects of control.

Furthermore we have seen that oscillation characterizes some types of both open- and closed-loop control systems. Many cyclic or periodic phenomena in geology, such as sedimentary cyclothems, periodic pulses of tectonic activity, and the advance and retreat of polar ice caps, may ultimately be explainable in terms of control theory. Geological simulation models of oscillatory phenomena must either assume that regular fluctuations of exogenous variables are responsible for these oscillations (such as sunspot cycles to explain ice ages or sea-level fluctuations to explain cyclothems) or explain the rhythmic behavior as an internal feature of the system (such as the transgressive–regressive cycle demonstrated in the sedimentary basin model at the beginning of this chapter). Control theory provides the means for developing models that exhibit oscillation as part of the system, and the use of formal control theory for such models may prove to be highly rewarding.

Annotated Bibliography

Carss, B. W., 1968, "In Search of Geological Cycles Using a Technique from Communications Theory," in Colloquium on Time Series Analysis, Computer Contribution 18, Kansas Geological Survey, pp. 51–56.

The author discusses the use of autocorrelation functions and spectral analysis for recognizing the presence of cyclicity in sedimentary sequences. The spacing between the harmonics describing the periodicity may represent the maximum thickness of sediment that the crust can support by its own strength, before isostatic readjustment restores equilibrium.

Carss, B. W., and Neidell, N. S., 1966, "A Geological Cyclicity Detected by Means of Polarity Coincidence Correlation," *Nature*, v. 212, pp. 136–137.

Regular cyclicity in Lower Carboniferous marine limestone, shale, sandstone, and coal from Britain was detected by using techniques from communication theory. The authors speculate that the oscillations can be explained by an isostatic mechanism of readjustment.

Grodins, F. S., 1963, *Control Theory and Biological Systems*, Columbia University Press, New York, 205 pp.

Although this book is concerned primarily with applications of control theory to biological systems, Chapters 1 to 3 provide an introduction to the mathematics of linear open-loop control, without delving too deeply into technicalities.

Himmelblau, D. M., and Bischoff, K. B., 1968, *Process Analysis and Simulation: Deterministic Systems*, John Wiley and Sons, New York, 348 pp.

A highly technical book, written from the chemical engineer's point of view; contains a wealth of ideas potentially useful for geological simulation.

Lajoy, M. H., 1954, *Industrial Automatic Controls*, Prentice-Hall, Englewood Cliffs, N.J., 276 pp.

A clearly-written book dealing with control theory as applied to technological systems. For the most part the treatment assumes a modest mathematical background, unlike most books on this subject.

Milhorn, H. T., 1966, *The Application of Control Theory to Physiological Systems*, W. B. Saunders, Philadelphia, 386 pp.

Although somewhat more technical than Grodins' book, several chapters are of particular interest. Chapter 4 describes how differential equations for physiological systems are derived, with special reference to fluid flow, diffusion, chemical kinetics, and population growth. Chapter 8 contains an introduction to physiological control systems. The topics covered, such as regulation of acid–base balance and regulation of electrolyte concentration, may give useful clues for those interested in modeling the control aspects of geochemical reactions.

Nye, J. F., 1965, "The Frequency Response of Glaciers," *Journal of Glaciology*, v. 5, pp. 567–587.

The author employs linear open-loop-control theory to describe the frequency response of glaciers to simple harmonic fluctuations of accumulation rate of snow. Ice discharge and ice thickness (the responses) are calculated as a function of time for various perturbations of snow-accumulation rate (fluctuating input). Theoretical results are compared with values obtained from glaciers in Sweden and the United States.

Oertel, G., and Walton, E. K., 1967, "Lessons from a Feasibility Study for Computer Models of Coal-Bearing Deltas," *Sedimentology*, v. 9, pp. 157–168.

The authors suggest that a deltaic system possessing multiple feedback loops will exhibit oscillation without direct control by such external factors as tectonic, eustatic, or climatic rhythms.

Schwarzacher, W., 1966, "Sedimentation in Subsiding Basins," *Nature*, v. 210, No. 5043, pp. 1349–1350.

A short note describing a model for equilibrium conditions in subsidence–sedimentation relationships, employing a first-order linear differential model.

Schwarzacher, W., 1967, "Some Experiments To Simulate the Pennsylvanian Rock Sequence of Kansas," in Colloquium on Time Series Analysis, Computer Contribution 18, Kansas Geological Survey, pp. 5–14.

Two models are developed to explain periodicity in geological processes. The first model assumes that a section of crust is supported by a central pivot. When the crust is disturbed from equilibrium by loading at one side of the pivot, it will oscillate around the pivot, the oscillations being damped with time. If the disturbances are random, the following stochastic differential equation may be applicable:

$$\frac{d^2x_t}{dt^2}+b_1\frac{dx_t}{dt}+b_2x_t=\epsilon_t$$

where x_t = magnitude of vertical displacement at some fixed distance from pivot at time t

b_1, b_2 = constants

ϵ_t = value of random variable at time t (≡ disturbing force).

Note that this equation is an example of a linear second-order open-loop control system, similar to the second-order equation developed for the spring system (Equation 7-10).

In the second model an area of uplift is assumed to be separated by rigid crust from an area of subsidence. Loading by sedimentation on the area of subsidence is followed by subcrustal flowage and subsequent tectonic upwarp in the area of uplift. Loading (and subsidence) is not followed immediately by uplift, however, there being an appreciable time lag due to the strength of the crust and very slow subcrustal flowage. The author suggests that such phenomena may be suitably described by a differential-difference equation of the form

$$\frac{dx_t}{dt} + c_1 x_t + c_2 x_{t-1} = \epsilon_t$$

where x_t = magnitude of vertical displacement at time t

x_{t-1} = similar quantity for time $t-1$, where time is divided into increments whose magnitude is equivalent to the time lag

c_1, c_2 = constants

ϵ_t = value of random variable at time t (≡ disturbing force)

Note that in this equation the first term (derivative) deals with time as a continuous function, whereas the second and third terms treat time as a discrete function.

CHAPTER 8

Optimization

Optimization is implicit in any simulation process. The basic steps in simulation are (a) observation of the real system, (b) construction of a model, (c) operation of the model, and (d) comparison of the model's behavior with that of the real system. The last step involves measuring, either quantitatively or qualitatively, the degree to which the model deviates from the real system. The general goal is to minimize these deviations, so that the model behaves most like the real system. The adjustment of the model to minimize the deviations is an optimization process.

Initially let us consider an actual application of optimization in simulation. The first model described in Chapter 9 is concerned with the rate of overturn of sedimentary material in the earth's crust throughout geologic time and was devised by Garrels and Mackenzie. The goal is to find a rate of recycling or overturn of sedimentary material that will best account for the observed mass distribution of sedimentary material in the earth's crust. Figure 8-1 is a histogram showing estimates of the mass of sedimentary material according to geologic age. The model involves the assumption that sedimentary material has undergone continual recycling or overturn, old sedimentary rocks being uplifted, eroded, and the material transformed into new sedimentary material. The rate of recycling, α, which can be expressed as the half-life or *half-mass* of the sedimentary mass, is

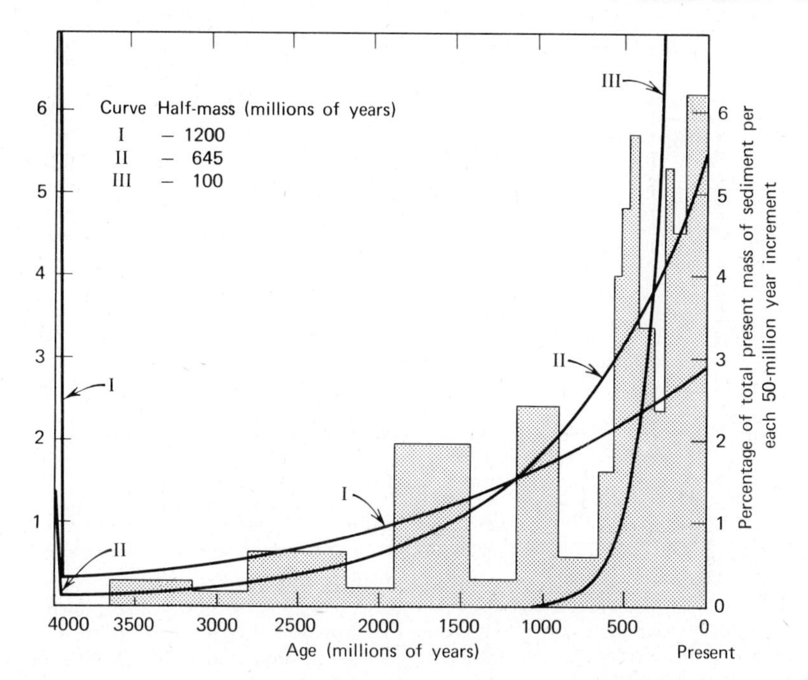

Figure 8-1 Distribution of mass of sedimentary rocks as function of age. Curves are based on models that assume total mass of sediments existing today has remained constant throughout geologic time. Curves I, II, and III represent age distributions of a mass of sediments assuming half-mass ages of 1200, 645, and 100 million years, respectively. Shaded histogram is estimate of actual mass distribution based on observed occurrence, adapted from Garrels and Mackenzie (1969).

critical to the model. A computer program (Table 9-1) representing the model provides for input of different values of α. The problem is to manipulate the value of α so as to obtain the best fit of the resulting theoretical curve to the histogram (Figure 8-1). The best fit can be defined in various ways. We shall define it as the minimum value of the sum of the absolute deviations. Thus, obtaining the optimum fit involves adjusting the value of α until the sum of the absolute deviations is the least possible. Figure 8-2 is a graph relating the rate of sedimentary overturn versus sum of absolute deviations. The optimum rate of overturn can be read directly from the curve.

In the sedimentary recycling model only one input parameter was adjusted in the trial-and-error optimization process. More complex simulation models may involve two or more input parameters that vary simultaneously. Under these conditions trial-and-error optimization becomes extremely time consuming and tedious. More powerful

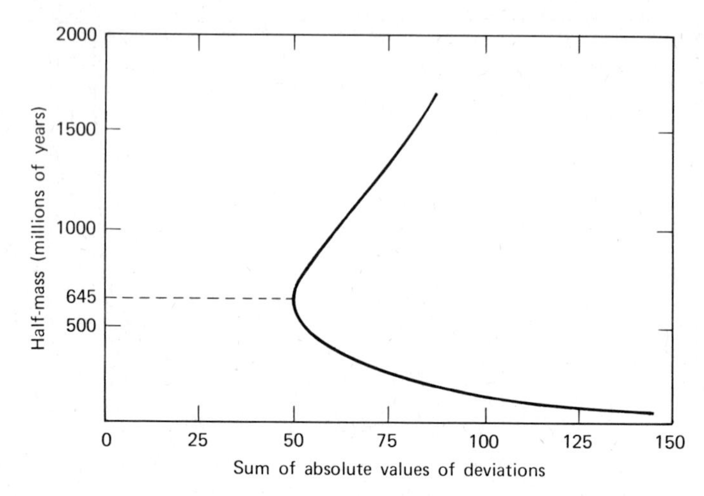

Figure 8-2 Graph of half-mass age in millions of years versus sum of absolute values of deviations. Curve is fitted to a large number of individual points, each one obtained by running simulation program, setting rate of sediment overturn (half-mass), and determining deviation of calculated-sediment-distribution curve from observed-distribution histogram (Figure 8-1). Calculations involve division of 4 billion years of geological time into 80 discrete intervals, each 50 million years long. Deviations are minimized at 645 million years.

and direct optimization techniques are needed. This chapter is concerned with some of the available techniques.

FUNDAMENTAL CONCEPTS

Analytical methods for optimizing systems have grown rapidly since World War II. Most optimization methods are applied in man-made systems in which some optimum situation is the goal. What is "best" is generally a human value judgment. A common goal in many man-made systems (management of a profit-seeking corporation for example) is to maximize profits. Lesser goals may be to minimize costs or time. Analytical optimization methods, in general, are used with the goal of either maximizing or minimizing a sharply defined *objective function*, such as profit or cost.

Any system invariably operates with constraints. For example, a bridge designer generally wishes to design a bridge that costs as little as possible. His design, however, must consider the maximum intended load of the bridge, and furthermore it is subject to the constraint that the overall strength of the bridge be greater than or equal to some minimum safety factor. Thus the cost of the bridge is the *objective function*, whereas the minimum safety factor is a *constraint*. By designing a bridge that is stronger than the minimum safety requirements, the safety constraint has been observed, but the cost of the bridge is no longer minimal. Similarly, if the cost of the bridge is reduced too much, the minimum safety limits will have been violated. Clearly the optimum design for a bridge of specified carrying capacity, given this single constraint, is one that meets the minimum safety requirements and costs the least.

Many other forms of constraints are possible. For example, in a bridge design the time required for construction may be an important consideration. During war the objective function for a bridge to carry advancing troops may be minimum construction time, rather than cost. Similarly on the battlefield a military tactician may wish to maximize the size of territory captured from the enemy, subject to the constraint of losing no more than a certain number of his troops.

In the examples above each optimum is a human value judgment that involves either a maximum or minimum. The concept of optimization can be broadened, however, to include the search for maxima or minima, regardless of whether a human value judgment is involved. Given this broader view, we can regard optimization as an aspect of many natural processes in which maxima and minima are obtained. At the same time we can retain its meaning as a class of mathematical methods used to search for maxima and minima. For example, plants and animals that evolve within a highly complex system are continually being optimized by natural selection. The objective function in biological evolution might be defined as fitness of each type of organism for its environment, assuming that natural selection works to promote the "survival of the fittest." The result is that each type of organism tends to evolve in such a way that it exhibits maximum fitness for its environment, subject to constraints imposed by prey–predator relationships, physiological and chemical laws, energy supply, living space, and many other biological and physical aspects of environment.

Some of the important ideas of optimization have developed out of observations of natural processes. The well-known mathematician Leonhard Euler (1707–1783) wrote that "since the fabric of the world is the most perfect and was established by the wisest Creator, nothing happens in this world in which some reason of maximum or minimum would not come to light" [quoted by Wilde and Beightler (1967) in *Foundations of Optimization*]. Wilde and Beightler trace some of the historical landmarks in optimization theory and go on to point out that the earliest optimum principles concerned the behavior of light.

Around 100 B.C. Heron of Alexandria asserted that light travels between two points by the shortest path, thus minimizing the distance traveled. This principle leads to the verifiable fact that light rays are straight lines unless reflected, refracted, or diffracted. Fermat's more general principle, made in 1657, that light travels between two points in the *least time* rather than the least distance, allows for Snell's law of refraction without contradicting Heron's principle.

In 1829 Gauss stated the "principle of least restraint" from which could be deduced the equality of internal to external forces in statics. Light and mechanics were unified by a single principle suggested in 1834 by Hamilton, which may be stated as follows: "of all the possible paths along which a dynamic system may move from one point to another within a specified time interval (consistent with any constraints) the actual path followed is that which minimizes the time integral of the difference between kinetic and potential energies." From this principle could be obtained all the optical and mechanical laws then known, and it forms one of the basic elements of wave mechanics and relativity.

Optimization theory bears an important relationship to the system-control ideas discussed in Chapter 7. In a technological control system the criteria that govern the overall performance of the system are established by a human operator. For example, in an on–off thermostat-controlled steam-heating system, as described in Chapter 7, the desired temperature is preset by the human operator. The objective function of the control system, with its negative-feedback loop, is to keep the room temperature close to the preset temperature. Stated differently, the control system seeks to minimize the difference between the actual and the desired room temperature, subject to constraints imposed by the properties of the system, including the magnitude of the differential gap and the transfer lag. From an optimization-theory point of view the control system can be regarded as possessing an objective function and various constraints. To make a distinction, however, between control theory and optimization theory, we consider system control as principally concerned with the response of a system to fluctuating inputs, whereas optimization is concerned with the definition and properties of an objective function, which in turn can serve as the criterion for control.

A number of optimization methods are generally grouped under the term *mathematical programming*, the word "program" being synonymous with "plan." It is important, of course, to distinguish between computer programming and mathematical programming, since the two have quite independent meanings. It is true, however, that most mathematical-programming techniques involve the use of computers.

Mathematical programming has several main branches, which include linear, nonlinear, and dynamic programming. All are generally concerned with maximizing or minimizing objective functions. Linear programming is the only aspect of mathematical programming that we are concerned with in this book.

BASIC OPTIMIZATION THEORY

Despite the wide variety of methods available for optimization, the basic concepts and terminology are more or less common to all. The objective function is the central concept. The objective function is a dependent variable, often denoted by z, whose value depends on one or more independent variables ($x_1, x_2, \cdots, x_n$). In some systems there are in turn two types of independent variables, *decision variables* (alternatively termed *control variables*) and *state variables.* State variables reflect the condition of the system being optimized, but they do not exert control over the system, as do the decision variables. Regardless of whether state variables are distinguished, all optimization techniques are concerned with maximizing or minimizing the objective function. This is achieved by finding the values of the independent variables for which the objective function is maximal or minimal.

$$z_{\max} = f(x_1, x_2, \cdots, x_n) \tag{8-1}$$

or

$$z_{\min} = f(x_1, x_2, \cdots, x_n) \tag{8-2}$$

To illustrate these optimization ideas, let us consider a system only two independent variables, x_1 and x_2, which will permit us to portray the relationships graphically. Since three variables (including z) are involved, we can represent z as the height of a contoured surface extending over an area defined by x_1 and x_2. Assume that we seek the maximum.

If the objective function is linear and involves two independent variables it may be written

$$z = ax_1 + bx_2 \tag{8-3}$$

where a and b are coefficients. The equation is that of a plane (Figure 8-3a). If no constraints are established, there is no point on the surface at which z is maximized, because z will tend toward infinity as x_1 and x_2 are increased indefinitely. Furthermore we make the assumption that x_1 and x_2 cannot be negative.

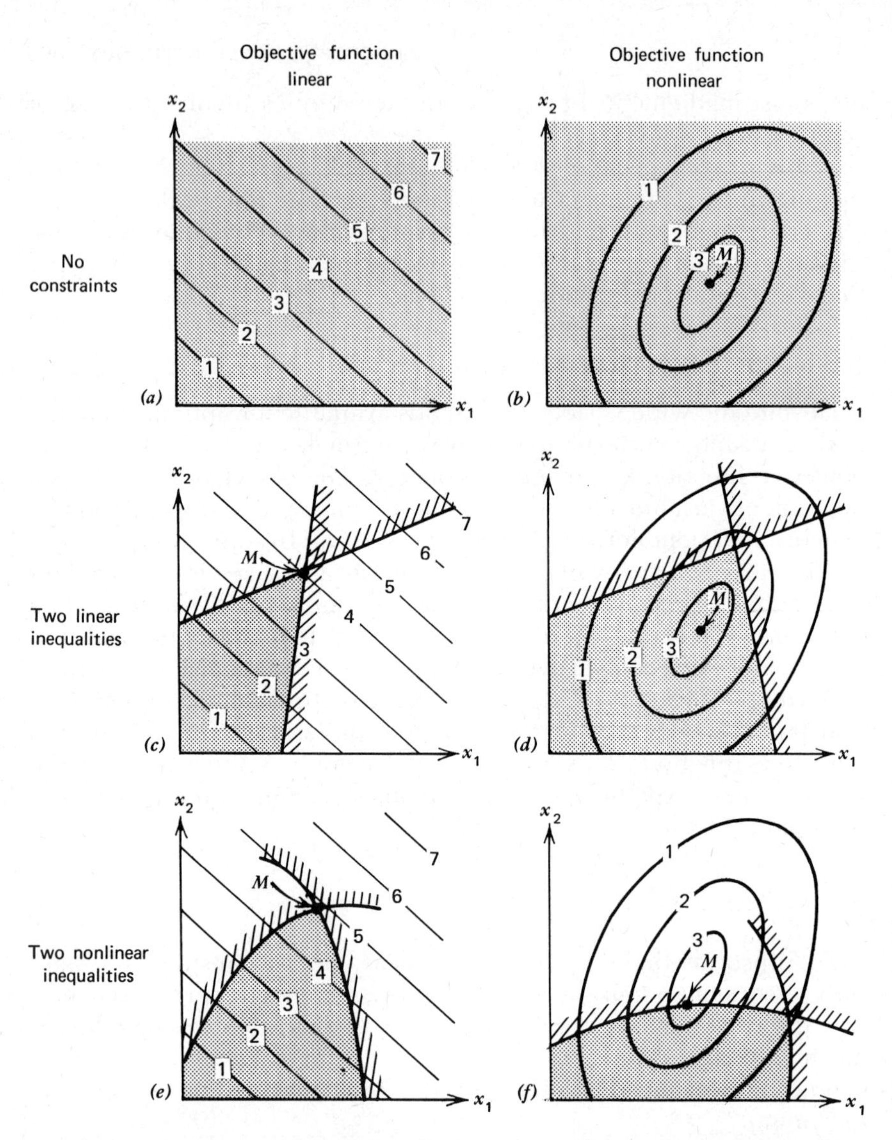

Figure 8-3 Graphic representation of major types of optimization relationships. Objective function z is represented by contour lines on surfaces. Surfaces in diagrams on left are planes representing linear function of x_1 and x_2, whereas surfaces in diagrams on right represent nonlinear function. Shaded part of each surface denotes area of feasible solution, it being assumed that x_1 and x_2 cannot take on negative values. Entire surface in (a) and (b) is within area of feasible solution because there are no constraints. Two linear inequalities have been added as constraints in (c) and (d), whereas two nonlinear inequalities provide constraints in (e) and (f). Maximum value of objective function falling within area of feasible solution is labeled M.

If the objective function is nonlinear, local maxima or minima may be present. For example, suppose that we define z as a general second-degree polynomial,

$$z = ax_1 + bx_2 + cx_1^2 + dx_1x_2 + ex_2^2 \tag{8-4}$$

where a, b, c, d, and e are coefficients. Equation 8-4 yields a parabolic surface (Figure 8-3*b*). In this case a maximum exists at point M, whose location is specified by appropriate values of x_1 and x_2.

In Figure 8-3 we follow the convention of shading that part of each surface for which a solution is feasible—in other words, for solutions that do not violate any constraints. Thus all points on the area of surfaces in Figure 8-3*a* and *b* yield feasible solutions because there are no constraints. This assertion is not contrary to our earlier statement that there is no definable maximum for surface *a*, and a single maximum for surface *b*.

Now let us add constraints to these two functions. If the constraints are linear, they can be represented by straight lines (Figure 8-3*c* and *d*) that represent linear inequalities (we discuss the meaning of an inequality later in this chapter). The area of feasible solution is reduced, being defined as the area represented by the function that lies "inside" the inequality lines. With the linear objective function (Figure 8-3*c*) the optimum solution is at the point defined by the intersection of the two inequality lines, for here z attains its maximum value without going beyond the constraints. The maximum for the nonlinear function, given the two linear inequalities as constraints (Figure 8-3*d*), is at point M, and the highest point on the objective function lies within the area of feasible solution. Finally, if we impose constraints in the form of nonlinear inequalities (Figure 8-3*e* and *f*), the optimum occurs at the highest point on each surface within the area of feasible solution.

Inequalities

As we have pointed out, constraints can be represented by inequalities. Before defining and illustrating an inequality, let us recall means for graphically representing straight lines. Figure 8-4 is a graph of two straight lines, one represented by the equation

$$2x_1 + 3x_2 = 100 \tag{8-5}$$

and the other by

$$x_1 + x_2 = 60 \tag{8-6}$$

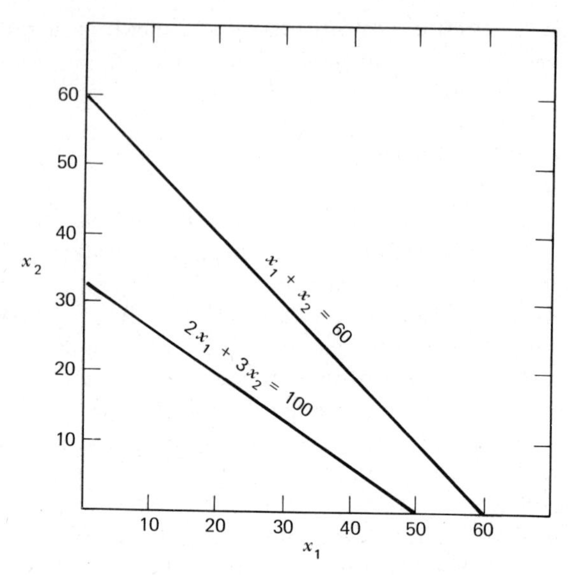

Figure 8-4 Graph of two linear equations involving independent variables x_1 and x_2.

Graphing these equations is simply a matter of setting x_1 to zero and solving for x_2 and vice versa, and then connecting the points. If the lines are extended so that they intersect off the present graph, a unique solution is obtained, the same values of x_1 and x_2 satisfying both equations.

Instead of representing Equation 8-6 in its conventional form, we can express it as an inequality, or inequation:

$$x_1 + x_2 \leq 60 \tag{8-7}$$

Stated in words, this expression implies that the value of $x_1 + x_2$ is, under all conditions, either equal to or less than 60. Since it is an inequality, however, we cannot express it as a line. Instead it can be expressed as an area (Figure 8-5) lying beneath the line for Equation 8-6. The axes in themselves can be regarded as inequalities, $x_1 \geq 0$ and $x_2 \geq 0$, whose values are equal to or greater than zero. Any point within the area bounded by these three inequalities yields a feasible solution, there being no unique solution for x_1 and x_2.

Given this understanding of inequalities, let us return to Figure 8-3*c*, where the problem is to find the values of x_1 and x_2 where

$$z_{\max} = ax_1 + bx_2 \tag{8-8}$$

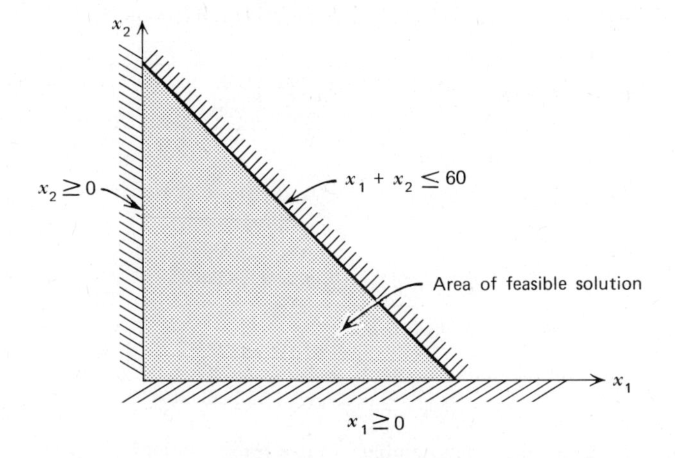

Figure 8-5 Graphic representation of area of feasible solution for three inequalities.

subject to the constraints defined by the four inequalities that bound the area of feasible solution

$$cx_1 + dx_2 \leqslant e$$
$$fx_1 + gx_2 \leqslant h$$
$$x_1 \geqslant 0$$
$$x_2 \geqslant 0$$

where a, b, c, d, e, f, g, and h are known constants. The optimum, or maximum in this example, occurs at the intersection of the two lines representing the principal inequalities. Figure 8-3*c* represents relationships typical of all linear programming methods in that both the objective function and all constraints are linear. Although the graphic solution is intuitively obvious, most applications of linear programming deal with more than two independent variables and therefore cannot be treated graphically.

CLASSIFICATION OF OPTIMIZATION METHODS

It is now appropriate to consider the various methods for searching for optima, keeping in mind the fact that all methods involve finding either the maximum or minimum value of the objective function within the area of feasible solution, as bounded by inequalities. Table 8-1 provides a classification of methods, identifying those treated in this book. The book by Wilde and Beightler (1967) is recommended for its thorough, overall treatment of optimization methods.

TABLE 8-1 Classification of Principal Optimization Methods[a]

General Class	*Specific Method*
Direct-search methods:	
Elimination methods	Sequential search[b] Contour tangents[b]
Climbing methods	Steepest ascent[b]
Indirect-search methods:	
Calculus methods	Maxima and minima[b] Variational calculus
Mathematical programming	Linear programming[b] Nonlinear programming Dynamic programming

[a]After Wilde and Beightler (1967).
[b]Specific methods treated in this volume.

Optimization methods fall into two broad classes, *direct-search* and *indirect-search* methods. In turn the direct-search methods may be subdivided into *elimination* methods and *climbing* methods. The indirect-search methods may be subdivided into methods employing calculus and into those employing mathematical programming.

The traditional approach to optimization involves classical calculus, in which the extreme value of the objective function is obtained by setting the first partial derivative of z with respect to each independent variable to zero. Given n independent variables, n simultaneous equations are obtained that can then be solved. The fitting of functions by least squares is an example of this method.

Variational calculus provides useful optimizing methods where the objective function varies in space or in time. A classical problem in variational calculus involves finding the maximum area that can be enclosed by a line of given length. Variational-calculus methods are not treated in this book.

The use of calculus optimization methods is fraught with a number of difficulties. The most obvious difficulties, as discussed by Bellman and Dreyfus (1962) are (a) the presence of multiple local maxima and minima on the objective function, within its area of feasible solution; (b) the effect of constraints that often place the maximum or minimum on a boundary, where they are not detected by setting the first derivative to zero; and (c) the oversensitivity of calculus methods to small errors in the data.

Mathematical programming methods overcome many of these problems. These are iterative methods, well suited for digital computation. Linear programming, as we have emphasized, is applicable where the objective function is a linear combination of the independent variables and the constraints are linear inequalities. Nonlinear programming, as its name implies, is applicable where either a nonlinear objective function or nonlinear constraints are employed. Finally, dynamic programming is a very powerful class of methods for treating both linear and nonlinear multistage decision processes, but it is restricted to cases with a relatively small number of variables. Both dynamic programming and nonlinear programming may ultimately prove to be important techniques in geological simulation, but we have not treated them in this book.

Direct-search methods provide the only feasible optimization technique when an analytical expression for the objective function is not available. In addition they are useful when the objective function is too complex for the indirect methods of calculus and mathematical programming to be applied unless simplifying approximations are made. As mentioned, direct-search methods can be grouped under two general categories, elimination methods and climbing methods. These methods approach the peak of the objective function by a direct ascent, as would a mountain climber, instead of "parachuting" to the summit (indirect method). We turn to the direct-search methods first for detailed treatment.

DIRECT-SEARCH METHODS

Elimination Methods

To illustrate the concepts of direct search let us use data from a hypothetical geochemical prospecting survey. The data (Table 8-2) consist of values of copper in parts per million in soil samples and pairs of coordinate values specifying the geographic location of each sample within a rectangular area 8 by 10 miles in extent (Figure 8-6). Our problem is to find point M, where the copper values are highest. Although we can find this point simply by inspecting the contour map, let us define the problem by employing formal optimization terminology. Copper concentration z is the objective function, and x_1 and x_2 are the decision, or control, variables. Thus our purpose is to find the values of x_1 and x_2, such that

$$z_{\text{max}} = f(x_1, x_2) \tag{8-9}$$

TABLE 8-2 Hypothetical Geochemical Prospecting Data Consisting of Copper Values and Geographic-Coordinate Values Pertaining to the Area Shown in Figure 8-6

Sample Locality	z*(Copper in ppm)*	x_1*(miles)*	x_2*(miles)*
1	34	1.5	6.7
2	39	5.1	7.2
3	40	9.0	7.3
4	47	9.2	6.2
5	62	3.5	5.6
6	61	1.7	4.7
7	63	2.8	4.7
8	62	8.0	5.2
9	50	8.2	4.4
10	78	5.5	4.2
11	47	1.4	3.2
12	59	3.1	3.2
13	62	7.9	5.2
14	70	6.3	5.6
15	41	5.1	2.0
16	53	6.7	5.5
17	31	7.8	1.9
18	17	6.8	0.6
19	28	2.8	0.8
20	30	1.5	1.1

subject to the constraints

$$0 \leqslant x_1 \leqslant 10; \qquad 0 \leqslant x_2 \leqslant 8$$

which define the boundaries of the map area (Figure 8-6). If we knew the function that relates z to x_1 and x_2, we could use indirect calculus methods to find the maximum point by setting the partial derivatives to zero. One way of obtaining such a relationship is to use regression methods, which involve the fitting of a polynomial or some other function by least squares. The least-squares fitting process is in itself a form of optimization since it involves minimizing the sum of the squared deviations. Using either a manually fitted contour surface, as in Figure 8-6, or an appropriate function fitted by least squares, we find the maximum copper concentration to be about 82 parts per million, with coordinate values of $x_1 = 5.9$ and $x_2 = 4.8$ miles. The three direct-search methods listed in Table 8-1 are illustrated below, employing the hypothetical copper data described previously.

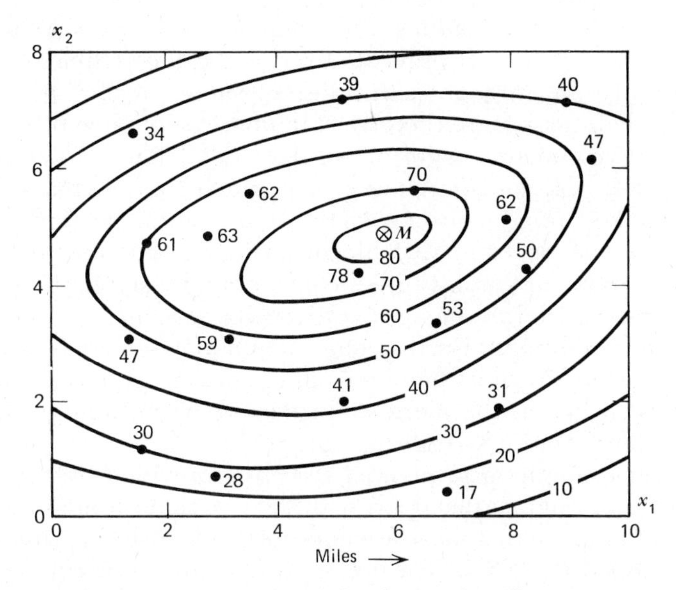

Figure 8-6 Contoured surface $z = f(x, y)$, fitted to copper values in parts per million. Point M is specified by values of x_1 and x_2 where z is maximum.

Sequential Search

The most elementary approach would consist of scanning each copper value and in turn eliminating all those except the largest. This would be extremely straightforward by hand or with a computer and would result in the selection of $z = 78$ at $x_1 = 5.5$, $x_2 = 4.2$. The disadvantage in this method lies in the absence of any interpolation between data points, restricting the selection of x_1 and x_2 corresponding to z_{max} to the coordinate values of one of the data points. Clearly this does not give the true maximum in our example, because the maximum M does not coincide with any of the data values. This disadvantage could be partly overcome by using nonlinear interpolation between data points, yielding a larger number of points on which a sequential search could be made. For example, we could overlay a gridwork of cells on the data from Figure 8-6 and obtain values of z for the center of each cell by interpolation. Simple linear interpolation between surrounding points, however, would be unsuitable because the newly-derived points would never exceed the highest original data-point values and therefore would not yield the true maximum.

Despite the computational inefficiency of sequential searching of the objective function, this method is probably the most practical for many simulation applications. For example, the curve shown in Figure 8-2 was obtained by repeatedly running the sedimentary-rock-mass simulation program with different values of the one decision variable (half-mass age), thereby obtaining a sequence of values for the objective function (sum of absolute deviations). In addition to locating the optimum (minimum in this case) sequential searching produces the curve (or surface) of the objective function for a range of settings of the decision variable or variables. Examination of this *response* curve (or surface) gives considerable insight into the sensitivity of the model to changes in decision variables, under a wide variety of conditions. Furthermore sequential search is extremely easy to perform with a computer (although it is expensive for large problems), whereas the more sophisticated optimization techniques may be difficult or impossible to apply.

Another example of sequential search is provided by a simulated grid-drilling program that deals with the search for mineral deposits. We might imagine that there are four buried deposits in a given hypothetical area (Figure 8-7). The deposits are unknown, of course, before their discovery. Each deposit is circular in area and can be discovered by a single drill hole. If we adopt a grid-drilling method, what grid spacing will be most profitable to us? Assume that each hole costs 100,000 dollars and each deposit yields a profit, after mining, of one million dollars. The objective is to maximize profits. If we simulate various grid patterns (assuming knowledge of the deposits), we can tabulate the net profit or loss for each pattern (Table 8-3), plotting (Figure 8-8) the financial outcome (objective function) versus the number of holes (decision variable) for the different patterns. As Figure 8-8 reveals, the most profitable course to follow is to use the 25-hole drilling pattern, which results in the discovery of three of the four deposits. The 36-hole pattern results in the discovery of all four deposits, but the increase in drilling costs causes the net profit to be lower than that for the 25-hole pattern.

It can be argued that the grid-drilling example is extremely artificial. Mineral deposits that are circular in shape and of uniform diameter are geologically unrealistic. Furthermore the use of rectangular grid-drilling patterns is uncommon. It has been shown, however, that simulated grid-drilling programs applied as "hindsight" over large areas containing many deposits make it possible to draw curves showing what would have been an optimum drilling plan had it been undertaken before exploration was begun. Papers by Drew (1966, 1967), Drew and Griffiths (1965), Griffiths (1966*a*, *b*), and Griffiths and Drew (1964), are particularly notable in this respect.

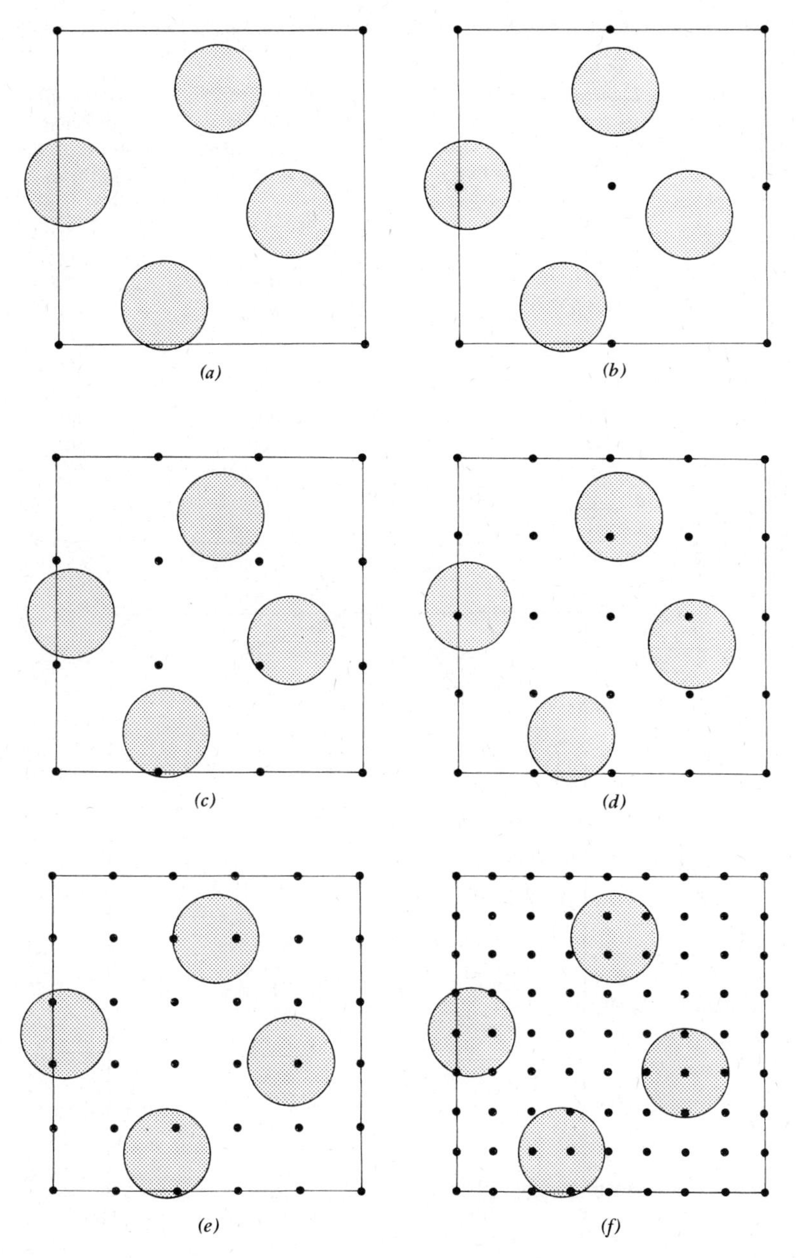

Figure 8-7 Simulation of different spacing patterns in grid-drilling program for hypothetical mineral deposits. Deposits are represented by shaded large circles, and drill holes by small solid circles.

TABLE 8-3 Results for Hypothetical Grid-Drilling Programs of Figure 8-7[a]

Number of Holes	*Total Drilling Costs (thousands of dollars)*	*Number of Discoveries*	*Profit from Mining[b] (thousands of dollars)*	*Net Profit or Loss (thousands of dollars)*
0	0	0	0	0
4	400	0	0	−400
9	900	1	1000	100
16	1600	2	2000	400
25	2500	3	3000	500
36	3600	4	4000	400
49	4900	4	4000	−900
64	6400	4	4000	−2400
81	8100	4	4000	−4100

[a]Each hole is presumed to cost 100,000 dollars, and each deposit yields a profit of one million dollars, exclusive of drilling costs.

[b]Excluding drilling costs.

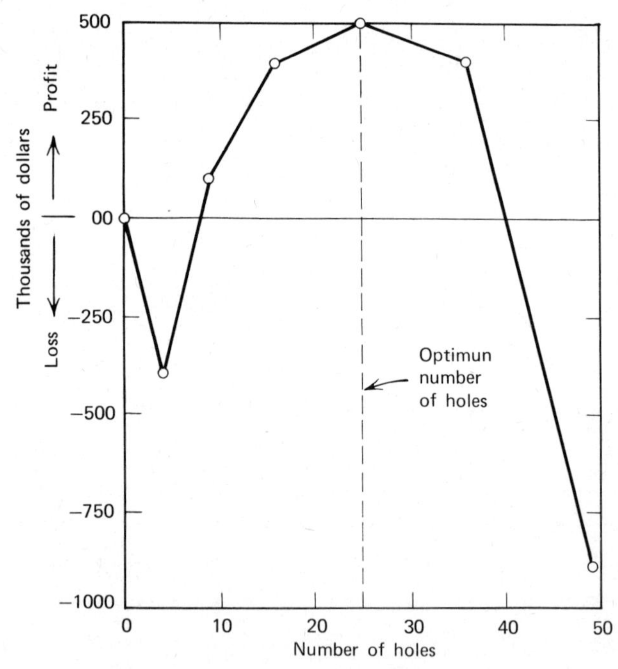

Figure 8-8 Plot of profit or loss versus number of drill holes for hypothetical drilling patterns of Figure 8-7. Optimum grid spacing corresponds to 25 holes, where profit is maximized at 500,000 dollars.

Figure 8-9 is a curve, taken from Drew (1966, p. 50), relating the expected net value versus grid spacing in miles for hypothetical grid-drilling programs for oil in the Michigan basin. Preparation of the curve involves a number of assumptions, including an elliptical shape of the oil fields in plan view, uniform-frequency distribution of the orientation of the long axes of the oil fields, detection of each field encountered by a single exploratory hole, and random uniform frequency distribution of the centers of the oil fields throughout the grid area. The highest point on the curve, about 850 million dollars, corresponds to a grid spacing of about 2.5 miles. Of course a grid spacing smaller than this would result in additional discoveries, but the progressive increase in drilling costs would reduce the net yield.

Although the grid-drilling example illustrated here has been simulated by hand, a computer program has been developed by Drew (1966) for performing experiments of this type; it incorporates a number of sophisticated features. Besides using regularly gridded holes, a random-sampling scheme could also be developed, employing random numbers for selecting the geographic coordinates of drill holes. Another computer model of this general type has been developed by Miesch, Connor, and Eicher (1964); it deals with the simulation of sampling designs for geochemical surveys.

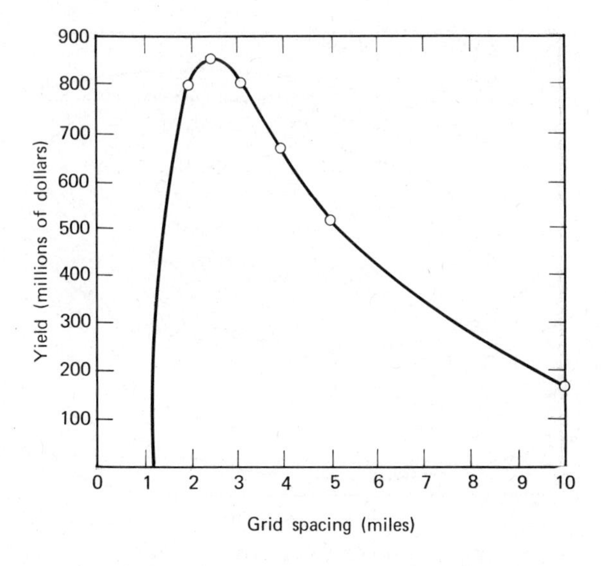

Figure 8-9 Curve relating yield in dollars versus grid spacing for simulated grid-drilling program in the Michigan basin. After Drew (1966, p. 50).

Method of Contour Tangents

The method of contour tangents is a simple and powerful type of elimination technique. Its application is restricted, however, in that it is useful only in dealing with a strongly unimodal objective function. In principle it consists of applying a sequence of tangents to a contoured surface, gradually converging on either a single maximum or minimum. The contoured surface of Figure 8-6 has been reproduced in Figure 8-10, providing an example of its application. Points, labeled 1, 2, · · ·, n are located successively at random on the surface representing the function. At point 1, which has been randomly chosen, a line that is a tangent to the contours is drawn so that it extends across the entire region. If the function is unimodal, the maximum must lie on the upper side of the tangent. Thus all of the area on the lower side of the tangent line can be eliminated from further consideration. The second point is picked at random within the remainder of the region, and a second line is drawn tangent to the contours of the surface at this point. As before, all the area on the lower side of the tangent line is omitted from further consideration. The third point at random is selected within the available remaining "uphill" area, and so on, as successive tangents at randomly chosen points are fitted. Clearly this method is fast and direct, the search narrowing rapidly to the maximum.

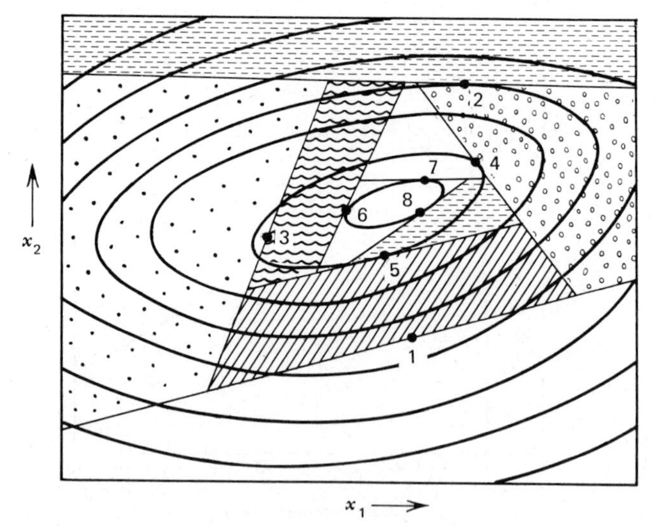

Figure 8-10 Contour-tangent method of elimination. Numbered points denote locations where tangents to contours are fitted. In seeking maximum each successive point will be located on "uphill' side of previously located tangent line.

Climbing Methods

If a man is climbing a mountain, he is unlikely to take a route that follows the steepest line unless he is particularly energetic. If, however, the mountain is shrouded in mist and he is unable to see more than a few feet in front of him, he may wish to take the steepest possible grade, reasoning that this must eventually lead him to the summit. This is also the rationale behind the optimization method of steepest ascent. The method of steepest ascent, however, tends to be unsatisfactory unless the objective function is unimodal, without subsidiary peaks.

The method of steepest ascent is illustrated in Figure 8-11. Starting at any point on the margin of the area of feasible solution, the slope of the objective function is evaluated for the neighboring region, and the direction of steepest ascent is found. The direction of steepest ascent is always perpendicular to the contour tangent. A small step is then made in this direction to a new point. The procedure is then repeated until the summit is reached. To find the minimum of a unimodal objective function the procedure is that of steepest descent.

For finding the line of steepest ascent a circle of specified diameter can be centered about the point from which the initial upward movement is to be made (Figure 8-12). Using the original data points that

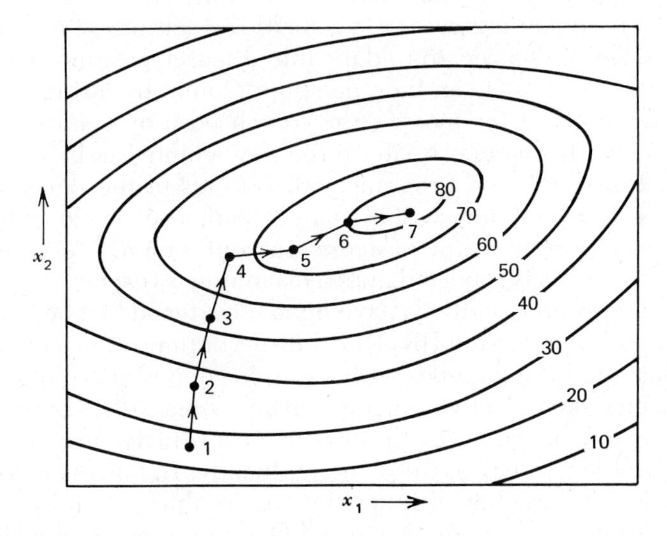

Figure 8-11 Path of steepest ascent in reaching maximum of unimodal function. Starting at point 1, move is made along route of steepest ascent to point 2 which is separated from point 1 by an increment. Process is repeated until maximum, or "summit," is reached.

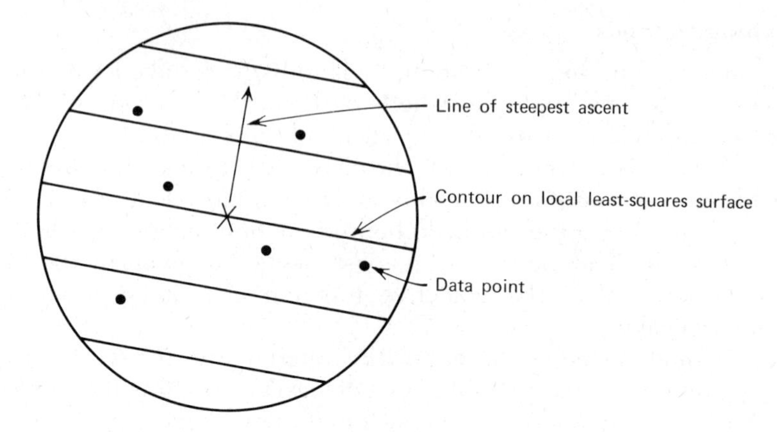

Figure 8-12 Method of obtaining line of steepest ascent (or descent) by fitting contoured surface by least squares to data points lying within circle. Line of steepest ascent extending from center of circle is perpendicular to contour lines.

fall within the circle, a plane surface can be fitted by least squares. The steepest ascent (or descent) is in turn perpendicular to the contours on the fitted surface.

In situations in which values of the objective function z are output from a computer-simulation model, the line of steepest ascent is evaluated by obtaining values of z for three settings of the decision variables, closely spaced around the initial search position. The direction of steepest ascent can then be simply found by fitting a plane to the three points. After moving the search position a small distance upward along the steepest path, three new points can be calculated, running the simulation program with settings of the decision variables selected randomly within some specified distance of the new search position. A new line of steepest ascent is then determined, and the whole process is repeated until a maximum is reached.

The direct-search methods have been introduced first because they are more easily understood than the indirect optimization methods. We have illustrated the methods with a simple copper-distribution problem in which the objective function relates copper distribution to the two independent variables or two map-coordinate directions. The copper problem is trivial to solve graphically. If, however, we were to increase the number of independent variables, even by one, the problem would become much more difficult to solve graphically. The direct-search methods can be readily extended to multidimensional problems, whereas our ability to solve problems graphically when more than two independent variables are involved is very limited.

We might also mention that the geochemical survey example that we first considered did not include constraints on the independent variables, except the map boundaries. It would not be difficult, however, to include inequalities as constraints which would simply narrow the region of search to a smaller area of feasible solution. In fact the contour-tangent method of elimination can be thought of in terms of the successive addition of linear inequalities (the contour-tangent lines) that progressively decrease the area of feasible solution until the "peak" is all that remains.

INDIRECT-SEARCH METHODS

Calculus of Maxima and Minima

The most elementary indirect-search methods involve the calculus of maxima and minima. The concepts of the calculus of maxima and minima can be illustrated with the behavior of a hypothetical river channel. Let us suppose that the shape of a river channel in cross section is always rectangular and that the breadth and depth of the channel form such a ratio that the wetted perimeter is always minimized for a given cross-sectional area. From Figure 8-13 we can see that the wetted perimeter is equal to breadth plus two times depth and that the cross-sectional area is equal to breadth times depth. The problem is to find those values of x_1 and x_2 where

$$z_{\max} = f(x_1, x_2) = 2x_1 + x_2 \tag{8-10}$$

subject to

$$x_1x_2 = k; \qquad x_1 \geqslant 0; \qquad x_2 \geqslant 0$$

where z = wetted perimeter
x_1 = depth of channel
x_2 = breadth of channel
k = cross-sectional area

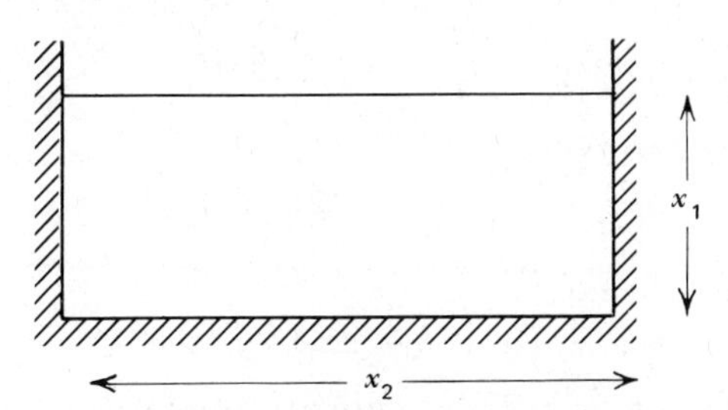

Figure 8-13 Rectangular channel of depth x_1 and breadth x_2 Wetted perimeter z is equal to $2x_1 + x_2$, and cross-sectional area equals x_1x_2.

This problem, then, consists of minimizing a linear objective function subject to one nonlinear equality constraint and two linear inequality constraints.

Because the nonlinear constraint is an equation, we can use it to eliminate by substitution one of the independent variables from the objective function expression, as follows:

$$x_1 x_2 = k; \qquad x_2 = \frac{k}{x_1} \tag{8-11}$$

In turn, substituting for x_2 in the objective function yields

$$z = 2x_1 + \frac{k}{x_1} \tag{8-12}$$

If we graph z and x_1 for various values of k (e.g., $k = 20$ and $k = 2$ square feet), we obtain the relationships shown in Figure 8-14. On this graph the depth x_1 for $k = 20$ that minimizes z can be read off directly as approximately 3, and for $k = 2$ the depth is approximately 1.

Analytically we can arrive at a precise relationship between x_1, x_2, and k by simple calculus methods. The first derivative of z with respect to x_1, dz/dx_1, is the slope of the curves shown in Figure 8-14 taken at any point along their lengths. This slope varies of course, and the minimum value of z for each curve is the point at which the slope is zero. Thus, if we take the derivative and set it to zero, we can solve for x_1 as follows:

$$\frac{dz}{dx_1} = 0 \tag{8-13}$$

Substituting into Equation 8-12, we obtain

$$\frac{dz}{dx_1} = \frac{d}{dx_1}\left(2x_1 + \frac{k}{x_1}\right) = 0 \tag{8-14}$$

Therefore

$$\frac{dz}{dx_1} = 2 - \frac{k}{x_1^2} = 0 \tag{8-15}$$

On solving for x_1 we obtain

$$x_1 = \sqrt{k/2} \tag{8-16}$$

or for k we have

$$k = 2x_1^2 \tag{8-17}$$

But we know that $x_1 x_2 = k$. Thus, substituting for k, we obtain

$$x_1 x_2 = 2x_1^2 \tag{8-18}$$

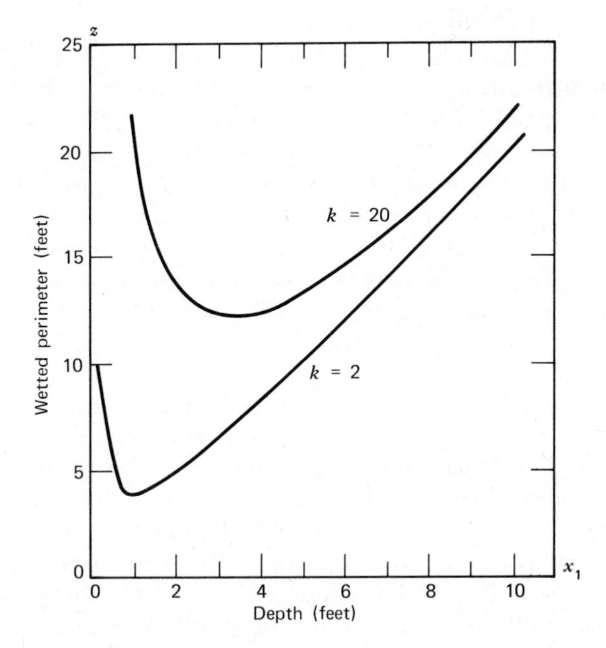

Figure 8-14 Graph of channel depth x_1 versus wetted perimeter z, for two values of rectangular cross-sectional area, $k = 20$ and $k = 2$. Problem is to find channel depth that minimizes wetted perimeter for given cross-sectional area.

and, solving for x_2, we find

$$x_2 = 2x_1 \tag{8-19}$$

Thus we have shown that, if the wetted perimeter of a rectangular channel is to be minimized for any fixed cross-sectional area, the channel shape must be such that it is twice as broad as it is deep.

We can also calculate precise values of x_1 and x_2 for any given values of k. For $k = 20$ we obtain

$$\begin{aligned} x_1 &= \sqrt{k/2} = \sqrt{20/2} = 3.1623 \\ x_2 &= 2x_1 = 6.3246 \\ z_{\min} &= 2x_1 + x_2 = 12.6492 \end{aligned}$$

For $k = 2$ we find that

$$\begin{aligned} x_1 &= \sqrt{2/2} = 1 \\ x_2 &= 2 \\ z_{\min} &= 4 \end{aligned}$$

Least-Squares Method

Let us turn now to the method of least squares, as it applies to the problem of fitting functions to raw data. The least-squares method is exceedingly widely used and well known. We view it here as an optimization method. To illustrate the least-squares method assume that a carbonate petrologist has estimated the proportions of various fossils in thin sections of limestone by point counting. Percentages of crinoids and ostracods in the limestone bear an inverse relation to one another (Table 8-4), as is revealed by a graph of crinoid x versus ostracod percentages y (Figure 8-15). A straight line fitted to the points is described by the equation

$$y = a + bx \tag{8-20}$$

If we decide to estimate the proportions of ostracods on the basis of the proportion of crinoids, we can define y as the dependent variable (percentage of ostracods) and x as the independent variable (percentage of crinoids); a and b are constants that represent the intercept and slope, respectively. By adjusting the values of a and b, any straight line could be produced on the graph. Our problem is to find those values of a and b (the decision variables) that describe the line of best fit by minimizing the sum of the squared deviations (the objective function).

TABLE 8-4 Proportions of Crinoids and Ostracods Measured from 14 Hypothetical Limestone Specimens

Specimen Number	*Percentage of Crinoids, x*	*Percentage of Ostracods, y*
1	1	18
2	2	16
3	5	15
4	4	14
5	7	13
6	8	11
7	9	10
8	9	9
9	11	8
10	14	6
11	14	4
12	17	2
13	18	2
14	18	1

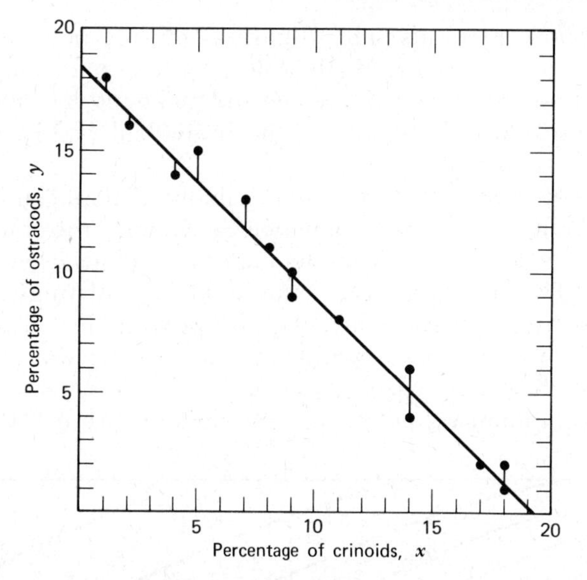

Figure 8-15 Graph of x (crinoids) versus y (ostracods) based on data listed in Table 8-4. Straight line $y = 18.61 - 0.96x$ minimizes sum of squared deviations of predicted values of y from observed values of y.

The required line could be fitted by eye, but we use the criterion of least squares, which can be defined as

$$z_{\min} = \sum (y_{\text{obs}} - y_{\text{trend}})^2 \tag{8-21}$$

where y_{obs} = observed data value
y_{trend} = calculated value on the best-fit line.

We may define y_{trend}, however, in terms of x, a, and b:

$$y_{\text{trend}} = a + bx \tag{8-22}$$

Substituting into the above expression for z, we obtain

$$z = \sum (y_{\text{obs}} - a - bx)^2 \tag{8-23}$$

From here on y_{obs} is written as y, so

$$z = f(a, b) = \sum (y - a - bx)^2 \tag{8-24}$$

The optimization problem is to find those values of a and b that will minimize z, where

$$z = \sum (y - a - bx)^2 \tag{8-25}$$

subject to no constraints. The slightly confusing point here is that, whereas a and b were originally defined as constants characterizing a straight line, now a and b are the independent decision variables of our optimization problem, and the individual x and y data values are treated as constants.

The role of least squares as an optimizing method can be clarified with a diagram (Figure 8-16) in which the value of the objective function z with respect to the two independent variables a and b has been contoured. This diagram was constructed by calculating z for a series of values of a and b, using the computer program listed in Table 8-5, and then fitting contours to points obtained in this way. A sample of output from the program is shown in Table 8-6. By referring to Figure 8-16, the minimum value of z is at point A, where $a = 18.6$ and $b =$

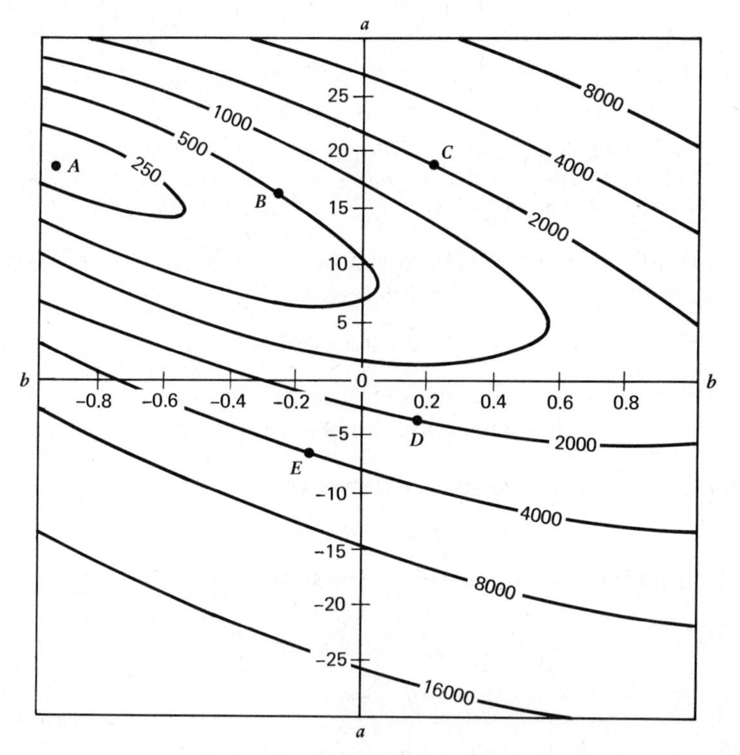

Figure 8-16 Contoured surface portraying values of objective function z with respect to coefficients a and b. Values of z represent sum of squared deviations of observed values from range of possible straight lines, defined by coefficients a and b that have been fitted to hypothetical ostracod–crinoid data of Table 8-4. Points labeled B through E refer to Figure 8-17, where they have been chosen to illustrate fit of lines as defined by coefficient values. Point A refers to Figure 8-15.

TABLE 8-5 FORTRAN Program for Calculating the Sum of Squared Deviations *z* Corresponding to Various Values of Coefficients *a* and *b* in the Straight-Line Equation

```
      DIMENSION X(14), Y(14)
    1 FORMAT(I5, 15F5.0/16F5.0)
    2 FORMAT(1H /7X, 'A', 9X, 'B', 9X, 'Z')
    3 FORMAT(1H , 3F10.3)
      READ(5,1) N, (X(I), I=1,N), (Y(I), I=1,N)
      WRITE(6,2)
      A=-32.0
      DO 20 I=1,15
      A=A+4.0
      B=-1.2
      DO 20 J=1,11
      B=B+0.2
      Z=0.0
      DO 10 M=1,14
      YCALC=A+B*X(M)
   10 Z=Z+(YCALC-Y(M))**2
   20 WRITE(6,3) A, B, Z
      RETURN
      END
```

−0.96. The effects of pairs of coordinate values, *a* and *b* at five points on the contoured surface, labeled A through *E*, are illustrated in Figures 8-15 and 8-17 by plots showing the position of the fitted line corresponding to each pair of values. For example, at point *B* on

TABLE 8-6 Extract from Output Generated by Program Listed in Table 8-5 and Used To Construct Figure 8-16

A	B	Z
12.000	-1.000	693.998
12.000	-0.800	375.239
12.000	-0.600	198.160
12.000	-0.400	162.760
12.000	-0.200	269.040
12.000	0.000	517.000
12.000	0.200	906.640
12.000	0.400	1437.958
12.000	0.600	2110.957
12.000	0.800	2925.637
12.000	1.000	3882.000
16.000	-1.000	134.000
16.000	-0.800	34.440
16.000	-0.600	76.560
16.000	-0.400	260.360
16.000	-0.200	585.840
16.000	0.000	1053.000
16.000	0.200	1661.837
16.000	0.400	2412.357
16.000	0.600	3304.556
16.000	0.800	4338.434
16.000	1.000	5514.000
20.000	-1.000	22.000
20.000	-0.800	141.640
20.000	-0.600	402.960
20.000	-0.400	805.959
20.000	-0.200	1350.637

Figure 8-16, where $z = 500$ corresponds to $a = 16$ and $b = -0.25$, the graph of the line in Figure 8-17 reveals that it is a poor fit to the data points. However, at points C, D, and E he fit is even poorer—for example, at point D, where $z = 2000$, $a = -3.5$, and $b = 0.15$ (Figure 8-17). The graph for point A where z is minimized is shown in Figure 8-15.

In producing Figure 8-16, using the program in Table 8-5, we have essentially used a direct-search technique to find those values of a and b that minimize z. Although useful for illustrative purposes, it is extremely inefficient. Alternatively we can obtain the optimum (minimum value of z) much more directly by employing calculus,

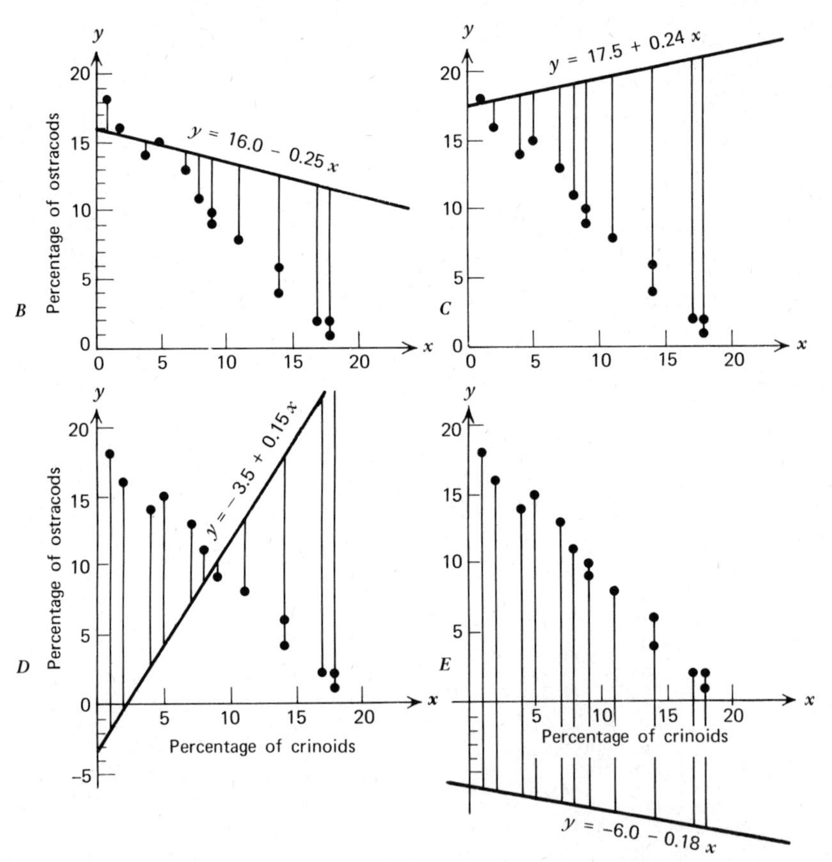

Figure 8-17 Series of graphs B to E relating fit of various straight lines as defined by different values of coefficients a and b. Sum of squared values of z, corresponding to pairs of coefficient values, is shown at points labeled A to E in Figure 8-16. Graph for point A is shown in Figure 8-15.

in which the objective function is differentiated with respect to the two independent variables. This gives rise to two partial derivatives, which are set to zero to obtain the values of a and b.

$$\frac{\partial z}{\partial a}=\frac{\partial z}{\partial b}=0 \tag{8-26}$$

In turn the partial derivatives of Equation 8-25 are

$$\frac{\partial z}{\partial a}=\sum 2(y-a-bx)(-1)=0 \tag{8-27}$$

$$\frac{\partial z}{\partial b}=\sum 2(y-a-bx)(-x)=0 \tag{8-28}$$

which yield on simplification

$$-\sum y+an+b\sum x=0 \tag{8-29}$$

$$-\sum xy+a\sum x+b\sum x^2=0 \tag{8-30}$$

where n is the number of data points. From the first of these equations, a may be expressed in terms of b

$$a=\frac{\sum y-b\sum x}{n} \tag{8-31}$$

Substituting this expression for a in Equation 8-30 yields

$$-\sum xy+\frac{\sum y-b\sum x}{n}\sum x+b\sum x^2=0 \tag{8-32}$$

which can be solved for b as follows:

$$-n\sum xy+\sum x\sum y-b\left(\sum x\right)^2+nb\sum x^2=0$$

$$b\left[n\sum x^2-\left(\sum x\right)^2\right]=n\sum xy-\sum x\sum y$$

$$b=\frac{n\sum xy-\sum x\sum y}{n\sum x^2-\left(\sum x\right)^2} \tag{8-33}$$

Those familiar with simple linear regression will recognize Equation 8-33 as the formula for the slope of the regression line of the dependent variable y on the independent variable x. The coefficient b is the slope of the fitted straight line, and a is the intercept on the y-axis.

The fitting of a straight line to satisfy the least-squares criterion is the simplest of all regression applications involving a single independent variable. The principles of any regression application are identical, regardless of the number of independent variables involved and whether the fitted functions are linear or nonlinear. Thus regression methods can be regarded as a special class of indirect optimization techniques.

Regression applications other than simple linear regression generally involve the use of simultaneous equations expressed in matrix form. Harbaugh and Merriam (1968, pp. 64–66) illustrate the steps in fitting a plane by least squares by creating a matrix equation. The Appendix of the present volume provides a brief introduction to the solution of simultaneous linear equations in matrix form, and most modern general textbooks on statistics deal with regression methods.

LINEAR PROGRAMMING

Optimization problems in which the objective function is a linear combination of the decision variables and is subject to linear inequality constraints can be solved by linear programming methods. In linear programming the problem is to find $x_1, x_2, \cdots, x_n$ so that the linear function

$$z = c_1x_1 + c_2x_2 + \cdots + c_nx_n \tag{8-34}$$

is either maximized or minimized, subject to the constraints

$$\begin{aligned} a_{11}x_1 + a_{12}x_2 + \cdots + a_{1n}x_n &\leq b_1 \\ a_{21}x_1 + a_{22}x_2 + \cdots + a_{2n}x_n &\leq b_2 \\ &\vdots \\ a_{m1}x_1 + a_{m2}x_2 + \cdots + a_{mn}x_n &\leq b_m \end{aligned}$$

and

$$x_1 \geq 0, x_2 \geq 0, \cdots, x_n \geq 0$$

where a_{ij}, b_i, and c_j are given constants.

Let us illustrate the manner in which linear programming can be employed with a hypothetical problem. Let us suppose that a company manufactures ornamental concrete slabs used as facing material by the building industry. These ornamental slabs are made by mixing a concrete aggregate of a certain composition, allowing it to harden, and then cutting and polishing it. Further let us assume that this company makes three different types of slab, *A*, *B*, and *C*, and that the most expensive constituents of these slabs are pebbles of jasper and agate. However, the amounts of jasper and agate pebbles that can be obtained per week are limited. The data for the problem are listed in Table 8-7. Each slab of type *A* requires 30 pounds of jasper pebbles

TABLE 8-7 Data for Linear Programming Problem Involving the Manufacture of Ornamental Slabs

	Jasper Pebbles (lb)	*Agate Pebbles (lb)*	*Profit per Slab*
Slab type *A*	30	10	\$30
Slab type *B*	40	30	\$60
Slab type *C*	10	20	\$20
Maximum pebbles available per week	≤ 2000	≤ 1000	

and 10 pounds of agate pebbles and yields a profit of 30 dollars per slab. Slab type *B*, on the other hand, requires 40 pounds of jasper pebbles and 30 pounds of agate pebbles, yielding a 60-dollar profit. Slab type *C* requires 10 pounds of jasper pebbles and 20 pounds of agate pebbles and yields a 20-dollar profit. The problem facing the company is to decide how many slabs of each type should be made during a week in which 2000 pounds of jasper pebbles and 1000 pounds of agate pebbles are available, so as to maximize profit.

The decision variables are the number of slabs of type *A* (x_1), type *B* (x_2), and type *C* (x_3). Since the objective is to maximize profit z, the problem consists of finding the values of x_1, x_2, and x_3 that maximize z where

$$z = 30x_1 + 60x_2 + 20x_3 \tag{8-35}$$

and where the coefficients 30, 60, and 20 are the profits per slab of type *A*, *B*, and *C*, respectively. The constraints may be expressed as

$$30x_1 + 40x_2 + 10x_3 \leq 2000$$

$$10x_1 + 30x_2 + 20x_3 \leq 1000$$

$$x_1 \geq 0; \qquad x_2 \geq 0; \qquad x_3 \geq 0$$

The first of these constraints pertains to jasper pebbles. Table 8-7 shows that the number of pounds of jasper pebbles needed in slab types *A*, *B*, and *C* are 30, 40, and 10, respectively, but not more than 2000 pounds are available during the week. The second constraint pertains to agate pebbles, and the other constraints dictate that a negative number of slabs is impossible.

Given this combination of decision variables and constraints, there is a unique solution to this problem. To maximize profits the company must manufacture 40 type *A* slabs, 20 type *B* slabs, and no type *C* slabs, yielding a profit of 2400 dollars for the week. With these values of x_1, x_2, and x_3 no surplus jasper or agate pebbles remain. This solution was obtained by using the simplex method, which we discuss next.

SIMPLEX METHOD

The simplex method is the major method for solving linear programming problems. A clear introduction to the method is provided by Hillier and Lieberman (1967, pp. 138–160). The principles of the method can be illustrated with an example involving two decision variables, permitting a graphic solution. Consider the equation

$$z = 2x_1 + 3x_2 \tag{8-36}$$

where z is to be maximized subject to the restrictions

$$x_1 \leqslant 5; \qquad x_2 \leqslant 4$$

and

$$4x_1 + 3x_2 \leqslant 24; \qquad x_1 \geqslant 0; \qquad x_2 \geqslant 0$$

If we graph x_1 versus x_2 (Figure 8-18), the inequality constraints can be plotted as lines that bound the area of feasible solution, forming an irregular polygon. Contoured values of the objective function are plotted in Figure 8-19. The values of $z = 0$, 6, 12, and 18 correspond to contour lines

$$z = 2x_1 + 3x_2 = 0 \tag{8-37}$$

$$z = 2x_1 + 3x_2 = 6 \tag{8-38}$$

$$z = 2x_1 + 3x_2 = 12 \tag{8-39}$$

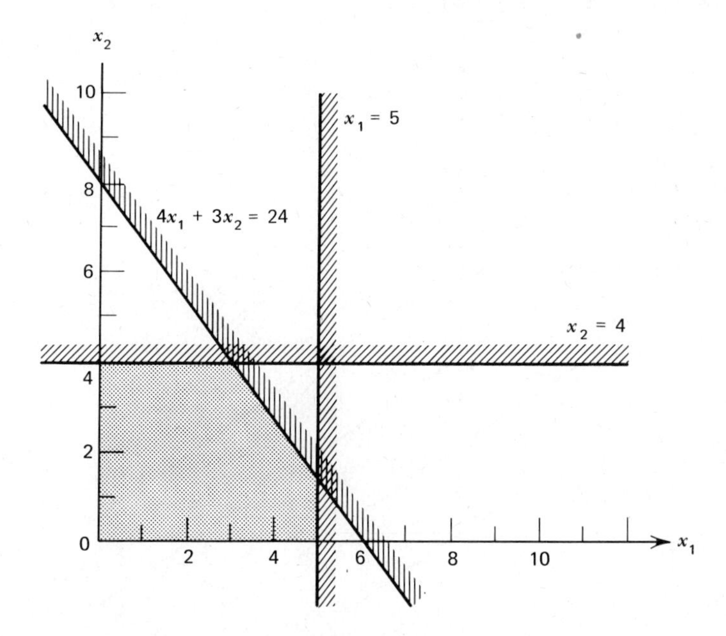

Figure 8-18 Area of feasible solution (shaded) is defined by linear boundaries corresponding to linear constraints. Permissible values of x_1 and x_2 lie inside shaded area.

$$z = 2x_1 + 3x_2 = 18 \tag{8-40}$$

The contours must be straight lines because the objective function is linear. It is clear from the diagram that the optimum solution is defined as the point (x_1,x_2) at which z is a maximum and which falls within the area of feasible solution. Inspection of the graph (Figure 8-19) reveals that this point is at the apex of the shaded area where $x_1 = 3$, $x_2 = 4$. In fact, with a linear objective function and linear constraints, the optimum solution must always lie at one of the apices of the polygonal area of feasible solution. Each of these apex points is known as a *basic feasible solution.*

The simplex method employs a recursive algebraic procedure that searches the basic feasible solutions until it finds the optimum solution and then stops. At each stage a basic feasible solution is selected and z is calculated. The neighboring basic feasible solutions, reached by moving along an edge of the polygonal area, are progressively searched as long as the value of z continues to increase (Figure 8-19). If in passing from one solution to the next a smaller value of z is obtained, the search ends, because the solution obtained previously is necessarily the optimum. This follows because the collection of

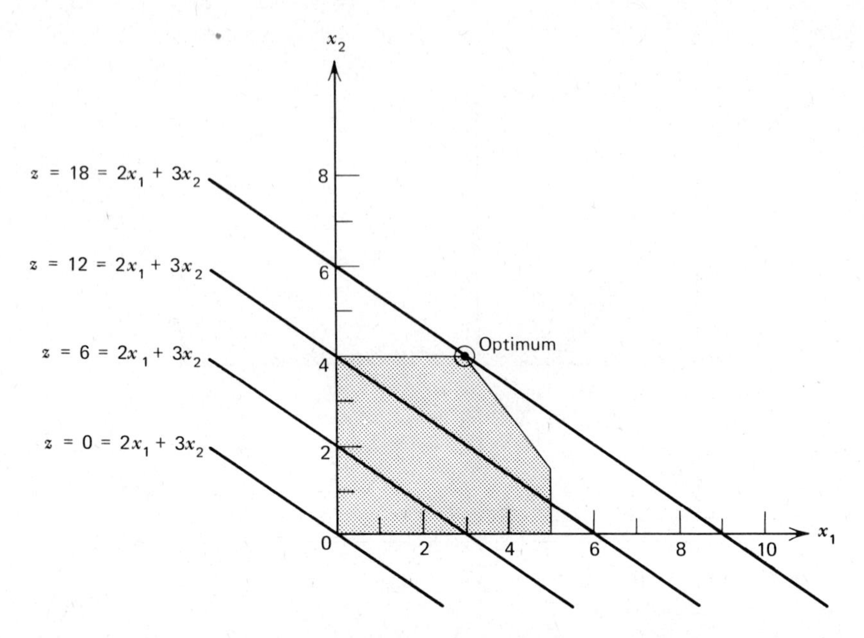

Figure 8-19 Contours of linear objective function z superimposed on area of feasible solution (shaded) defined by constraints shown in Figure 8-18.

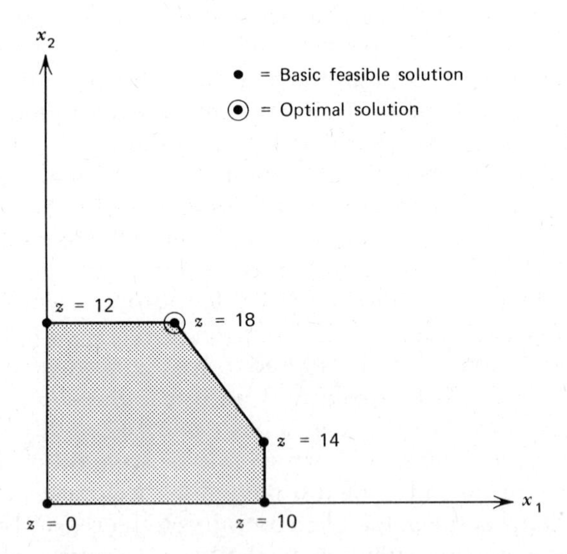

Figure 8-20 Five basic feasible solutions at points on the perimeter of area of feasible solution. Points lie at apices of a polygon which is sequentially searched for optimum solution.

basic feasible solutions constitutes a *convex set*, a rather elusive mathematical concept that is precisely described in an appendix in the book of Hillier and Lieberman (1967). This concept can be graphically illustrated with reference to Figure 8-20. By following along any edge of the shaded area from point to point in the convex set, the optimal solution is eventually reached and can be recognized by noticing that the two neighboring points in the set have lower values of z. In other words, the optimum solution is reached by following a sequence of line segments joined together that are continuously convex outward, having no inflection points and no subsidiary peaks. This concept of a convex set holds true for every linear programming problem, irrespective of the number of decision variables (or dimensions) involved. Figure 8-21 represents how a set of basic feasible solutions might appear to be partly concave (Figure 8-21*a*). A set cannot be concave, however, if the area of feasible solution is correctly recognized. The presence of superfluous constraints gives rise to the seeming concavity. If the superfluous constraints are ignored, the area of feasible solution forms a convex set.

Algebraic Representation of the Simplex Method

Thus far we have attempted to explain the principles of the simplex method by geometrical means. Although geometric representation is intuitively appealing, we are incapable of representing relationships with more than two independent variables (decision variables) by strictly geometric means. Consequently we must employ an algebraic representation, where we can deal in principle with any number of decision variables. The problem is centered about the algebraic formulation of the basic feasible solutions.

The first step is to translate the inequalities into equations by introducing artificial or *slack* variables. For example, the inequality

$$x_1 \leqslant 5 \tag{8-41}$$

can be transformed to an equation by adding the slack variable x_3 to the left-hand side; thus

$$x_1 + x_3 = 5 \tag{8-42}$$

Of course $x_3 = 0$ when $x_1 = 5$, but otherwise x_3 must have a positive value. Similarly the inequality

$$x_2 \leqslant 4 \tag{8-43}$$

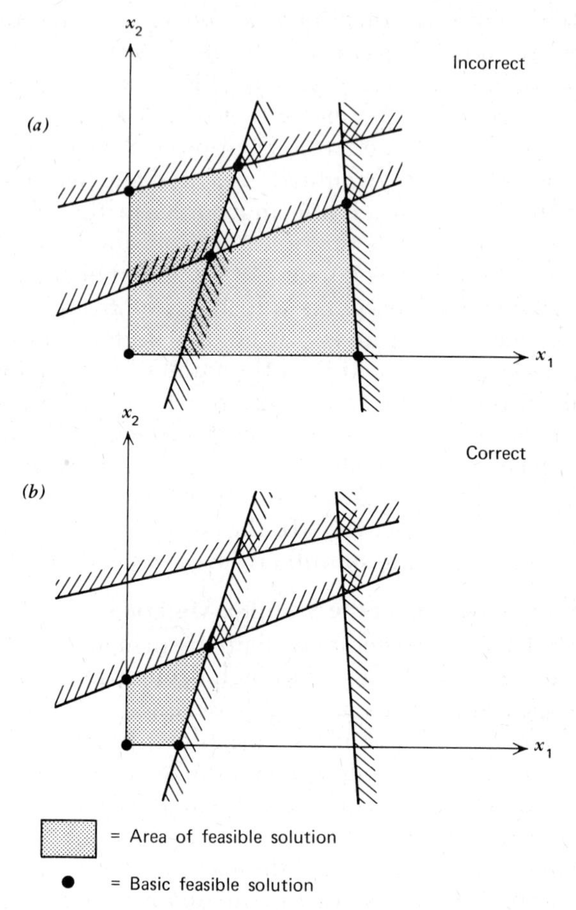

Figure 8-21 Diagrams portraying incorrect and correct solutions in linear programming applications: (*a*) area of feasible solution with seemingly concave reentrant that has been incorrectly specified; (*b*) correctly formulated area of feasible solution is fully convex, indicating that two constraints are superfluous.

can be transformed by adding

$$x_2 + x_4 = 4 \tag{8-44}$$

The constraint

$$4x_1 + 3x_2 \leqslant 24 \tag{8-45}$$

can be similarly altered to

$$4x_1 + 3x_2 + x_5 = 24 \tag{8-46}$$

The optimization problem now can be stated as finding those values of x_1 and x_2 that maximize z where

$$z = 2x_1 + 3x_2 + 0x_3 + 0x_4 + 0x_5 \tag{8-47}$$

subject to

$$\begin{aligned} 1x_1 + 0x_2 + 1x_3 + 0x_4 + 0x_5 &= 5 \\ 0x_1 + 1x_2 + 0x_3 + 1x_4 + 0x_5 &= 4 \\ 4x_1 + 3x_2 + 0x_3 + 0x_4 + 1x_5 &= 24 \end{aligned}$$

and

$$x_j \geq 0 \quad \text{for} \quad j = 1, 2, \cdots, 5$$

where x_3, x_4, and x_5 are the slack variables. The constraints thus form three simultaneous linear equations with five unknowns.

To solve linear simultaneous equations we need to have as many equations as unknowns. The only way of obtaining solutions to the set of three equations above is to assign zero values to two of the unknowns and solve for the remaining three. If there are m equations and n unknowns ($n \geq m$), a *basic* solution is one obtained by solving for m variables and setting ($n - m$) *nonbasic* variables to zero.

For the linear programming example above let us make x_4 and x_5 nonbasic, setting them to zero. The basic variables, x_1, x_2, and x_3, now must be solved. The equations are

$$\underbrace{\begin{bmatrix} 1x_1 + 0x_2 + 1x_3 \\ 0x_1 + 1x_2 + 0x_3 \\ 4x_1 + 3x_2 + 0x_3 \end{bmatrix}}_{\text{basic variables}} \begin{matrix} + \\ + \\ + \end{matrix} \underbrace{\begin{bmatrix} 0x_4 + 0x_5 \\ 1x_4 + 0x_5 \\ 0x_4 + 1x_5 \end{bmatrix}}_{\substack{\text{set nonbasic} \\ \text{variables to zero}}} \begin{matrix} = 5 \\ = 4 \\ = 24 \end{matrix} \qquad \begin{matrix} (8\text{-}48) \\ (8\text{-}49) \\ (8\text{-}50) \end{matrix}$$

which may be written as

$$\begin{aligned} x_1 \qquad\quad + x_3 &= 5 &\quad (8\text{-}51) \\ x_2 \qquad &= 4 &\quad (8\text{-}52) \\ 4x_1 + 3x_2 \qquad &= 24 &\quad (8\text{-}53) \end{aligned}$$

The value of x_2 (=4) can be read directly from Equation 8-52. Substituting Equation 8-52 into Equation 8-53, we find that

$$4x_1 + (3 \times 4) = 24 \tag{8-54}$$

or

$$x_1 = 3$$

Now substituting this value into Equation 8-51, we can solve for x_3:

$$3 + x_3 = 5 \tag{8-55}$$

or

$$x_3 = 2$$

Thus a basic feasible solution has been obtained by making x_4 and x_5 equal to zero, so that $(x_1, x_2, x_3, x_4, x_5) = (3, 4, 2, 0, 0)$. In doing so we have by chance selected the optimal solution directly, as this point corresponds to the maximum value of $z(= 18)$ shown in Figure 8-19. The method by which we solved the simultaneous equations was that of elimination. By setting up the equations in a form where the number of unknowns equal the number of equations, we can always obtain a basic feasible solution by elimination.

Methods for the solution of simultaneous linear equations in matrix form by a technique known as Gaussian elimination are briefly introduced in the Appendix. An appreciation of the basic principles of Gaussian elimination is very helpful for understanding the simplex method.

Returning to our example, the next step is to decide how to choose the *initial* basic feasible solution. Given the initial basic solution, succeeding basic feasible solutions are then sought. In the present example one way would be simply to select any three variables and solve for them. Although this would always give a basic solution, it would not necessarily provide one that was feasible because the variables could be negative. A better procedure is to select the slack variables as the initial basic variables because these will always provide a feasible solution. We have done so in the example below:

$$\underbrace{\begin{bmatrix} 1x_1 + 0x_2 \\ 0x_1 + 1x_2 \\ 4x_1 + 3x_2 \end{bmatrix}}_{\substack{\text{nonbasic} \\ \text{variables} \\ \text{set to zero}}} \begin{matrix} + \\ + \\ + \end{matrix} \underbrace{\begin{bmatrix} 1x_3 + 0x_4 + 0x_5 \\ 0x_3 + 1x_4 + 0x_5 \\ 0x_3 + 0x_4 + 1x_5 \end{bmatrix}}_{\substack{\text{slack variables} \\ \text{made basic}}} \begin{matrix} = 5 \\ = 4 \\ = 24 \end{matrix} \qquad \begin{matrix} (8\text{-}56) \\ (8\text{-}57) \\ (8\text{-}58) \end{matrix}$$

which simplifies to

$$x_3 = 5; \qquad x_4 = 4; \qquad x_5 = 24$$

The initial basic feasible solution is therefore the point $(x_1, x_2, x_3, x_4, x_5) = (0, 0, 5, 4, 24)$ that corresponds to the point at the origin of the graph in Figure 8-19. The value of z is found by substituting into the objective function:

$$z = 2x_1 + 3x_2 + 0x_3 + 0x_4 + 0x_5 \tag{8-59}$$

where we notice that the coefficients of the slack variables are all zero, as they are not contributing to z in the initial form. Thus $z = (2 \times 0) + (3 \times 0) + (0 \times 5) \times (0 \times 4) + (0 \times 24) = 0$.

The simplex method now searches the neighboring basic feasible solutions to see if the value of z can be improved. In practice the next step consists of selecting one of the nonbasic variables in the initial solution as the *entering* variable, making it basic, and making one of the initial basic variables nonbasic. The problem is first to select which nonbasic variable to bring into the solution. This is done by choosing the nonbasic variable that is most likely to make the greatest improvement in the value of z. In the present case

$$z = \underbrace{[2x_1 + 3x_2]}_{\text{nonbasic}} + \underbrace{[0x_3 + 0x_4 + 0x_5]}_{\text{basic}} \tag{8-60}$$

and it seems most likely that x_2 would increase the value of z more rapidly than x_1, since x_2 increases z at the rate of 3 per unit increase of x_2, whereas the rate for x_1 is only 2. Therefore x_2 is selected as the entering variable.

One basic variable is now chosen as the leaving variable. This is done by finding which of the basic variables is first reduced to zero by increasing the value of the entering variable, which is x_2 in this example. This permits the greatest increase in z by increasing x_2, without actually producing a negative value for variable x_3, x_4, or x_5, which is not permissible. This can be best explained by the following table:

Equation	*Maximum Increase in x_2*
$x_3 = 5 - x_1$	No limit
$x_1 = 4 - x_2$	$x \leqslant 4$
$x_5 = 24 - 4x_1 - 3x_2$	$x_2 \leqslant 8$

In the first equation in the table x_2 does not appear, so increasing x_2 has no effect. If we increase x_2 in the second equation in the table, we find that $x_2 = 4$ is the maximum value possible without making x_4 negative (and the solution therefore infeasible). In the third equation the maximum increase in x_2 is 8, in the case of $x_1 = 0$, as this would make $x_5 = 0$. Any larger value of x_2, irrespective of the values of x_1, would give negative values to x_5. Thus x_4 is chosen as the leaving variable, as the maximum permissible increase in x_2 for this variable is 4, and this is smaller than the maximum permissible values for x_3 (no limit) and x_5 (8).

The new nonbasic variables may now be isolated by brackets:

$$1x_1 \quad 0x_2 + 1x_3 + 0x_4 + 0x_5 = 5 \tag{8-61}$$

$$0x_1 + 1x_2 + 0x_3 + 1x_4 + 0x_5 = 4 \tag{8-62}$$

$$4x_1 + 3x_2 + 0x_3 + 0x_4 + 1x_5 = 24 \tag{8-63}$$

or

$$z - [2x_1 + 0x_4] - [3x_2 + 0x_3 + 0x_5] = 0 \tag{8-64}$$

$$[1x_1 + 0x_4] + [0x_2 + 1x_3 + 0x_5] = 5 \tag{8-65}$$

$$[0x_1 + 1x_4] + [1x_2 + 0x_3 + 0x_5] = 4 \tag{8-66}$$

$$[4x_1 + 0x_4] + [3x_2 + 0x_3 + 1x_5] = 24 \tag{8-67}$$

nonbasic variables   basic variables

where the objective function has also been included as Equation 8-64. By Gaussian elimination, x_2 (the entering variable) is now eliminated from every equation except Equation 8-66. Equation 8-64 is altered by adding three times Equation 8-66, and Equation 8-67 is altered by subtracting three times Equation 8-66. This yields

$$z - [2x_1 + 3x_4] + [0x_2 + 0x_3 + 0x_5] = 12 \tag{8-68}$$

$$[1x_1 + 0x_4] \quad [0x_2 + 1x_3 + 0x_5] = 5 \tag{8-69}$$

$$[0x_1 + 1x_4] + [1x_2 + 0x_3 + 0x_5] = 4 \tag{8-70}$$

$$[4x_1 - 3x_4] + [0x_2 + 0x_3 + 1x_5] = 12 \tag{8-71}$$

nonbasic variables   basic variables

In this form the values of the basic variables and z can be read directly as $z = 12$, $x_2 = 4$, $x_3 = 5$, and $x_5 = 12$. This point $(x_1, x_2, x_3, x_4, x_5) = (0, 4, 5, 0, 12)$ can be identified on Figure 8-19 as the point on the x_2-axis where $z = 12$.

In order to ascertain algebraically whether this solution is optimal we need to examine the objective function. We shall not refer to it in its original form, but in its new form as Equation 8-68. We need to decide whether an increase in either x_1 or x_4 will produce an increase in z. Rearranged, Equation 8-68 can be written as

$$z = 12 + 2x_1 - 3x_4 \tag{8-72}$$

If $x_1 = 0$, any increase in x_4 will decrease the value of z. With $x_4 = 0$, however, an increase in x_1 will produce an increase in z; thus the present solution is not optimal, and x_1 is the next entering variable and is now made basic.

To select the leaving variable we ascertain which basic variable is reduced to zero most rapidly by increasing x_1, the entering variable.

Equation	*Maximum Increase in* x_1
$x_3 = 5 - x_1$	$\leqslant 5$
$x_2 = 4 - x_4$	No limit
$x_5 = 12 - 4x_1 + 3x_4$	$\leqslant 3$

Therefore x_5 is selected as the leaving variable. The new arrangement of basic and nonbasic variables is

$$z + [3x_4 + 0x_5] - [2x_1 + 0x_2 + 0x_3] = 12 \tag{8-73}$$

$$[0x_4 + 0x_5] + [1x_1 + 0x_2 + 1x_3] = 5 \tag{8-74}$$

$$[1x_4 + 0x_5] + [0x_1 + 1x_2 + 0x_3] = 4 \tag{8-75}$$

$$-[3x_4 + 1x_5] + [4x_1 + 0x_2 + 0x_3] = 12 \tag{8-76}$$

nonbasic variables — basic variables

By using Equation 8-76, x_1 can be eliminated from Equations 8-73 and 8-74. Equation 8-73 is changed by adding Equation 8-76 divided by 2, and Equation 8-74 is changed by subtracting Equation 8-76 divided by 4; Equation 8-75 remains unchanged. This yields

$$z + [\tfrac{3}{2}x_4 + \tfrac{1}{2}x_5] + [0x_1 + 0x_2 + 0x_3] = 18 \tag{8-77}$$

$$[\tfrac{3}{4}x_4 - \tfrac{1}{4}x_5] + [0x_1 + 0x_2 + 1x_3] = 2 \tag{8-78}$$

$$[1x_4 + 0x_5] + [0x_1 + 1x_2 + 0x_3] = 4 \tag{8-79}$$

$$-[\tfrac{3}{4}x_4 + \tfrac{1}{4}x_5] + [1x_1 + 0x_2 + 0x_3] = 3 \tag{8-80}$$

nonbasic variables — basic variables

As before, we can now read off the basic feasible solution from the equations arranged in this way. This solution is the point $(x_1, x_2, x_3, x_4, x_5) = (3, 4, 2, 0, 0)$ which we determined previously as being optimal, at a value of $z = 18$.

We can further check that this is indeed optimal by examining the new form of the objective function

$$z = 18 - \tfrac{3}{2}x_4 - \tfrac{1}{2}x_5 \tag{8-81}$$

If we set $x_4 = 0$, any increase in x_5 reduces z from 18. Similarly, if we set $x_5 = 0$, any increase in x_4 will decrease z. Thus we have confirmed that this is an optimal solution, and the simplex iterations are stopped.

We may summarize the steps of the simplex method as follows:

Step 1. Introduce the slack variables. Obtain the initial feasible solution by making the slack variables basic.

Step 2. Select the new entering basic variable from the previous nonbasic variables. Choose the variable that increases z at the fastest rate in the objective-function equation.

Step 3. Select the leaving variable from the old basic variables. This is the basic variable that is reduced to zero first by increasing the new entering variable.

Step 4. Determine the new basic feasible solution by solving for the basic variables by Gaussian elimination.

Step 5. Check to see if the solution is optimal. This will be the case if the sign of the coefficients of the nonbasic variables in the objective function are all positive (or all negative if the variables are on the right-hand side). If the solution is not optimal, return to step 2 and repeat.

Simplex Tableau

The simplex tableau is a useful way of displaying the essential details of calculations pertaining to the simplex method in shorthand form. For the initial representation of the equation in the present example the simplex tableau is arranged as follows:

Basic Variable	*Row Number*	*Coefficient of* z	x_1	x_2	x_3	x_4	x_5	*Right Side of Equation*
z	0	1	−2	−3	0	0	0	0
x_3	1	0	1	0	1	0	0	5
x_4	2	0	0	1	0	1	0	4
x_5	3	0	4	3	0	0	1	24

We can now use this concisely expressed information to proceed as before, starting at step 2:

Step 2. Inspecting row 0 and noticing that the values of x_1, x_2, x_3, and x_4 are all written as if they were on the left side of the equation, we select as the entering variable the one with the smallest coefficient, which in this case is x_2 with a value of −3.

Step 3. To find the leaving variable inspect the x_2 column, omitting row 0. Calculate the ratio of the right-hand side of the equation to the coefficient of x_2 for each row and select the smallest. In this case $4/1 \leqslant 24/3 \leqslant 5/0$, so the leaving variable is selected as x_4.

Step 4. The next simplex tableau is constructed by reorganizing the basic variable column to include x_2 in place of x_4. Since 1 is the coefficient of x_2 in row 2, (the row vacated by the leaving variable x_4), that entire row is divided by 1. Then, since −3, 0, and 3 are the coefficients of x_2 in rows 0, 1, and 3, subtract −3, 0, and 3 times row 2 from those rows, respectively. This gives the following tableau:

Basic Variable	*Row Number*	Coefficient of						*Right Side of Equation*
		z	x_1	x_2	x_3	x_4	x_5	
z	0	1	−2	0	0	3	0	12
x_3	1	0	1	0	1	0	0	5
x_2	2	0	0	1	0	1	0	4
x_5	3	0	4	0	0	−3	1	12

Step 5. This basic feasible solution is checked for optimality by scanning row 0 for negative values of the x coefficients. Since x_1 is negative, the solution is not optimal. Another iteration using steps 2 to 5 is required, using the same shorthand tableau notation. This is left to the reader as an exercise.

Computer Program

A computer program for solving small linear programming problems is listed in Table 8-8. This particular program is restricted to 12 variables. Many computation centers have linear programming packages in their libraries that can handle several hundred variables and are equipped to deal with unusual conditions that may sometimes arise in the solution. The algorithm in Table 8-8 is very straightforward and simply represents a coding of the rules outlined in the preceding section.

A sample problem has been worked to illustrate the input and output from the problem. The problem is to find the values of x that maximize z, where

$$z = -3x_1 - x_2 + 5x_3 + 2x_4 \tag{8-82}$$

subject to the constraints

$$\begin{aligned} x_1 + 5x_2 + 2x_3 - x_4 &\leqslant 2 \\ 2x_1 - x_2 + 4x_3 + 3x_4 &\leqslant 5 \\ 6x_1 + 2x_2 + x_3 + 3x_4 &\leqslant 3 \\ x_i \geqslant 0; \qquad i &= 1,2,3 \end{aligned}$$

TABLE 8-8 FORTRAN Program for Linear Programming by the Simplex Method

```
C.....SIMPLEX METHOD OF LINEAR PROGRAMMING
C          NVAR            NO. OF VARIABLES
C          NEQU            NO. OF EQUATIONS
C          COEF(I,J)       COEFFS ON LEFT SIDE OF CONSTRAINT EQUATIONS
C          RTSID(I)        TERMS ON RIGHT SIDE OF CONSTRAINT EQUATIONS
C          WEIGHT(I)       COEFFS ON LEFT SIDE OF OBJECTIVE FUNCTION EQUAT
      DIMENSION COEF(12,12), WEIGHT(12), IBASIC(12), RTSID(12)
    1 FORMAT(16I5)
    2 FORMAT(10F8.0)
    3 FORMAT(1H /////' SIMPLEX TABLEAU NUMBER', I5// 3X, 'BASIC', 4X,
     1 'EQUATION', 4X, 'COEFFICIENTS'/ 2X, 'VARIABLE', 3X, 'NUMBER',
     2 6X, 'Z', 12I8)
    4 FORMAT(1H , 3X, 'Z', 10X, '0', 9X, '1', 12F8.1)
    5 FORMAT(1H , 2X, I2, I11, 9X, '0', 12F8.1)
    6 FORMAT(1H // ' OBJECTIVE FUNCTION =', F10.3)
    7 FORMAT(1H ,' ENTERING VARIABLE =',I5, 5X, 'LEAVING VARIABLE =',I5)
    8 FORMAT(1H , 'THIS SOLUTION IS OPTIMAL')
      READ(5,1) NVAR, NEQU, NTIM
      READ(5,2) (WEIGHT(J), J=1,NVAR), OBJFUN
      DO 15 I=1,NEQU
   15 READ(5,2) (COEF(I,J), J=1,NVAR), RTSID(I)
      READ(5,1) (IBASIC(I), I=1,NEQU)
C.....BEGIN MAJOR ITERATIVE CYCLE
      DO 90 NT=1, NTIM
      WRITE(6,3) NT, (J,J=1,NVAR)
C.....FIND VARIABLE WITH SMALLEST WEIGHT IN OBJECTIVE FUNCTION. THIS
C.....BECOMES ENTERING VARIABLE JENTER
      SMALL1=WEIGHT(1)
      JENTER=1
      DO 20 J=2,NVAR
      IF (WEIGHT(J).GE.SMALL1) GO TO 20
      SMALL1=WEIGHT(J)
      JENTER=J
   20 CONTINUE
C.....FIND LEAVING VARIABLE ILEAVE
      SMALL2=999999.9
      DO 30 I=1,NEQU
      IF (COEF(I,JENTER).LE.0.000) GO TO 30
      RATIO=RTSID(I)/COEF(I,JENTER)
      IF (RATIO.GE.SMALL2) GO TO 30
      SMALL2=RATIO
      ILEAVE=I
   30 CONTINUE
C.....WHICH VARIABLE CORRESPONDS TO ILEAVE-TH EQUATION?
      JLEAVE=IBASIC(ILEAVE)
C.....WRITE OUT SIMPLEX TABLEAU
      WRITE(6,4) (WEIGHT(J), J=1,NVAR), OBJFUN
      DO 40 I=1, NEQU
   40 WRITE(6,5) IBASIC(I), I, (COEF(I,J), J=1,NVAR), RTSID(I)
      WRITE(6,6) OBJFUN
C.....IF SOLUTION IS OPTIMAL, STOP ITERATIONS
      IF (SMALL1.GE.0.0.OR.SMALL2.EQ.999999.9) GO TO 100
      WRITE(6,7) JENTER, JLEAVE
C.....PERFORM GAUSSIAN ELIMINATION TO OBTAIN NEXT FEASIBLE SOLUTION
C.....DIVIDE ILEAVE-TH ROW BY COEF(ILEAVE,JENTER)
      CON=COEF(ILEAVE,JENTER)
      DO 50 J=1,NVAR
   50 COEF(ILEAVE,J)=COEF(ILEAVE,J)/CON
      RTSID(ILEAVE)=RTSID(ILEAVE)/CON
```

TABLE 8-8 (*contd.*)

```
C.....ELIMINATE JENTER-TH VARIABLE FROM REMAINING EQUATIONS
      DO 70 I=1,NEQU
      IF (I.EQ.ILEAVE) GO TO 70
      CON=COEF(I,JENTER)
      DO 60 J=1,NVAR
C.....SUBTRACT COEF(I,JENTER) TIMES COEF(ILEAVE,J) FROM COEF(I,J)
   60 COEF(I,J)=COEF(I,J)-CON*COEF(ILEAVE,J)
      RTSID(I)=RTSID(I)-CON*RTSID(ILEAVE)
   70 CONTINUE
C.....PERFORM ELIMINATION PROCEDURE ON OBJECTIVE FUNCTION TOO
      CON=WEIGHT(JENTER)
      DO 80 J=1,NVAR
   80 WEIGHT(J)=WEIGHT(J)-CON*COEF(ILEAVE,J)
      OBJFUN=OBJFUN-CON*RTSID(ILEAVE)
C.....CORRECT IBASIC LIST
      IBASIC(ILEAVE)=JENTER
   90 CONTINUE
      RETURN
  100 WRITE(6,8)
      RETURN
      END
```

Introducing slack variables and reorganizing the equations, we obtain

$$z+3x_1+1x_2-5x_3-2x_4+0x_5+0x_6+0x_7=0 \qquad (8\text{-}83)$$
$$1x_1+5x_2+2x_3-1x_4+1x_5+0x_6+0x_7=2 \qquad (8\text{-}84)$$
$$2x_1-1x_2+4x_3+3x_4+0x_5+1x_6+0x_7=5 \qquad (8\text{-}85)$$
$$6x_1+2x_2+1x_3+3x_4+0x_5+0x_6+1x_7=3 \qquad (8\text{-}86)$$
$$x_i \geqslant 0; \qquad i=1,2,\cdots,7$$

The data input for this example is listed in Table 8-9 and consists of the following cards or lines:

Line 1: Number of variables (7)
Number of constraint equations (3)
Maximum number of simplex iterations (20)

Line 2: Coefficients of x on the left side of the objective-function equation (3, 1, −5, −2, 0, 0, 0)
Value of constant on the right side of the objective-function equation (0)

Line 3: Coefficients of x on the left side of the first-constraint equation (1, 5, 2, −1, 1, 0, 0)
Value of constant on the right side of the first-constraint equation (2)

Line 4: Same as line 3, but with values for the second-constraint equation

Line 5: Same as line 3, but with values for the third-constraint equation

Line 6: Indices of the variables to be used in the initial basic feasible solution, normally the slack variables (5, 6, 7)

The program output using data of Table 8-9 is listed in Table 8-10. In each simplex tableau there are columns for the basic variables, equation numbers, and coefficients of z and $x_1, x_2, \cdots, x_7$; the final column contains the term from the right side of the equations. The

TABLE 8-9 Data Input for Example Used To Illustrate Computer Program for the Simplex Method

7	3	20						
	3	1	-5	-2	0	0	0	0
	1	5	2	-1	1	0	0	2
	2	-1	4	3	0	1	0	5
	6	2	1	3	0	0	1	3
5	6	7						

value of the objective function corresponds to the value of the term on the right side of the objective-function equation. The solution at any one stage of the calculation is read directly by inspecting the basic variable index in the first column and reading its value in the last column. For example, in the first tableau we find that $x_5 = 2$, $x_6 = 5$, and $x_7 = 3$. These are the slack variables that have been made the initial basic variables, and $x_1 = x_2 = x_3 = x_4 = 0$.

In the second tableau the previously basic variable x_5 has been replaced by the previously nonbasic variable x_3. Again we can read directly the values of the basic variables as $x_3 = 1$, $x_6 = 1$, $x_7 = 2$ at a value of $z = 5.0$. Noticing that the coefficient of x_4 in the objective-function equation is still negative, we recognize that this solution is not optimal. By substituting x_4 for x_6 for the next basic solution, we obtain the third tableau. This solution is $x_3 = 1.1$, $x_4 = 0.2$, $x_7 = 1.3$ at $z = 5.9$. Since there are no negative coefficients in the objective function equation, this solution is optimal. The values of x for the optimal solution are therefore $(x_1, x_2, x_3, x_4, x_5, x_6, x_7) = (0, 0, 1.1, 0.2, 0, 0, 1.3)$.

Difficulties in the Use of the Simplex Method

A number of difficulties may arise in applying the simplex method, and readers are referred to one of the many books that deal solely with linear programming (such as that by Smythe and Johnson, 1966) for more complete treatment. Two problems are briefly mentioned here. The first is related to inequality constraints where the left side

TABLE 8-10 Output from Computer Run Illustrating Use of Program for the Simplex Method Listed in Table 8-8, Employing Input Data of Table 8-9

```
SIMPLEX TABLEAU NUMBER    1

 BASIC     EQUATION    COEFFICIENTS
VARIABLE    NUMBER      Z      1      2      3      4      5      6      7      .
   Z          0         1    3.0    1.0   -5.0   -2.0    0.0    0.0    0.0    0.0
   5          1         0    1.0    5.0    2.0   -1.0    1.0    0.0    0.0    2.0
   6          2         0    2.0   -1.0    4.0    3.0    0.0    1.0    0.0    5.0
   7          3         0    6.0    2.0    1.0    3.0    0.0    0.0    1.0    3.0

OBJECTIVE FUNCTION =    0.000
 ENTERING VARIABLE =    3      LEAVING VARIABLE =    5


SIMPLEX TABLEAU NUMBER    2

 BASIC     EQUATION    COEFFICIENTS
VARIABLE    NUMBER      Z      1      2      3      4      5      6      7
   Z          0         1    5.5   13.5   -0.0   -4.5    2.5   -0.0   -0.0    5.0
   3          1         0    0.5    2.5    1.0   -0.5    0.5    0.0    0.0    1.0
   6          2         0   -0.0  -11.0   -0.0    5.0   -2.0    1.0   -0.0    1.0
   7          3         0    5.5   -0.5   -0.0    3.5   -0.5   -0.0    1.0    2.0

OBJECTIVE FUNCTION =    5.000
 ENTERING VARIABLE =    4      LEAVING VARIABLE =    6


SIMPLEX TABLEAU NUMBER    3

 BASIC     EQUATION    COEFFICIENTS
VARIABLE    NUMBER      Z      1      2      3      4      5      6      7
   Z          0         1    5.5    3.6   -0.0   -0.0    0.7    0.9   -0.0    5.9
   3          1         0    0.5    1.4    1.0   -0.0    0.3    0.1   -0.0    1.1
   4          2         0    0.0   -2.2    0.0    1.0   -0.4    0.2    0.0    0.2
   7          3         0    5.5    7.2   -0.0   -0.0    0.9   -0.7    1.0    1.3

OBJECTIVE FUNCTION =    5.900
THIS SOLUTION IS OPTIMAL
```

of the inequality is *greater than* the right side (previously we always discussed *less than* cases). The second concerns the problem of *equality* rather than inequality constraints.

At first it may seem that neither of these conditions is problematical. However, let us examine a very simple case in which the constraint is an inequality going the "wrong" way. Let us maximize

$$z = 3x_1 + 2x_2 \tag{8-87}$$

subject to

$$x_1 + 5x_2 \geqslant 6; \qquad 2x_1 - 4x_2 \leqslant 8; \qquad x_i \geqslant 0; \qquad i = 1, 2$$

Introducing slack variables these conditions may be rewritten as

$$z - 3x_1 - 2x_2 + 0x_3 + 0x_4 = 0 \tag{8-88}$$
$$1x_1 + 5x_2 - 1x_3 + 0x_4 = 6 \tag{8-89}$$
$$2x_1 - 4x_2 + 0x_3 + 1x_4 = 8 \tag{8-90}$$

In order to make the inequality $x_1 + 5x_2 \geqslant 6$ into an equation a slack variable is subtracted from the left side, to give Equation 8-89. However, this is no longer a basic feasible solution, as the coefficient of the slack variable must be +1. Furthermore we cannot avoid the problem by changing all the signs since this would yield

$$-1x_1 - 5x_2 + 1x_3 + 0x_4 = -6 \tag{8-91}$$

Although this is a legitimate *basic initial* solution, it is not feasible as $x_3 = -6$, and one of the constraints of the simplex method is that all variables be nonnegative.

One way of avoiding this problem is to add still another slack variable (x_5), so that the equations now read

$$z - 3x_1 - 2x_2 + 0x_3 + 0x_4 + 0x_5 = 0 \tag{8-92}$$
$$1x_1 + 5x_2 - 1x_3 + 0x_4 + 1x_5 = 6 \tag{8-93}$$
$$2x_1 - 4x_2 + 0x_3 + 1x_4 + 0x_5 = 8 \tag{8-94}$$

The initial basic feasible solution to this new problem is obtained by making x_4 and x_5 basic, so the solution is $(x_1, x_2, x_3, x_4, x_5) = (0, 0, 0, 8, 6)$ and the rules against nonnegativity are obeyed. However, by adding another slack variable, the constraint

$$x_1 + 5x_2 - x_3 + x_5 = 6$$

has become meaningless, since x_3 and x_5 could take on any values without violating the original constraint that $x_1 + 5x_2 \geqslant 6$.

The way around this dilemma is to make certain that the extra slack variable x_5 is always equal to zero in the optimal solution. This can be achieved by driving the value to zero by the "big M" method, as discussed by Hillier and Lieberman (1967), where M is some large positive number. The objective-function equation is rewritten

$$z = 3x_1 + 2x_2 - Mx_5 \tag{8-95}$$

and the maximum value of z must occur when $x_5 = 0$ (negativity is not permitted). However, x_5 is used in the initial basic feasible solution, so it must be eliminated from the objective function before testing for optimality and bringing in a new entering variable. In this example it can be accomplished as follows:

$$z - 3x_1 - 2x_2 + 0x_3 + 0x_4 + Mx_5 = 0 \tag{8-96}$$
$$1x_1 + 5x_2 - 1x_3 + 0x_4 + 1x_5 = 6 \tag{8-97}$$
$$2x_1 - 4x_2 + 0x_3 + 1x_4 + 0x_5 = 8 \tag{8-98}$$

Subtracting M times Equation 8-97 from Equation 8-96 gives a new objective function with a zero coefficient for x_5:

$$z - (3 + \mathrm{M})x_1 - (2 + 5M)x_2 + Mx_3 + 0x_4 + 0x_5 = -6M \tag{8-99}$$

If we make $M = 1000$, the equations become

$$z - 1003x_1 - 5002x_2 + 1000x_3 + 0x_4 + 0x_5 = -6000 \tag{8-100}$$
$$1x_1 + 5x_2 - 1x_3 + 0x_4 + 1x_5 = 6 \tag{8-101}$$
$$2x_1 - 4x_2 + 0x_3 + 1x_4 + 0x_5 = 8 \tag{8-102}$$

which is solved in the normal way, making x_4 and x_5 the initial basic variables. The computer output shown in Table 8-11 represents the three iterations needed to reach an optimal solution for this problem. The optimal solution is $(x_1, x_2, x_3, x_4, x_5) = (4.6, 0.3, 0, 0, 0)$ at $z = 14.3$. Although there is still a negative coefficient in the objective-function equation (the coefficient for $x_3 = -1.1$), if either x_1 or x_2 is excluded from the basis and x_3 is entered, x_3 takes on a negative value; this is not legal, and thus the simplex iterations are stopped.

The second type of difficulty arises when equality constraints are used instead of inequality constraints. Again the problem is in obtaining an initial basic feasible solution. Let us take the problem of minimizing z where

$$z = -3x_1 - 2x_2 \tag{8-103}$$

TABLE 8-11 Computer Solution to Linear Programming Problem Involving Inequality Going the "Wrong" Way

SIMPLEX TABLEAU NUMBER 1

BASIC	EQUATION	COEFFICIENTS						
VARIABLE	NUMBER	Z	1	2	3	4	5	
Z	0	1	-1003.0	-5002.0	1000.0	0.0	0.0	-6000.0
5	1	0	1.0	5.0	-1.0	0.0	1.0	6.0
4	2	0	2.0	-4.0	0.0	1.0	0.0	8.0

OBJECTIVE FUNCTION = -6000.000
ENTERING VARIABLE = 2 LEAVING VARIABLE = 5

SIMPLEX TABLEAU NUMBER 2

BASIC	EQUATION	COEFFICIENTS						
VARIABLE	NUMBER	Z	1	2	3	4	5	
Z	0	1	-2.6	-0.0	-0.4	-0.0	1000.4	2.4
2	1	0	0.2	1.0	-0.2	0.0	0.2	1.2
4	2	0	2.8	-0.0	-0.8	1.0	0.8	12.8

OBJECTIVE FUNCTION = 2.398
ENTERING VARIABLE = 1 LEAVING VARIABLE = 4

SIMPLEX TABLEAU NUMBER 3

BASIC	EQUATION	COEFFICIENTS						
VARIABLE	NUMBER	Z	1	2	3	4	5	
Z	0	1	-0.0	-0.0	-1.1	0.9	1001.1	14.3
2	1	0	-0.0	1.0	-0.1	-0.1	0.1	0.3
1	2	0	1.0	0.0	-0.3	0.4	0.3	4.6

OBJECTIVE FUNCTION = 14.285
THIS SOLUTION IS OPTIMAL

subject to the constraints

$$x_1 + 5x_2 = 6; \qquad 2x_1 - 4x_2 \leqslant 8; \qquad x_i \geqslant 0; \qquad i = 1, 2$$

Minimization problems are exactly comparable to maximization, except that the sign of the objective function is changed. In other words, minimizing z in Equation 8-103 is equivalent to maximizing z where

$$z = -(-3x_1 - 2x_2) \tag{8-104}$$

or

$$z = 3x_1 + 2x_2 \tag{8-105}$$

The signs of the coefficients in the objective-function equation should be adjusted according to whether the problem is one of maximization or minimization. Then the computer program in Table 8-8, which is designed for maximization problems only, can be directly applied.

Writing out the equation by including a slack variable for the inequality, we obtain

$$z - 3x_1 - 2x_2 + 0x_3 = 0 \tag{8-106}$$
$$x_1 + 5x_2 + 0x_3 = 6 \tag{8-107}$$
$$2x_1 - 4x_2 + 1x_3 = 8 \tag{8-108}$$

There is only one slack variable, but there are a total of three unknown variables and two equations, so we need two basic variables to obtain a basic feasible solution. The problem can be resolved by adding another slack variable, x_4, to the equality constraint and once again using the "big M" method to ensure that this extra slack variable is not a basic variable when the optimum is reached. By doing this we can not only obtain an initial basic feasible solution but also ensure that the contribution of the slack variable x_4 to the equality (Equation 8-107) is zero at the optimal solution. The equations now become

$$z - 3x_1 - 2x_2 + 0x_3 + Mx_4 = 0 \tag{8-109}$$
$$x_1 + 5x_2 + 0x_3 + 1x_4 = 6 \tag{8-110}$$
$$2x_1 - 4x_2 + 1x_3 + 0x_4 = 8 \tag{8-111}$$

Eliminating x_4 from the objective-function equation by subtracting M times Equation 8-110 yields

$$z - (3 + M)\,x_1 - (2 + 5M)\,x_2 + 0x_3 + 0x_4 = -6M \tag{8-112}$$

If we set $M = 1000$, the equations for the simplex solution are

$$z - 1003x_1 - 5002x_2 + 0x_3 + 0x_4 = -6000 \tag{8-113}$$
$$1x_1 + 5x_2 + 0x_3 + 1x_4 = 6 \tag{8-114}$$
$$2x_1 - 4x_2 + 1x_3 + 0x_4 = 8 \tag{8-115}$$

This problem was solved by using the computer program listed in Table 8-8, with output listed in Table 8-12. The optimum solution is identical to that found for the preceding problem at the point $(x_1, x_2, x_3, x_4) = (4.6, 0.3, 0, 0)$.

TABLE 8-12 Output from Linear Programming Problem Involving One Equality Constraint Instead of Inequality Constraint

```
SIMPLEX TABLEAU NUMBER     1

 BASIC      EQUATION     COEFFICIENTS
VARIABLE     NUMBER      Z        1         2         3         4
   Z           0         1 -1003.0 -5002.0       0.0       0.0 -6000.0
   4           1         0     1.0     5.0       0.0       1.0     6.0
   3           2         0     2.0    -4.0       1.0       0.0     8.0

OBJECTIVE FUNCTION = -6000.000
 ENTERING VARIABLE =     2       LEAVING VARIABLE =       4


SIMPLEX TABLEAU NUMBER     2

 BASIC      EQUATION     COEFFICIENTS
VARIABLE     NUMBER      Z        1         2         3         4
   Z           0         1     -2.6     -0.0     -0.0   1000.4      2.4
   2           1         0      0.2      1.0      0.0      0.2      1.2
   3           2         0      2.8     -0.0      1.0      0.8     12.8

OBJECTIVE FUNCTION =      2.398
 ENTERING VARIABLE =     1       LEAVING VARIABLE =       3


SIMPLEX TABLEAU NUMBER     3

 BASIC      EQUATION     COEFFICIENTS
VARIABLE     NUMBER      Z        1         2         3         4
   Z           0         1     -0.0     -0.0      0.9   1001.1     14.3
   2           1         0     -0.0      1.0     -0.1      0.1      0.3
   1           2         0      1.0      0.0      0.4      0.3      4.6

OBJECTIVE FUNCTION =     14.285
THIS SOLUTION IS OPTIMAL
```

Mining Application

Let us now employ the simplex method to solve a linear programming problem in mining. A hypothetical mining company owns two very small mines, both of which produce the same kinds of ore. The mines are located in different areas, and the production capacities and production costs are different at the two mines. At each mine the ore is graded in three classes: high, medium, and low grade. Furthermore the mining company is under contract to supply a milling and smelting company with a certain tonnage of each grade of ore each week. The data are listed in Table 8-13.

Thus mine *A* produces 6 tons per day of high-grade ore, 2 tons per day of medium-grade ore, and 4 tons per day of low-grade ore, at a total cost of 200 dollars per day. At mine *B* the daily production is 2, 2, and 12 tons for high-, medium-, and low-grade ore, respectively,

Table 8-13 Hypothetical Data for Mining Application of Linear Programming[a]

	High-Grade Ore	*Medium-Grade Ore*	*Low-Grade Ore*	*Daily Cost* ($)
Daily production in mine *A* (tons)	6	2	4	200
Daily production in mine *B* (tons)	2	2	12	160
Weekly tonnage requirements	$\geqslant 12$	$\geqslant 8$	$\geqslant 24$	

[a]Adapted from Kemeny et al. (1966), with permission of Prentice-Hall Inc.

at a cost of 160 dollars per day. The mill requires a weekly supply of 12 tons of high-grade ore, 8 tons of medium-grade ore, and 24 tons of low-grade ore. The mining company wishes to know how many days each mine should be operated in order to minimize costs. The problem can be stated in linear programming terms as

$$\text{minimize} \qquad z = 200x_1 + 160x_2 \tag{8-116}$$

$$\text{subject to} \qquad \begin{aligned} 6x_1 + 2x_2 &\geqslant 12 \\ 2x_1 + 2x_2 &\geqslant 8 \\ 4x_1 + 12x_2 &\geqslant 24 \\ x_1 \geqslant 0; \quad x_2 &\geqslant 0 \end{aligned}$$

where x_1 = number of days of production per week at mine *A*
x_2 = number of days of production per week at mine *B*
z = objective function (cost per week)

In order to use our simplex algorithm we must first express the objective function in such a form that it is maximized instead of minimized. Next we must add two slack variables to each equation, driving one of them to zero by using the big *M* method. Our equations are now written as follows:

$$z + 200x_1 + 160x_2 + 0x_2 + 0x_4 + 0x_5 - Mx_6 - Mx_7 - Mx_8 = 0 \tag{8-117}$$
$$6x_1 + 2x_2 - 1x_3 + 0x_4 + 0x_5 + 1x_6 + 0x_7 + 0x_8 = 12 \tag{8-118}$$
$$2x_1 + 2x_2 + 0x_3 - 1x_4 + 0x_5 + 0x_6 + 1x_7 + 0x_8 = 8 \tag{8-119}$$
$$4x_1 + 12x_2 + 0x_3 + 0x_4 - 1x_5 + 0x_6 + 0x_7 + 1x_8 = 24 \tag{8-120}$$

where $z = -$(cost), the objective function to be maximized
x_3, x_4, x_5 = slack variables introduced to transform the inequalities to equalities

$x_6, x_7, x_8 =$ slack variables added for obtaining an initial basic feasible solution, but which must be excluded from the optimal solution

$M =$ a large number (say 1000)

The slack variables x_6, x_7, and x_8 are to be driven to zero by using the big M method. Accordingly these variables must now be eliminated from the objective function as already described:

$$z+(200-6M-2M-4M)x_1+(160-2M-2M-12M)x_2 \\ +Mx_3+Mx_4+Mx_5=-(12M+8M+24M)$$

If we set $M = 1000$, this expression simplifies to

$$z-11800x_1-15840x_2+1000x_3+1000x_4+1000x_5=-44000 \quad (8\text{-}122)$$

The data input and line-printed output for this problem are listed in Tables 8-14 and 8-15, respectively. We observe that the minimum operating cost is 680 dollars per week, and it is achieved by operating mine A for 1 day per week and mine B for 3 days per week. The objective-function values in this case are negative, because the problem has been formulated in terms of maximizing the negative cost, which is equivalent to minimizing the positive cost. Notice that this problem could also be solved graphically, since there are only two decision variables.

TABLE 8-14 Input Data for Mine-Scheduling Problem

```
     8      3     20
-11800 -15840        1000      1000      1000         0         0         0  -44000
     6      2          -1         0         0         1         0         0      12
     2      2           0        -1         0         0         1         0       8
     4     12           0         0        -1         0         0         1      24
     6      7      8
```

Problems

Problem 8-1

For a rectangular river channel of depth x_1 and width x_2, what ratio of depth to breadth will maximize the cross-sectional area for a given wetted perimeter? Formulate this problem in terms of an objective function and constraints. Solve the problem by using both differential calculus and graphical means.

TABLE 8-15 Output for Mine-Scheduling Problem[a]

SIMPLEX TABLEAU NUMBER 1

BASIC VARIABLE	EQUATION NUMBER	COEFFICIENTS Z	1	2	3	4	5	6	7	8	
Z	0	1	-11800.0	-15840.0	1000.0	1000.0	1000.0	0.0	0.0	0.0	-44000.0
6	1	0	6.0	2.0	-1.0	0.0	0.0	1.0	0.0	0.0	12.0
7	2	0	2.0	2.0	0.0	-1.0	0.0	0.0	1.0	0.0	8.0
8	3	0	4.0	12.0	0.0	0.0	-1.0	0.0	0.0	1.0	24.0

OBJECTIVE FUNCTION =-44000.000
ENTERING VARIABLE = 2 LEAVING VARIABLE = 8

SIMPLEX TABLEAU NUMBER 2

BASIC VARIABLE	EQUATION NUMBER	COEFFICIENTS Z	1	2	3	4	5	6	7	8	
Z	0	1	-6520.0	-0.0	1000.0	1000.0	-320.0	-0.0	-0.0	1320.0	-12320.0
6	1	0	5.3	-0.0	-1.0	-0.0	0.2	1.0	-0.0	-0.2	8.0
7	2	0	1.3	-0.0	-0.0	-1.0	0.2	-0.0	1.0	-0.2	4.0
2	3	0	0.3	1.0	0.0	0.0	-0.1	0.0	0.0	0.1	2.0

OBJECTIVE FUNCTION =-12320.000
ENTERING VARIABLE = 1 LEAVING VARIABLE = 6

SIMPLEX TABLEAU NUMBER 5

BASIC VARIABLE	EQUATION NUMBER	COEFFICIENTS Z	1	2	3	4	5	6	7	8	
Z	0	1	-0.0	-0.0	10.0	70.0	-0.0	990.0	930.0	1000.0	-680.0
1	1	0	1.0	-0.0	-0.2	0.2	-0.0	0.2	-0.2	-0.0	1.0
5	2	0	0.0	0.0	2.0	-8.0	1.0	-2.0	8.0	-1.0	16.0
2	3	0	-0.0	1.0	0.2	-0.7	-0.0	-0.2	0.7	-0.0	3.0

OBJECTIVE FUNCTION = -680.003
THIS SOLUTION IS OPTIMAL

[a]Tableaux from iterations 3 and 4 have been omitted for the sake of brevity. Minimum weekly cost ($680) is possible by operating mine *A* for 1 day per week and mine *B* for 3 days per week.

Problem 8-2

a. Given the following data on percentages of algae and corals in a limestone, compute the intercept and slope of the straight line that best fits these data, employing Equations 8-31 and 8-33, derived from the least-squares criterion.

Specimen Number	*Algae*	*Corals*
1	0.8	3.1
2	1.9	3.2
3	2.7	4.3
4	4.0	6.1
5	4.7	6.2
6	4.7	8.0
7	5.5	4.6
8	5.8	6.7
9	8.8	7.6
10	7.8	9.0

b. Write a computer program to determine by sequential evaluation the shape of the sum of squared deviations surface, as a function of intercept and slope of straight lines fitted to the data points. This problem is similar to the case illustrated in Figure 8-16 and Tables 8-5 and 8-6.

Problem 8-3

Modify the data in Table 8-7 and then calculate the number of rock slabs of each type that will maximize profit. Do this first by hand, using the simplex method. Second, obtain the solution by computer, using the program listed in Table 8-8.

Problem 8-4

Using the mine-production and mill-requirement data of Table 8-13, compute an optimum solution assuming that the daily operating costs differ from those listed in Table 8-13 as follows: mine *A* is 300 dollars per day and mine *B* is 200 dollars per day.

Annotated Bibliography

Bellman, R. E., and Dreyfus, S. E., 1962, *Applied Dynamic Programming*, Princeton University Press, Princeton, N.J., 363 pp.

Chapter 1 contains a readable discussion of difficulties encountered in using calculus methods for optimization. The other topics covered in this book—such as optimal search techniques, feedback-control processes, and Markovian decision processes—are highly relevant, but are treated at a mathematical level that is too advanced for the average geological reader.

Clyde, C. G., Jensen, B. C., and Milligan, J. H., 1967, "Optimizing Conjunctive Use of Surface Water and Groundwater," in *Proceedings of Symposium on Groundwater Development in Arid Basins*, Utah State University, pp. 59–86.

The authors show how linear programming can be used in the design of water resource systems.

Dougherty, E. L., and Smith, S. T., 1966, "The Use of Linear Programming to Filter Digitized Map Data," *Geophysics*, v. 31, pp. 253–259.

A polynomial trend surface can be fitted to map data using linear programming to determine the coefficient values that minimize the sum of the *absolute* values of the deviations. This contrasts with the usual least-squares approach, in which the coefficients are found by setting their partial derivatives with respect to each map coordinate to zero. Points which lie far from the fitted plane receive higher weights than points close to the plane when squared deviations are used.

Drew, L. J., 1966, *Grid-Drilling Exploration and Its Application to the Search for Petroleum*, Ph.D. Thesis, Pennsylvania State University, 141 pp.

Pioneer work dealing with the statistics of simulated grid-drilling programs.

Drew, L. J., 1967, "Grid-Drilling Exploration and Its Application to the Search for Petroleum," *Economic Geology*, v. 62, pp. 698–710.

Summarizes work presented in thesis (Drew, 1966).

Drew, L. J., and Griffiths, J. C., 1965, "Size, Shape, and Arrangement of Some Oilfields in the U.S.A.," *Short Course and Symposium on Computers and Computer Applications in Mining and Exploration*, University of Arizona, v. 3, pp. FF1–FF31.

Statistics of the geometry of oil fields is an important factor in devising systematic exploration programs.

Garrels, R. M., and Mackenzie, F. T., Jr., 1969, "Sedimentary Rock Types; Relative Proportions as a Function of Geological Time," *Science*, v. 163, pp. 570–571.

Presents histogram showing the distribution of presently preserved masses of sedimentary rocks according to geologic age.

Greenwood, H. J., 1967, "The N-Dimensional Tie Line Problem," *Geochimica et Cosmochimica Acta*, v. 31, pp. 465–490.

The point at which several lines intersect in an *N*-dimensional phase diagram can be determined by solving the system of linear algebraic equations that defines the lines. Such unique points are difficult to obtain in phase diagrams of naturally occurring mineral assemblages, because of errors of measurement and the lack of equilibrium conditions during mineral formation. The author shows how such a problem can be tackled using linear programming. Intersection of compositional lines occurs over an *N*-dimensional area (area of feasible solution), within which a unique point can be found by minimizing an objective function.

Griffiths, J. C., 1966*a*, "Exploration for Natural Resources," *Operations Research*, v. 14, pp. 189–209.

Discusses probabilistic search model involving a grid-drilling program in oil exploration.

Griffiths, J. C., 1966*b*, "Grid Spacing and Success Ratios in Exploration for Natural Resources," *Symposium and Short Course on Computers and Operations Research in Mineral Industries*, Pennsylvania State University, Mineral Industries Exploration Station Special Publication, v. 1, 24 pp.

Discusses the frequency distribution of "successes" in simulated grid-drilling program in oil exploration.

Griffiths, J. C., and Drew, L. J., 1964, "Simulation of Exploration Programs for Natural Resources by Models," *Quarterly of the Colorado School of Mines*, v. 59, pp. 187–206.

Describes a theoretical exploration system model that involves drilling on a grid basis, followed by statistical analysis of the data. Experiments with the model have been conducted in the Ohio–Pennsylvania–West Virginia oil-producing region.

Harbaugh, J. W., and Merriam, D. F., 1968, *Computer Applications in Stratigraphic Analysis*, John Wiley and Sons, New York, 282 pp.

Pages 62 to 66 briefly treat regression methods, showing how a matrix equation is derived by using the least-squares criterion for fitting polynomials to map data.

Hillier, F. S., and Lieberman, G. J., 1967, *Introduction to Operations Research*, Holden-Day, San Francisco, 639 pp.

A clearly written book that contains pertinent material in Chapters 5, 6, 7, and 8. Chapters 13 and 14 on Markov chains and simulation, respectively, are also of interest.

Horowitz, I., 1965, *An Introduction to Quantitative Business Analysis*, McGraw-Hill, New York, 270 pp.

Chapter 6 contains a readable introduction to linear programming. Chapter 7 discusses integer programming, similar to linear programming but dealing with cases where integers are used instead of floating-point numbers. In Chapter 8 a highly important application of linear programming, the so-called transportation problem, is clearly presented.

Jacobs, O. L. R., 1967, *An Introduction to Dynamic Programming; the Theory of Multistage Decision Processes*, Chapman and Hall, London, 126 pp.

A compact book dealing with an important optimization technique, not covered by the present chapter.

Kemeny, J. G., Snell, L., and Thompson, G. L., 1966, *Introduction to Finite Mathematics*, (Second edition), Prentice-Hall, Englewood Cliffs, N.J., 465 pp.

Chapter 6 provides a good introduction to linear programming and game theory. The mining application described above is solved graphically instead of by the algebraic simplex method.

Miesch, A. T., Connor, J. J., and Eicher, R. N., 1964, "Investigation of Geochemical Sampling Problems by Computer Simulation," *Quarterly of the Colorado School of Mines*, v. 59, pp. 131–148.

Discusses a model for simulating different sampling designs for mapping geochemical data. The model is concerned with finding the optimal sampling design, consisting of the design that uncovers sufficient information at minimal cost. The approach is a direct-search method involving simulation.

Rosen, R., 1967, *Optimality Principles in Biology*, Butterworths, London, 198 pp.

Discusses the relevance of optimization theory to biology. The author describes some mathematical techniques (mainly calculus methods and open-loop control systems), and treats such topics as the allometric law, homeostasis, regulators, and adaptive systems. There is much of interest for geologists in this book.

Shaffer, L. R., Ritter, J. B., and Meyer, W. L., 1965, *The Critical Path Method*, McGraw-Hill, New York, 216 pp.

The critical path method (CPM) is an optimization technique for aiding the planning, scheduling, and operation of industrial projects. It is a numerical technique widely used by operations-research specialists.

Smythe, W. R., and Johnson, L. A., 1966, *Introduction to Linear Programming*, Prentice-Hall, Englewood Cliffs, N.J., 221 pp.

A good elementary book on linear programming. It is especially useful for its chapter on topics dealing with linear algebra, which gives a firm foundation for understanding the Gauss–Jordan method of elimination for solving simultaneous linear equations.

Vajda, S., 1961, *Mathematical Programming*, Addison-Wesley, Reading, Mass., 310 pp.

This book not only provides a thorough treatment of linear programming but also introduces nonlinear programming. Numerous examples from operations research show how linear programming can be applied in business and industry.

Wilde, D. J., 1964, *Optimum Seeking Methods*, Prentice-Hall, Englewood Cliffs, N.J., 202 pp.

Much of the material in this book is included in Wilde and Beightler's more comprehensive book listed below. Wilde's book contains an excellent discussion of direct-search methods.

Wilde, D. J., and Beightler, C. S., 1967, *Foundations of Optimization*, Prentice-Hall, Englewood Cliffs, N.J., 480 pp.

Although this book is highly mathematical, it provides an excellent overview of optimization techniques, classifying them according to similarities and differences.

CHAPTER 9

Sedimentation Applications

This chapter deals principally with five types of sedimentation-simulation applications:

1. A sedimentary-rock-mass model
2. A generalized model dealing with gross aspects of clastic sedimentation in sedimentary basins
3. An evaporite-basin model
4. A deltaic sedimentation model
5. A carbonate sedimentation and ecology model

Models 3, 4, and 5 deal with three spatial dimensions and geologic time. Model 2 is concerned with two spatial dimensions through time, being confined to a vertical cross section through a sedimentary basin. Model 1 does not consider spatial relationships, being concerned only with variations in time. The five models are described in detail, whereas other pertinent sedimentation applications are briefly treated in the annotated bibliography.

SEDIMENTARY-ROCK-MASS MODEL

Changes in the mass or volume of material incorporated as sedimentary rock in the earth's crust through geologic time pose a challenging problem. A major question is whether the mass of material has remained more or less constant through geologic time or whether the mass has increased in some systematic manner. Garrels and Mackenzie* have approached the problem by developing two simple mass-balance models that attempt to explain the distribution through time of sedimentary rock masses in the earth's crust. Figure 9-1 is a histogram showing rough estimates of the masses of sedimentary rock, classified according to geologic age, that are present in the earth's

*Much of the information in this section dealing with sedimentary-rock-mass models was derived from the manuscript for the book *Evolution of Sedimentary Rocks* by R. M. Garrels and F. T. Mackenzie, Jr., to be published by Norton and Company. Their concepts and models have been adapted to our purpose, showing how these models can be programmed for computers.

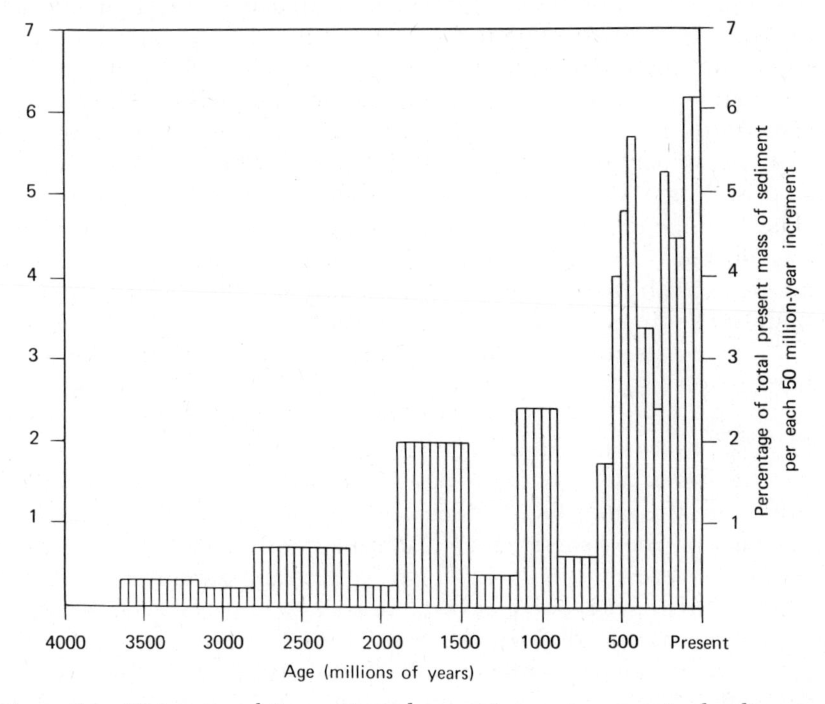

Figure 9-1 Histogram relating estimated present mass, in percent, of sedimentary rocks ranked according to geologic age. Age in millions of years is represented on horizontal scale. Vertical scale represents percentage of total present mass of sedimentary rocks, in 50 million year units (each unit is represented by width of a narrow column). Sum of all of narrow columns equals 100 percent. Metasedimentary rocks that are sufficiently similar to sedimentary rocks are classified as sedimentary for construction of histogram. Adapted from Garrels and Mackenzie (1969). Copyright by the American Association for the Advancement of Science.

crust today. Metasedimentary rocks that are sufficiently similar in chemical composition and other aspects to sedimentary rocks are classified as sedimentary for the construction of this histogram. On the other hand, metasedimentary rocks with a composition in the range of igneous rocks are assumed to have been destroyed and are classified as igneous, thereby being excluded from the histogram.

This histogram is based on geologic data from many parts of the world. Estimates of the masses of sedimentary rocks are difficult to make, inasmuch as they include assumptions as to the thicknesses and areal extent of sedimentary rock masses. In places these rock masses are deeply buried and relatively inaccessible for direct scrutiny. The irregularities in the histogram reflect these uncertainties, but the histogram shows that there is a general exponential increase in the mass of preserved sedimentary material with decreasing geologic age. The gross changes are due to the fact that previously deposited sediment is continually subject to erosion and redeposition as newly deposited sediment, as well as being subject to destruction by metamorphism and melting. It is clear that a large proportion of the sediment of any age must consist of recycled material. The rates of recycling and of sediment deposition are critical in interpreting the histogram.

The average depositional rate for sediments that are younger than Precambrian is estimated to be about 5×10^{15} grams per year. If this rate were assumed to apply over the past 4 billion years, the total amount of sediment deposited during this time span would be about 200×10^{23} grams, or 200,000 geograms (one geogram = 10^{20} grams). This amount is four or five times greater than the estimated present mass of sediment, which is on the order of 50,000 geograms. If this rate of deposition is assumed, it is essential also to assume a recycling rate that is compatible with the depositional rate.

Although the large fluctuations in the histogram reflect difficulties in estimating rock masses, they also represent secular fluctuations in the rates of erosion, metamorphism, and granitization. Factors that include climate, vegetative cover, and rock type are clearly important in affecting the rate of erosion. Gregor (1968) has suggested that the post-Precambrian part of the histogram reflects two well-defined rates of erosion. The more rapid rate pertains to the pre-Carboniferous interval before the widespread appearance of land plants, and the slower rate pertains to the interval from the start of the Carboniferous to the present. Garrels and Mackenzie (1969) show that the mass–age distribution of evaporites, carbonates, and clastics, considered separately, strongly suggests different rates of erosion and recycling, as one would expect from differences in their solubility.

If the fluctuations in the histogram are ignored, thus assuming a constant rate of erosion through time, two simple theoretical models may be constructed and used to experiment with different rates of erosion in an attempt to match the observed time distribution of sediment masses. The two models are (a) a constant-mass model and (b) a linear-accumulation model. Below we develop the mathematics of the models and illustrate their manipulation with a FORTRAN program.

Development of Models

The major processes affecting the total sedimentary rock mass are shown diagrammatically in Figure 9-2. Sedimentary rock material is continually being cycled by erosion and deposition. Newly deposited sediment is partly derived from older sediments and partly from the

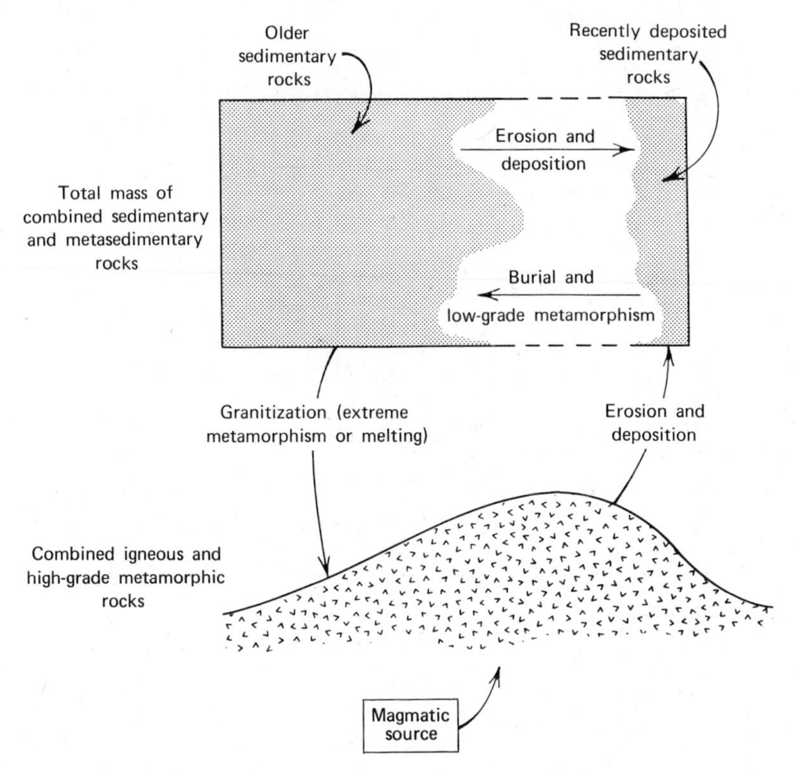

Figure 9-2 Schematic diagram illustrating exchanges of material between mass of sedimentary rocks and mass of combined igneous and high-grade metamorphic rocks. Igneous rocks are presumed to be derived from magmatic sources as well as from melting of sedimentary rocks.

weathering of undifferentiated igneous and metamorphic rocks. On the other hand, some of the sediment mass is continually subject to obliteration by melting and high-grade metamorphism ("granitization"), causing sedimentary material to be returned to the mass of undifferentiated igneous and high-grade metamorphic rocks.

The two models reflect the balance between the granitization of sedimentary rocks and the erosion of igneous rocks (Figure 9-3). The

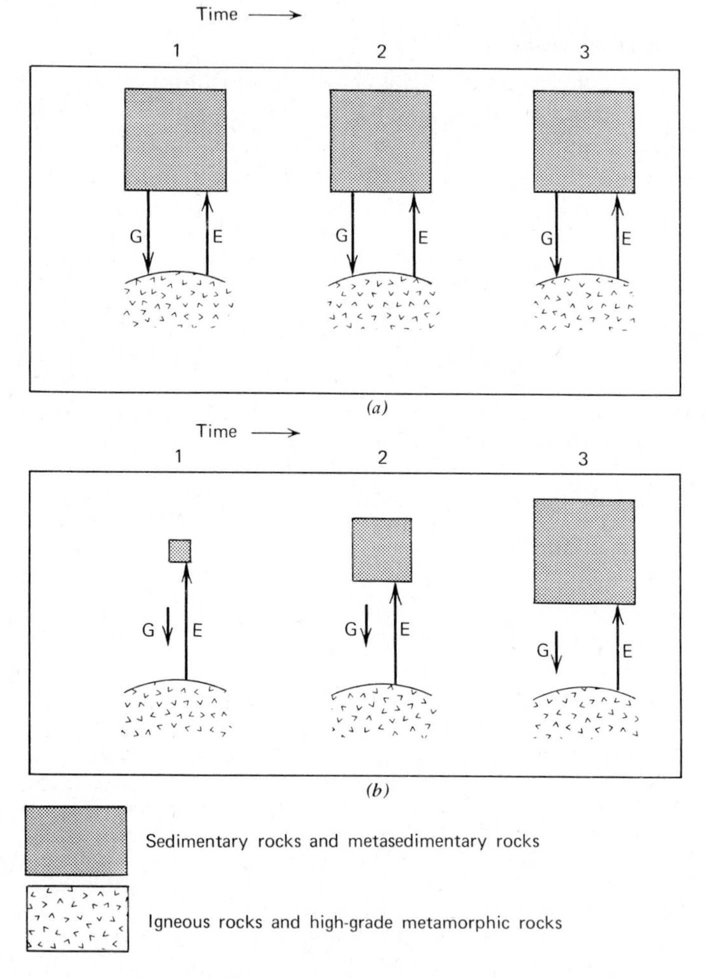

Figure 9-3 Diagrams contrasting constant-mass model (*a*) with linear-accumulation model (*b*). Melting and erosion rates balance in constant-mass model, whereas erosion rate exceeds melting rate in linear-accumulation model. G = granitization and melting of sedimentary rocks; E = erosion.

constant-mass model assumes that the granitization of sedimentary rocks and the erosion of igneous rocks of primary or secondary origin have been balanced through time, so that the total mass of sediments has remained constant. This implies in turn that an initial mass of sediment was formed very rapidly in the early stages of the earth's history, following cooling and degassing, and that this constant mass has been repeatedly recycled in the crust. The linear-accumulation model, on the other hand, assumes that the rate of erosion of igneous rocks has exceeded granitization and remelting of sediments, resulting in a progressive increase in the total mass of sedimentary rocks. A linear increase for the accumulation model has been assumed for the sake of simplicity in the model, but a nonlinear function could be employed instead. The accumulation model assumes that the total sedimentary mass began at zero and grew linearly to its present value.

Constant-Mass Model

In developing the constant-mass model let M be the total mass of sedimentary rock. At time 0 the age of the total mass is 0. During a finite time interval Δt the mass of sediment deposited is equal to the mass eroded from older sedimentary rocks, plus the mass eroded from igneous rocks. During the same time interval the mass of old sediments is reduced by obliteration, partly by erosion and partly by melting. We assume that the mass obliterated by melting equals the sediment mass recently derived from the erosion of igneous rocks. Thus the sediment mass deposited during any time interval equals the sediment mass obliterated. We can therefore state that during any Δt the amount of sediment deposited is constant and equal to some fixed fraction of the total sedimentary mass M. Thus at time 1 there will be a small mass of newly deposited sediment aged 0 and the remaining original mass will have advanced in age to time increment 1 (Figure 9-4). During the second time interval obliteration by erosion and melting will remove a fraction of material from rocks of each age, the fraction being a constant proportion of the mass of each age. In turn the mass of newly deposited sediment will equal the mass of obliterated sediment. By repeating the process, we can generate a mass–age histogram similar to the gross shape of the histogram based on observations (Figure 9-1).

The process of continuing obliteration and deposition is an illustration of the law of growth and decay, or exponential law. The decay of radioactive material and the growth of daughter products (e.g., uranium to lead and helium) is a familiar geological manifestation of this law. The law states that the amount of material lost (or,

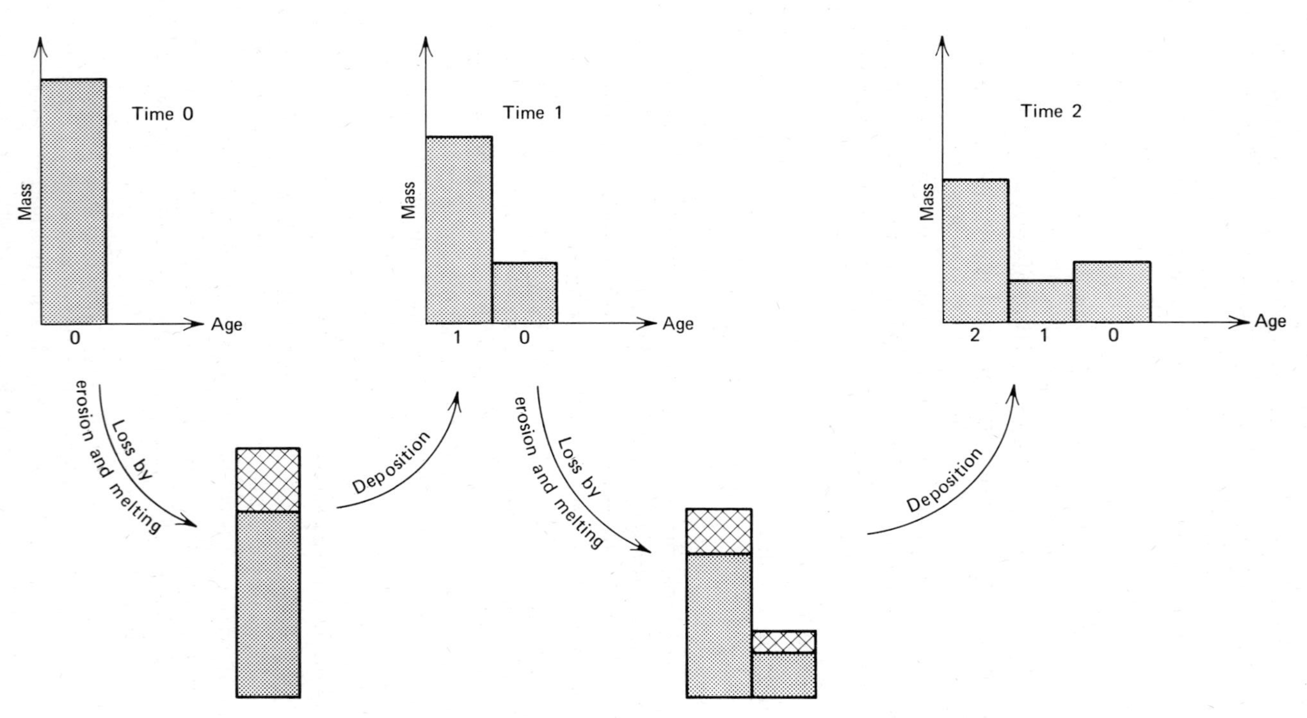

Figure 9-4 Schematic diagram illustrating method of generating histogram for constant-mass model through time. Model is shown advancing from time 0 to time 2. Cross-hatching denotes proportion of mass eroded from previously deposited sediment and from igneous rock. Mass contributed from weathered igneous rock is equal to mass of sedimentary rock "lost" through melting. Therefore total old sedimentary mass lost by both erosion and melting equals mass deposited as new sediment during each time increment.

alternatively, produced) per unit time is proportional to the amount present. The basic equation in differential form is

$$-\frac{dM}{dt} = \alpha M \tag{9-1}$$

Where M = mass
t = time
α = decay constant

The negative sign is employed when a relationship involving decay or loss is implied.

If Equation 9-1 is integrated and applied to our sediment-mass model, we obtain

$$M_t = M_0 e^{-\alpha t} \tag{9-2}$$

where M_t = mass remaining at time t
M_0 = mass at time 0

In dealing with the law of growth and decay it is convenient to express decay rates in half-lives. In a sediment-mass model, however, we can express a half-life as a *half-mass*, defined as the amount of time required for half of the material initially present to have been lost by erosion and melting. The relationship between half-life and the decay constant is

$$\frac{M_t}{M_0} = \tfrac{1}{2} = e^{-\alpha t_{0.5}} \tag{9-3}$$

where $t_{0.5}$ is the half-mass in time units. If we solve for $t_{0.5}$, we obtain

$$t_{0.5} = \frac{-\log_e 0.5}{\alpha} = \frac{0.69315}{\alpha}$$

or

$$\alpha = \frac{0.69315}{t_{0.5}}$$

Calculation of the amount of mass transferred in sedimentary overturn employs an adaptation of Equation 9-2. During time Δt the total mass M is reduced by erosion and melting to $Me^{-\alpha\Delta t}$, and the amount of material deposited during the interval is equivalent to the amount removed. Therefore the net amount deposited M_d during each time increment is equal to the total mass M, minus the amount that remains, $Me^{-\alpha\Delta t}$

$$M_d = M - Me^{-\alpha\Delta t} = M(1 - e^{-\alpha\Delta t}) \tag{9-4}$$

We can employ this equation in a geological time sequence by dividing the total time span into N time increments, each Δt long, and labeling each time increment i, where $i = 1, 2, \cdots, N$, with $i = 1$ as the oldest and $i = N$ as the youngest. Then the present age of rocks deposited during the ith time increment is $(N - i)\Delta t$. The mass of sediment deposited during any time increment is M_d. Applying Equation 9-2, the mass remaining today, M_r, from sediments deposited in the ith time increment is given by

$$M_r = M_d \, e^{-\alpha(N-i)\Delta t} \tag{9-5}$$

Rearranging superscripts and substituting for M_d, we obtain

$$M_r = M(1 - e^{-\alpha\Delta t})e^{\alpha(i-N)\Delta t} \tag{9-6}$$

This is the basic equation employed in our model. Given M, α, Δt, N, and e, we can obtain mass remaining M_r for values of i ranging from 1 to N. This is readily accomplished in a simple DO-loop in the FORTRAN program of Table 9-1. We also need to calculate the amount of the initial mass that remains, M_{r0}

$$M_{r0} = Me^{-\alpha N\Delta t} \tag{9-7}$$

and the number of times the sedimentary mass has been completely recycled or turned over, R,

$$R = N(1 - e^{-\alpha\Delta t}) \tag{9-8}$$

Linear-Accumulation Model

Now let us examine an alternative model that assumes that the total mass of sediment has accumulated linearly through time. If we use M to denote the present total mass of sediment, the total mass M_i present at time i in the past is given by

$$M_i = i\frac{M}{N} \tag{9-9}$$

This relationship implies that the mass varies linearly from 0 at $i = 0$ to M at $i = N$, increasing by the fraction M/N with a unit increase in i. The amount of sediment deposited during any time increment will also increase with increasing i, no longer being equal to the constant M_d but instead corresponding to a variable M_{di}. Thus

$$M_{di} = M_{i-1}(1 - e^{-\alpha\Delta t}) + \frac{M}{N} \tag{9-10}$$

This equation states that the mass deposited during the ith time increment equals the amount derived from the erosion of the total sedimentary mass during the $(i-1)$ time increment plus the fraction M/N derived from the erosion of igneous rock and newly added to the mass of sedimentary rock. Substituting for M_{i-1}, we obtain

$$M_{di} = (i-1)\frac{M}{N}(1-e^{-\alpha\Delta t}) + \frac{M}{N} \tag{9-11}$$

Finally the amount remaining at present of sediments deposited during the ith time interval is

$$M_r = M_{di}e^{-\alpha(N-i)\Delta t} \tag{9-12}$$

Substituting for M_{di} and rearranging, we obtain

$$M_r = \left\{(i-1)\frac{M}{N}(1-e^{-\alpha\Delta t}) + \frac{M}{N}\right\}e^{\alpha(i-N)\Delta t} \tag{9-13}$$

This is the basic equation for the linear-accumulation model. Once more we can obtain values of M_r for each time step i, where $i = 1, 2, \cdots, N$. This can be accomplished in the same FORTRAN DO-loop that was employed in the constant-mass model.

Computer Program

For the computer program listed in Table 9-1 a geological time span of 4 billion years has been subdivided into 80 steps. Each Δt is thus defined as spanning 50 million years but for simplicity is regarded as unity in the equations. The present total mass of sediment has been converted to 100 arbitrary rock-mass units. Thus in making calculations the total present mass of sediment must sum to 100 as a matter of definition. The program provides for calculating both the constant-mass model and the linear-accumulation model.

Output from a computer run employing an assumed half-mass of 912.5 million years is shown in Table 9-2. As we show subsequently, this rate is optimal for the linear-accumulation model, giving the best fit to the observed mass–age distribution of Figure 9-1. Values of mass present are listed only for every fourth time increment to produce a more compact table. The column totals are obtained by summing mass values for all 80 time increments and confirm that the total present mass is 100 units. In addition the fit of the calculated values to the observed values has been calculated by summing the absolute values

TABLE 9-1 FORTRAN Program Representing Both Constant-Mass and Linear-Accumulation Sedimentary-Cycle Models

```
C.....CONSTANT MASS AND LINEAR ACCUMULATION SEDIMENTATION MODELS
C.....INSPIRED BY GARRELS AND MACKENZIE(1969)
C.....    AK    HALF-MASS FOR SEDIMENT DESTRUCTION IN UNITS OF 50 M.Y.
      DIMENSION TITLE(20), CMMASS(80), ACMASS(80), RMASS(80)
      DATA  RMASS/ 7*0.0,10*0.25,7*0.15,12*0.65,6*0.20,9*1.95,6*0.30,
     1  5*2.40,5*0.55,2*1.6,4.0,4.8,5.7,2*3.35,2.35,5.3,2*4.45,2*6.2/
    1 FORMAT(20A4/F10.0)
    2 FORMAT(1H1, 20A4/1X,'RATE OF DESTRUCTION - HALF-MASS IN 50 M.Y. UN
     1ITS =', F10.3/ 23X, 'EQUIVALENT DECAY CONSTANT  =', 2X,  F8.3/
     2 1X, 'NO. OF SEDIMENT TURNOVERS FOR CONSTANT MASS MODEL=',  F9.2/
     3 /// 1X, 'TIME' 16X, 'MASS REMAINING'/ 8X,'CONSTANT MASS',
     4     2X, 'LINEAR ACCUMULATION',  2X, 'OBSVERVED'/  3X, '0', 5X,
     5  F7.2, 13X, '0.00', 15X, '0.00')
    3 FORMAT(1H ,I3,5X,F7.2,10X,F7.2,12X,F7.2)
    4 FORMAT(1H / ' TOTALS', 2X,F7.2,10X,F7.2,13X, '100.00'/
     1 ' FIT', 5X, F7.2, 10X, F7.2)
C.....READ HALF-MASS FOR DESTRUCTION RATE
  999 READ(5,1) TITLE, AK
      NPRINT=1
      ALPHA=0.69315/AK
      C1=1.0-EXP(-ALPHA)
      C2=1.25
      TURNOV=80.0*C1
C.....DETERMINE AMOUNT OF STARTING MASS THAT REMAINS AT PRESENT
      SMASS=100.0*EXP(-80.0*ALPHA)
      DEVCM=SMASS
      DEVAC=0.0
      CMTOT=SMASS
      ACTOT=0.0
      WRITE(6,2) TITLE, AK, ALPHA, TURNOV, SMASS
C.....DETERMINE MASS REMAINING FOR ROCKS DEPOSITED DURING EACH TIME STEP
      DO 10 I=1,80
      RI=I
      C3=EXP((RI-80.0)*ALPHA)
C.....CONSTANT MASS MODEL
      CMMASS(I)=100.0*C1*C3
      DEVCM=DEVCM+ABS(CMMASS(I)-RMASS(I))
      CMTOT=CMTOT+CMMASS(I)
C.....LINEAR ACCUMULATION MODEL
      ACMASS(I)=((RI-1.0)*C2*C1+C2)*C3
      DEVAC=DEVAC+ABS(ACMASS(I)-RMASS(I))
      ACTOT=ACTOT+ACMASS(I)
   10 IF (MOD(I,NPRINT).EQ.0) WRITE(6,3) I,CMMASS(I),ACMASS(I),RMASS(I)
      WRITE(6,4) CMTOT, ACTOT, DEVCM, DEVAC
      GO TO 999
      END
```

of the deviations represented by the differences between the calculated and observed values. Small values of "FIT" indicate a close concordance between theoretical and observed distributions.

Output from several computer runs has been graphed for comparison purposes in Figure 9-5. The observed distribution is shown as a histogram (shaded), and the theoretical distributions are curves that have been superimposed. Three experiments are shown for the constant-mass model (Figure 9-5*a*). Curve I was produced by a relatively slow sedimentary turnover involving a half-mass of 1200 million

TABLE 9-2 Example Output from Sedimentary-Rock-Mass Model Program[a]

```
SEDIMENTARY CYCLE MODEL
RATE OF DESTRUCTION - HALF-MASS IN 50 M.Y. UNITS =      18.250
                      EQUIVALENT DECAY CONSTANT   =      0.038
NO. OF SEDIMENT TURNOVERS FOR CONSTANT MASS MODEL=       2.98
```

TIME	MASS REMAINING		
	CONSTANT MASS	LINEAR ACCUMULATION	OBSVERVED
0	4.79	0.00	0.00
4	0.21	0.08	0.00
8	0.24	0.10	0.25
12	0.28	0.13	0.25
16	0.33	0.17	0.25
20	0.38	0.22	0.15
24	0.44	0.28	0.15
28	0.52	0.35	0.65
32	0.60	0.44	0.65
36	0.70	0.54	0.65
40	0.82	0.67	0.20
44	0.95	0.83	1.95
48	1.11	1.02	1.95
52	1.29	1.25	0.30
56	1.50	1.53	0.30
60	1.74	1.87	2.40
64	2.03	2.28	0.55
68	2.36	2.77	1.60
72	2.75	3.36	5.70
76	3.20	4.08	5.30
80	3.73	4.93	6.20
TOTALS	100.00	100.00	100.00
FIT	57.86	48.90	

[a]Rate of sedimentary overturn is given as half-mass (here equal to $50 \times 18.25 = 91.25$ million years). Time 0 represents the youngest rocks. Each time increment represents 50 million years, but only every fourth increment is printed out. Left column is index of time increments. Other columns, from left to right, give proportions, as percentages per each time increment, of the present total mass of sediment (a) for constant-mass model, (b) for linear-accumulation model, and (c) mass values supplied as observations. Concordance of calculated masses with observed mass distribution (labeled FIT) represents sum of absolute values of deviations for 80 time increments. Notice that time 0 for constant-mass model contains 4.79 percent of the total sedimentary mass. This is the proportion of the original mass that has never been eroded under assumed conditions. A faster erosion rate would reduce or eliminate this residual mass.

years. At this rate a total mass of sediment 2.28 times the present total sedimentary mass would have been recycled. This assumption implies that a sizeable proportion of the original sedimentary mass has never been eroded, as shown by the tail at the left side of the graph. Curve III involves a half-mass of 100 million years, which is equivalent to 23.43 complete sedimentary turnovers. If the turnover were this rapid, there would be no rocks older than about 1300 million years, and the bulk of all sedimentary rocks would necessarily be

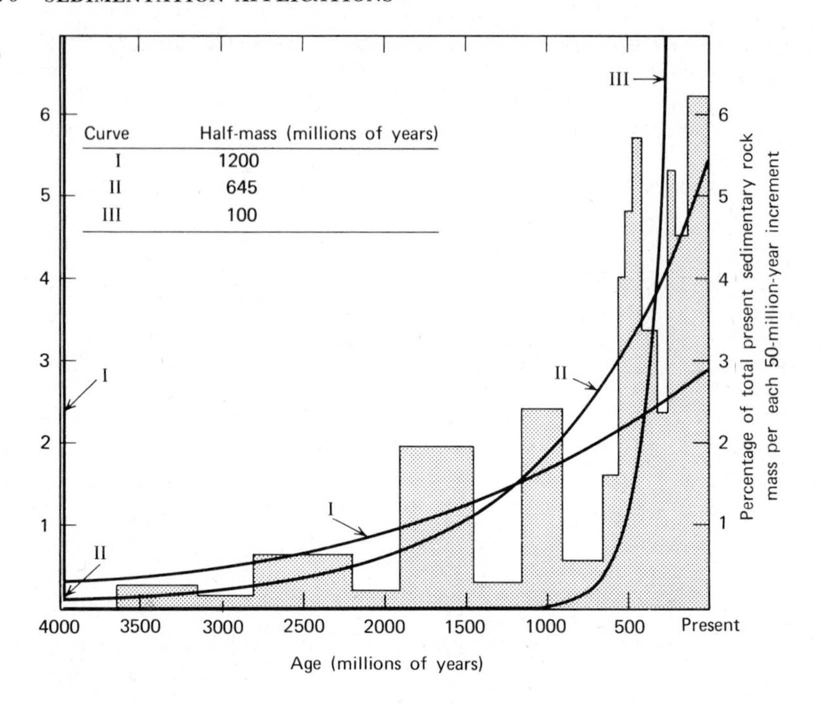

Figure 9-5a Curves yielded by sedimentary-mass model superimposed on histogram of estimated age distribution of sedimentary rock masses (shaded). (*a*) Constant-mass model;

younger than about 400 million years. Curve II was obtained by selecting a half-mass that optimized the fit of the theoretical curve to the observed data. A half-mass of 645 million years implies that about half the mass of sedimentary rocks deposited since the beginning of the Cambrian has been recycled. Only a small tail representing the original sedimentary mass that is 4 billion years old remains today.

The results for the linear-accumulation model (Figure 9-5*b*) may be interpreted similarly. Curves I and III represent slow and fast assumed turnover rates, with half-masses of 1200 and 100 million years, respectively. The optimum rate of turnover, yielding the curve with the best fit, involves a half-mass of 912 million years—a turnover rate that is considerably slower than that for the constant-mass model.

Optimization of Models

It is surprising that two models with such different basic assumptions produce curves that are so similar. Our choice between the two

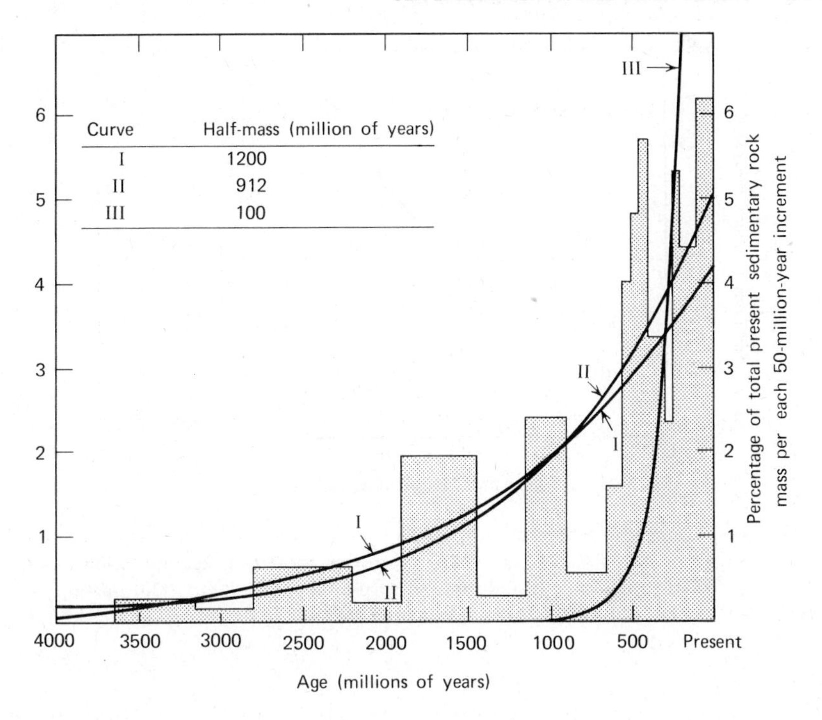

Figure 9-5b Linear-accumulation model. Histogram after Garrels and Mackenzie (1969). Copyright by the American Association for the Advancement of Science.

models is complicated by the fact that we are also faced with choosing a half-mass value. We can better compare the two models by analyzing the results obtained over a range of half-masses. We can do this by calculating the goodness of fit of each model to the observed data for a range of half-mass values (Figure 9-6). The curves shown in Figure 9-6 were generated by a modification of the computer program in Table 9-1, employing a DO-loop that successively generates curves for sedimentary half-masses ranging from 100 to 2000 million years in steps of 100 million years. This process is equivalent to optimization by direct enumeration, as discussed in Chapter 8, with FIT being the objective function and half-mass being the decision variable.

The curves in Figure 9-6 reveal that the sum of the absolute values of the deviations is minimized at about 900 million years for the linear-accumulation model and at about 650 million years for the constant-mass model. There are several interesting features in this diagram. First we notice that there are two regions where the curves lie very close to each other. Where the half-mass is less than about

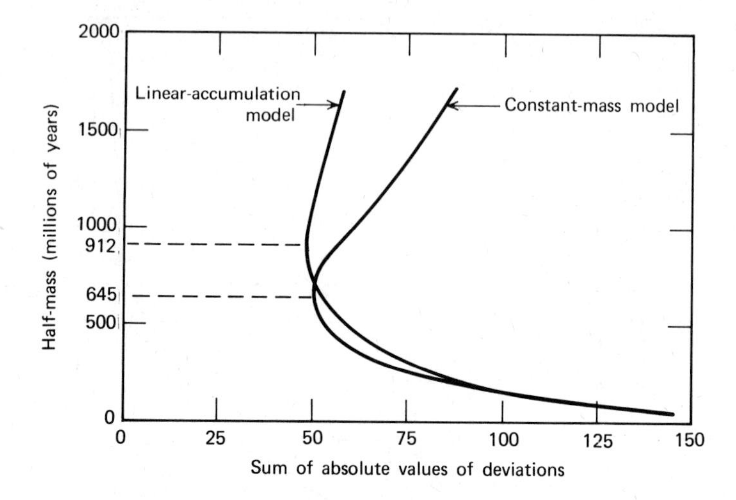

Figure 9-6 Curves showing the response of the constant-mass and linear-accumulation models to various half-mass values. Note that the linear-accumulation model is very insensitive to changes in half-mass between about 500 and 1000 million years. Minima for the constant-mass and linear-accumulation curves lie at 645 and 912 million years, respectively.

200 million years, the two curves nearly coincide and become increasingly close with further decrease in half-mass. At zero half-mass (instant erosion) the curves of the models coincide. The curves also cross each other at a half-mass of approximately 700 million years, close to the optimum for the constant-mass model. Another interesting feature concerns the shape of the curves on either side of the minima. The goodness of fit of the curves on the rapid turnover (short half-mass) side of the minima rapidly deteriorates with decreasing half-mass. Both models are almost equally sensitive to half-mass variations in this respect, and it seems clear that, if the models are valid, half-masses less than 400 million years are unlikely, regardless of inaccuracies in the observed mass–age distribution of sediments.

Estimates of present day erosion rates do not accord with erosion rates inferred from either of the models. Garrels and Mackenzie estimate that roughly 24×10^{15} grams of sediment are carried by streams per year at present. If the total sedimentary rock mass is about 50×10^{23} grams, the half-mass at the present erosion rate is $\frac{1}{2} \times (50 \times 10^{23})/(24 \times 10^{15}) \simeq 10^{8}$, or 100 million years. The curves of Figure 9-6 suggest that such a short half-mass probably has not prevailed over much of geologic time, and the present rate of erosion is abnormally high. Alternatively, if the present erosion rate is representative of that of the geologic past, then both the constant-mass and the linear-accumulation models are invalid. On the other hand, secular variations in the rates of erosion are probable, and the present high rate of erosion may be related to the rugged topography over much of the earth's land area, coupled with climatic extremes.

Returning to Figure 9-6, we notice that both curves are relatively insensitive to changes in half-mass longer than about 500 million years. In particular this applies to the fit of the linear-accumulation model, which varies only slightly for half-masses between 600 and 1000 million years. Although the optimum for the linear-accumulation model differs from that of the constant-mass model, the low sensitivity of both models for half-masses between 500 and 1000 million years is such that this difference probably has only minor significance, particularly when one considers the error in the observed data. Furthermore, as the half-mass equivalent to the present rate of erosion is about 100 million years, as compared with the optimum of 645 million years for the constant-mass model and 912 million years for the linear-accumulation model, an optimum half-mass of about 650 million years may be more likely than one of 900 million years, and this is close to the point where the responses of the constant-mass and linear-accumulation models overlap.

We conclude that present erosion rates are much greater than the average erosion rates of the geologic past and that the average rate of erosion throughout the past 4 billion years corresponds to a half-mass age between 500 and 1000 million years, with the most probable rate corresponding to a half-mass age of roughly 650 million years. Finally, we cannot decide on present evidence whether the constant-mass model or the linear-accumulation model is the more reasonable.

SEDIMENTARY-BASIN MODEL[a]

The sedimentary-rock-mass model that we have just discussed is an example of a simple conceptual and mathematical model in which spatial relationships are not represented and in which time is the only independent variable. Most stratigraphic interpretation, however, involves models that embrace either two or three spatial dimensions, as well as time. Interpretation of lithofacies and biofacies generally

[a]Reprinted (with permission) from the Systematics Association Volume *Data Processing for Geology and Biology*, edited by J. L. Cutbill, to be published by Academic Press. Originally entitled "Stratigraphic Modeling by Computer Simulation," the paper was presented orally at the Systematics Association Symposium held at Cambridge, England, September 24–26, 1969.

involves construction of geometrical depositional models for a succession of moments in time. In turn construction of these models commonly involves various assumptions concerning the depositional environment—such as depth of water, wave and current action, distance from shore, types and sources of sediment, tectonic warping, and fluctuations of sea level. Regardless of the uncertainties in interpreting these variables, they should not be ignored. The difficulty in interpreting them is compounded by the fact that they are strongly interdependent. Below we describe a simple mathematical sedimentation model that provides a formal method for quantitatively treating several geological variables in sedimentary basins. The model was briefly described in Chapter 7 in a discussion of elementary system-control concepts and is an extension of a conceptual model developed by Sloss (1962).

The Sloss model deals principally with the deposition of clastic sediments on continental shelves. Sediments are delivered to the edge of the sea by streams flowing across a coastal plain. When brought to the sea, each sediment particle is transported until it finds a position of rest and becomes available for incorporation into the sedimentary sequence. The position of rest is a function of kinetic energy (waves and currents), material (composition and particle size), and boundary conditions (bottom slope and roughness). The interaction of these factors produces an equilibrium surface, base level, above which a particle cannot come to rest and below which deposition and burial are possible. At any instant in time, given an adequate supply of sediment, the interface between water and sediment tends to coincide with base level. Successive interfaces representing successive instants of time can be interpreted as a record of the relative rate of subsidence of the depositional basin.

The gross geometry or shape S of a body of sedimentary rocks varies as a function of the quantity Q of material supplied to the depositional site, the rate of subsidence R at the site, the rate of dispersal D, and the nature of the material supplied M:

$$S = f(Q, R, D, M) \tag{9-14}$$

In this expression R is a measure of the rate of subsidence expressed as a receptor value, defined as the available volume below base level created per unit time by subsidence. The proportion of various particle sizes of sedimentary material M is assumed to be constant. This assumption implies that weathering and erosion in a heterogeneous source area yield coarse, medium, and fine clastic particles to the depositional area in unchanging proportions.

With these assumptions, Sloss developed the process elements and the resulting stratigraphic responses in a series of diagrams. Figure 9-7 shows hypothetical responses when the quantity Q of material supplied and receptor value R are varied. In Figure 9-7*a* more material is supplied than can be handled by dispersal and subsidence ($Q > R$), resulting in deposition of a regressive sequence. The results where $Q = R$ and $Q < R$ are illustrated in Figure 9-7*b* and *c*, respectively.

Computer Model

The computerized extension of the Sloss model that we have developed incorporates the following factors:

1. Quantity of material supplied, which may include from one to five different sediment-size fractions.
2. Initial geometry of the sedimentary basin, expressed as water depth.
3. Tectonic warping (subsidence) through time and from place to place in the basin.
4. Position of base level or equilibrium surface defined with respect to sea level for each particle-size class.

The model treats only two spatial dimensions, representing a vertical section through a sedimentary basin. The other horizontal dimension is not represented, but could be incorporated if a more advanced version were to be developed. As in the rock-mass model, time is divided into discrete steps. Space is represented by a sequence of columns, each of unit width, which represent water depth and thickness increments of various sediment types (Figure 9-8). Subsidence (or uplift) is represented by sliding the columns up or down relative to sea level. FORTRAN arrays are used to store water depths and sediment thicknesses for each vertical column for each time increment, in a manner similar to that described in Chapter 2. Each column of sediment and water is displayed as a row of symbols on the computer printer, and the resulting sequence of rows forms a cross section through the sedimentary basin (see Figure 9-10).

Sediment Transport and Deposition

Transport and deposition of sediment are treated heuristically in the computer model. It is assumed that during any time interval, a certain increment of sediment (the sediment "load") enters the basin from a source area on one side of the basin and is then transported from column to column. Deposition may take place in each column,

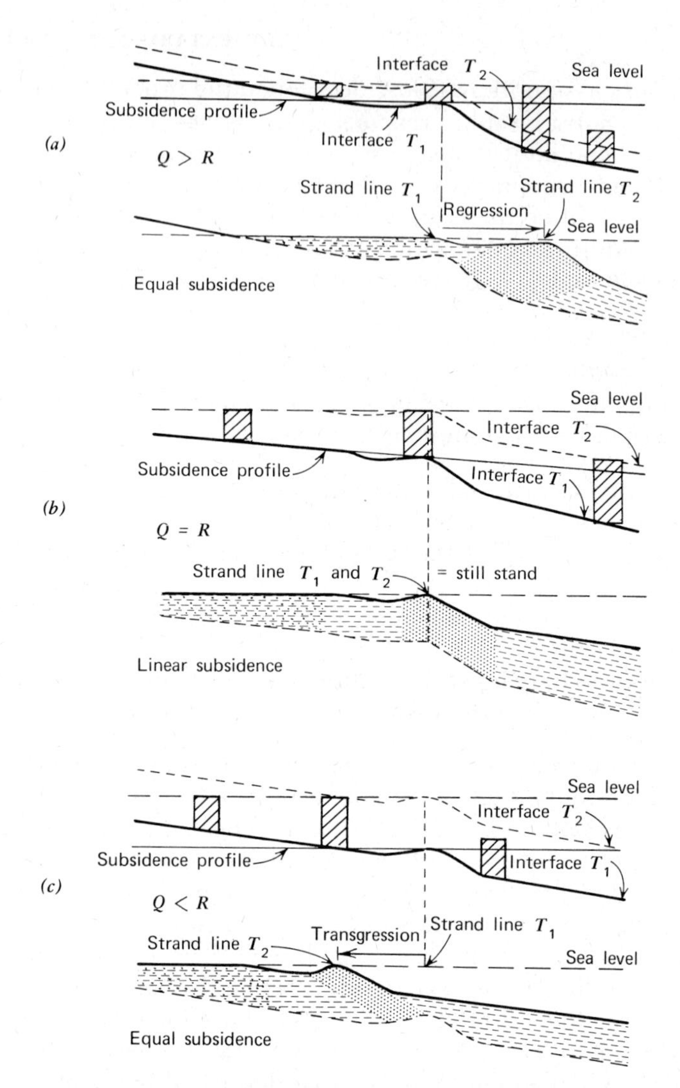

Figure 9-7 Sequence of cross sections that illustrate hypothetical responses of the Sloss (1962) model under three different conditions. Upper part of each diagram shows interfaces at times T_1 and T_2. Diagonally ruled bars represent amount of sediment supplied during interval between T_1 and T_2, whereas receptor value is denoted by lowering of interface with respect to sea level between T_1 and T_2, as shown by subsidence profile. Lower part of each diagram shows facies that will result. (*a*) When quantity Q of sediment supplied is greater than receptor R value ($Q > R$), coupled with uniform subsidence, model responds by deposition of regressive facies sequence. (*b*) When $Q = R$, coupled with subsidence rate that increases linearly seaward, model responds by producing facies that have vertical boundaries. (*c*) When $Q < R$, coupled with uniform subsidence, model yields transgressive sequence. Reproduced with permission from the *Journal of Sedimentary Petrology*.

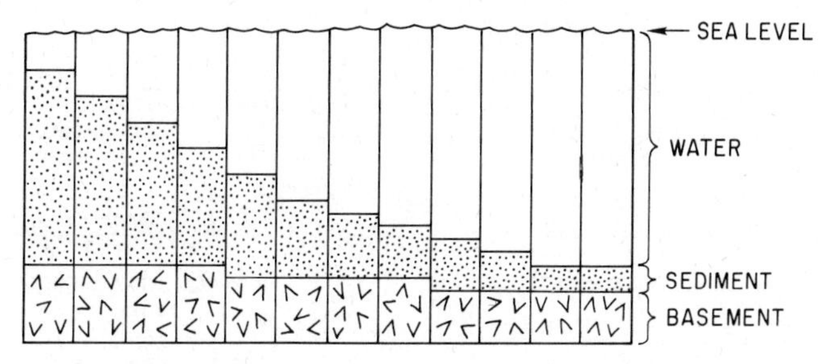

Figure 9-8 Subdivision of two-dimensional sedimentary basin into series of discrete vertical columns representing water, sediment, and basement. From Bonham-Carter and Harbaugh (1970).

the amount deposited being debited from the sediment load and the remainder passed onto the next column, where the process is repeated. The sequence proceeds from the sediment-source side of the basin toward the seaward side. The amount of sediment deposited in a particular column depends on (a) the amount of sediment available for deposition and (b) the water depth in that column in relation to base level. Mass balance is observed by accounting for all sedimentary materials as they move through the system.

The rules governing transportation and deposition of sediment are extremely simple and are outlined below. Part of the sediment load reaching a particular column is deposited if the water depth is greater than depth to base level. In columns that contain water sufficiently deep so that base level exerts no control the proportion of sediment deposited for each particle-size class is represented by a curve that declines exponentially toward the seaward side of the basin (Figure 9-9*a* and *b*). The accounting system arithmetic involved in this process can be envisioned as follows. Let the sediment load entering the basin be L and let the proportion of this load that is deposited in the first column be k. Thus the amount deposited is kL, and the remaining load that is shunted on to the next column in the sequence is $L-kL$, or $L(1-k)$. The quantities deposited in successive columns are listed in Table 9-3.

The amount deposited in the second column is found by multiplying the remaining load by k, giving $kL(1-k)$. In column 3 the remaining load is $L(1-k)^2$, and the amount deposited is $kL(1-k)^2$. Subsequent columns are treated similarly. Thus we may generalize: for the nth column the sediment load remaining is $L(1-k)^{n-1}$ and the amount deposited is $kL(1-k)^{n-1}$. This relationship is an adaptation of the familiar law of growth and decay in which k is the decay constant, and is similar to the relationship used in the sedimentary mass model.

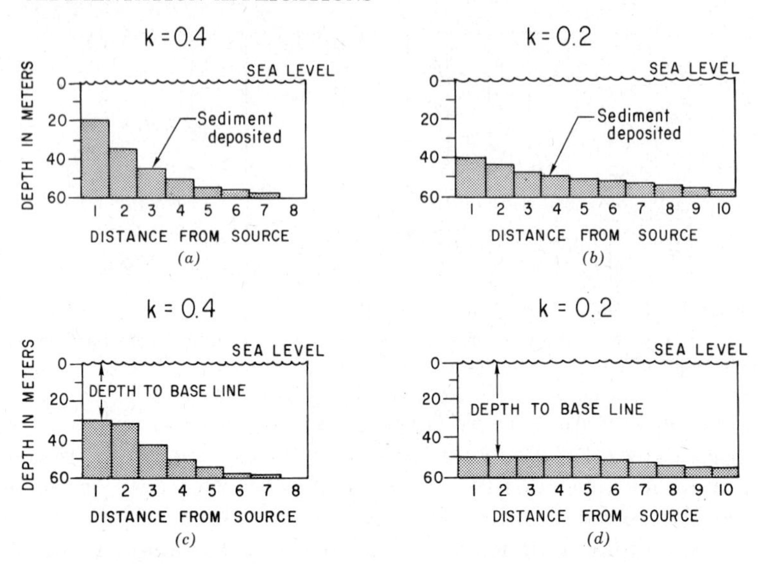

Figure 9-9 Four diagrams representing vertical sections through sedimentary basin into which uniform volume of sediment is supplied from source at left. Initial water depth is uniform. Diagrams (*a*) and (*b*) assume that base level does not exert influence and illustrate effects of varying the "decay constant" k. Diagrams (*c*) and (*d*) illustrate the effects of two different base levels, coupled with values of k equivalent to those of diagrams (*a*) and (*b*), respectively. Volume of sediment introduced in all four diagrams is equivalent to a column 100 meters high. From Bonham-Carter and Harbaugh (1970).

The significance of different values of k is open to interpretation. It is clear from Figure 9-9*a* and *b* that k pertains to slope of deposits. If slope were solely a function of grain size, larger values of k might correspond to coarse sediment capable of reposing on steeper slopes and being less mobile than fine sediment. Maximum slope angle and

TABLE 9-3 Quantities of Sediment Deposited in Successive Columns

Column	*Sediment Load*	*Sediment Deposited*
1	L	kL
2	$L(1-k)$	$kL(1-k)$
3	$L(1-k)^2$	$kL(1-k)^2$
4	$L(1-k)^3$	$kL(1-k)^3$
5	$L(1-k)^4$	$kL(1-k)^4$
.	.	.
.	.	.
.	.	.
n	$L(1-k)^{n-1}$	$kL(1-k)^{n-1}$

mobility, however, do not bear a simple relationship to grain size. Thus it is difficult to interpret k in terms of a simple physical relationship; instead it should be regarded as a parameter used in conjunction with a relationship that is largely heuristic.

If the model dealt only with the transport of sediment according to the growth-and-decay law, the calculations would be simple and could be readily carried out by hand. If we introduce constraints, however, the solutions are not simple. If we introduce base level, above which sediment of a particular particle size cannot come to rest, we must continually check the elevation of the sediment–water interface in each column to ensure that it is not above base level. Figure 9-9*c* and *d* illustrates the effect of base level as a constraint, employing specified values of k and holding other factors constant.

The model incorporates various decision rules that govern sediment deposition. We can distinguish three situations:

1. Where the sediment–water interface is above or equal to base level (as specified for a particular particle-size class), deposition is not possible and all the load is shunted on to the next column.
2. Where the sediment–water interface is slightly below base level, only part of the quantity of sediment that would otherwise be deposited is accommodated. Sufficient sediment is deposited to bring the column to base level, and the remainder is passed on to the next column.
3. Where the sediment–water interface is sufficiently far below base level, all of the sediment available for deposition is accommodated.

These relationships can be expressed algebraically. If water depth in a column is D and depth to base level is B, then the amount deposited S is given by one of the following relationships: If water depth is less than or equal to depth to base level, no sediment is deposited:

$$S = 0; \qquad D \leqslant B \tag{9-15}$$

If water depth minus the quantity kL is less than or equal to depth to base level, the amount deposited is equal to depth minus base level:

$$S = D - B; \qquad (D - kL) \leqslant B \tag{9-16}$$

Otherwise the amount deposited equals the quantity kL, and base level has no influence on sedimentation; thus

$$S = kL \tag{9-17}$$

Figure 9-9*c* and *d* illustrates these algebraic base-level control relationships. In Figure 9-9*c* depth to base level is 30 meters. Without base-level control, of the total of 100 meters of sediment supplied, 40 meters would have been deposited in the first column, reducing its water depth to 20 meters. But 20 meters is shallower than base level. Thus deposition of $D - B = 60 - 30 = 30$ meters of sediment takes place in column 1, and the remaining load of 70 meters of sediment passes on to column 2. The sediment load reaching column 2 is equivalent to $100 - 30 = 70$ meters of sediment. The new value of kL is thus $0.4 \times 70 = 28$. This time, if all 28 meters of sediment are deposited, water depth is given by $D - kL = 60 - 28 = 32$, which is greater than depth to base level. Thus 28 meters are deposited in column 2.

Similar calculations are carried out for columns 3 to 8. In Figure 9-9*d*, depth to base level is 50 meters and $k = 0.2$. Base level imposes a depth limit for sediment in the first five columns. In columns 6 to 10 the amount deposited in each column dies away exponentially.

The treatment of more than one grain size further complicates the calculations. Each grain size is assigned (a) a specific value of k, (b) an initial sediment load to be released from the source for each time increment, and (c) a depth to base level. In a computer run many columns receive sediment of more than one grain size during a time increment. Mixtures of sediment of varying size in a particular column are graphically represented by using a symbol that represents the particle-size class that forms the largest proportion of the volume deposited during the particular time increment.

Crustal Subsidence

The effect of subsidence or, conversely, eustatic changes in sea level can be represented by adding or subtracting values to the water depth in each column. Increasing water depth in a single column by some number of meters has the effect of depressing the entire column downward by the same number of meters. In this model there are two ways in which water-depth values can be changed. First, provision is made in the program for adding a specific increment to water depth in each column during each time increment. The increment to water depth may vary from column to column but is constant for each time interval. This simulates uniform subsidence with time and also may be used to represent changes in sea level. Alternatively subsidence can be related to deposition by a simple proportionality constant. Each column can be depressed by an amount equal to the thickness of sediment deposited multiplied by a proportionality constant F. At one extreme, if F equals one, the amount of subsidence equals the amount of deposition. At the other extreme subsidence does not occur if F equals zero.

Subsidence need not be instantaneous with deposition. As an alternative subsidence may lag behind deposition by some whole number of time increments. For example, if the lag length is three time increments, subsidence will occur at the end of every third time increment, and the amount of subsidence in a particular column is obtained by multiplying F by the total quantity of sediment deposited in that cell during the previous three time increments. The lag cannot be made shorter than one time increment because of the division of time into discrete steps in the model. Unit lag causes virtually instant response. It could be argued that there may be an appreciable lag between the loading of the actual crust and its subsequent subsidence. The model makes it convenient to experiment with the effects of different lag lengths.

Experiments with the Model

The computer program representing the model is listed in Table 9-4, and an example of input is shown in Table 9-5. The response of the model, consisting of a sequence of stratigraphic cross sections, is shown in Figure 9-10. Other output is listed in Table 9-6. As the cross sections reveal, without subsidence and with only a single grain size, the model produces a "deltaic" deposit that grows progressively out into deep water, forming a series of "foreset" beds that dip progressively more steeply as the delta builds outward.

A similar experiment, in which crustal subsidence occurs, is shown in Figure 9-11. The sediment–water interface remains at a constant elevation after the first time increment, each column subsiding by an amount equal to the thickness of sediment deposited immediately before. Under these conditions the response of the crust is to subside the most where the maximum amount of sediment has been deposited. The overall form of the deposits is that of a lens, reminiscent of the lens-shaped mass of Cenozoic sediments of the Gulf Coast.

Three other experiments with the model are illustrated in Chapter 7. In Figure 7-7 a regressive sequence has been produced by introducing two grain sizes and making subsidence equal to zero. At time 1 sand is deposited close to shore, and silt is deposited farther from shore. Beginning at time 2 and continuing during subsequent time increments, sand begins to build out over silt; the effect of base level (set to three depth units for sand) is to cause the zone of maximum deposition to migrate progressively seaward. Note that time lines

TABLE 9-4 FORTRAN Program for Two-Dimensional Sedimentary-Basin Model[a]

```
C.....SIMPLE TWO-DIMENSIONAL SEDIMENTARY BASIN MODEL
C.....  NTIM          NO. OF TIME INCREMENTS
C.....  NCOLS         NO. OF COLUMNS OF SEDIMENT
C.....  NFRACT        NO. OF SEDIMENT SIZE FRACTIONS
C.....  SED(NT,I,L) NO. OF SEDIMENT UNITS DEPOSITED IN N-TH TIME INCREM.
C                          IN I-TH COLUMN, AND L-TH SIZE FRACTION
C.....  SEDINP(L)     NO. OF SED UNITS IN INITIAL SED LOAD (L-TH SIZE FR)
C.....  SEDIN(L)      TEMPORARY ACCOUNT FOR L-TH SIZE FRACTION
C.....  DEPTH(I)      WATER DEPTH IN I-TH COLUMN
C.....  SUBSID(I)     AMOUNT OF EXOGENOUS SUBSIDENCE (OR COULD BE USED TO
C                          REPRESENT SEA LEVEL CHANGES) ADDED TO WATER DEPTH
C                          IN I-TH COLUMN EVERY LAG-TH TIME INCREMENT
C.....  SUBFAC        SUBSIDENCE FACTOR RELATING SUBSIDENCE TO DEPOSITION
C.....  LAG           TIME LAG IN TIME INCREMENTS BETWEEN DEPOSITION AND
C                          RESULTING SUBSIDENCE (LAG=1 IMPLIES INSTANT RESP)
C.....  EQUIL(L)      EQUILIBRIUM DEPTH BELOW WHICH L-TH SIZE FRACTION
C                     NOT BE DEPOSITED
C.....  CON(L)        CONSTANT DETERMINING RATE OF DEPOSITION FOR L-TH SIZ
C.....  KPRINT        PRINTED OUTPUT EVERY KPRINT-TH TIME INCREMENT
C.....  NPLOT         X-SECTION PLOTTED EVERY NPLOT-TH TIME INC
C.....  KPLOT         TIME LINE PLOTTED EVERY KPLOT-TH TIME INC ON X-SECT
      DIMENSION SUBSID(30),SEDIN(5),SEDINP(5),EQUIL(5),CON(5),A(5),D(5)
      COMMON DEPTH(30), SED(30,30,5), TITLE(18), NCOLS, NFRACT
    1 FORMAT(18A4)
    2 FORMAT(7I5, F5.0)
    3 FORMAT(16F5.0)
    4 FORMAT(1H1, 18A4/ 5X, 'TIME INCREMENT ', I5/ ' COLUMN', 3X,
     1 'DEPTH', 3X, 5('SIZE', I2, ' DEPOS  LEFT    '))
    5 FORMAT(1H , I4, F10.1, 3X, 5(6X,2F6.1,4X))
C.....READ INPUT PARAMETERS
  999 READ(5,1) TITLE
      READ(5,2) NTIM,NCOLS,NFRACT,KPRINT,NPLOT,KPLOT,LAG,SUBFAC
      IF (LAG.LT.1) LAG=1
      READ(5,3) (SEDINP(L), L=1,NFRACT)
      READ(5,3) (EQUIL(L), L=1,NFRACT)
      READ(5,3) (CON(L), L=1,NFRACT)
      READ(5,3) (DEPTH(I), I=1,NCOLS)
      READ(5,3) (SUBSID(I), I=1,NCOLS)
C.....PLOT INITIAL CROSS SECTION BEFORE SEDIMENTATION
      CALL CROSEC(0,KPLOT)
C.....BEGIN MAJOR DO-LOOP, ONCE THRU PER TIME INCREMENT
      DO 110 NT=1,NTIM
      IF (MOD(NT,KPRINT).EQ.0) WRITE(6,4) TITLE, NT, (L, L=1,NFRACT)
C.....ALLOCATE SEDIMENT LOAD TO BE DEPOSITED THIS TIME INC
      DO 40 L=1,NFRACT
   40 SEDIN(L)=SEDINP(L)
C.....BEGIN LOOP, ONCE THRU PER SED COLUMN
      DO 80 I=1,NCOLS
C.....CALCULATE AMOUNT OF SED OF EACH SIZE READY FOR DEPOSITION (A(LLL))
      DO 50 L=1,NFRACT
      SED(NT,I,L)=0.0
   50 A(L)=SEDIN(L)*CON(L)
C.....ENTER LOOP ONCE THRU PER EQUILIB DEPTH STARTING WITH DEEPEST
      DO 70 LL=1,NFRACT
      L=NFRACT-LL+1
      ATOT=0.0
      DO 55 LLL=1,L
```

```
   55 ATOT=ATOT+A(LLL)
      IF (ATOT.LE.0.01) GO TO 80
C.....CALCULATE TOTAL AVAILABLE SPACE FOR DEPOSITION. IF 0 GO TO 70
      B=DEPTH(I)-EQUIL(L)
      IF (B.LE.0.0) GO TO 70
C.....DETERMINE ACTUAL AMOUNT OF DEPOSITION. DEPOSIT SIZE FRACTIONS
C       IN PROPORTION TO THEIR TOTALS IN THE LOAD
      DEPOS=AMIN1(ATOT,B)
      Z=DEPOS/ATOT
      DO 60 LLL=1,L
C......AMOUNT OF SED OF SIZE LLL DEPOSITED IS D(LLL), ADDED TO SED
C       ARRAY, SUBTRACTED FROM A(LLL)
      D(LLL)=A(LLL)*Z
      SED(NT,I,LLL)=SED(NT,I,LLL)+D(LLL)
      SEDIN(LLL)=SEDIN(LLL)-D(LLL)
   60 A(LLL)=A(LLL)-D(LLL)
      ATOT=ATOT-DEPOS
      DEPTH(I)=DEPTH(I)-DEPOS
   70 CONTINUE
   80 IF (MOD(NT,KPRINT).EQ.0) WRITE(6,5) I,DEPTH(I), (SED(NT,I,L),
     1  SEDIN(L), L=1,NFRACT)
C.....IF NT(MODULO NPLOT) EQUALS ZERO PLOT CROSS SECTION
      IF (MOD(NT,NPLOT).LT.1) CALL CROSEC(NT,KPLOT)
C.....ADD SUBSIDENCE TO DEPTH IF NT(MODULO LAG) EQUALS ZERO
      IF (MOD(NT,LAG).GT.0) GO TO 110
      DO 100 I=1,NCOLS
      SUM=0.
      IF (SUBFAC.LT.0.00001) GO TO 100
C.....DETERMINE AMOUNT OF SEDIMENT LOADED ONTO COLUMN SINCE PREVIOUS
C       SUBSIDENCE ADJUSTMENT
      DO 90 LG=1,LAG
      INDEX=NT-LG+1
      DO 90 L=1,NFRACT
   90 SUM=SUM+SED(INDEX,I,L)
      SUM=SUM*SUBFAC
  100 DEPTH(I)=DEPTH(I)+SUBSID(I)+SUM
  110 CONTINUE
      GO TO 999
      END
      SUBROUTINE CROSEC(NT,KPLOT)
C.....SUBROUTINE FOR DRAWING GRAPHIC SECTIONS
      DIMENSION PLOT(120), PLOT1(120), SYMBOL(5)
      COMMON DEPTH(30), SED(30,30,5), TITLE(18), NCOLS, NFRACT
      DATA SYMBOL,DOT,EYE,BLANK,RLT/'O','$','A','B','C','.','I',' ','<'/
    1 FORMAT(1H1, 18A4/ 5X, 'TIME INCREMENT ', I5/)
    2 FORMAT(1H , 120A1)
    3 FORMAT(1H0)
      WRITE(6,1) TITLE, NT
C.....FOR EVERY COLUMN IN VERTICAL SECTION, DO DOWN TO 80
      DO 80 II=1,NCOLS
      I=NCOLS-II+1
C.....SET BLANKS INTO ALL POSITIONS OF PLOT ARRAYS
      DO 10 K=1,120
      PLOT1(K)=BLANK
   10 PLOT(K)=BLANK
C.....INDEX DENOTES POSITION IN PLOTTING ARRAY BELOW SEA LEVEL - ONLY
C       POSITIVE VALUES ARE PRINTED. USE I'S FOR WATER
```

TABLE 9-4 *(contd.)*

```
      NDEP=ABS(DEPTH(I))+0.5
      INDEX=0
      IF (DEPTH(I).LT.0) INDEX=-NDEP
      DEV=ABS(DEPTH(I))-NDEP
      IF (NDEP.LT.1) GO TO 30
      DO 20 K=1,NDEP
      INDEX=INDEX+1
   20 IF (INDEX.GE.1) PLOT(INDEX)=EYE
C.....FILL SEDIMENT POSITIONS WITH APPROPRIATE SYMBOLS
   30 IF (NT.EQ.0) GO TO 55
C.....FOR EACH TIME INCREMENT DO DOWN TO 50
      DO 50 NN=1,NT
      N=NT-NN+1
C.....SUM SED FRACTIONS. FIND DOMINANT SED FRACTION AND ALLOCATE SYMBOL
      SUM=0.0
      LBIG=0
      BIG=0.0
      DO 35 L=1,NFRACT
      IF (BIG.GT.SED(N,I,L)) GO TO 35
      BIG=SED(N,I,L)
      LBIG=L
   35 SUM=SUM+SED(N,I,L)
      NSUM=SUM+0.5+DEV
      DEV=SUM+DEV-NSUM
      IF (NSUM.LT.1) GO TO 50
      DO 40 K=1,NSUM
      INDEX=INDEX+1
      IF (INDEX.LT.1) GO TO 40
      PLOT(INDEX)=SYMBOL(LBIG)
      IF (K.EQ.NSUM.AND.MOD(N,KPLOT).EQ.0) PLOT1(INDEX)=DOT
   40 CONTINUE
   50 CONTINUE
C.....INSERT 'LESS THAN' SIGNS FOR BASEMENT ROCKS
   55 DO 60 K=1,2
      INDEX=INDEX+1
   60 IF (INDEX.GE.1) PLOT(INDEX)=RLT
      IF (INDEX.LT.1) GO TO 70
      WRITE(6,2) (PLOT(K), K=1,INDEX)
      WRITE(6,2) (PLOT1(K), K=1,INDEX)
      GO TO 80
   70 WRITE(6,3)
   80 CONTINUE
      RETURN
      END
```

[a]About two-thirds of total program consists of subroutine CROSEC, which is used for plotting stratigraphic cross sections with line printer.

TABLE 9-5 Input for Program in Table 9-4 To Produce Output Shown in Figure 9-10

```
TEST RUN 1, ONE GRAIN SIZE, NO SUBSIDENCE
   10   20    1    1    1    3    1    0
   10
    3
   .5
    1    2    3    4    5    6    7    8    9   10   11   12   13   14   15   16
   17   18   19   20
    0    0    0    0    0    0    0    0    0    0    0    0    0    0    0    0
    0    0    0    0
```

TABLE 9-6 Example of Tabular Output from Run of Sedimentary-Basin Model[a]

TEST RUN 1, ONE GRAIN SIZE, NO SUBSIDENCE
TIME INCREMENT 1

COLUMN	DEPTH	SIZE 1 DEPOS	LEFT
1	1.0	0.0	10.0
2	2.0	0.0	10.0
3	3.0	0.0	10.0
4	3.0	1.0	9.0
5	3.0	2.0	7.0
6	3.0	3.0	4.0
7	5.0	2.0	2.0
8	7.0	1.0	1.0
9	8.5	0.5	0.5
10	9.7	0.3	0.3
11	10.9	0.1	0.1
12	11.9	0.1	0.1
13	13.0	0.0	0.0
14	14.0	0.0	0.0
15	15.0	0.0	0.0
16	16.0	0.0	0.0
17	17.0	0.0	0.0
18	18.0	0.0	0.0
19	19.0	0.0	0.0
20	20.0	0.0	0.0

[a]Output is based on input data listed in Table 9-5 and pertains to time increment 1. Each row pertains to an individual column in cross section. Column labeled DEPTH refers to depth after deposition in arbitrary units, column labeled DEPOS refers to amount of sediment deposited in equivalent arbitrary units, and column labeled LEFT pertains to the load of sediment that remains to be transported to the next column. Column DEPOS thus accounts for all sediment supplied (10.0 units) during time increment (that which is deposited is debited from that which remains, and what is left is shunted to column LEFT).

intersect the facies boundary. Irregularities in the facies boundary are produced by numerical rounding associated with the use of whole numbers of graphic symbols on the computer printout.

In the experiment shown in Figure 7-9 the same controlling parameters are employed as in Figure 7-7, except that rate of subsidence is made equal to the rate of sedimentation, with a lag of one time increment. As in Figure 9-11, the configuration of the sediment–water interface remains constant through time, and the site of maximum deposition remains in one place. Because there are two facies, however, we note that vertical boundaries separate the sand and silt facies.

The effect of lag is introduced in the experiment shown in Figure 7-10. A time lag of three time increments results in interfingering of the sand and silt layers. At times 1, 2, and 3 a regressive sequence is formed, similar to that shown in Figure 7-7. At the end of the third time increment, however, each column subsides by an amount equal to the total thickness of sediment deposited in the column during the preceding three time increments. During increments 4, 5, and 6 a new

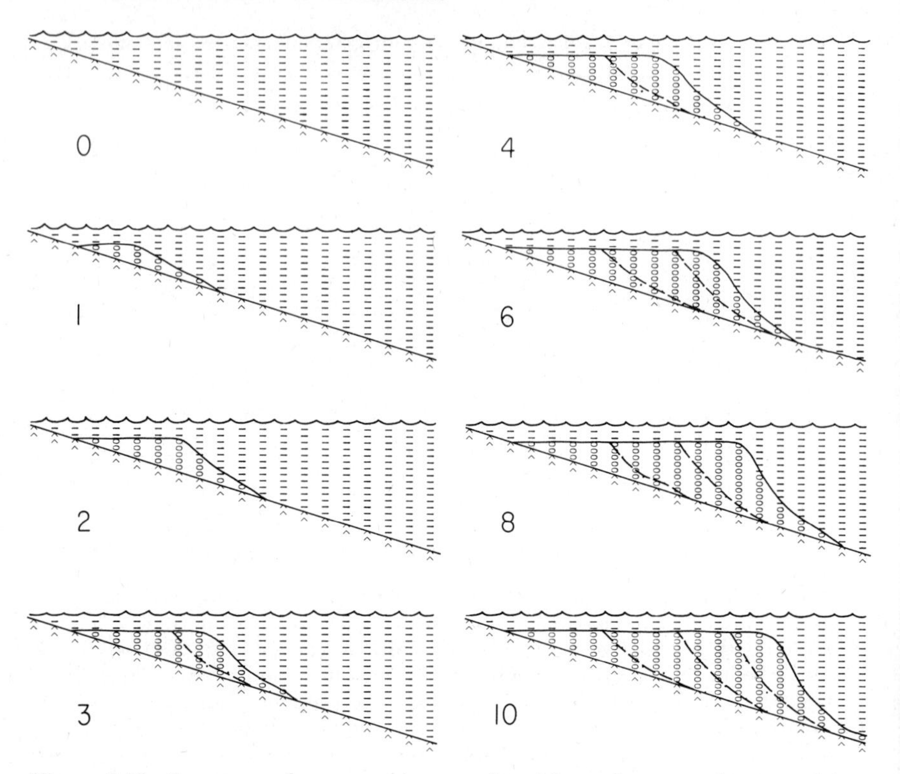

Figure 9-10 Sequence of cross sections produced by sedimentary-basin model in experimental run in which crustal subsidence does not occur and which involves sediment of a single size class. Time increments 0, 1, 2, 3, 4, 6, 8, and 10 are shown. Water is symbolized by prostrate I's, sand by O's, silt by $'s. Model responds by producing a regressive sequence of deltaic deposits. Slope of deposits is affected by initial slope of sea floor and by "decay constant" k, which has been set at 0.5. Sediment–water interface is plotted every third time increment, thus forming series of stratigraphic time lines. From Bonham-Carter and Harbaugh (1970).

regressive sequence builds out over the previous one, causing the facies to interfinger. In the experiment shown in Figure 9-12 the basin slope has been reduced, but otherwise the controlling parameters are similar to those used to produce the output in Figure 7-10. As a result the region of interfingering is more pronounced, the effects of transgression and regression being more widespread.

The sedimentary-basin model in its present form can be used for making a wide variety of simple sedimentation–simulation experiments. The program can be readily employed to simulate eustatic changes in sea level and could also be modified to incorporate additional factors that affect the deposition of sediment. The usefulness of

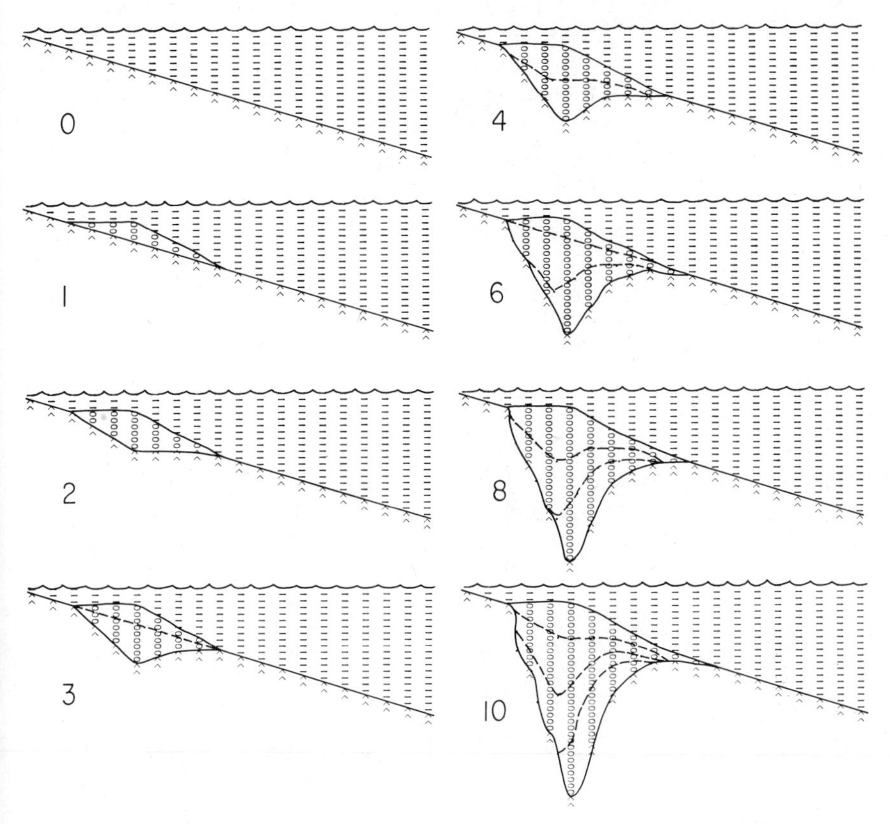

Figure 9-11 Response of model when controlling parameters are similar to those in experiment shown in Figure 9-10, except that subsidence takes place in each column in proportion to volume of sediment deposited in that column. Each increment of subsidence lags behind deposition by one time increment. Note that configuration of the sediment–water interface is unchanged after the first time increment. Greatest amount of subsidence occurs where maximum quantity of sediment is deposited. From Bonham-Carter and Harbaugh (1970).

such a model lies in its ability to test the effects of simple assumptions and to formalize concepts that are qualitative and geometric.

EVAPORITE-BASIN MODEL

The sedimentary-basin model that we have discussed emphasizes the importance of materials balance. The model employs a simple accounting method that debits and credits sedimentary material from column to column within the two-dimensional vertical cross sections that represent the system. The evaporite-basin model that we discuss

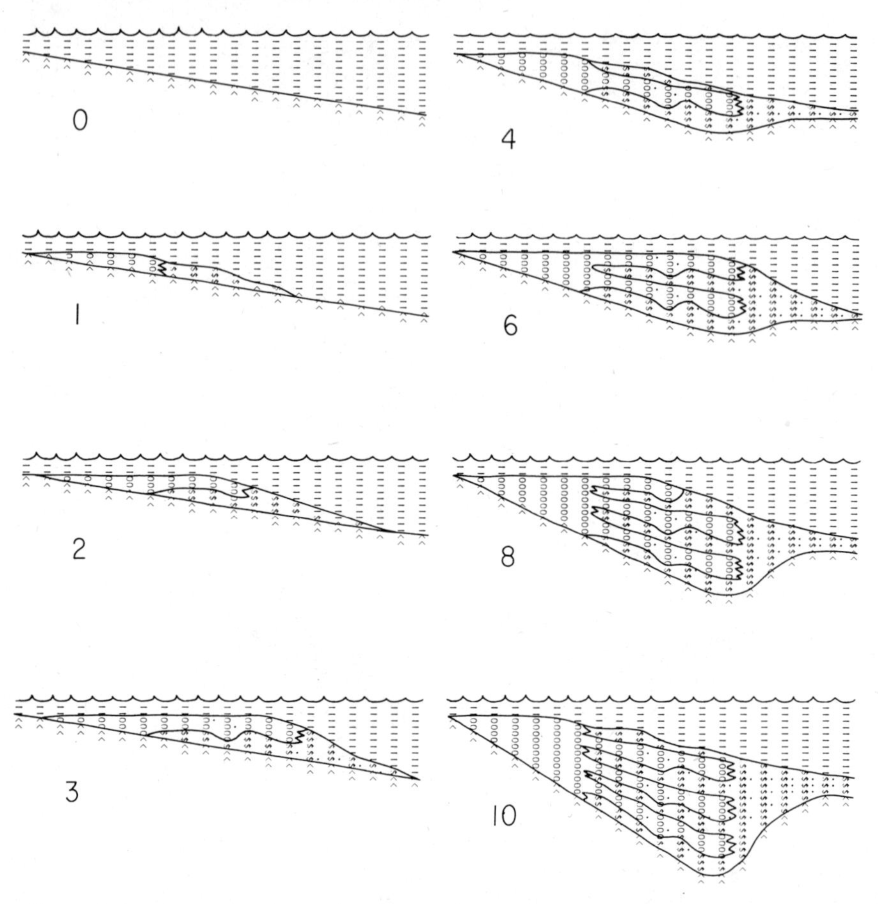

Figure 9-12 Response of model similar to that shown in Figure 7-10, with two grain sizes and with subsidence lagging behind deposition by three time increments and occurring every third time increment. Experiment differs from that shown in Figure 7-10, having a more gentle initial slope of basin floor, and larger sedimentary load supplied during each time increment. From Bonham-Carter and Harbaugh (1970).

next also emphasizes materials balance, but the accounting method is different. The evaporite model makes use of the mass-conservation principles of the continuity equation, from which the equations for potential flow are derived.

Actual evaporite basins provide good examples of the conservation of matter. Although evaporites can be produced under a variety of circumstances, we can envision a relatively simple evaporite-basin configuration in which a barrier, or shallowly submerged sill, isolates the evaporite basin from the open sea (Figure 1-4). The barrier permits ordinary seawater to flow into the basin, but it prevents the more dense brine from flowing back to the open sea. The water that flows into the basin goes out by evaporation, and the dissolved solids are precipitated as salts. If the volumetric rate of flow into the basin is known, the quantity of salts that are precipitated can be readily calculated.

Briggs and Pollack (1967) and Pollack (1967) have adapted a potential-flow model to simulate the deposition of salt in the Late Silurian Salina Formation in the Michigan basin. Their model was briefly described in Chapter 1. The evaporite model that we describe incorporates the concepts outlined in their papers and employs the principles of steady-state potential flow that are treated in Chapter 6. The model can be described in terms of its three principal components: (a) fluid flow, (b) transport of dissolved salt, and (c) precipitation of salt.

Seawater is of course, a brine solution, and with the passage of time and continuous evaporation from the surface, the salt concentration in the brine will increase. By increasing concentration the density of the upper brine layer will also increase. This will have two effects on the fluid flow. First, the rate of surface evaporation will change with density, highly concentrated brine evaporating less slowly than weakly concentrated brine. Although this factor could be considered in the model, the computation time would increase and the overall result is unlikely to be greatly altered. Second, it is assumed in deriving the basic Laplace equation that the fluid is incompressible and remains of constant density. Referring to Chapter 6, notice that Equation 6-18 contains density (ρ) as a variable, which is subsequently eliminated from each term in the equation. In the evaporite model, this incompressibility assumption is still made, although the fluid is clearly of variable density and one would therefore expect small changes in the pattern of fluid flow. Again, although the full equation (6-18), could have been used, computation time would have increased enormously, and the overall configuration of fluid flow (and subsequent distribution of precipitated salt) is believed to be only slightly altered.

In Chapter 6 we adapted the Laplace equation for potential flow. For two dimensions the Laplace equation is

$$\frac{\partial^2 \phi}{\partial x^2} + \frac{\partial^2 \phi}{\partial y^2} = 0 \tag{9-18}$$

which in turn is derived from the continuity equation

$$\frac{\partial v_x}{\partial x} + \frac{\partial v_y}{\partial y} = 0 \tag{9-19}$$

and incorporates the relation between velocity and potential:

$$v_x = \frac{\partial \phi}{\partial x}; \qquad v_y = \frac{\partial \phi}{\partial y} \tag{9-20}$$

In these equations x and y represent the two geographic dimensions. If we are to represent evaporation, we must add the third, or vertical, dimension z. We wish to transport water by evaporation from the upper boundary of the layer, keeping the lower boundary as one of "no flow." This can be accomplished by adding a term of the form $\partial v_z/\partial z$ to the continuity equation. During an increment of time Δt a certain vertical "thickness" E of water will evaporate, E being the evaporation rate, analogous to and having the same dimensions as, velocity. If we let the depth of water be h instead of ∂z, we can assume that $\partial v_z/\partial z \simeq E/h$ because of the no-flow condition across the lower boundary. Substituting into the continuity equation (Equation 9-19), we obtain

$$\frac{\partial v_x}{\partial x} + \frac{\partial v_y}{\partial y} + \frac{E}{h} = 0 \tag{9-21}$$

which yields

$$\frac{\partial^2 \phi}{\partial x^2} + \frac{\partial^2 \phi}{\partial y^2} = -\frac{E}{h} \tag{9-22}$$

This modified form of the Laplace equation is called a Poisson equation, because of the constant term $-E/h$. Although Equation 9-22 at first appears to violate the principle of continuity, the constant evaporation term really represents flow in the vertical, or z, direction, and, as described above, continuity is maintained.

It is now a relatively simple matter to replace this equation by a finite-difference expression, as described in Chapter 6. If we let $\Delta x = \Delta y$, the potential at any cell (i, j) is given by

$$\phi_{i,j} = \frac{\phi_{i+1,j} + \phi_{i,j-1} + \phi_{i-1,j} + \phi_{i,j+1} + k}{4.0} \tag{9-23}$$

where $k = E(\Delta x)^2/h$, an accumulation term that consolidates the constants.

The computer program representing the evaporite model (Table 9-7) contains subroutine POTFLO, which is similar in principle to the relatively simple program listed in Table 6-5 for two-dimensional steady-state potential flow, except that subroutine POTFLO employs an accumulation term, ACCUM, which has been added to the velocity-potential equations. The program adheres to all the provisions of potential flow, including continuity of fluid mass over the entire system. This requires a balance between total inflow and total outflow, which is accomplished by summing water lost through evaporation on a cell-by-cell basis (ignoring obstacle cells that do not contain water), adding water lost by flow through an outlet or outlets if these are present, and scaling the input velocities so that the inflow and outflow quantities are equal.

The method of designating the shape of the basin, including obstacles, and positions of inlets and outlets is identical to that shown in Figures 6-9 and 6-11.

Transport of Dissolved Salt

The method of representing the transportation of dissolved salt is related to that of the fluid itself, although there are important differences. The model considers only the transportation of a single salt, sodium chloride, but other salts (e.g., calcium sulfate or calcium carbonate) could readily be represented. It is clear that if sodium chloride is progressively concentrated in each grid cell, then the continuity equation for the dissolved material must contain an accumulation term:

$$\text{input} - \text{output} = \text{accumulation}$$

The concentration field thus changes progressively with time as salt accumulates and the brine becomes more dense. Thus we need an equation with a time-dependent term, instead of the steady-state equation that we have used for fluid motion. A non-steady-state equation can be derived for our purposes by using a two-dimensional version of the continuity equation shown in Equation 6-18:

$$\frac{\partial(\rho v)_x}{\partial x} + \frac{\partial(\rho v)_y}{\partial y} = \frac{\partial \rho}{\partial t} \tag{9-24}$$

where ρ is density, defined as mass per unit volume. We are concerned with the concentration of salt, however, and we can replace ρ with c, where c has the same dimensions as ρ and represents the concentration of sodium chloride in grams per liter. Thus we can write

$$\frac{\partial(cv)_x}{\partial x} + \frac{\partial(cv)_y}{\partial y} = \frac{\partial c}{\partial t} \tag{9-25}$$

where the term cv pertains to mass flux.

Equation 9-25 states that the change of concentration with time (right-hand side) is equal to the change of mass flux with respect to the

TABLE 9-7 FORTRAN Program for Evaporite-Basin Model[a]

```
C.....EVAPORITE BASIN MODEL AFTER BRIGGS-POLLACK
C.....CONC(I,J)        CONCENTRATION ARRAY IN GMS/LITER
C.....VELX(I,J)        VELOCITY ACROSS LEFT FACE OF CELL I,J IN CMS/SEC
C.....VELY(I,J)        VELOCITY ACROSS TOP FACE OF CELL I,J IN CMS/SEC
C.....VELPOT(I,J)      VELOCITY POTENTIAL ARRAY
C.....LABEL(I,J)       CELL 'TYPE' ARRAY
C.....THICK(I,J)       THICKNESS OF PRECIPITATED SALT
C.....DTYR             LENGTH OF TIME INCREMENT IN YEARS
C.....DXM              LENGTH OF CELL SIDE IN METERS
C.....EVCMYR           EVAPORATION RATE IN CMS/YEAR
C.....DEPCM            DEPTH OF BRINE LAYER IN CMS
C.....SATGPL           SATURATION POINT FOR HALITE  IN GRMS/LITER
C.....DNGCM3           DENSITY OF SALT PRECIPITATE IN GRMS PER CC
C.....IOXIN            NO. OF BOUNDARY CELLS WITH INPUT VELS. IN X-DIRN
C.....IOYIN            NO. OF BOUNDARY CELLS WITH INPUT    VELS. IN Y-DIRN
C.....IOXOUT           NO. OF BOUNDARY CELLS WITH OUTPUT VELS. IN X-DIRN
C.....IOYOUT           NO. OF BOUNDARY CELLS WITH OUTPUT VELS. IN Y-DIRN
C.....ITMAX            MAX NO OF ITERATIONS FOR VEL. POT. SOLUTION
C.....TOL              PERMISSIBLE TOLERANCE FOR VEL. POT. SOLUTION
C.....CSTART            INITIAL CONCENTRATION OF SEA WATER IN GRMS/LITER
C.....CON              CUTOFF POINT FOR STEADY STATE CONDITION OF CONCCENTR
C                      FIELD. ITERATIONS STOP WHEN MAX((NEW CONC-OLD CONC)
C                      *100/OLD CONC).LE.CON
    1 FORMAT(20A4/9I5/9F5.0)
    2 FORMAT(16I5/16I5)
    3 FORMAT(2I5,F5.0)
    4 FORMAT(1H1, 20A4/' EVAPORATION RATE IN CMS PER YEAR ', F10.3/
     1 ' DEPTH OF BRINE LAYER ', F10.3/ ' SATURATION OF HALITE IN GRMS P
     2ER LITER ', F10.3/ ' LENGTH OF TIME STEP IN YEARS', F10.3/
     3  ' LENGTH OF CELL SIDE IN METERS', F10.3/' TOTAL VOLUME LOST BY E
     4VAPORATION IN CCS/SC ', E12.3/' VOLUME LOST BY OUTFLOW IN CCS/SC '
     5, E12.3/ ' TOTAL THROUGHPUT OF WATER IN CCS/SC', E12.3//
     6 ' EQUIVALENT VELOCITY ASSUMING SINGLE INLET 1 CELL WIDE',E12.3)
    5 FORMAT(1H1, 20A4/' ERROR',F10.3,'  NO. OF ITERATIONS', I5/
     1 ' VELOCITY POTENTIAL'/)
    6 FORMAT(1H1, 20A4/' VELOCITY COMPONENT IN X-DIRECTION'/)
    7 FORMAT(1H1, 20A4/' VELOCITY COMPONENT IN Y-DIRECTION'/)
    8 FORMAT(1H1, 20A4/ ' TIME INCREMENT ', I5/
     1 ' CONCENTRATION AFTER ', I10, ' YEARS'/)
    9 FORMAT(1H1, 20A4/ ' TIME INCREMENT ', I5/' SALT THICKNESS AFTER ',
     1 I10, ' YEARS'/)
      COMMON TITLE(20), THICK(20,20), CONC(20,20), CONC1(20,20),
     1 VELPOT(20,20), VELX(20,21), VELY(21,20), LABEL(20,20),
     2 RESVEL(20,20), VELDIR(20,20)
      DIMENSION INX(50), JNX(50), IOUTY(50), JOUTY(50)
C.....READ INPUT PARAMETERS
      READ(5,1) TITLE, NTIM, NROWS, NCOLS, ITMAX, NPRINT, IOXOUT,IOYOUT,
     1 IOXIN, IOYIN,  DEPCM, SATGPL, DNGCM3, TOL, CSTART, EVCMYR,DTYR,
     2 DXM, CON
C.....SCALE DIMENSIONS OF SOME PARAMETERS
      DXCM=DXM*100.0
      DTSEC=DTYR*31471200.0
      DTX=DTSEC/DXCM
      EVCMSC=EVCMYR/31471200.0
      ACCUM=(EVCMSC*DXCM*DXCM)/DEPCM
      NRP1=NROWS+1
      NCP1=NCOLS+1
      DUM=DEPCM*31471200.0/DXCM
      FACTOR=DEPCM/(DNGCM3*100.0)
C.....READ LABEL CELLS
      ISUM9=0
      DO 20 I=1,NROWS
      READ(5,2) (LABEL(I,J), J=1,NCOLS)
```

```
      DO 20 J=1,NCOLS
      IF (LABEL(I,J).EQ.9) ISUM9=ISUM9+1
   20 CONTINUE
C.....CALCULATE TOTAL CCS. LOST BY EVAPORATION PER SEC
      SUMCEL=NROWS*NCOLS-ISUM9
      TEVAP=EVCMSC*DXCM*DXCM*SUMCEL
C.....INITIALIZE ARRAYS
      CSTART=CSTART/1000.
      SATGPL=SATGPL/1000.
      DO 25 I=1,NROWS
      DO 25 J=1,NCP1
   25 VELX(I,J)=0.0
      DO 26 I=1,NRP1
      DO 26 J=1,NCOLS
   26 VELY(I,J)=0.0
      DO 30 I=1,NROWS
      DO 30 J=1,NCOLS
      VELPOT(I,J)=0.0
      THICK(I,J)=0.0
      CONC1(I,J)=0.0
   30 CONC(I,J)=CSTART
C.....READ VELOCITIES AT OUTLETS (IF ANY)
      SUMOUT=0.0
      IF (IOXOUT.LT.1) GO TO 32
      DO 31 IO=1,IOXOUT
      READ(5,3) I,J,VELX(I,J)
   31 SUMOUT=SUMOUT+ABS(VELX(I,J))
   32 IF (IOYOUT.LT.1) GO TO 34
      DO 33 IO=1,IOYOUT
      READ(5,3) I,J,VELY(I,J)
   33 SUMOUT=SUMOUT+ABS(VELY(I,J))
   34 SUMOUT=SUMOUT*DXCM*DEPCM
      TOTOUT=TEVAP+SUMOUT
      VELOC1=TOTOUT/(DXCM*DEPCM)
C.....READ VELOCITIES AT INLETS
      SUMIN=0.0
      IF (IOXIN.LT.1) GO TO 50
      DO 40 IO=1,IOXIN
      READ(5,3) I, J, VELX(I,J)
      INX(IO)=I
      JNX(IO)=J
   40 SUMIN=SUMIN+ABS(VELX(I,J))
   50 IF (IOYIN.LT.1) GO TO 61
      DO 60 IO=1,IOYIN
      READ(5,3) I, J, VELY(I,J)
      IOUTY(IO)=I
      JOUTY(IO)=J
   60 SUMIN=SUMIN+ABS(VELY(I,J))
C.....SCALE INLET VELOCITIES TO BALANCE OUTPUT
   61 IF (IOXIN.LT.1) GO TO 63
      DO 62 IO=1,IOXIN
      I=INX(IO)
      J=JNX(IO)
   62 VELX(I,J)=VELX(I,J)*VELOC1/SUMIN
   63 IF (IOYIN.LT.1) GO TO 65
      DO 64 IO=1,IOYIN
      I=IOUTY(IO)
      J=JOUTY(IO)
   64 VELY(I,J)=VELY(I,J)*VELOC1/SUMIN
   65 WRITE(6,6) TITLE
      CALL PRINT(NROWS,NCP1,VELX,20,21,100000.)
      WRITE(6,7) TITLE
      CALL PRINT(NRP1,NCOLS,VELY,21,20,100000.)
```

```
      WRITE(6,4) TITLE, EVCMYR, DEPCM, SATGPL, DTYR, DXM,TEVAP,SUMOUT,
     1TOTOUT, VELOC1
C.....CALCULATE VELOCITY POTENTIALS AND VELOCITY COMPONENTS
      CALL POTFLO(NROWS,NCOLS,ACCUM,TOL, ITMAX, DXCM, ERROR, IT, SUMCEL,
     1 DUM)
      WRITE(6,5) TITLE, ERROR, IT
      CALL PRINT(NROWS, NCOLS, VELPOT,20,20,10.0)
      WRITE(6,6) TITLE
      CALL PRINT(NROWS, NCP1, VELX,20,21,100000.)
      WRITE(6,7) TITLE
      CALL PRINT(NRP1,NCOLS,VELY,21,20,100000.)
      CALL VECTOR(RESVEL,VELDIR,NROWS,NCOLS)
C.....FOR EACH TIME INCREMENT CALCULATE SALT CONCENTRATION AND DETERMINE
C.....THICKNESS OF PRECIPITATED SALT
      DO 70 NT=1,NTIM
      ITIM=FLOAT(NT)*DTYR
      CALL CONCEN(DTX,NROWS,NCOLS,CSTART,NT,DXCM,DEPCM,DIFF)
      CALL PRECIP(NROWS,NCOLS,SATGPL,FACTOR)
      IF (DIFF.LE.CON) GO TO 80
      IF (MOD(NT,NPRINT).GT.0) GO TO 70
      WRITE(6,8) TITLE, NT, ITIM
      CALL PRINT(NROWS,NCOLS,CONC,20,20,1000.)
      WRITE(6,9) TITLE, NT, ITIM
      CALL PRINT(NROWS,NCOLS,THICK,20,20,100.)
   70 CONTINUE
   80 WRITE(6,9) TITLE, NT, ITIM
      CALL PRINT(NROWS,NCOLS,CONC,20,20,1000.)
      WRITE(6,9) TITLE, NT, ITIM
      CALL PRINT(NROWS,NCOLS,THICK,20,20,100.)
      RETURN
      END
      SUBROUTINE POTFLO(NROWS,NCOLS,ACCUM,TOL, ITMAX,DXCM,ERROR,
     1 IT, SUMCEL, DUM)
      DIMENSION ACHECK(20,20)
      COMMON TITLE(20), THICK(20,20), CONC(20,20), CONC1(20,20),
     1 VELPOT(20,20), VELX(20,21), VELY(21,20), LABEL(20,20),
     2 RESVEL(20,20), VELDIR(20,20)
C.....USE GAUSS-SEIDEL ITERATIVE METHOD TO SOLVE POTENTIAL FIELD
      DO 215 IT=1,ITMAX
      ERROR=0.0
      DO 210 I=1,NROWS
      DO 210 J=1,NCOLS
C.....CALCULATION DEPENDS ON TYPE OF CELL LABEL
      L=LABEL(I,J)
      GO TO (110,120,130,140,150,160,170,180,190,100,191,192), L
C.....NORMAL CELL (L=10)
  100 DUMMY=(VELPOT(I+1,J)+VELPOT(I-1,J)+VELPOT(I,J+1)+VELPOT(I,J-1)
     1  +ACCUM)/4.0
      GO TO 200
C.....UPPER LEFT CORNER CELL (L=1)
  110 DUMMY=(VELPOT(I+1,J)+VELPOT(I,J+1)-(VELY(I,J)+VELX(I,J))*DXCM
     1 +ACCUM)/2.0
      GO TO 200
C.....UPPER BOUNDARY CELL (L=2)
  120 DUMMY=(VELPOT(I+1,J)+VELPOT(I,J-1)+VELPOT(I,J+1)-VELY(I,J)*DXCM
     1 +ACCUM)/3.0
      GO TO 200
C.....UPPER RIGHT CORNER (L=3)
  130 DUMMY=(VELPOT(I+1,J)+VELPOT(I,J-1)-(VELY(I,J)-VELX(I,J+1))*DXCM
     1 +ACCUM)/2.0
      GO TO 200
C.....RIGHT BOUNDARY (L=4)
  140 DUMMY=(VELPOT(I+1,J)+VELPOT(I,J-1)+VELPOT(I-1,J)+(VELX(I,J+1)*DXCM
     1  )+ACCUM)/3.0
      GO TO 200
C.....LOWER RIGHT CORNER (L=5)
```

TABLE 9-7 *(contd.)*

```
  150 DUMMY=(VELPOT(I,J-1)+VELPOT(I-1,J)+(VELX(I,J+1)+VELY(I+1,J))*DXCM
     1  +ACCUM)/2.0
      GO TO 200
C.....LOWER BOUNDARY (L=6)
  160 DUMMY=(VELPOT(I,J-1)+VELPOT(I-1,J)+VELPOT(I,J+1)+VELY(I+1,J)*DXCM
     1  +ACCUM)/3.0
      GO TO 200
C.....LOWER LEFT CORNER (L=7)
  170 DUMMY=(VELPOT(I-1,J)+VELPOT(I,J+1)-(VELX(I,J)-VELY(I+1,J))*DXCM
     1 +ACCUM)/2.0
      GO TO 200
C.....LEFT BOUNDARY (L=8)
  180 DUMMY=(VELPOT(I-1,J)+VELPOT(I,J+1)+VELPOT(I+1,J)-(VELX(I,J)*DXCM)
     1  +ACCUM)/3.0
      GO TO 200
C.....INTERIOR OBSTACLE CELL (L=9)
  190 DUMMY=VELPOT(I,J)
      GO TO 200
C.....L=11
  191 DUMMY=(VELPOT(I,J-1)+VELPOT(I,J+1)+(VELY(I+1,J)-VELY(I,J))
     1 *DXCM+ACCUM)/2.0
      GO TO 200
C.....L=12
  192 DUMMY=(VELPOT(I-1,J)+VELPOT(I+1,J)-(VELX(I,J)-VELX(I,J+1))
     1 *DXCM+ACCUM)/2.0
      GO TO 200
C.....TEST FOR CONVERGENCE
  200 VP=ABS(VELPOT(I,J))
      IF (VP.LE.0.0000000001) GO TO 205
      DIFF=ABS(DUMMY-VELPOT(I,J))/VP
      IF (DIFF.GT.ERROR) ERROR=DIFF
  205 VELPOT(I,J)=DUMMY
  210 CONTINUE
      IF (IT.LE.5) GO TO 215
      IF ((ERROR*100.).LE.TOL) GO TO 220
  215 CONTINUE
C.....CALCULATE VELOCITY COMPONENTS ACROSS EACH CELL BOUNDARY
  220 DO 260 I=1, NROWS
      DO 260 J=1,NCOLS
      LL=LABEL(I,J)
C.....FIRST IN X-DIRECTION
      GO TO (240,230,230,230,230,230,240,240,240,230,230,240), LL
C.....CALCULATE VELOCITY ON LEFT SIDE OF CELL
  230 VELX(I,J)=(VELPOT(I,J)-VELPOT(I,J-1))/DXCM
C.....NOW IN Y-DIRECTION
  240 GO TO (260,260,260,250,250,250,250,250,260,250,260,250), LL
C.....CALCULATE VELOCITY ON UPPER SIDE OF CELL
  250 VELY(I,J)=(VELPOT(I,J)-VELPOT(I-1,J))/DXCM
  260 CONTINUE
C.....CHECK ACCUMULATION TERMS
      DO 270 I=1,NROWS
      DO 270 J=1,NCOLS
      ACHECK(I,J)=(VELX(I,J)+VELY(I,J)-VELX(I,J+1)-VELY(I+1,J))*DUM
  270 CONTINUE
      WRITE(6,1)  TITLE
      CALL PRINT(NROWS,NCOLS,ACHECK,20,20,1.)
    1 FORMAT(1H1, 20A4/ ' EVAPORATION CHECK')
C.....CALCULATE VELOCITY VECTORS
      DO 290 I=1,NROWS
      DO 290 J=1,NCOLS
      VX=(VELX(I,J+1)+VELX(I,J))/2.0
      VY=(VELY(I+1,J)+VELY(I,J))/2.0
      RESVEL(I,J)=SQRT(VX*VX+VY*VY)
      IF (ABS(RESVEL(I,J)).LT.0.000000001) GO TO 280
      VELDIR(I,J)=ATAN2(VY,VX)*180.0/3.14159+90.0
      GO TO 290
```

TABLE 9-7 *(contd.)*

```
  280 VELDIR(I,J)=0.0
  290 CONTINUE
      RETURN
      END
      SUBROUTINE CONCEN(DTX,NROWS,NCOLS,CSTART,NT,DXCM,DEPCM,BIG)
      COMMON TITLE(20), THICK(20,20), CONC(20,20), CONC1(20,20),
     1 VELPOT(20,20), VELX(20,21), VELY(21,20), LABEL(20,20),
     2 RESVEL(20,20), VELDIR(20,20)
C.....CALCULATE NEW CONCENTRATION VALUES
      BIG=0.0
      DO 28 I=1,NROWS
      DO 28 J=1,NCOLS
      L=LABEL(I,J)
C.....ASSIGN CONCENTRATION VALUES FOR CELL(I,J) AND 4 CLOSEST CELLS
      GO TO (1,2,3,4,5,6,7,8,9,10,11,12), L
C.....TOP LEFT CORNER
    1 CA=CSTART
      CB=CSTART
      CC=CONC(I,J+1)
      CD=CONC(I+1,J)
      GO TO 13
C.....TOP MARGIN
    2 CA=CONC(I,J-1)
      CB=CSTART
      CC=CONC(I,J+1)
      CD=CONC(I+1,J)
      GO TO 13
C.....TOP RIGHT CORNER
    3 CA=CONC(I,J-1)
      CB=CSTART
      CC=CSTART
      CD=CONC(I+1,J)
      GO TO 13
C.....RIGHT MARGIN
    4 CA=CONC(I,J-1)
      CB=CONC(I-1,J)
      CC=CSTART
      CD=CONC(I+1,J)
      GO TO 13
C.....BOTTOM RIGHT CORNER
    5 CA=CONC(I,J-1)
      CB=CONC(I-1,J)
      CC=CSTART
      CD=CSTART
      GO TO 13
C.....LOWER MARGIN
    6 CA=CONC(I,J-1)
      CB=CONC(I-1,J)
      CC=CONC(I,J+1)
      CD=CSTART
      GO TO 13
C.....LOWER LEFT CORNER
    7 CA=CSTART
      CB=CONC(I-1,J)
      CC=CONC(I,J+1)
      CD=CSTART
      GO TO 13
C.....LEFT MARGIN
    8 CA=CSTART
      CB=CONC(I-1,J)
      CC=CONC(I,J+1)
      CD=CONC(I+1,J)
      GO TO 13
C.....OBSTACLE
    9 CA=CSTART
```

TABLE 9-7 *(contd.)*

```
      CB=CSTART
      CC=CSTART
      CD=CSTART
      GO TO 13
C.....'NORMAL' CELL
   10 CA=CONC(I,J-1)
      CB=CONC(I-1,J)
      CC=CONC(I,J+1)
      CD=CONC(I+1,J)
      GO TO 13
C.....L=11
   11 CA=CONC(I,J-1)
      CB=CSTART
      CC=CONC(I,J+1)
      CD=CSTART
      GO TO 13
C.....L=12
   12 CA=CSTART
      CB=CONC(I-1,J)
      CC=CSTART
      CD=CONC(I+1,J)
C.....CALCULATE AMOUNT ENTERING OR LEAVING EACH CELL
   13 IF (VELX(I,J)) 14,14,15
   14 CA=CONC(I,J)*VELX(I,J)
      GO TO 16
   15 CA=CA*VELX(I,J)
   16 IF (VELY(I,J)) 17,17,18
   17 CB=CONC(I,J)*VELY(I,J)
      GO TO 19
   18 CB=CB*VELY(I,J)
   19 IF (VELX(I,J+1)) 20,20,21
   20 CC=CC*VELX(I,J+1)
      GO TO 22
   21 CC=CONC(I,J)*VELX(I,J+1)
   22 IF (VELY(I+1,J)) 23,23,24
   23 CD=CD*VELY(I+1,J)
      GO TO 27
   24 CD=CONC(I,J)*VELY(I+1,J)
   27 CONC1(I,J)=CONC(I,J)+DTX*(CA+CB-CC-CD)
      DIFF=ABS(CONC1(I,J)-CONC(I,J))/CONC1(I,J)
      IF (DIFF.GT.BIG)BIG=DIFF
   28 CONTINUE
      BIG=BIG*100.
      RETURN
      END
      SUBROUTINE PRECIP(NROWS, NCOLS,SATGPL, FACTOR)
C.....CALCULATE THICKNESS OF PRECIPITATED SALT
      COMMON TITLE(20), THICK(20,20), CONC(20,20), CONC1(20,20),
     1 VELPOT(20,20), VELX(20,21), VELY(21,20), LABEL(20,20),
     2 RESVEL(20,20), VELDIR(20,20)
      DO 10 I=1,NROWS
      DO 10 J=1,NCOLS
      CONC(I,J)=CONC1(I,J)
      IF (CONC(I,J).LT.SATGPL) GO TO 10
      THICK(I,J)=THICK(I,J)+(CONC(I,J)-SATGPL)*FACTOR
      CONC(I,J)=SATGPL
   10 CONTINUE
      RETURN
      END
```

[a]Use of the program requires that subroutines PRINT and VECTOR, which are listed in Tables 6-3 and 6-4, respectively, be added to the end of the program.

x-direction plus the change of mass flux with respect to the y-direction (left-hand side). Values for the velocity terms v_x and v_y are provided by the previous velocity-field calculations, and initial-concentration values of sodium chloride in each cell of the grid must be supplied as input data.

Transformation of Equation 9-25 to finite-difference form is relatively straightforward. For cell (i, j) at time t, using time interval Δt and assuming that $\Delta x \simeq \partial x = \partial y$, we can write

$$\frac{\Delta[(cv)_x]_{ij} + \Delta[(cv)_y]_{ij}}{\Delta x} = \frac{c_{t,i,j} - c_{t-1,i,j}}{\Delta t} \tag{9-26}$$

where the c in the (cv) terms refers to concentration at time $t-1$. Rearranging so as to solve for concentration at time t, we obtain

$$c_{t,i,j} = c_{t-1,i,j} + \frac{\Delta t}{\Delta x}\{\Delta[(cv)_x]_{ij} + \Delta[(cv)_y]_{ij}\} \tag{9-27}$$

Thus the "new" concentration at cell (i, j) equals the "old" concentration at cell (i, j) plus the change in mass flux across the cell boundaries in the x- and y-directions, multiplied by a constant $(\Delta t/\Delta x)$.

The mass-flux terms can be calculated from the concentration and velocity data already known. Referring to Figure 9-13, we can see that the change in mass flux across cell (i, j) will depend on the flux across the four faces A, B, C, and D. For the x-direction

$$\Delta[(cv)_x]_{ij} = (cv)_C - (cv)_A \tag{9-28a}$$

and for the y-direction

$$\Delta[(cv)_y]_{ij} = (cv)_D - (cv)_B \tag{9-28b}$$

The velocity components for each cell face (A, B, C, and D) are known, as are the concentration values for each cell center. The actual concentration values employed at each cell face will depend, however, on the sign on the velocity component. For example, let us assume that at face A the velocity component in the x-direction is positive, implying flow from left to right into cell (i, j). The velocity at A, say v_A, must therefore be multiplied by $c_{i,j-1}$ as this is the concentration in the cell immediately on the left of face A. If the velocity is negative, then v_A is multiplied by c_{ij}, as this is the concentration in the cell immediately to the right of face A, implying flow from right to left. Thus we assume that

$$(cv)_A = c_{i,j-1}v_A \qquad \text{for} \quad v_A > 0 \tag{9-29a}$$

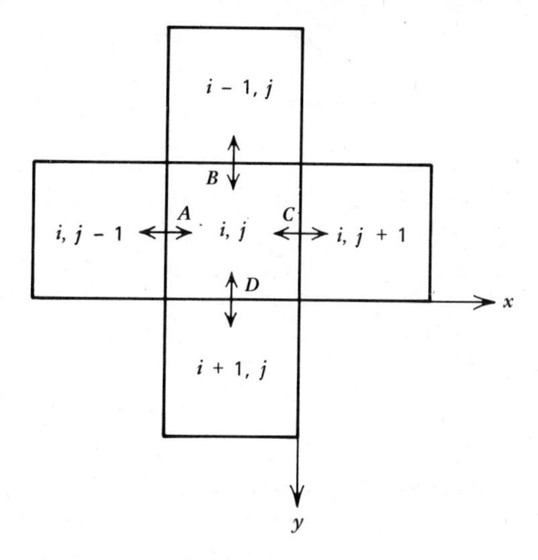

Figure 9-13 Diagram to illustrate calculation of dissolved salt flux across four faces of cell (i, j).

or

$$(cv)_A = c_{ij}v_A \qquad \text{for} \quad v_A < 0 \tag{9-29b}$$

Similarly at face B we assume that

$$(cv)_B = c_{i-1,j}v_B \qquad \text{for} \quad v_B > 0 \tag{9-30a}$$

$$(cv)_B = c_{i,j}v_B \qquad \text{for} \quad v_B < 0 \tag{9-30b}$$

and similar relationships hold for faces C and D.

These computations are carried out in the FORTRAN program (Table 9-7). Subroutine CONCEN in the program is used to calculate new concentration values from old concentration values, given the velocity components, and the length of the time interval Δt. Each time this subroutine is used the concentration field is progressively modified. The concentration of salt (here assumed to be sodium chloride) rises steadily in all cells except close to inlets where the water is constantly refreshed from the open sea. Starting with normal seawater, which contains 27 grams of sodium chloride per liter, the saturation concentration is reached at 311 grams per liter. The choice of Δt is critical. If Δt is too large, concentration gradients become unnaturally steep at inlets, with subsequently unrealistic results. If Δt is too small, so many time increments are required to produce significant changes in the concentration field that the computing time and costs are excessive. Experimentation is generally necessary to select a value for Δt that is appropriate for use with the other conditions assumed in the simulation run.

Precipitation of Salt

Sodium chloride precipitates when a concentration of 311 grams per liter is reached. At each time increment salt is removed from seawater by precipitation whenever this concentration is reached, so that the concentration never exceeds 311 grams per liter. Precipitation calculations are handled by subroutine PRECIP, employing the following equation:

$$S_{ij} = (c_s - c_{ij}) \frac{h}{\rho_s} \tag{9-31}$$

where S_{ij} = thickness of salt deposited in cell (i, j)
c_s = saturation concentration of sodium chloride set at 311 grams per liter at standard temperature and pressure (or to other values for different salts)
c_{ij} = concentration of sodium chloride in cell (i, j) in grams per liter
h = depth of water in the surface layer of brine
ρ_s = bulk density of sedimentary halite (density of combined salt particles and pore space)

Computer Program

The computer program in Table 9-7 involves the following principal steps:

1. Input parameters are read, and inflow and outflow velocities (including outflow by evaporation) are scaled so that continuity of mass is preserved.
2. The velocity field is calculated, employing boundary velocity conditions to obtain velocity potentials by Gauss–Seidel iteration (Chapter 6). Velocity vectors are calculated and plotted by using subroutine VECTOR (Table 6-4). The potential-flow calculations are carried out in subroutine POTFLO.
3. The rate of water "accumulating" in each cell is checked. The rate of accumulation should equal exactly the rate of evaporation, which is assumed to be constant and is supplied as an input parameter. If the accumulation terms are not all equal to the assigned constant, it probably means that either too few Gauss–Seidel iterations have been made or that an error has been made in assigning labels to boundary cells, as described in Chapter 6 and illustrated below.

4. Salt-concentration values are calculated for each cell in the grid.

5. If the salt concentration is greater than 311 grams per liter in any cell, the excess salt is converted to an equivalent thickness of deposited sediment.

6. Steps 4 and 5 are repeated iteratively, each step representing the advance of time by Δt, until the concentration field has stabilized. After a certain number of iterations, the concentration values will remain essentially unchanged, although the precipitation of salt will continue in those cells in which the saturation point of 311 grams per liter has been reached.

Experimental Run

Use of the computer program is illustrated with the hypothetical evaporite basin shown in Figure 9-14*a*. The shape of the basin is defined by obstacle and boundary cells, and encloses an area 550 by 500 meters in extent. A single inlet 100 meters wide admits normal seawater, and there are no other inlets or outlets. Thus all the water flowing into the basin through the inlet is necessarily lost through evaporation. The basin is represented by using a grid 11 columns wide and 10 rows deep, each cell being 50 meters square. Cells labeled with 9's (shaded) are obstacle cells and outline the shoreline of the basin. The other numbers in the grid denote types of cells, which are either boundary cells (diagonally ruled) or interior cells (labeled 10). The values for inflow velocity are given as relative values. In this example both inlet cells have been assigned a value of unity. The actual velocity values are subsequently obtained automatically by the program which calculates the total loss by evaporation from each cell (except for obstacle cells) and then scales inflow velocities so as to balance total outflow.

Input to the program for this experimental run is listed in Table 9-8. The significance of each item in the input data is described below.

Line 1: Alphanumeric title=HYPOTHETICAL EVAPORITE BASIN

Line 2: Number of time increments, NTIM=1500
Number of rows in grid, NROWS=10
Number of columns in grid, NCOLS=11
Number of Gauss–Seidel iterations, ITIM=500
Frequency of printed output, NPRINT=100
(value of NPRINT, k, causes output to be printed every kth time increment)
Number of outlets for flow in x-direction, IOXOUT=0

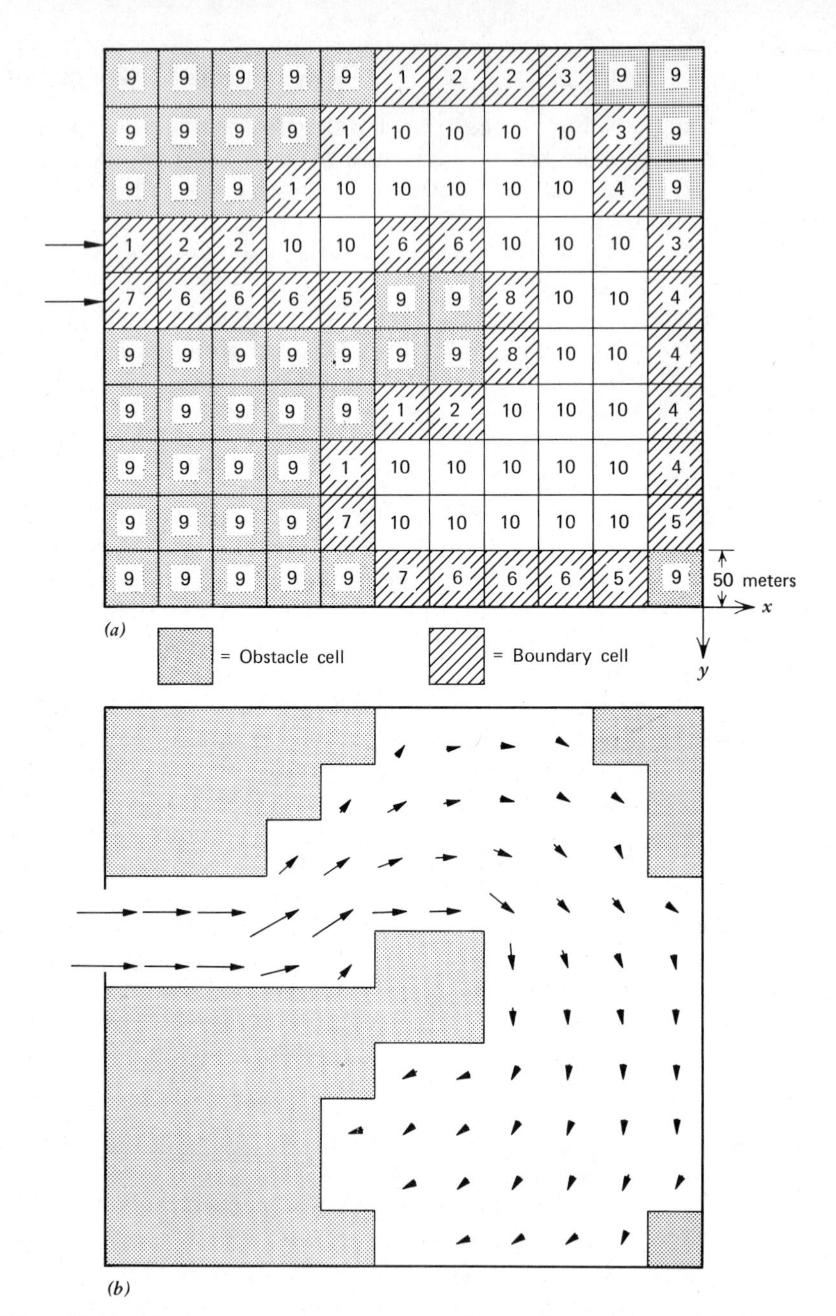

Figure 9-14 Maps pertaining to experimental run of evaporite model. (*a*) Labels applied to cells in grid to establish role of each cell in run (method is explained in conjunction with Figure 6-9). Obstacle cells (labeled with 9's and shaded) define shoreline. Cells that serve as boundaries are diagonally ruled. (*b*) Velocity-vector map representing magnitude and direction of flow.

TABLE 9-8 Input Data for Evaporite-Basin Program Used to Produce Output Shown in Figures 9-14 and 9-15

```
HYPOTHETICAL EVAPORITE BASIN
 1500   10   11  500  100    0    0    2    0
  400  311    2  .01   27   40   .3   50   .1
    9    9    9    9    9    1    2    2    3    9    9
    9    9    9    9    1   10   10   10   10    3    9
    9    9    9    1   10   10   10   10   10    4    9
    1    2    2   10   10    6    6   10   10   10    3
    7    6    6    6    5    9    9    8   10   10    4
    9    9    9    9    9    9    9    8   10   10    4
    9    9    9    9    9    1    2   10   10   10    4
    9    9    9    9    1   10   10   10   10   10    4
    9    9    9    9    7   10   10   10   10   10    5
    9    9    9    9    9    7    6    6    6    5    9
    4    1    1
    5    1    1
```

Number of outlets for flow in y-direction, IOYOUT=0
Number of inlets for flow in x-direction, IOXIN=2
Number of inlets for flow in y-direction, IOYIN=0

Line 3: Depth of surface layer of seawater in centimeters, DEPCM=400.0
Saturation point of sodium chloride in grams per liter, SATGPL=311.0
Density of resulting salt sediment in grams per cc. DNGCM3=2.0
Tolerance limit for iterative velocity-potential calculation, TOL=0.01 (see Chapter 6 for explanation)
Initial concentration of sodium chloride in grams per liter, CSTART=27.0
Evaporation rate in centimeters per year, EVCMYR=40.0
Length of time step in years, DTYR=0.3
Length of cell side in meters, DXM=50.0
Automatic cutoff limit for time-step loop when concentration field reaches steady state, CON=0.1 (calculations terminate where maximum difference between successive concentration values, divided by previous concentration, is less than or equal to CON)

Lines 4 to 13: Integer array denoting cell types (as shown in Figure 9-14a)

Lines 14 and 15: Positions of inlets and relative velocity magnitudes for flow in x-direction (as shown in Figure 9-14a)

Response of the program in terms of evaporation and precipitation is shown in Figures 9-15 and 9-16. Figure 9-15*a* consists of an array representing a map showing the depth of water in centimeters lost through evaporation per cell. The map confirms that each nonobstacle cell has lost 40 centimeters of water per year, which accords with the value assigned as input. Figure 9-15*b* is a map of velocity-potential values. The actual inflow velocities are very small. The velocity at the two inlet cells, which allow sufficient seawater to enter so as to balance total loss by evaporation, is 5.2×10^{-4} centimeters per second. The flow vectors in Figure 9-14*b* show the progressive decline in velocities away from the inlet.

The gradual concentration and subsequent precipitation of salt are illustrated with maps (Figure 9-16) for time increments 50, 100, and 200, which represent 15, 30, and 60 scale years, respectively. At the outset of the simulation run the concentration in all cells is set at 27 grams per liter, which is the salinity of ordinary seawater. As the model is moved forward through time, the concentration increases most rapidly in those cells that are most distant from the inlets. By the 15th scale year of operation concentration values range up to 118 grams per liter, and by the 30th year concentrations have reached the maximum, or saturation, point of 311 grams per liter in some cells. Salt is deposited in each cell in which the concentration reaches 311 grams per liter. The concentration is less than the maximum in the cells between the inlet and the far interior end of the basin. In the intermediate part of the flow region continuous inflow of ordinary

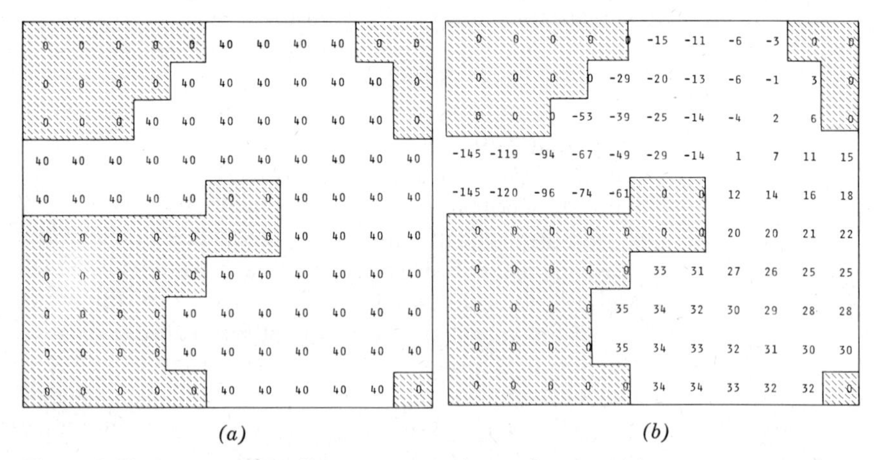

Figure 9-15 Maps produced in experimental run of evaporite-basin program to show (*a*) depth of layer of water, in centimeters, removed by evaporation per year in each cell, and (*b*) velocity-potential values multiplied by 10. Shading represents land area.

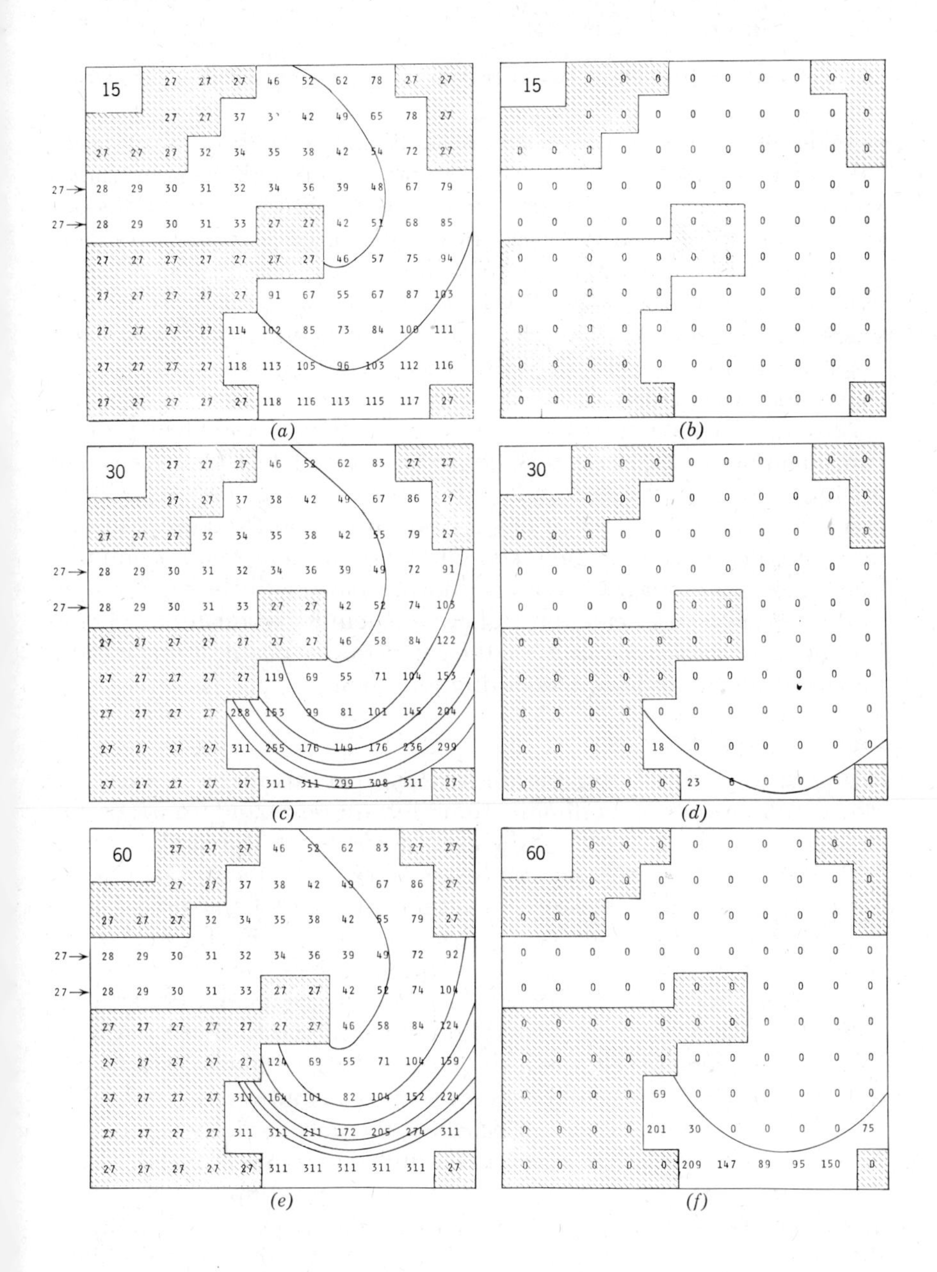

Figure 9-16 Maps produced in experimental run of simulation program showing progressive changes in sodium chloride concentration in grams per liter (*a*, *c*, *e*) and thickness of halite in centimeters deposited (*b*, *d*, *f*) at 15, 30, and 60 scale years of model operation.

seawater prohibits the concentration from reaching the saturation point.

At the 60th year the concentration field appears to be almost identical to that shown for the 30th year. In other words, the concentration values, ranging from 27 grams per liter at the inlet to 311 grams per liter in the most restricted part of the basin, have reached a stable condition. Continued inflow of seawater causes precipitation of salt in cells with a concentration of 311 grams per liter, and concentration values in all cells remain unchanged. The distribution of salt deposits is shown in Figure 9-16, deposition occurring adjacent to the shoreline in that part of the basin farthest from the inlet.

Although we have illustrated the model with only a single computer run, a number of parameters could be altered to test a variety of assumptions. The configuration of the basin and the position of inlets could be changed. Depth of the surface brine layer and the evaporation rate also have a pronounced effect on the rate of salt accumulation. Materials other than sodium chloride may readily be represented in the program (e.g., calcium sulfate or calcium carbonate) by changing the seawater-concentration, the saturation-concentration, and the bulk-density values for the resulting sediment.

Limitations of Model

In case we should give the impression that this model is the "last word" on evaporite sedimentation, let us point out some of the limitations of the model in its present form. First, the model is concerned only with evaporite deposition in restricted marginal marine basins; for example, it does not attempt to model the tidal flat or "sabka" evaporite mechanism (Kinsman, 1969). Furthermore the whole topic of evaporite models is one on which several respected researchers have widely differing viewpoints (Schmalz, 1969; Sloss, 1969), and simulation models could be developed to represent their hypotheses. Second, although the Briggs–Pollack model is dynamic in the sense that the salt-concentration distribution is continually altered until an equilibrium is reached, the flow model is static only. In other words, the velocity field is assumed to remain constant while deposition occurs. In reality the salt would build up in some areas and alter the velocity field. For example, the problem of simulating the deposition of a large thickness of salt in the center of the Michigan basin (see Chapter 1) with only small thicknesses at the margins might be overcome by using a flow model in which the depth of water changes with deposition. Salt deposited in relatively shallow water near shore would progressively fill the basin from the outer edge inward, gradually confining the flow region to a small zone in the basin center. Thus the distribution of salt thickness would tend to reflect the three-dimensional configuration of the basin floor, the greatest thicknesses of salt being deposited in deeper parts of the basin, and the shallow areas receiving lesser thicknesses. If subsidence and compaction were added to the model, its behavior might then be highly realistic, although difficult to predict in advance.

Finally, the flow pattern of the model that we illustrate does not consider the effect of varying density of brine in the uppermost layer. Once evaporation has produced concentration gradients over the flow region, density gradients would also occur, and one would expect the flow pattern to be somewhat altered, particularly in cases where the uppermost layer was of appreciable thickness. This effect could be included in the model by retaining the density parameter in the basic continuity equation (Equation 6-18).

DELTA MODEL

A simulation model for deltaic deposition at river mouths has been developed by Bonham-Carter and Sutherland (1968). Basically the model is concerned with tracing the movement of sediment particles as they leave a river mouth, determining where they will settle, and calculating the rate of sediment accumulation in front of the mouth (Figure 2-11). The principal factors that affect the size, shape, and composition of river-mouth deltas are (a) density differences between inflow and basin water, (b) water discharge from the river (c) composition and quantity of sediment load of the river, (d) geometry of the basin, (e) tectonic behavior of the crust underlying the basin, (f) offshore wave-energy conditions, (g) tides, and (h) climate. The Bonham-Carter–Sutherland model considers only the first four of these factors. Although the model does not have immediate application to most real deltas—because few, if any, deltas are formed without modification by offshore energy factors—it provides a method for exploring the interrelationships between the variables important in the constructive phase of delta building. This is an essential step in the development of more realistic and complex models of deltaic sedimentation.

Although the mathematical basis for this model involves the calculation of a velocity field and the transport of material in flowing water, the techniques employed are quite different from those used in the evaporite-basin model. Instead of using potential flow, as in the evaporite model, the delta model combines equations for flow in an open channel with equations for jet flow. Three important differences between potential-flow models and the delta model should be noted. First, potential-flow equations are derived from basic theory that employs continuity of mass, whereas equations for open-channel and jet flow are largely empirical, being obtained from laboratory and field observations. Second, empirical flow relationships as used in the delta model incorporate the effects of viscosity and turbulence, which potential-flow theory does not consider. Third, the empirical flow equations in the delta model refer only to the velocity component parallel to the main axis of channel flow. This means that the channel-velocity terms in the delta model are not directly comparable to vector quantities defined by potential-flow equations. For this reason velocity-vector maps are not plotted for the delta model as in the evaporite model.

Another important difference between the evaporite and delta models concerns the transport of sediment. In the evaporite model salt is transported in solution, whereas in the delta model sediment is transported principally by suspension, although transport by traction is implied but not explicitly represented. The paths, or trajectories, of suspended sediment particles as they leave the channel mouth in the delta model are calculated on the basis of current velocities in the front of the channel mouth, as well as on the rate of fall of the particles (Figure 2-11).

The formulation of the delta model is relatively abstract and simple. From a channel, assumed to be rectangular in cross section (Figure 9-17), a stream debouches into a body of salt water. As it leaves the channel mouth, the fresh water in the stream is assumed to spread out over the more dense salt water as a *plane jet*, forming two distinct layers (Figure 9-18). The model employs a rectangular coordinate system whose origin is the center of the channel floor at the mouth. The principal direction of river flow is the X-axis, the Y-axis is horizontal and transverse to the X-axis, and the Z-axis is vertical, with the positive direction upward.

Velocity Field

The river entering the basin is thus compared to a jet discharging from a rectangular slot of width B. This follows the suggestion by Bates (1953) that flow at the mouth of a river entering an ideal, tideless, currentless, marine basin of infinite size is similar to a horizontal two-dimensional plane jet. The fresh river water is considerably less dense than salty ocean water due to the difference in salinity. For most rivers, salinity is shown by Bates to be much more important

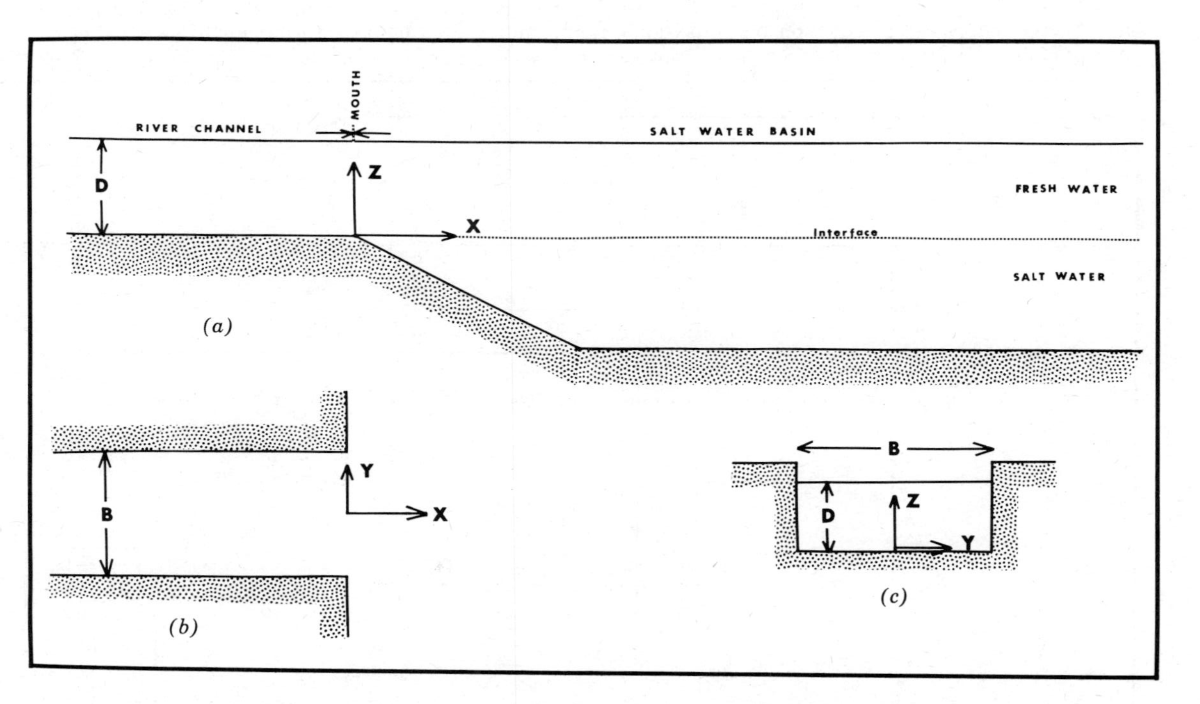

Figure 9-17 Diagram defining rectangular river channel leading into salt-water basin. Fresh water is assumed to be buoyed up by salt water without vertical mixing between layers. Elevation views are shown in (*a*) and (*c*), plan view in (*b*). From Bonham-Carter and Sutherland (1968).

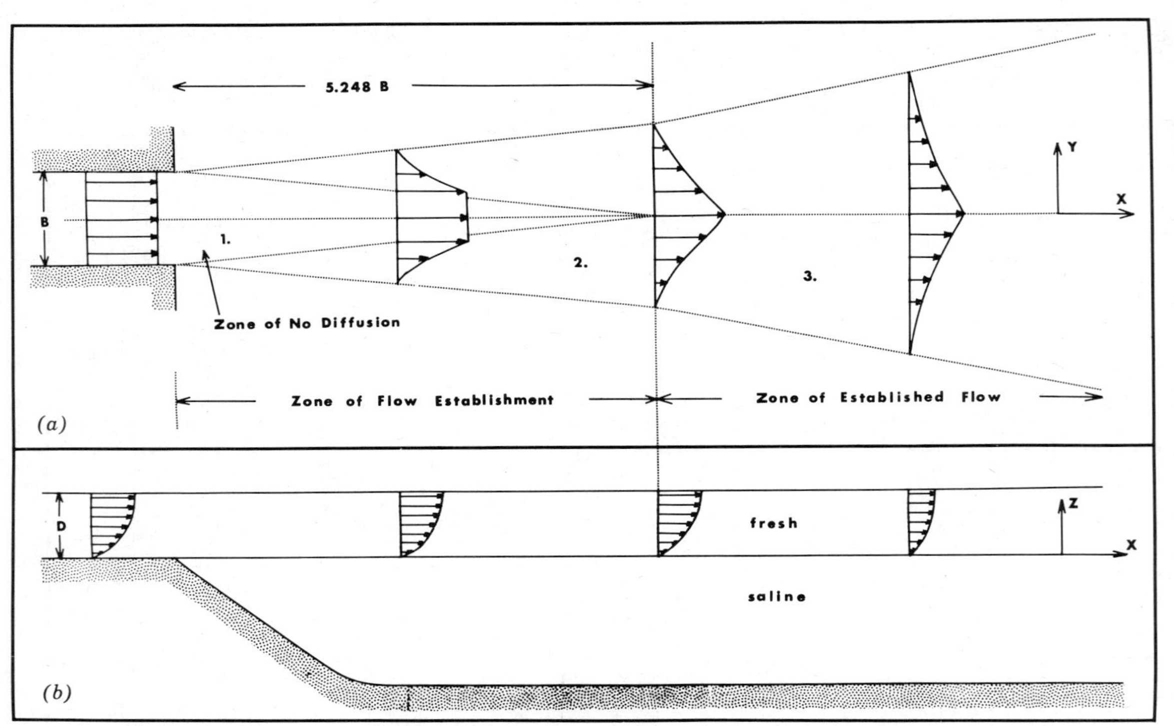

Figure 9-18 Plan and elevation view of flowfield. (*a*) Plan view depicts structure of plane-jet velocity field with three distinct flow regions. (*b*) Elevation view (center-line section) shows exponential decrease in velocity with depth, and illustrates assumption of two-layer system with fresh water buoyed up on more dense saline water. From Bonham-Carter and Sutherland (1968).

than temperature or the presence of sediment particles in affecting density. As a consequence the river water—even when muddy and cold compared with the salty ocean water—is buoyed up by the salt water, and vertical mixing between the two types of water is assumed to be negligible. Shearing and mixing occur in the horizontal plane along the margins of the plane jet, however, and this causes the velocity to be gradually reduced (Figure 9-19).

Construction of the model involves linking the equations for a horizontal two-dimensional plane jet, with variations in stream velocity in the third or vertical dimension. The current velocity in a stream is greatest near the top and decreases exponentially toward the bottom (Figure 9-20). The model provides for variations in vertical velocity by employing a sequence of plane jets, each occupying a thin layer, much like sheets of cardboard stacked one upon another. The velocity pattern in each layer is of the same general form (Figure 9-19), except that the initial velocity in each layer (at the channel mouth) depends on the velocity in the stream channel at that elevation. The velocities elsewhere in the plane jet are scaled to the initial velocity, as shown in Figure 9-19. For a given elevation the velocities at the channel mouth are assumed not to vary across the width of the channel.

We define the streamwise (X-direction) component of velocity for any point in front of the channel mouth as $U(X, Y, Z)$, where X, Y, and Z are the distances to the point from the center of the river mouth on the channel floor along the three coordinate axes. We shall show how the streamwise-velocity component can be determined given the mean streamwise velocity in the river $\overline{V}$ and values of X, Y, and Z.

The model is formulated in terms of dimensionless variables, using slot width B and mean velocity in the river $\overline{V}$ as reference quantities. This is a convenient approach because it permits any consistent system of measurement units to be employed. Capital letters have been employed for dimensional quantities and lower case letters for their dimensionless counterparts. Thus let $x = X/B$, $y = Y/B$, $z = Z/B$, $d = D/B$, $u_* = U_*/\overline{V}$, and $u = U/\overline{V}$, where U_* is the shear velocity in the river.

The velocity profile in the river $V(Z)$ is assumed to be that of an open channel, given by the equation

$$V(Z) = \overline{V} + \frac{U_*}{k}\left(1 + \log_e \frac{Z}{D}\right) \tag{9-32}$$

where $V(Z)$ = streamwise component of velocity at elevation Z above channel floor

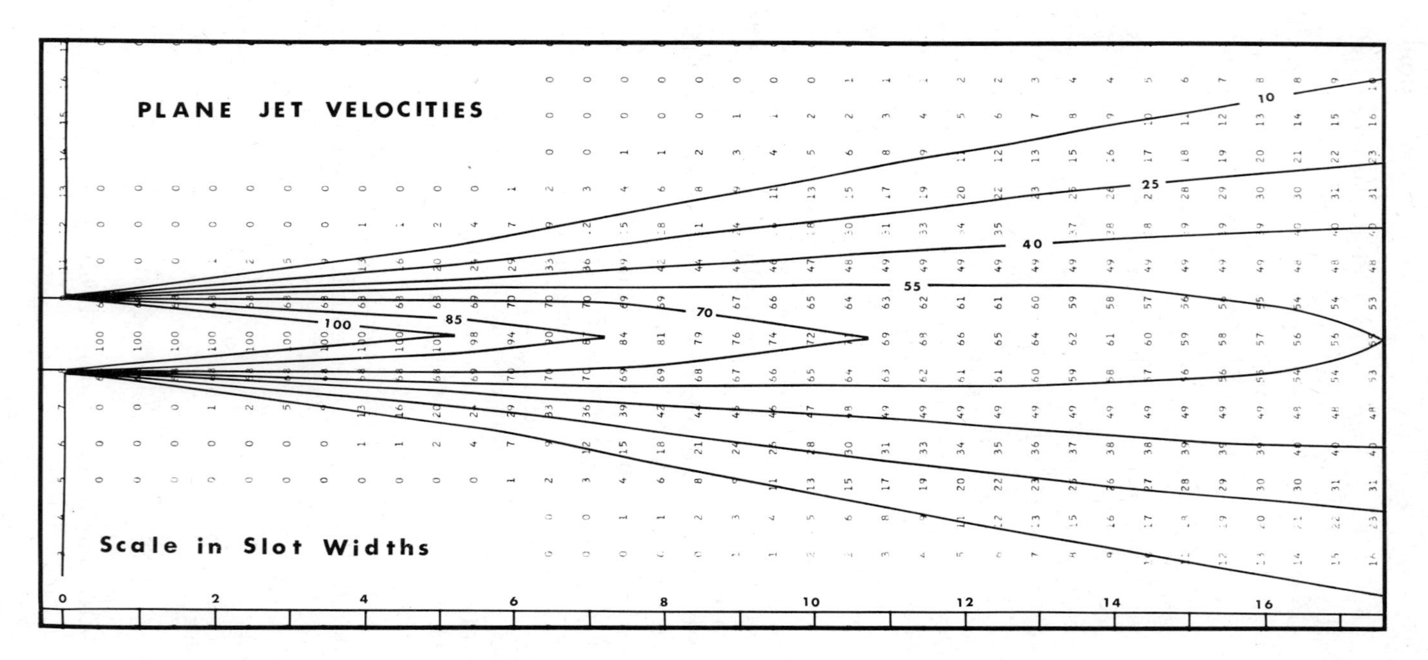

Figure 9-19a Contour map of relative velocities in plane-jet velocity field. Geographic scale is in unit "slot widths". Number printed in each cell represents velocity as a percentage of initial velocity (which is defined as 100 percent). With velocities expressed as percentages and geographic scale in terms of slot widths, velocity field is identical for any plane jet regardless of initial velocity or of actual width of slot.

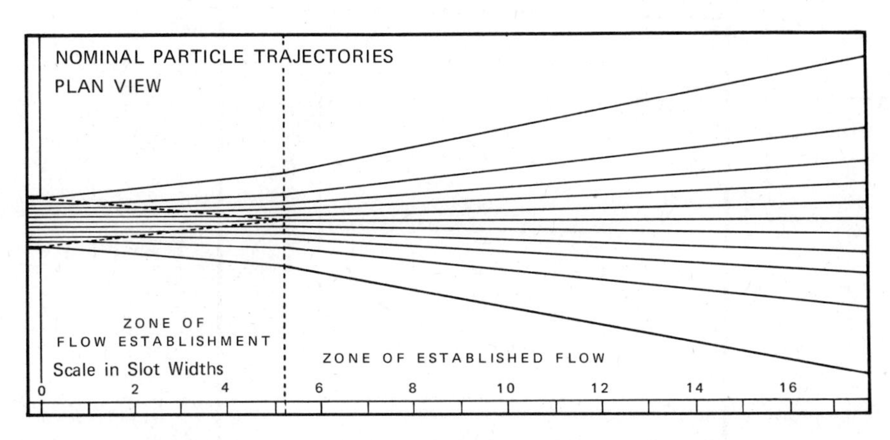

Figure 9-19b Map showing a number of representative particle paths or trajectories in map view. As in (*a*), geographic scale is in slot widths. From Bonham-Carter and Sutherland (1968).

$\overline{V}$ = mean streamwise-velocity component in river channel
U_* = shear velocity, a measure of the frictional characteristics of the channel floor
k = Von Kármán's constant, a numerical factor that varies with the amount of sediment in suspension and is here assumed to be 0.4
D = depth of channel

The equation states that the streamwise-velocity component equals the mean river velocity $\overline{V}$, plus an expression that accounts for the frictional characteristics of the channel floor, amount of sediment in suspension, depth of channel, and elevation above the channel floor. This is a well-known equation, described, for example, by Vanoni, Brooks, and Kennedy (1961). In converting Equation 9-32 to dimensionless form it is convenient to define a new velocity function $f_1(z)$ as

$$f_1(z) = \frac{V(Z)}{\overline{V}} \equiv 1 + \frac{u_*}{k}\left(1 + \log_e \frac{z}{d}\right) \tag{9-33}$$

As the current moves away from the channel mouth in a plane jet, the distribution of current velocity (Figure 9-18) is such that three zones of flow can be distinguished (Albertson and others, 1950):

1. A zone of no diffusion
2. A zone of flow establishment
3. A zone of established flow.

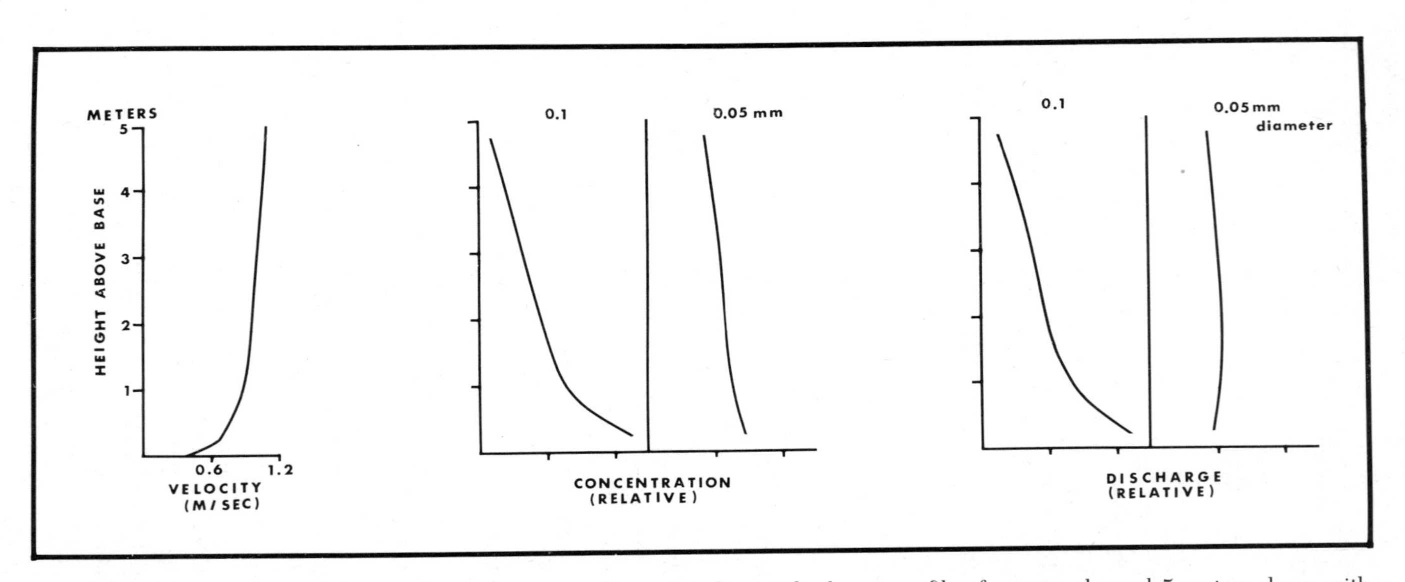

Figure 9-20 Theoretical velocity, sediment-concentration and sediment-discharge profiles for open channel 5 meters deep, with average water velocity of 1 meter per second. Ideal spherical quartz particles that are 0.1 and 0.05 millimeters in diameter are used for concentration and discharge curves. Note that fine material is more evenly distributed with depth than coarse material. From Bonham-Carter and Sutherland (1968).

The zone of no diffusion is an elongate triangular area that tapers away from the channel mouth. Within the zone of no diffusion the velocity remains constant. Beyond the edges of this zone, however, the velocities fall off sharply. In the zone of established flow the graphs of the distribution of velocity values measured along lines parallel to the Y-axis are similar to regular Gaussian frequency-distribution curves, each of which has a standard deviation that increases linearly with distance from the river mouth. Within the zone of flow establishment the velocity-distribution curves are also similar to Gaussian curves, but flattened at the center because the forward velocities do not vary within the zone of no diffusion. In the zone of established flow, which lies totally beyond the zone of no diffusion, the forward velocities are greatest at the center of the flow and decrease fairly rapidly to either side in the Y-direction.

The functional relationships for the streamwise-velocity component at any point in front of the river mouth in terms of x, y, and z are defined as follows. In the zone of no diffusion (labeled region 1 in Figure 9-18)

$$u(x, y, z) = f_1(z) \tag{9-34}$$

which simply says that the streamwise-velocity component equals the initial velocity at that elevation at the mouth. In region 2

$$u(x, y, z) = f_1(z) \exp\left\{-42.3\left[0.096 + \frac{y-0.5}{x}\right]^2\right\} \equiv f_1(z) f_2(x, y) \tag{9-35}$$

and in region 3 the zone of established flow, we may write

$$u(x, y, z) = f_1(z)\, x^{-1/2} \exp\left(0.828 - \frac{42.3y^2}{x^2}\right) \equiv f_1(z)\, f_3(x, y) \tag{9-36}$$

The relationships for $f_2(x, y)$ and $f_3(x, y)$ are from Albertson and others (1950). Thus for the three regions we may write

$$u(x, y, z) = f_1(z)\,[1, f_2(x, y), f_3(x, y)] \tag{9-37}$$

where the terms in square brackets apply to regions 1, 2, and 3, respectively. This equation is later used to calculate the settling trajectory of sedimentary particles issuing from the mouth, as shown below.

Equation 9-37 indicates that the streamwise velocity at any point in front of the river mouth can be calculated if the coordinates of the point along the x-, y-, and z-axes are given.

Sediment Discharge

The individual sediment particles transported by the river are subject to random motion due to the turbulent nature of the flow. Grains are continually plucked from the bed, bounced or rolled along the bottom, swept up into the turbulent water, and dropped again. Despite the random nature of this process, the gross behavior of the total sedimentary load of a river can be modeled deterministically. In the model the river is subdivided into a number of streamtubes, and the total sediment traveling along any tube is represented by the *nominal*, or *statistical*, particle at the center of the tube. Each nominal particle thus represents the average behavior of the real sediment grains transported in that part of the flow (Figure 2-11). The model is thus concerned with calculating the proportion of the total sediment load represented by each nominal particle and then tracking the movement of the particles as they settle in front of the mouth.

Although individual particles of suspended sediment move in a locally random fashion, due to turbulence, it is assumed that the nominal particles can be traced in the same general manner as ballistic missiles. Systematic decrease in forward velocity, coupled with systematic variations in sediment content with height above river bottom affect the distribution and subsequent deposition of sediment. Idealized geographic paths of nominal sediment particles are shown in Figure 9-19*b*. Shear action along the margins of the jet causes declining velocities beyond the zone of no diffusion, and the paths are progressively deflected away from the centerline of the jet. Variations in forward velocity with depth also have important bearing. Figure 2-11 shows the idealized paths of nominal particles of identical size that are suspended at different heights above the base of the stream as they emerge at the stream mouth. Note that highest particles travel the farthest by virtue of greater initial height and greater initial velocity. Each path curves downward with a continuously varying slope that depends on the settling velocity of the particle and the forward velocity at each point of descent. Finally, when the particles pass from the fresh-water jet to the quiet underlying seawater, their forward velocities are assumed to be zero, so they fall vertically downward.

The sediment transported along an open channel is distributed unevenly with depth, so the actual quantity of sediment represented by a nominal particle will depend on the height of the particle above the river bottom. For coarse-grained materials the bulk of the load is carried near the bottom. Fine-grained materials are more evenly distributed through the water column (Figure 9-20). The concentration of sediment at a particular depth in an open channel can be calculated from the equation

$$c_z = c_a\left(\frac{d-z}{z} \cdot \frac{a}{d-a}\right)^p \tag{9-38}$$

where c_z = concentration at elevation z above bottom
d = channel depth
c_a = reference concentration at elevation a
a = reference elevation
$p = w_s/ku_*$
w_s = fall velocity of sediment in still water

The qualities k and u_* have been already defined.

In order to find the proportion of the total load transported along a single streamtube it is necessary to calculate the sediment discharge per unit width s between the lower z_i and upper z_j boundaries of that tube. This is done by integrating the product of sediment concentration c times velocity $f_1(z)$ between elevations z_i and z_j. Thus, substituting Equation 9-38 for c, we obtain

$$s(z_i, z_j) = \int_{z_i}^{z_j} f_1(z)\, c_a\left(\frac{d-z}{z} \cdot \frac{a}{d-a}\right)^p dz \tag{9-39}$$

The total sediment discharge per unit width of channel can be defined as $s(\epsilon, d)$, where ϵ is a very small elevation value (four times particle diameter here) introduced because the velocity function $f_1(z)$ does not apply for flow very close to the channel bottom. We can now obtain the fraction F (the sediment load traveling along a single tube between elevations z_i and z_j divided by the total load) as

$$F(z_i, z_j) = \frac{s(z_i, z_j)}{s(\epsilon, d)} = \frac{\int_{z_i}^{z_j} f_1(z)[(d-z)/z]^p\, dz}{\int_{\epsilon}^{d} f_1(z)[(d-z)/z]^p\, dz} \tag{9-40}$$

Now given the total sediment load (g_s = mass/unit time), we can calculate the load represented by each nominal particle as $g_s\, F(z_i, z_j)/n$, where n is the number of streamtubes across the channel width.

The actual settling trajectory of a nominal particle in elevation view is obtained by considering the forces acting on the particle.

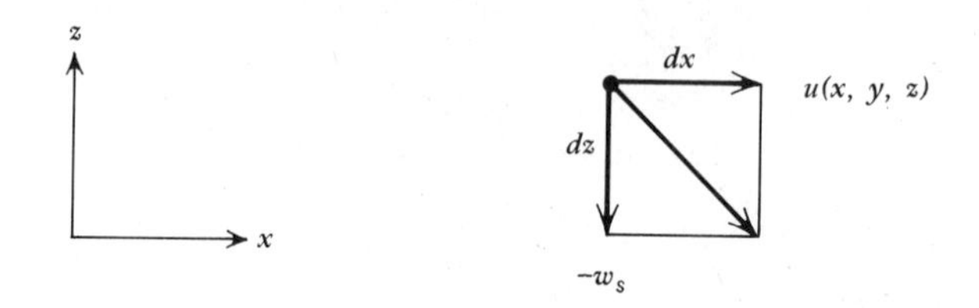

The fall velocity w_s acts vertically downward, and the streamwise component of velocity $u(x, y, z)$ acts in the x-direction. Thus at any point on the settling trajectory the slope dz/dx is assumed to be given by $-w_s/u(x, y, z)$, as shown above. The function $u(x, y, z)$ is known; given fall velocity w_s and the initial coordinates of the particle at the river mouth $(0, y_0, z_0)$, we can integrate over the length of the trajectory to determine the terminal coordinates $(x_t, y_t, 0)$. At this point the nominal particle passes through the boundary between fresh water and salt water, and is then assumed to settle vertically to the bottom. Substituting Equation 9-37, we obtain the slope of the trajectory:

$$\frac{dz}{dx} = -\frac{w_s}{u(x, y, z)} = -\frac{w_s}{f_1(z)[1, f_2(x, y), f_3(x, y)]} \tag{9-41}$$

Integration of Equation 9-41 over the length of the trajectory after cross-multiplication yields

$$\int_{\epsilon}^{z_0} f_1(z)\, dz = \int_0^{x_t} \frac{w_s\, dx}{[1, f_2(x, y), f_3(x, y)]} \tag{9-42}$$

Equation 9-42 is then solved for x_t. Bonham-Carter and Sutherland (1968) derive general solutions to obtain the terminal coordinates (x_t, y_t) for any nominal particle, considering lateral spread as well as vertical settling. For the simple case of particles traveling along the jet center line (i.e., at $y = 0$) Equation 9-42 yields the following solutions:

$$x_t = \frac{\alpha}{w_s} \quad \text{if} \quad x_t \leq 5.248 \quad \text{(region 1)} \tag{9-43}$$

and

$$x_t = \left[\frac{\alpha}{0.29 w_s} - 6.075\right]^{2/3} \quad \text{if} \quad x_t > 5.248 \quad \text{(region 3)} \tag{9-44}$$

where

$$\alpha = z_0 - \epsilon + \frac{u_*}{k}\left[z_0 \log_e\left(\frac{z_0}{d}\right) + \epsilon \log_e\left(\frac{d}{\epsilon}\right)\right] \tag{9-45}$$

For solutions with $y > 0$ readers are referred to Bonham-Carter and Sutherland's paper (1968). The parameters ϵ, u_*, k, w_s, and d are the same as those already defined.

Computer Program

The computer program for this model is not reproduced in this book, but it is listed in full by Bonham-Carter and Sutherland (1968). The computational steps are as follows:

1. The total sediment load (mass/unit time) is distributed among the streamtubes, so that the mass of sediment represented by each nominal particle is known.
2. The terminal coordinates for each nominal particle are calculated by using Equations 9-43 and 9-44 or those developed by Bonham-Carter and Sutherland.
3. An array representing a horizontal (x, y plane) grid is employed to account for the location and quantity of sediment represented by each nominal particle as it settles in front of the mouth (Figure 2-11). The accounting grid is represented by a four-dimensional FORTRAN array that has two dimensions for the horizontal coordinates, one dimension for grain size, and one for time. The mass of sediment represented by each nominal particle is converted to a stratigraphic thickness, given the dimensions of the accounting cell and the bulk density of the resulting sediment. Compaction is not considered.

The model can be both static (single time increment) and dynamic (several time increments). We shall first discuss some results obtained with the static model and then proceed to the dynamic model.

Table 9-9, 9-10, and 9-11 list the computer output from a run employing three grain-size fractions and a channel 30 meters wide and 10 meters deep. Table 9-9 lists the parameters describing the channel, which has a slope of 1 to 1000, a friction coefficient f for bottom roughness of .02, and a Von Kármán constant k of 0.4. The shear velocity U_* is calculcated from depth D and slope S by employing the relationship

$$U_* = \sqrt{gDS} \tag{9-46}$$

where g is the acceleration due to gravity (Vanoni, Brooks, and

TABLE 9-9 Computer Printout Showing Details of River Channel[a]

CHANNEL CHARACTERISTICS	SLOTS		MTRS		
DEPTH	0.10000		3.00000		
SLOPE					0.00100
DARCY WEISBACH FRICTION COEFFICIENT					0.02000
VON KARMAN'S CONSTANT					0.40000
NOMINAL LOWER BOUNDARY	0.133333E-03		0.40000E-02		
SHEAR VELOCITY	0.57184E-02	PER SEC	0.17155	PER SEC	
AVERAGE VELOCITY	0.11437E 00	PER SEC	3.43103	PER SEC	
WATER DISCHARGE	0.11437E-01	PER SEC	308.79272	PER SEC	

[a]Units are shown in slots (widths of channel at the mouth) and meters. Channel is 30 meters wide.

Kennedy, 1961). Then the average channel velocity is found by using the relation

$$\bar{V} = U_* \sqrt{8/f} \tag{9-47}$$

The sediment characteristics of each grain-size fraction are listed in Table 9-10. The distribution of sediment with depth is shown in Table 9-11 and clearly shows that the coarse grain-size fraction is transported predominantly near the channel floor, whereas the fine-grained fractions are more evenly distributed through the water column.

The contents of the accounting grid array are listed in Figure 9-21*a* through *c*. These tables represent map views of the thickness distributions of each grain size. Figure 9-22 shows the same information displayed graphically as a series of perspective diagrams drawn automatically by a CALCOMP digital plotter. These thickness-distribution surfaces are symmetrical about the axis of channel flow. Fine-grained material is carried much further from the channel mouth than the coarse material. The average slope on the "foreset" surface is approximately half a degree, assuming an initially flat basin floor, being greater for coarse sediment and smaller for fine sediment. Although the perspective plots show gross vertical exaggeration, the decrease in foreset angle with decrease in grain size is clearly visible. One feature brought out by the perspective diagrams reflects the distribution of the three sediment-size fractions in the total sediment load. In the input data size fraction 3 (0.25-mm diameter) is two times more abundant than size fraction 2 (0.5-mm diameter), which in turn is five times more abundant than size fraction 1 (1-mm diameter),

TABLE 9-10 Computer Printout Showing Sediment Characteristics for Three Grain-Size Fractions Used in This Experiment[a]

```
SEDIMENT CHARACTERISTICS
  NUMBER OF SIZE FRACTIONS    3

     FRACTION    1
         LOAD IN GRAMS PER YEAR                                                   0.74291E 11
         LOAD AS CELL SQUARE THICKNESS IN MTRS ACCUMULATED OVER 1.0 YEAR         79.30663
         PARTICLE DIAMETER                                                        1.00000 MM
         SETTLING VELOCITY                                                        0.98419E-01 MTRS PER SEC
         DENSITY                                                                  0.26600E 07 GRAMS PER CUBIC MTR
         POROSITY                                                                 0.20000

     FRACTION    2
         LOAD IN GRAMS PER YEAR                                                   0.37146E 12
         LOAD AS CELL SQUARE THICKNESS IN MTRS ACCUMULATED OVER 1.0 YEAR        396.53296
         PARTICLE DIAMETER                                                        0.50000 MM
         SETTLING VELOCITY                                                        0.62013E-01 MTRS PER SEC
         DENSITY                                                                  0.26600E 07 GRAMS PER CUBIC MTR
         POROSITY                                                                 0.20000

     FRACTION    3
         LOAD IN GRAMS PER YEAR                                                   0.74291E 12
         LOAD AS CELL SQUARE THICKNESS IN MTRS ACCUMULATED OVER 1.0 YEAR        793.06616
         PARTICLE DIAMETER                                                        0.25000 MM
         SETTLING VELOCITY                                                        0.32162E-01 MTRS PER SEC
         DENSITY                                                                  0.26600E 07 GRAMS PER CUBIC MTR
         POROSITY                                                                 0.20000
```

[a]Total sediment load is dominated by a large quantity of the fine-grained fraction. "Cell square thickness" pertains to the equivalent stratigraphic thickness in a single cell of the accounting grid.

TABLE 9-11 Computer Printout Listing Distribution of Sediment over One Vertical Column of Stream Tubes at the Channel Mouth[a]

```
SEDIMENT DISCHARGE PROFILE AT CHANNEL MOUTH

DISCHARGE EXPRESSED AS A/C GRID CELL THICKNESS IN MTRS PER  1.00 YEARS

HEIGHT ABOVE CHANNEL FLOOR EXPRESSED IN MTRS
```

ROW	HEIGHT	SIZE FRACTIONS 1	2	3
10	2.85019	0.36952E-03	0.54491E-01	0.84473E 00
9	2.55059	0.18546E-02	0.16182E 00	0.15498E 01
8	2.25100	0.45024E-02	0.28331E 00	0.20639E 01
7	1.95139	0.87826E-02	0.42976E 00	0.25435E 01
6	1.65180	0.15681E-01	0.61509E 00	0.30351E 01
5	1.35220	0.27243E-01	0.86338E 00	0.35768E 01
4	1.05260	0.48277E-01	0.12232E 01	0.42201E 01
3	0.75300	0.92821E-01	0.18138E 01	0.50655E 01
2	0.45340	0.22345E 00	0.30517E 01	0.63935E 01
1	0.15380	0.30484E 01	0.11351E 02	0.10359E 02
TOTAL FOR THE COLUMN		0.34714E 01	0.19848E 02	0.39652E 02
TOTAL FOR THE GRID		0.69427E 02	0.39695E 03	0.79303E 03

[a]There are 10 stream tubes (rows) per column and three grain-size fractions. Units represent stratigraphic thicknesses of sediment that would result if sediment was deposited in a single cell of the accounting grid. Note that the bulk of the coarse fraction travels near the channel bottom, but the fine fraction is more uniformly distributed according to height above bottom.

reflecting a nonlinear distribution of sediment abundance with sediment size. However, the thickness of sediment deposited in the cells closest to the river mouth (as reflected by the height of the peaks on the plots) is shown to be greatest for the middle-size fraction (Figure 9-22*b*), which initially seems surprising. In fact, if the initial sediment load contained equal proportions of all three size fractions, the peak would be highest for the coarse fraction (Figure 9-22*a*). However, the relative paucity of the coarse fraction in the total load produces a relatively small peak with a steep foreset slope, whereas the medium-grained material is sufficiently abundant close to the mouth to produce the highest peak.

Experiments with the Static Model

A series of experiments employing the model in static form were carried out to evaluate the sensitivity of the model to changes in input parameters. The results of these experiments are listed in compact form in Table 9-12, and are illustrated graphically in Figure 9-23. Four series of experiments illustrate the effect of changing four parameters—grain size, channel depth, channel slope, and channel width. By decreasing grain diameter progressively from

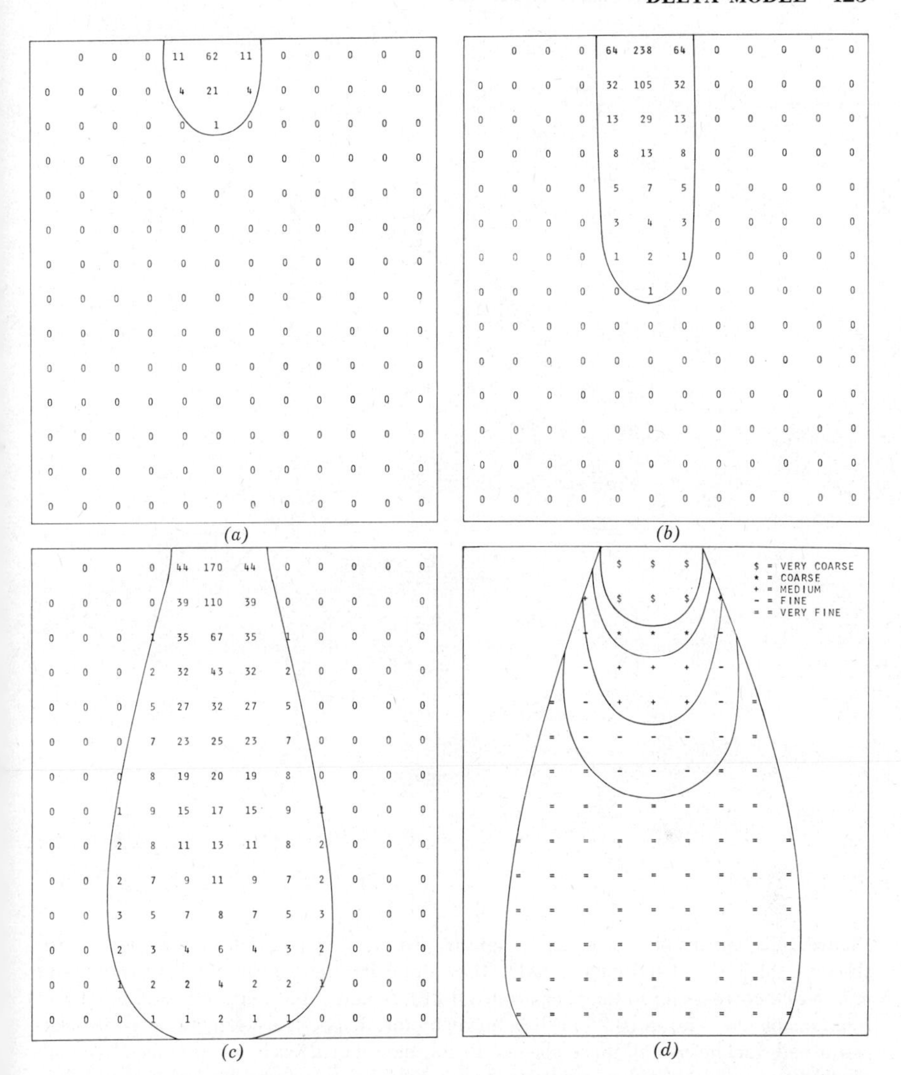

Figure 9-21 Maps printed by computer, showing thickness distribution of sediment-size fractions in simulated river-mouth delta. Size fraction in (*a*) has grains 1 millimeter in diameter, in (*b*) 0.5 millimeter, and in (*c*) 0.25 millimeter. Map with sediments classified into five sedimentary facies based on grain size is shown in (*d*).

0.3 to 0.2 and then to 0.1 mm (series I) and holding all other parameters constant, the depositional area responds by extending successively farther from the mouth in the direction of flow, but the lateral spread remains comparatively narrow and cigar shaped. The average

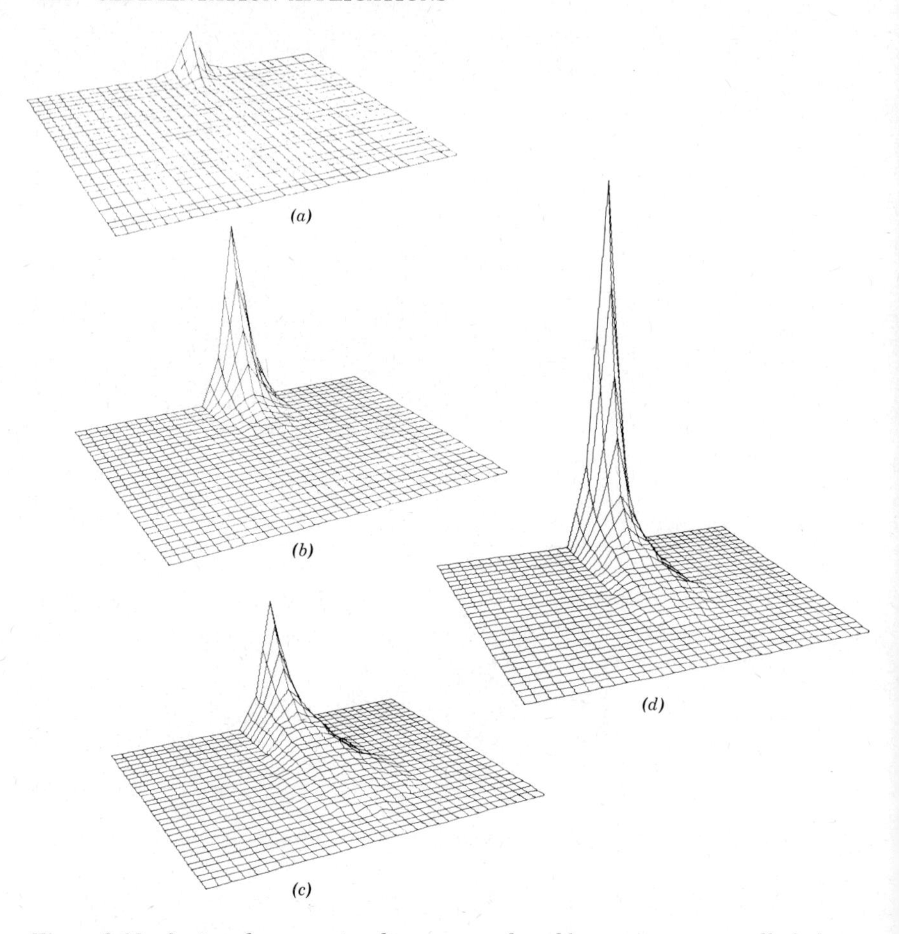

Figure 9-22 Series of perspective diagrams produced by computer-controlled plotter showing thickness of sediment-size fractions deposited in experimental run. Diagrams *a*, *b*, and *c* correspond to those of Figure 9-21*a*, *b*, and *c* and pertain to sediment-size fractions of 1.0-, 0.5-, and 0.25-millimeter diameter, respectively. Diagram *d* represents combined thickness of all three classes. Resolution of grid has been enhanced by quadrupling the number of data points by linear interpolation between previously calculated data points.

slope of the deposits in the elevation view is about half a degree. By increasing the channel slope (series II) and holding other parameters constant, the average channel velocity is increased from 2 to 3 and then to 4 meters per second. Comparison of series I and series II reveals that variation in grain size has considerably greater effect in modifying the size of the depositional area than variation in river slope over the range of values selected. Altering the channel depth

TABLE 9-12 Summary of Results Obtained from Four Series of Experiments with the Static Model[a]

Series	Run	*Depth D (meters)*	*Width B (meters)*	*Slope*	*Average Channel Velocity $\overline{V}$ (meters/sec)*	*Water Discharge Q (meters³/sec)*	*Shear Velocity U_* (meters/sec)*	*Von Kármán Constant k*	*Darcy–Weisbach Friction Factor f*	*Length of Cell Side (meters)*	*Grain Diameter (mm)*	*Fall Velocity W_s (mm/sec)*	*Sediment Discharge Q_s (tons × 10^6/year)*	*X_t max (meters)*	*Y_t max (meters)*
I	A	10.0	100.0	0.00010	2.00	2000.0	0.10	0.4	0.02	140.0	0.3	3.90	0.230	508.0	130
	B	10.0	100.0	0.00010	2.00	2000.0	0.10	0.4	0.02	140.0	0.2	2.40	0.239	794.2	140
	C	10.0	100.0	0.00010	2.00	2000.0	0.10	0.4	0.02	140.0	0.1	0.78	0.247	1890.0	300
II	A	10.0	100.0	0.00010	2.00	2000.0	0.10	0.4	0.02	140.0	0.3	3.90	0.230	508.0	130
	B	10.0	100.0	0.00023	3.00	3000.0	0.15	0.4	0.02	140.0	0.3	3.90	1.250	741.1	140
	C	10.0	100.0	0.00041	4.00	4000.0	0.20	0.4	0.02	140.0	0.3	3.90	4.030	941.8	150
III	A	10.0	100.0	0.00010	2.00	2000.0	0.10	0.4	0.02	140.0	0.3	3.90	0.230	508.0	130
	B	15.0	100.0	0.00010	2.43	3639.2	0.12	0.4	0.02	140.0	0.3	3.90	0.421	872.1	140
	C	20.0	100.0	0.00010	2.80	5602.9	0.14	0.4	0.02	140.0	0.3	3.90	0.660	1225.3	180
IV	A	10.0	100.0	0.00010	2.00	2000.0	0.10	0.4	0.02	140.0	0.3	3.90	0.230	508.0	130
	B	10.0	200.0	0.00010	2.00	4000.0	0.10	0.4	0.02	140.0	0.3	3.90	0.442	508.0	230
	C	10.0	300.0	0.00010	2.00	6000.0	0.10	0.4	0.02	140.0	0.3	3.90	0.662	508.0	350

[a]In series I grain size has been altered; in series II channel slope has been altered; in series III channel depth has been altered; and in series IV channel width has been altered. X_t max and Y_t max give the maximum size of the depositional area in planview for X and Y directions, respectively. From Bonham-Carter and Sutherland (1968).

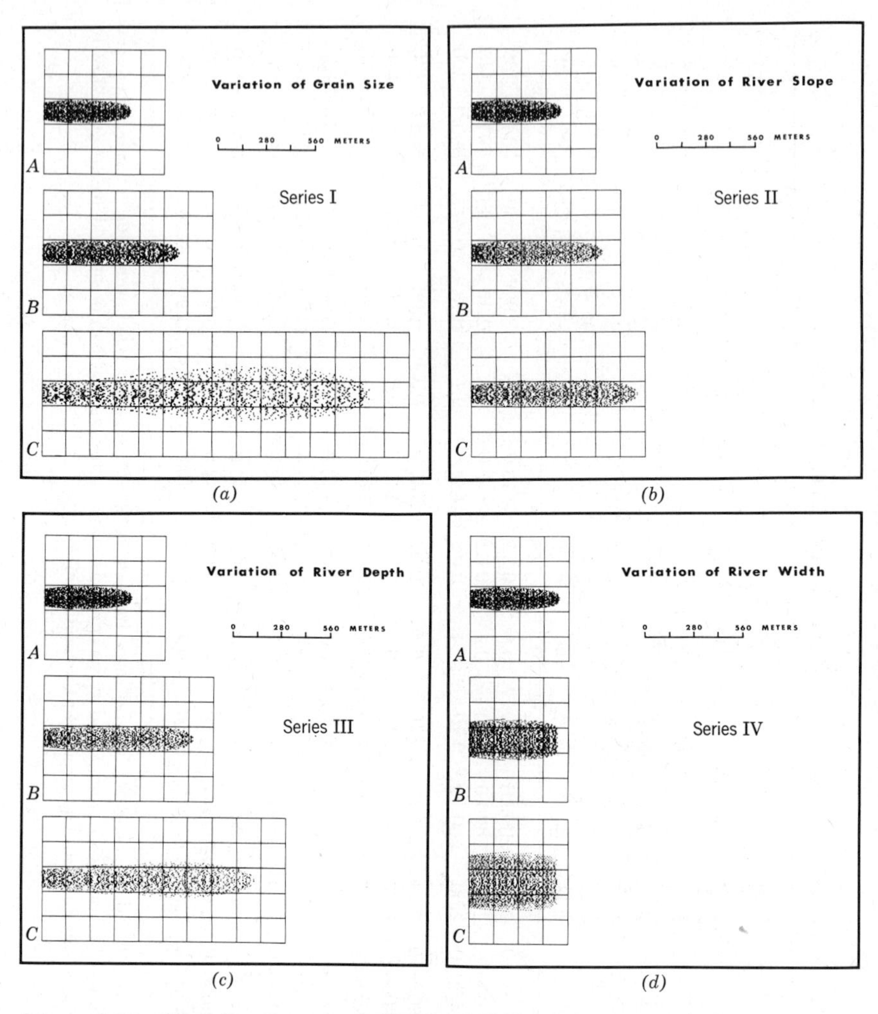

Figure 9-23 Plots showing output from four series of experiments with static model. (*a*) Series I pertains to variation in grain size only, (*b*) series II to variation in channel slope, (*c*) series III to variation in channel depth, and (*d*) series IV to variation in channel width. Diagrams depict depositional areas in map view. Each dot represents point at which nominal particle (statistical aggregate) settles. Details of parameters are summarized in Table 9-12. From Bonham-Carter and Sutherland (1968).

(series III) from 10 to 15 and then to 20 meters, holding other parameters constant, causes the average channel velocity to increase, despite the unaltered slope, and produces a substantial increase in the size of the depositional area. Comparison of the maps reveals, however, that grain size is more important than either channel slope or depth in governing the extent of the depositional area over the range of values tested in these experiments. The slope of the deposits varies slightly from one experiment to the next but is always about half a degree, a value that is reasonable by comparison with other major deltas. Van Straaten (1961) reports average delta foreset slopes for the Orinoco River delta of about 0.25 degree, for the Mississippi delta of about 1 degree, and for the Grand Rhone delta of about 2 degrees.

In series IV the channel width has been increased from 100 to 200 and then to 300 meters. Although the discharge is greatly increased, the average channel velocity is unchanged. As a result increase in channel width does not cause particles to be carried further from the mouth but simply produces a wider deposit.

An unrealistic feature of the static delta model is the narrowness of the depositional area. Actual deltas appear to fan out laterally more then those of the model. There are a number of possible reasons for this difference. At low river velocities potential flow might be a more reasonable flow model than the jet model because of the reduced effect of turbulent mixing. A potential-flow model would produce a radially symmetrical sediment-dispersal pattern more like those found in actual deltas. Another factor is the unstable nature of the channel axis. In real deltas it is likely that, even over small intervals of time, the main axis of flow will oscillate in direction. To demonstrate the effects of oscillation a simulation experiment was performed in which the x-axis was permitted to rotate with respect to the accounting grid coordinates. The angle of rotation for each fan position was generated by using a Gaussian random-number source, with zero mean and a specified standard deviation. The model responded by producing a fan-shaped deposit (Figure 9-24). If we consider actual deltas, it seems likely that the effect of offshore energy factors (waves, tides, currents), and the influence of river-mouth bars are probably significant factors in producing broad, rather than narrow, deposits. The offshore energy factors are totally ignored in the model, but river-mouth bars are considered in a dynamic version of the model described below.

Experiments with the Dynamic Model

In a paper dealing with the jet theory of delta formation Bates (1953) suggested that the structure of the jet velocity field could be used to explain the position of submerged levees at the margins of the flow and a lunate river-mouth bar transverse to the axis of channel

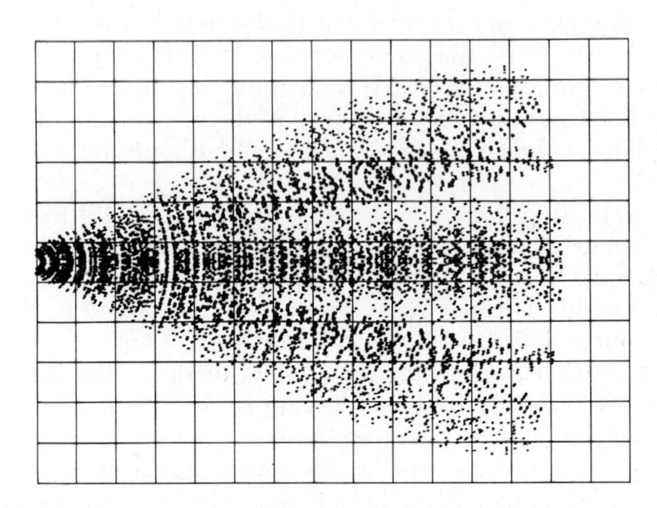

Figure 9-24 Depositional area resulting from experiment in which central axis of flow was permitted to oscillate from side to side. From Bonham-Carter and Sutherland (1968).

flow. These features are common to many deltas, and an effective model should account for their formation.

The static delta model produces a simple wedge of sediment, thick near the mouth and tapering basinward. There is no tendency for a bar at the mouth or submerged levees to form. It is clear, however, that such a model is unrealistic in that continued sedimentation would soon block the mouth and obstruct the channel. It is more reasonable to assume that sediment can be deposited so that it builds up to some limiting depth. Once this limiting depth is reached, nominal particles are swept along with horizontal, rather than sloping, trajectories until they encounter deeper water.

The provision for limiting depth has been incorporated somewhat heuristically into the dynamic model. Each cell in the accounting grid is permitted to fill to a limiting depth only, where the limiting depth is a function of the local streamwise-velocity component. The limiting depth surface (LDS) is calculated at the beginning of the run and is identical in shape to the jet velocity field. Figure 9-25 is a computer-drawn perspective diagram of this surface, which represents depth. At the mouth the LDS is equivalent to channel depth. The surface

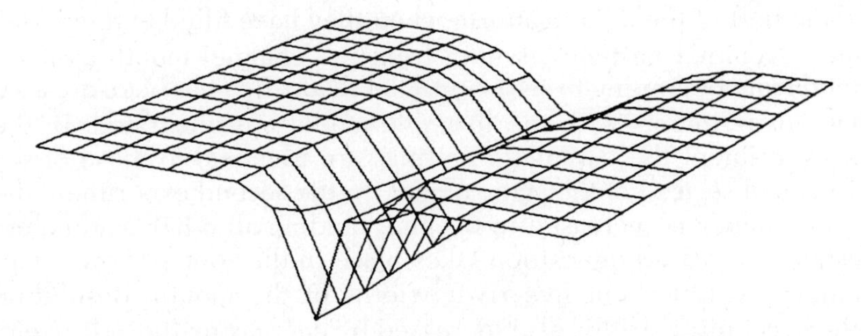

Figure 9-25 Computer-drawn perspective diagram of limiting depth surface (LDS). Channel mouth is at bottom-left edge of grid, where LDS value is same as river depth. LDS gradually becomes shallower in x-direction (towards rear right edge of diagram) and shallows rapidly in y-directions away from central axis.

becomes gradually shallower with an increase in distance in the longitudinal or x-direction and shallows abruptly on either side of the x-axis. The configuration of the LDS can be regarded as the mold that governs the limiting shape of natural levees and the river-mouth bar. Figure 9-26 illustrates diagramatically how the accounting-grid cells

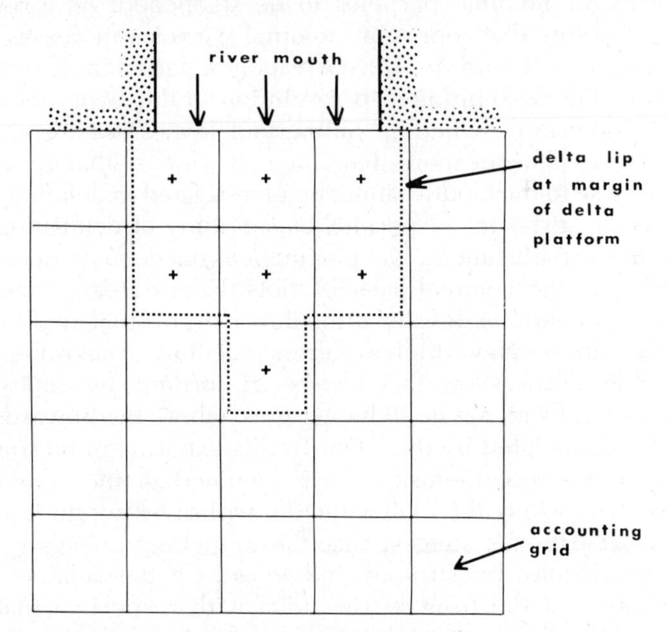

Figure 9-26 Diagram showing accounting grid and map of delta platform. Cells in which deposits have reached limiting depth become part of platform. Nominal particles do not settle until delta lip is crossed. From Bonham-Carter and Sutherland (1968).

become part of the delta platform when they have filled to their LDS values. Nominal particles issuing from the channel mouth are prevented from beginning to settle until the delta "lip" has been crossed.

Two experiments with the dynamic model are illustrated. In the first experiment the parameter settings are identical to those of the third run of series I of the static model. In the second experiment the grain diameter is increased to 0.4 mm, holding all other parameters constant, so that all deposition takes place in the zone of flow establishment, within about five river widths of the mouth. Resolution of the accounting grid is also increased by decreasing the cell size to one-fourth that of the river width. In this way deposition near the mouth can be more closely monitored.

The first experiment results in the deposition of two transverse "mouth bars" (Figure 9-27*b*). The bar closest to the mouth is produced by a gradual shallowing of the LDS from the end of the zone of flow establishment. Limiting depth values have restricted the upward growth in this region. The second bar is formed by an increase in the rate of deposition near the seaward end of the depositional area, where current velocities have decreased significantly, causing the trajectories for nominal particles to be steepened. As a result the density of points that represent nominal particles increases around the seaward end (Figure 9-27*a*), corresponding to an increased rate of deposition. The accounting-grid resolution in this experiment is too coarse to determine whether submerged levees are forming. The delta platform is never more than one cell wide, so that depositional variation close to the mouth cannot be investigated in detail.

The second experiment permits closer study of deposition in the zone of flow establishment. Development of the delta platform can be followed from the nominal-particle plots (Figure 9-28), which show no increase in particle density at the flow margins that might indicate levee formation. Nevertheless, cross sections transverse to the principal flow axis show that levees are forming by deposition of sediment in cells along the delta margins, where the upward growth of levees is regulated by the LDS. In this experiment no transverse bar formed because deposition was confined to the zone of flow establishment, where the LDS coincides with river depth.

These experiments suggest that the branching behavior of delta distributaries may be strongly influenced by the relative rate of sedimentation at the front of the delta with respect to that at the margins of the flow region. Rapid deposition of submerged levees, coupled with slow development of a mouth bar, causes the channel to be restricted and in turn brings about an increase in flow velocity over the seaward end, preventing the mouth bar from forming. If, on the other hand, the bar forms more rapidly than the levees, the channel will be obstructed, possibly leading to a split in the distributary. Further experiments may lead to a better understanding of the controls of this branching process.

The principal limitation of the dynamic model is its inability to consider the disruption of the velocity field by the formation of levees and mouth bars. Not only does the sediment behavior in relation to the water depth need to be dynamic but the velocity field needs to be progressively modified, too. If a dynamic velocity field were to be coupled with offshore energy factors, such as longshore drift and wind-driven currents, a more realistic delta model could be devised.

CARBONATE-ECOLOGY MODEL

The carbonate-ecology model has been designed primarily for simulation experiments involving interactions between carbonate-secreting marine-organism communities. In principle the model could be used for simulating the behavior of organism communities under a variety of conditions, ranging from bottom-dwelling marine organisms to aggregations of land plants. In practice the model has been applied to shallow-water carbonate-secreting marine-organism communities. The components of the model that involve interaction between organism communities are probabilistic and make extensive use of methods that involve Markov chains. The probabilistic mechanisms in the model are allied with triple-dependent, first-order Markov chains with nonstationary transition probabilities. The transition probabilities are recomputed at each step by methods that are simple in principle, although complicated in practice.

Representation of organism communities within the model is simple. The area that the organism communities occupy is divided into a grid with square cells of arbitrary size. Each cell contains only one organism community at a time. A community can be defined in the model as consisting of one or more species, but each community type is regarded as the basic unit, rather than species or individual organisms. In the computer programs representing the model (Harbaugh, 1966; Harbaugh and Wahlstedt, 1967), each organism community occupying a specific cell is symbolized by an integer (Figure 9-29). In turn the integers are stored in three-dimensional arrays. Two of the array dimensions pertain to geographic dimensions, and the third represents successive time planes (or successive strata, depending on the intended meaning).

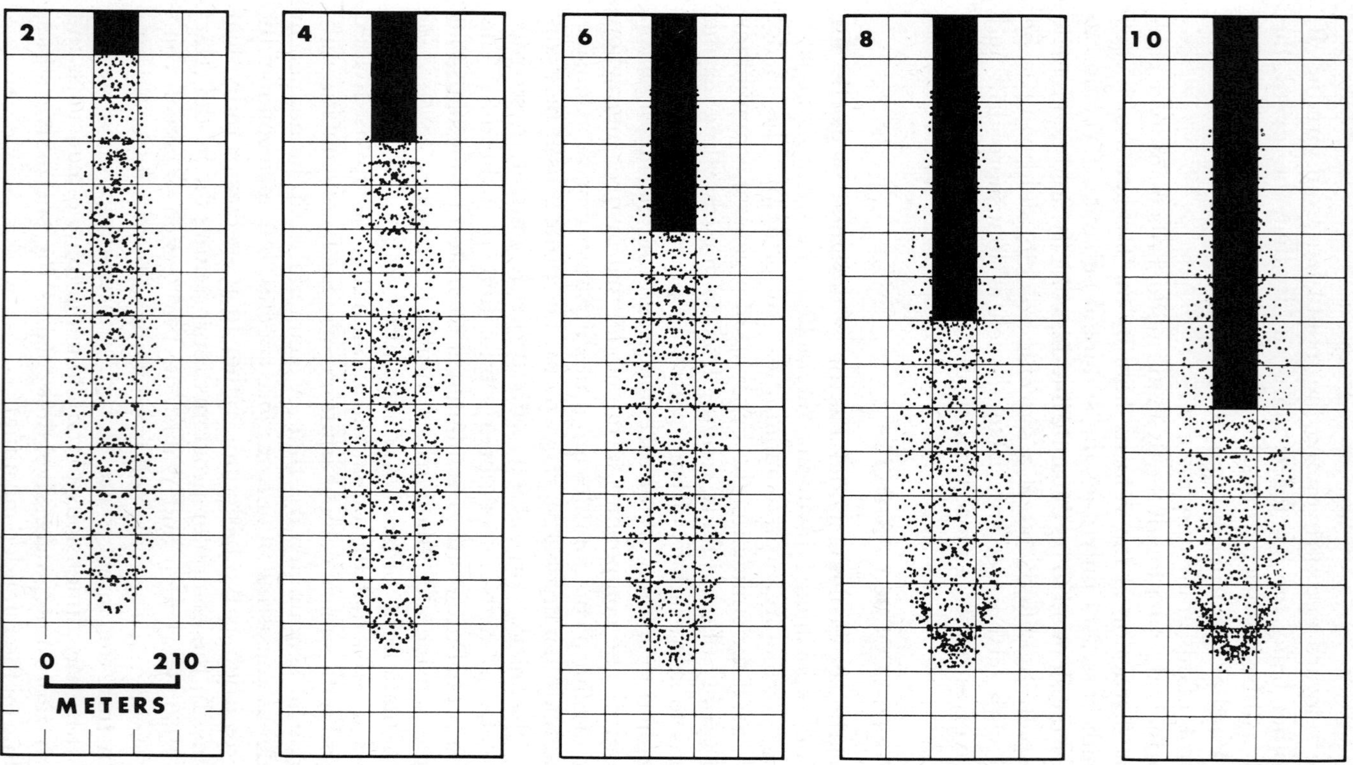

Figure 9-27a Output from first experiment with dynamic model. Particle maps from selected time increments show advance of delta platform (solid). Development of transverse bar is indicated by high density of points at seaward end of depositional area in later time increments. From Bonham-Carter and Sutherland (1968).

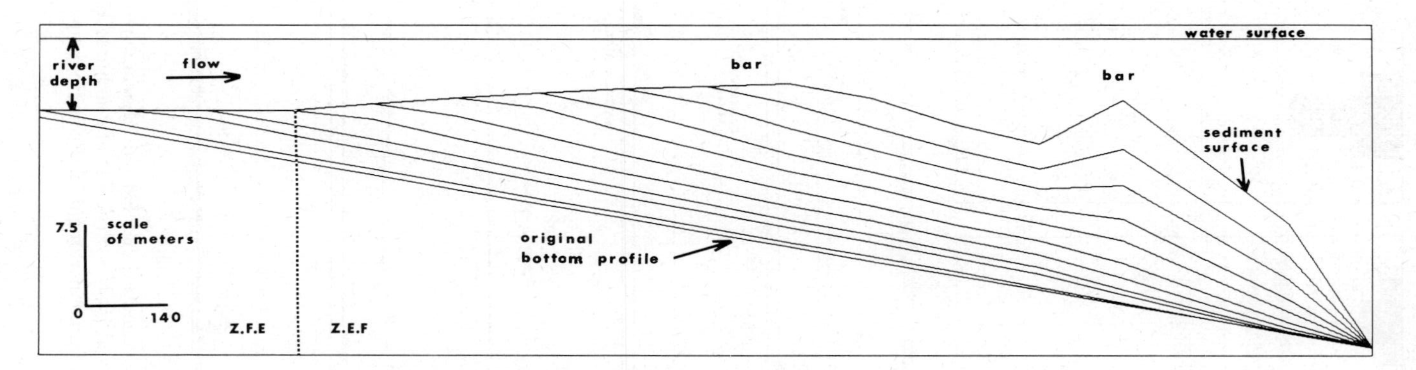

Figure 9-27b Vertical longitudinal stratigraphic section along middle of accounting grid after 10 time increments. Lower map edge is toward right. From Bonham-Carter and Sutherland (1968).

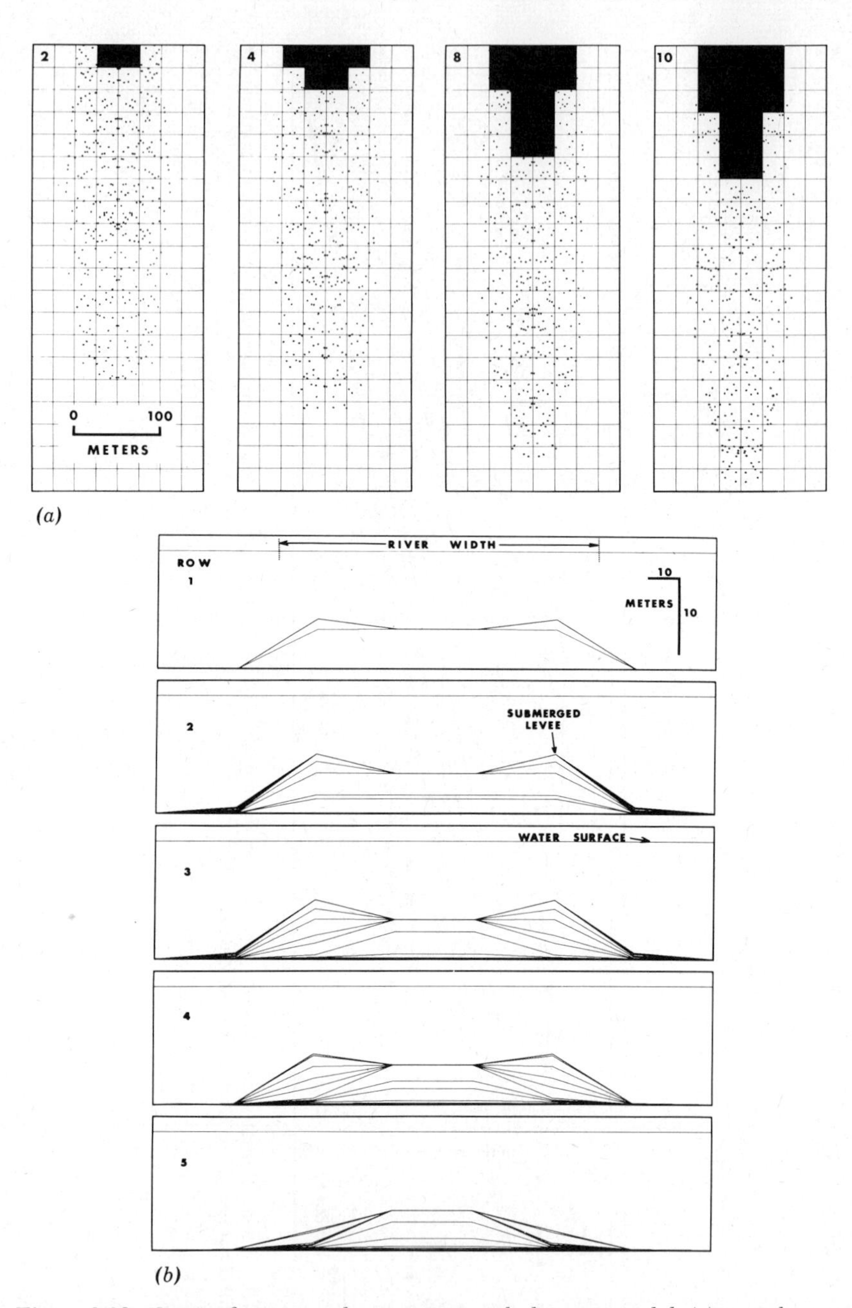

Figure 9-28 Output from second experiment with dynamic model: (*a*) particle maps for time increments 2, 4, 8, and 10; (*b*) vertical stratigraphic sections along rows 1, 2, 3, 4, and 5 of accounting grid of time increment 12. From Bonham-Carter and Sutherland (1968).

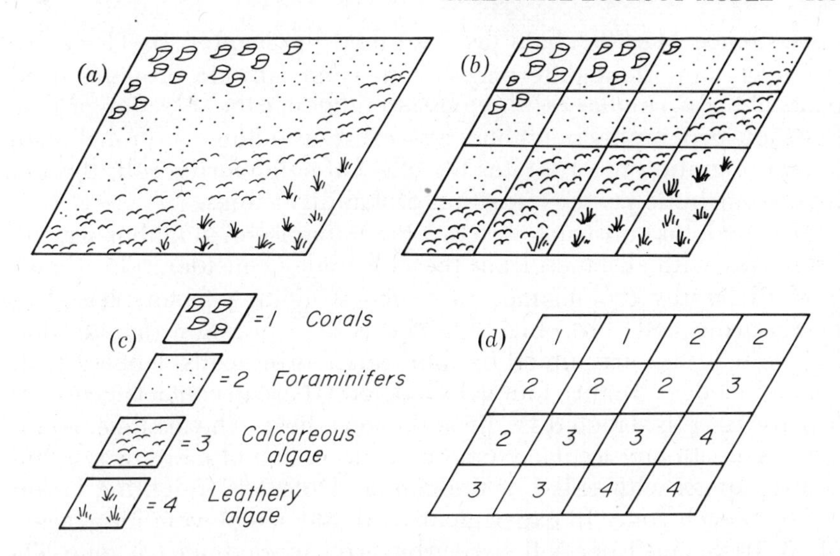

Figure 9-29 Means of representing geographic distribution of organism communities: (*a*) sea floor is populated by different communities that are continuous and are portrayed graphically; (*b*) sea floor has been divided into square cells; (*c*) graphic symbols are assigned numerical equivalents; (*d*) two-dimensional array of integers contains essentially same information as graphic symbols in (*a*) except for some loss of detail due to relative coarseness of discrete cells. From Harbaugh (1966).

Migration and Succession of Organism Communities

Migration and succession of organism communities in the model involves a method in which the different organism-community types are multiplied by weighting factors and then one is selected at random. The process might be likened to the process of placing balls with numbers painted on them in an urn and withdrawing one at random. The probability of drawing a ball with a particular number is proportional to the number of balls possessing that number with respect to the total number of balls in the urn. For example, if there are a total of 1000 balls in the urn and 150 balls have number 5 painted on them, the probability of drawing a ball with 5 painted on it is .150. The program representing the model, in effect, puts a succession of balls representing organism communities into an urn and then draws one at random for each cell during each time increment.

The probability pertaining to the selection of the next community to occupy a particular cell is influenced by preceding events and is therefore Markovian. The model assumes that three previous events influence the selection, each separated by a single time increment. Thus in some respects the selection process is very similar to a triple-dependence Markov chain (as described in Chapter 4). There are, however, important differences in that not only do previous occupants of the particular cell in question influence the selection process but their geographic neighbors also exert an influence. Probabilistic lateral migration incorporates the effect of neighboring cells through their contribution to the transition-probability values.

The model is constructed so that the influence of neighboring cells decreases with distance from the cell under consideration. Figure 9-30 illustrates the manner in which weighting factors based on neighboring cells are obtained. The cell in question (labeled I in Figure 9-30) is surrounded by four neighboring cells, labeled II, in contact at edges. In turn four cells labeled III are in contact at corners. Finally 12 cells, labeled IV, lie at the periphery. The arrangement of cells is an attempt to place them in a succession of more or less concentric zones with cell I at the center. Different weighting factors apply to each zone. In experiments with the model weighting factors of 50, 10, 3, and 1 per cell were arbitrarily applied in each zone. The weighting factors were then used to multiply the number of balls containing the particular type of community in a cell. For example, if organism community 3 occupies cell I, we can imagine that 50 balls

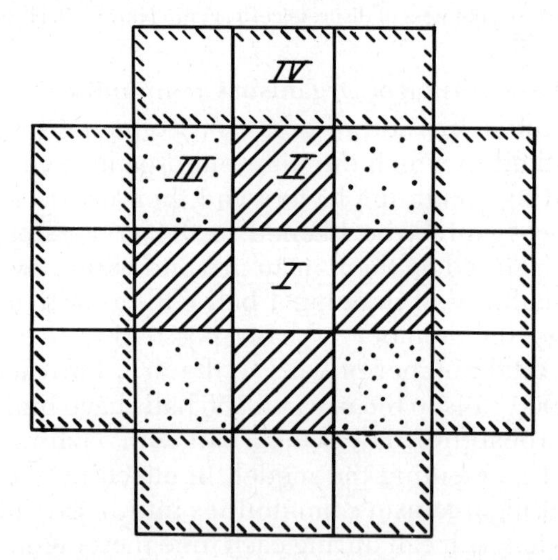

Figure 9-30 Method of representing influence of occupants of neighboring cells in carbonate-ecology model. Cell under consideration is labeled I. Surrounding cells are grouped into zones II, III, and IV, which lie at progressively increasing distances from central cell. The pattern is progressively moved from cell to cell within the rectangular meshwork in which the simulation experiment takes place. Special provisions must be observed at edges and corners of the meshwork because of boundary effects.

with this number will be placed in the urn. Since there are four cells in zone II, 10 balls will be added for each of the four cells, each ball being labeled with the integer representing the organism community occupying each cell. If all four cells in zone II are occupied by the same community, 40 balls with the same label will be added; if they are all different, 10 of each type will be added, and so on. Another way of thinking of the relative influence of the cells in the various zones would be to express their contribution as a percentage (Table 9-13).

TABLE 9-13 Relative Influence in Calculating the Transition Probabilities of Organism Communities That Occupy Neighboring Cells (Figure 9-30) During the Previous Time Increment

Zone	*Weighting Factor per Cell*	*Number of Cells*	*Total Weighting Factor*	*Influence of Cells in Each Zone (%)*
I	50	1	50	44.0
II	10	4	40	35.0
III	3	4	12	10.5
IV	1	12	12	10.5

The influence of neighboring cells applies to each particular time increment. In selecting a new occupant for a particular cell only the occupants of previous time increments are considered, not the neighbors for which the new occupants are being chosen. The reason for this is inherent in the construction of digital dynamic models. Because of the necessity of taking discrete steps in carrying out computing operations, simultaneous events cannot be directly represented in the model. In other words, we cannot carry out calculations that represent the behavior of the occupant of a particular cell simultaneously with those in other cells. Instead, we must perform all of way if thinking of the relative influence of the cells in the various the computing operations one by one that pertain to a particular time increment before we pass to the next time increment.

In selecting the occupant of each cell the model compares the occupants of the cell under consideration (cell I) for the three preceding time increments and in turn supplies an additional, or secondary, weighting factor that is used to further modify the contribution (as the imaginary numbered balls supplied to the urn). A secondary weighting factor is calculated for each of the three preceding time increments and is multiplied times the weighted contribution based on the neighbor-cell effect (or primary weighting factor), as illustrated in Figure 9-30. Figure 9-31 is an attempt to illustrate how the primary and secondary weighting factors pertaining to the three preceding time increments (at times $t-1$, $t-2$, and $t-3$) affect the selection of the occupant for a single cell at time t.

The secondary weighting factor serves the following purposes:

1. It permits the degree of persistence of an organism community to exert influence on succeeding events. For example, if a given type of organism community has occupied a given cell for three successive time increments, the probability of the same community occupying the same cell in the forthcoming time increment is greater than it would be if different communities had occupied the cell during the three preceding time increments. The effect of this may be likened to inertia in that long-established communities may tend to resist subsequent change much more than communities whose occupation has been brief.

2. It permits physical environmental factors to exert influence on succession. For example, to represent the influence of a favorable environmental factor a weighting factor greater than unity is used, whereas an unfavorable environmental factor is expressed through use of a weighting factor between 1.0 and 0.0. A weighting factor of zero would eliminate any contribution from the particular organism community for the particular time increment and therefore represents the effect of an environmental factor that is totally intolerable.

3. The secondary weighting factor permits influence to be exerted by the degree to which a succession of occupants of a particular cell approaches an ideal ecologic succession. The degree of "closeness" or "farness" in an ecologic succession can be expressed numerically. For example, if there are 12 communities that are symbolized by numbers such that community 1 is a pioneer community and community 12 is the climax community (for a given set of environmental conditions), given sufficient time increments, the pioneer community (No. 1) should be gradually replaced by communities symbolized by higher and higher numbers until the climax community (No. 12) is reached. Thus, although there is a tendency for the succession to be unidirectional (i.e., toward the climax), momentary reversals can occur as a result of random fluctuations, and major reversals can be produced by major changes in environmental factors (including catastrophic events). Some cyclically bedded sequences of fossiliferous strata, such as the marine Pennsylvanian strata of eastern Kansas and

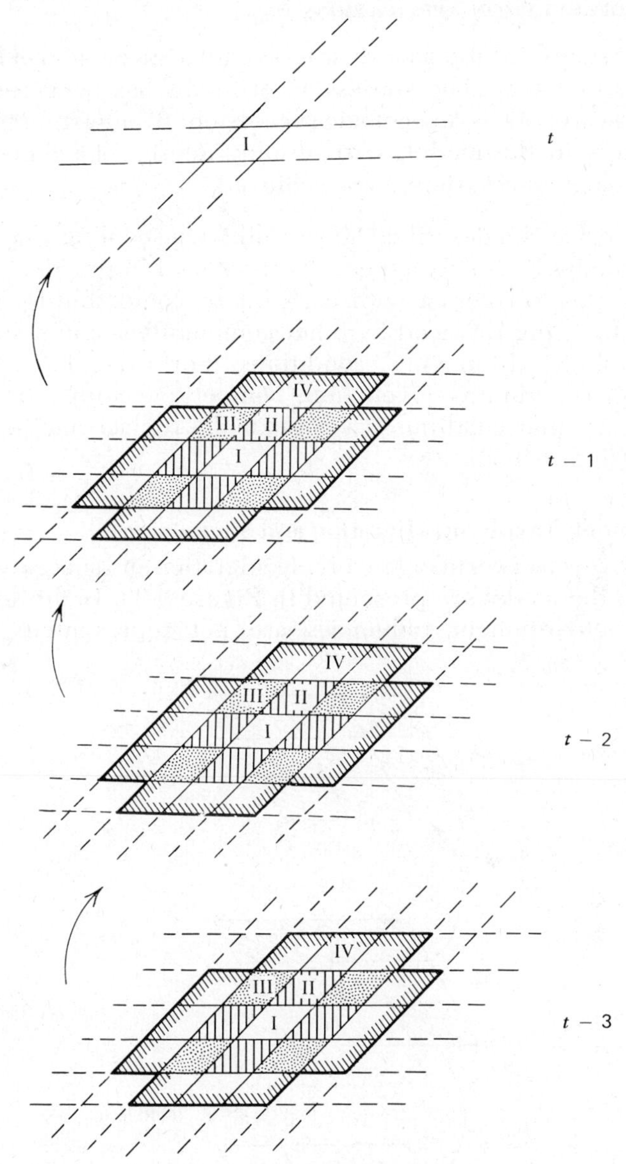

Figure 9-31 Diagram illustrating method of selection of organism community to occupy cell I at time t. During each of three preceding time increments (times $t-1$, $t-2$, and $t-3$) integers representing occupants of cells in zones I, II, III, and IV are multiplied by primary weighting factors to represent neighbor effect. Subsequently these integers for each time increment are again multiplied by secondary weighting factors that represent combined influence of degree of persistence of occupants through time, influence of physical environmental factors, and degree to which sequence adheres to a prescribed ideal ecologic succession.

adjacent states, exhibit a more or less regular succession of lithologies and fossil content. The succession of fossils has been regarded by some researchers as an ecologic succession. By adjusting the weighting factors in the model, virtually any degree of adherence to an ideal ecologic succession can be achieved.

The mechanism described above, although seemingly complex and almost totally heuristic, appears to work well in practice. It permits communities to interact with each other, competing for space, replacing or being replaced by other communities, and responding to changes in environmental conditions. For example, a particular organism community, given high competitive ability or "vitality" under particular conditions, will gradually replace another of lower vitality (Figure 9-32).

Experiments Involving Migration and Succession

Results of an experiment involving migration and succession carried out with the model are presented in Figure 9-33. In this experiment physical-environment influences are not represented. The only

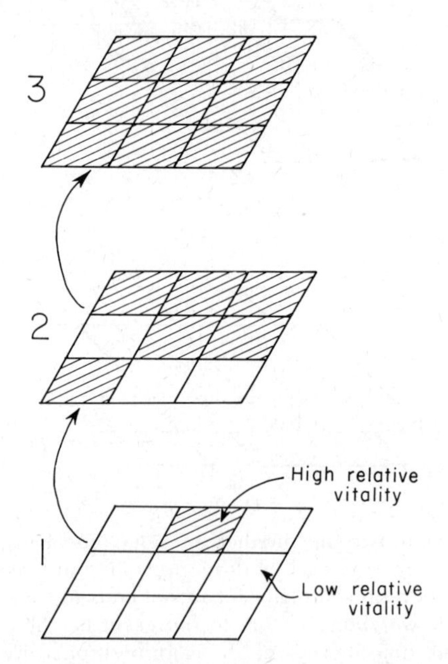

Figure 9-32 Replacement in simulation model of organism community of low relative vitality by one of high vitality through progressive time increments 1, 2, and 3. From Harbaugh (1966).

influences are the hypothetical organisms interacting with each other. Assumptions made in the experiment are as follows:

1. Five hypothetical organism communities are present symbolized by an asterisk, an equal sign, a plus sign, a slash, and a blank.
2. The organism communities form the following ecologic succession, ascending through time (we identify them by the names of their symbols): (a) asterisk community, (b) equal-sign community, (c) plus community, (d) slash community, and (e) blank community.
3. A rectangular grid containing 7 rows and 31 columns was initially populated entirely with the asterisk community.
4. The other four communities are within the system at the outset, potentially available for colonization but not populating the grid area at the start of the simulation run.
5. The tendency for one organism community to succeed another differs between the communities. The blank community has been arbitrarily endowed with a very strong tendency to replace its predecessor communities.

As Figure 9-33 reveals, during time increment 1 the grid area is entirely populated by the asterisk community. At time increments 3, 5, and 7 there is an interplay between the equal-sign, plus, and asterisk communities. By time increment 9, however, the blank community has established itself at several points, which thereafter serve as nuclei from which the blank community radiates during subsequent time increments. Toward the close of the run, at time increment 19, the grid area is mostly populated by blanks and slashes, an equilibrium between the blank and slash communities having been attained.

Although the experimental run shown in Figure 9-33 is of possible academic interest, it can be argued that it does not accord with any real situation because physical influences have been ignored. For example, the effect of wind-induced currents in seawater, which transport the larvae of bottom-dwelling marine organisms, would strongly influence the migration and succession patterns of certain marine organisms. The effect of a current can be imitated in the model by transporting organism communities from cell to cell, as illustrated in Figure 9-34.

The results of an experiment simulating the effect of wind-induced currents on a marine reef, bank, and shoal complex are shown in Figure 9-35. In this experiment an initial colony was "planted" at one side of the area and subjected to a steady transporting influence. As the run progressed colonies were progressively forced "down-

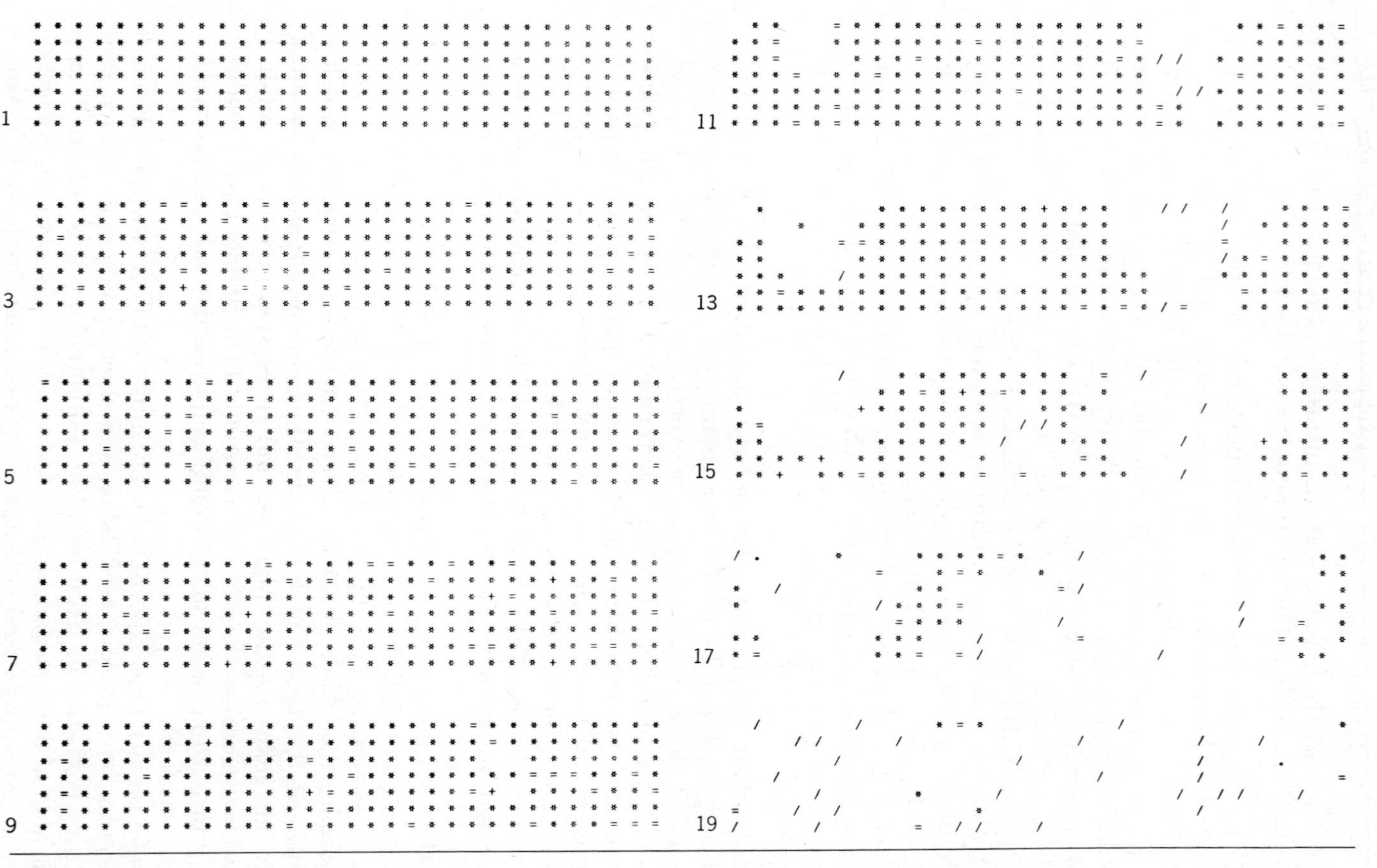

Figure 9-33 Results of experimental run with carbonate-ecology model in which influences consist solely of interactions between organism communities symbolized by asterisks, equal signs, plus signs, blanks, and slashes. Every other time increment is shown for time increments in simulation run extending from 1 to 19.

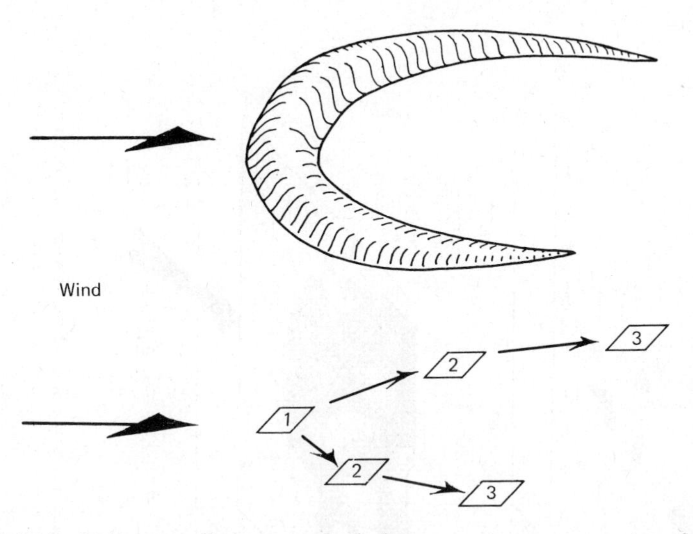

Figure 9-34 Diagram to illustrate method of representing transportation of organism communities from cell to cell to imitate effect of wind-induced current on distribution of marine-organism community.

wind," serving as nuclei around which new colonies grew and gradually coalesced. Toward the close of the run the initial colonizing communities were largely supplanted by a community late in the ecologic sequence.

Representation of Physical Factors

In dealing with physical factors in the model we may distinguish between the representation of the physical factors themselves and the representation of their effect on organism communities. The representation of some physical factors is simple, whereas that of others is more complex. For example, depth of water (or height above sea level) is represented by an array of numbers, each of which signifies elevation with respect to sea level for the cell that it represents. Changes from one time increment to the next can be represented by increments and decrements that represent tectonic movements and sedimentation. Each cell is treated independently for accounting purposes and may be likened to a prism, square in cross section, that is free to move up or down (Figure 9-36).

The deposition of clastic sediment in the model is represented by methods that are roughly similar to those used in the sedimentary-basin model described earlier in this chapter. The deposition of

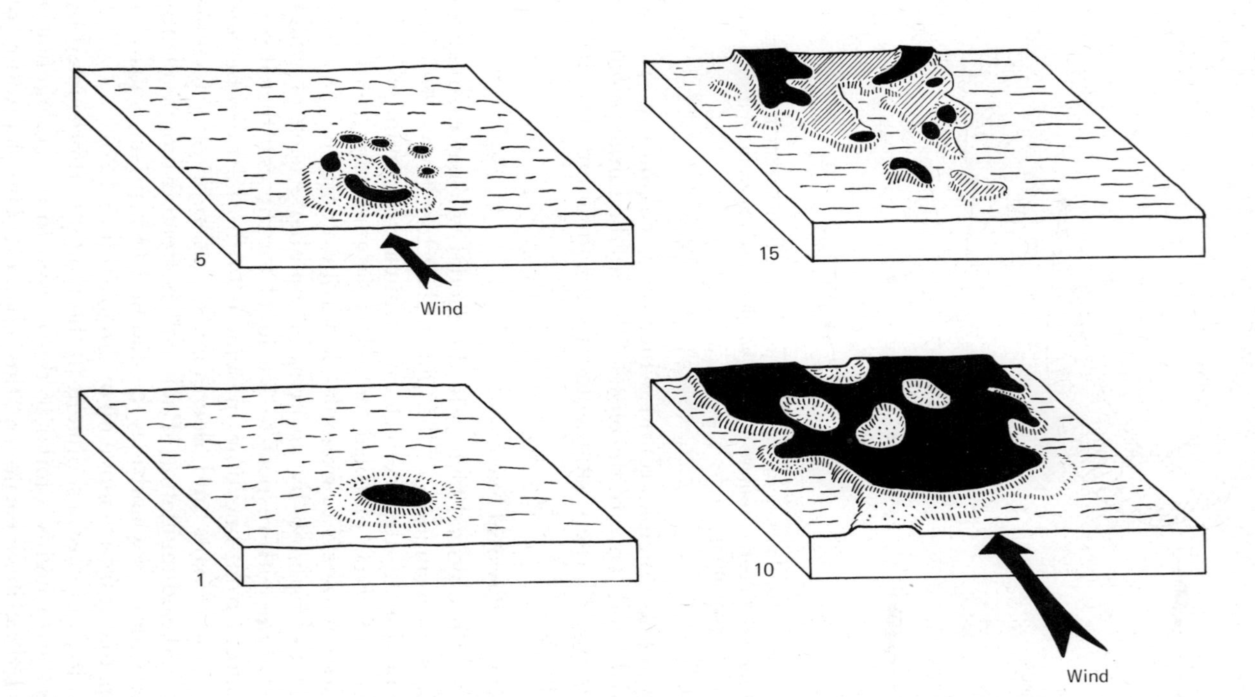

Figure 9-35 Series of block diagrams representing geologic interpretation of the effect of wind-induced currents on the ecologic succession of reef (black), bank (stippled), and shoal (wavy lines) communities. Community represented by diagonal ruled lines occurs late in ecologic succession. Status at time increments 1, 5, 10, and 15 is shown.

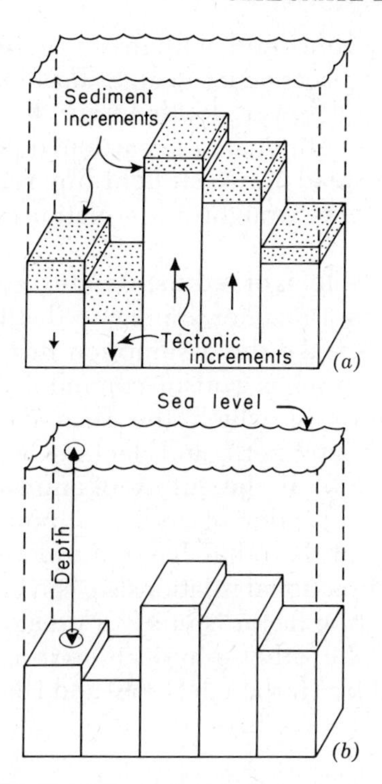

Figure 9-36 Relationship between (*a*) algebraically additive increments of tectonic warping and of sedimentation; (*b*) water depth as arithmetic complement. From Harbaugh (1966).

carbonate sediment is treated differently, however, in that carbonate sediment is assumed to be formed solely through secretion by organism communities. The production of carbonate sediment is treated as a function of community type and relative vitality. For example, a phylloid algal community could be assumed to produce more carbonate sediment per cell per time increment than a brachiopod community. The relative vitality of the community under the prevailing environmental conditions also has a large influence. A phylloid algae community on a shallowly submerged marine bank would have a high carbonate productivity, whereas the same community in deeper water might produce very little carbonate sediment.

The effect of depth variations on organism communities is treated as a secondary weighting factor. For each community an upper (shallow water) depth limit, a most favorable depth, and a lower (deeper water) limit is specified. Two linear relationships—one linking the upper limit with the most favorable depth and the other linking the most favorable depth with the lower depth limit—are used as components in the secondary weighting function (Figure 9-37). Beyond the upper and lower depth limits the weighting factor is zero, and its effect is to eliminate the particular organism community. The use of a curve instead of two straight-line relationships would be more realistic, but the straight-line relationships were used for simplicity.

The environmental effects of sediment influx, such as sand and mud, can be represented in a manner similar to that for representing the effect of depth. If we make the assumption that an influx of mud or sand is detrimental to an organism community, we can employ a functional relationship in which the secondary weighting factor ranges between unity and zero, and declines as mud or sand influx increases (Figure 9-38). If the influx of mud or sand reaches an intolerable level, a multiplier of zero is supplied to the secondary weighting factor. If, on the other hand, an increase in sand or mud influx is favorable, a functional relationship can be employed in which the secondary weighting factor is unity or greater, increasing as mud or sand increases. The other physical-environment factors in the model are described by Harbaugh (1966) and Harbaugh and Merriam (1968, pp. 246–251).

Simulation of Carbonate Marine Banks

The model was used in a series of experiments in which the goal was to simulate the environmental conditions in which certain Upper

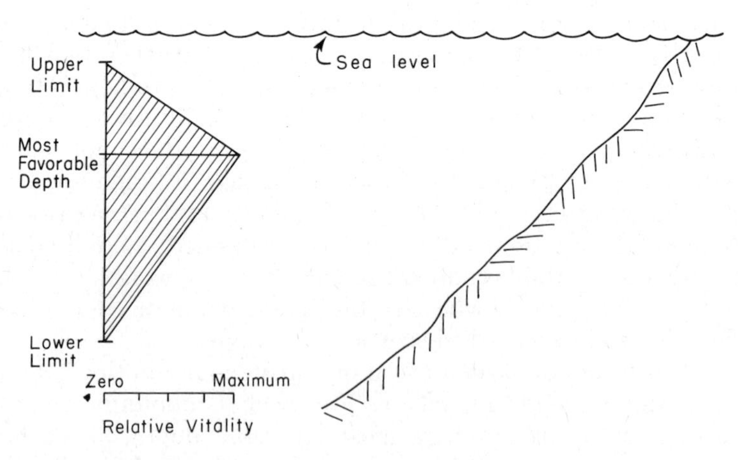

Figure 9-37 Effect of variation in depth on relative vitality of organism community. From Harbaugh (1966).

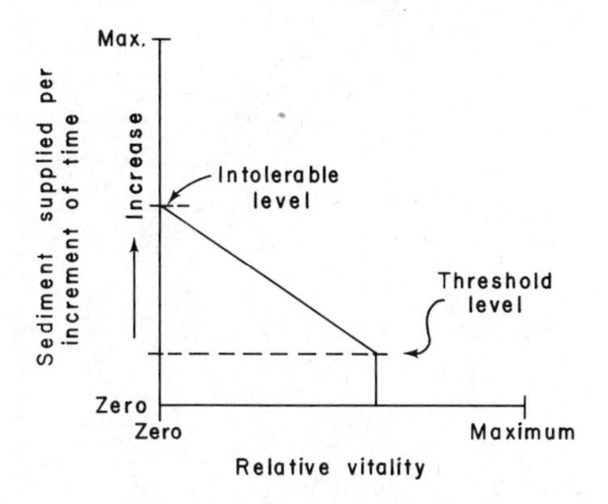

Figure 9-38 Effect of variations in rate of supply of terrestrially derived sediment on relative vitality of organism community. From Harbaugh (1966).

Pennsylvanian marine deposits were formed in southeastern Kansas. These deposits include locally thickened, lenslike masses of limestone, termed marine banks, that are composed principally of leaflike (phylloid) calcareous algae. One of the aspects of the problem of origin of the banks is understanding the factors that cause them to be localized. They appear to have been formed adjacent to an ancient shoreline, where the lobate, deltalike deposits formed as a result of debouching streams strongly influenced the position of the banks (Figure 9-39).

The environmental responses of a Pennsylvanian phylloid algae community are clearly difficult to ascertain and can only be inferred. In simulation experiments conducted with the model the problem was approached heuristically, with a great deal of trial-and-error probing, an example of which is shown in Figure 9-40. This figure contains three 40-mile-long, north–south geologic sections through an area in southeastern Kansas in which deposition of Pennsylvanian sediments has been simulated. The three sections show the response of the model to different "settings" that affect the vitality and sediment-contributing ability of phylloid algae. In Figure 9-40*a* the response of the algae is too weak, in Figure 9-40*b* it is too strong, and in Figure 9-40*c* it is similar to that inferred from observations of the actual deposits (Harbaugh, 1960). Thus the vitality factor and sediment-contributing ability of the algae have been adjusted in the model until the algae perform properly under the assumed environmental conditions.

In a series of simulation experiments with the model, dealing with

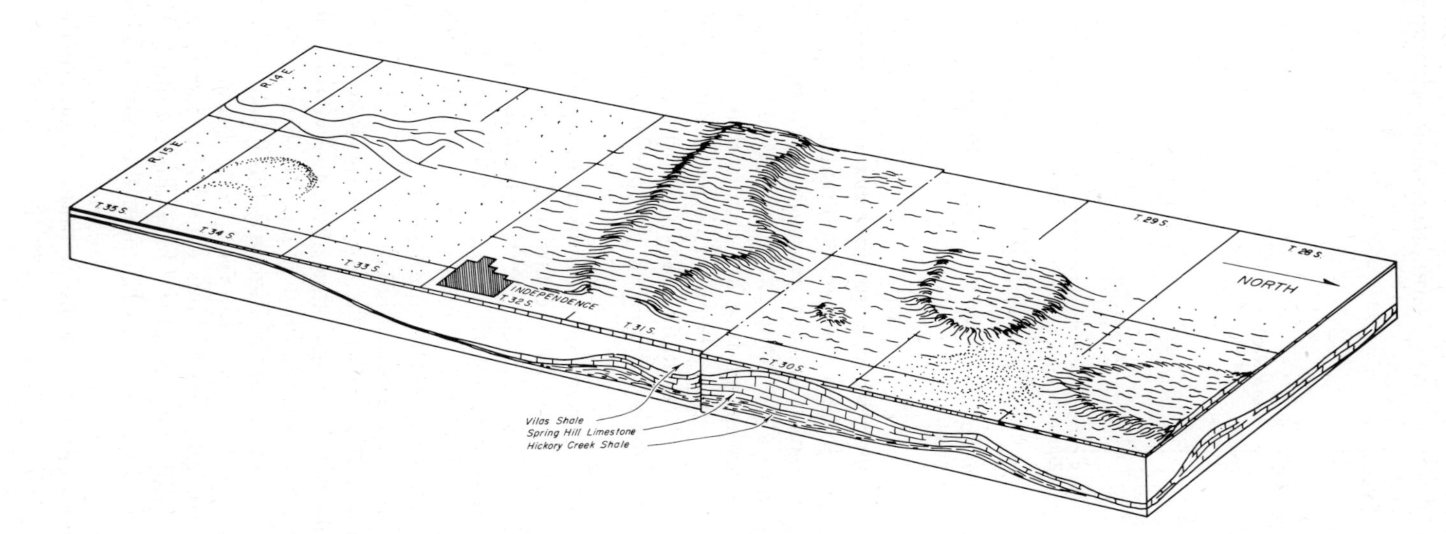

Figure 9-39 Block diagram showing interpretation of geography and geology during deposition of marine-bank deposits in Late Pennsylvanian time in southern Kansas. Block is about 40 miles long in north–south direction and 15 miles wide in east–west direction. From Harbaugh (1960).

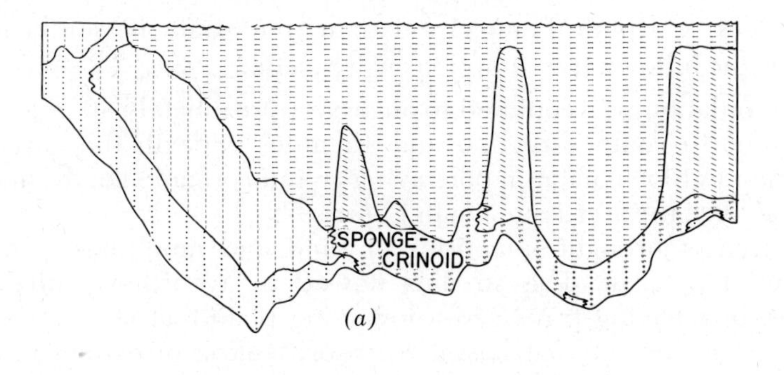

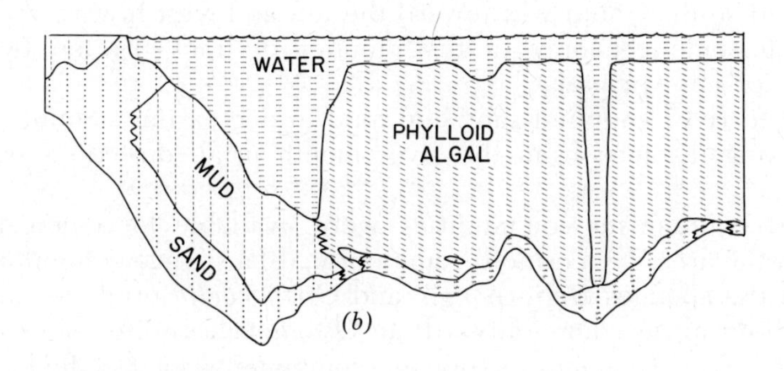

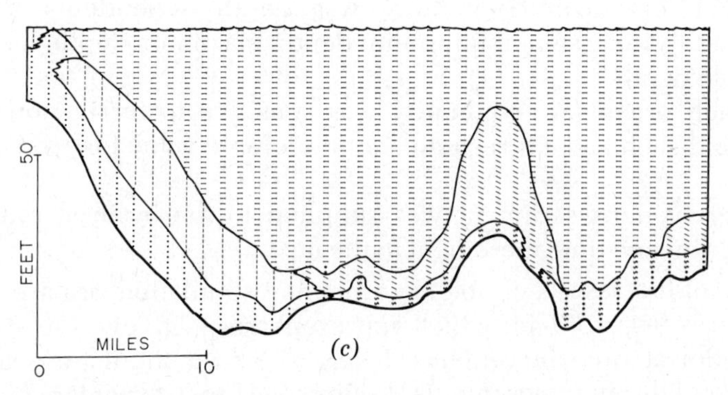

Figure 9-40 Series of geologic cross sections showing different responses of phylloid algae community to conditions assumed in experiments: (*a*) response of algae is too weak; (*b*) response is too strong; (*c*) response is appropriate. North is toward right. Except for lines and large identifying letters, symbols have been printed directly by computer's line printer. Symbol representing sediment type of greatest volumetric importance per time increment is printed out. Asterisks and dollar signs pertain to sediment contributed predominantly by crinoids and sponges, respectively; slash symbols pertain to phylloid algae; vertical dashes to mud; and prostrate I's to water.

the origin of Upper Pennsylvanian marine strata in southeastern Kansas, the following assumptions were made:

1. Simulation was confined to an area approximately 40 miles long in the north–south direction and 15 miles wide in the east–west direction. A block diagram portraying an interpretation of the ancient geography of the area is shown in Figure 9-39.

2. Streams brought quantities of sand and mud into a marine depositional basin. The streams flowed in a northerly direction, presumably having traversed a low-lying plain that lay marginal to the marine depositional basin. (In representing direction on maps produced by the computer in these experiments the maps were oriented so that south was toward the left and west toward the top, offering some economy in the number of lines printed by the computer's line printer.)

3. The river supplied sand and mud in decreasing amounts with increasing distance from the river mouth in all directions except south.

4. Five organism communities were available for colonization within the area: (a) a crinoid community, (b) a sponge community in which the sponges *Heliospongia* and *Girtyocoelia* predominate, (c) a phylloid algae community, (d) an *Osagia*-calcarenite community, and (e) a swamp community (biologic components not specified).

5. Depth varied from place to place initially.

6. Each organism community was depth dependent and was governed by an assumed minimum depth, maximum depth, and most favorable depth.

7. Each organism community was sensitive to terrigenous sediment, and each had a threshold and intolerable value for sand and for mud.

8. Terrigenous sediment was supplied to the basin at rates that varied from one time increment to the next.

Some of the results of one experimental simulation run are shown in Figures 9-41 to 9-44, which represent, respectively, the status of the model at time increments 2, 5, and 8 in a simulation run that spanned eight time increments. Figures 9-41 to 9-43 contain (a) a map showing depth of water, (b) a map showing sediment facies, and (c) a north–south geologic section. Figure 9-44 presents perspective diagrams of the topography of the sedimentary basin. These show the changes that took place during the simulation run. The following major changes occurred:

1. There was progressive filling of the southern two-thirds of the

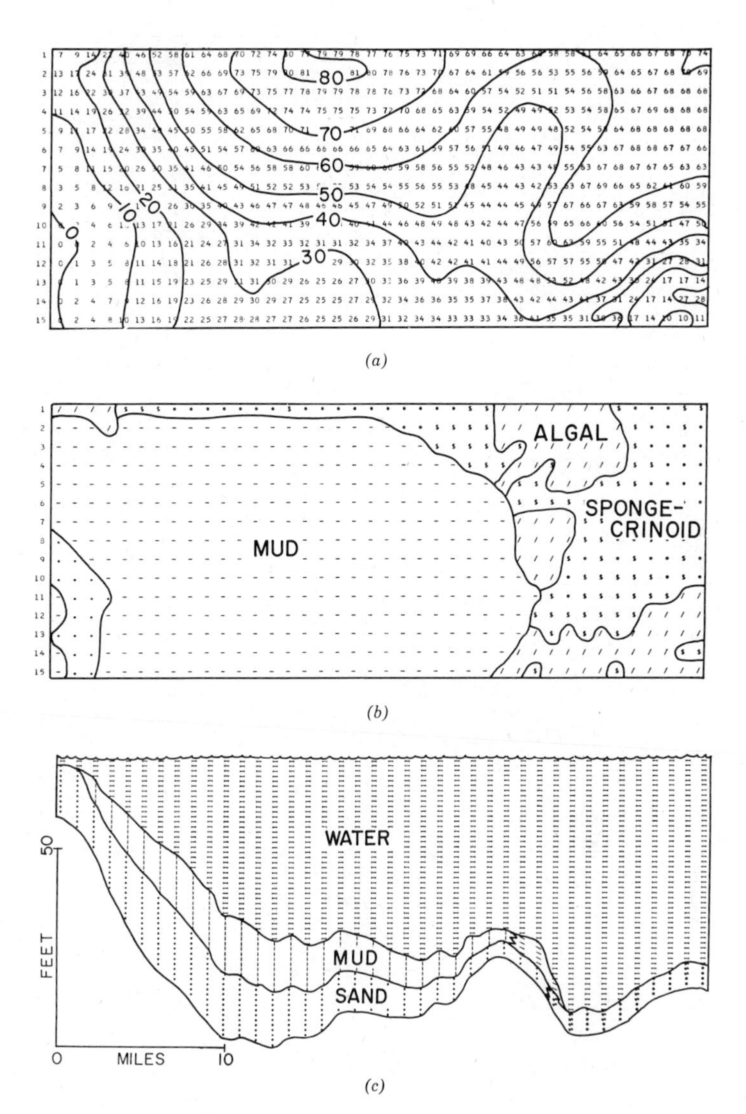

Figure 9-41 Depth map (*a*), facies map (*b*), and geologic cross section (*c*), showing status of simulation model at time increment 2 in experimental run. North is toward right. Depth values are in feet; negative values are above sea level. Except for lines and large identifying numbers and letters, symbols have been printed automatically by computer's line printer. Symbol representing sediment type (*b*) and (*c*) of greatest volumetric importance deposited in cell during time increment is printed. Dots indicate sand, dashes indicate mud, slashes indicate phylloid algae, asterisks indicate crinoids, and dollar signs indicate sponges. Cross section is along ninth row from top of map. From Harbaugh (1966).

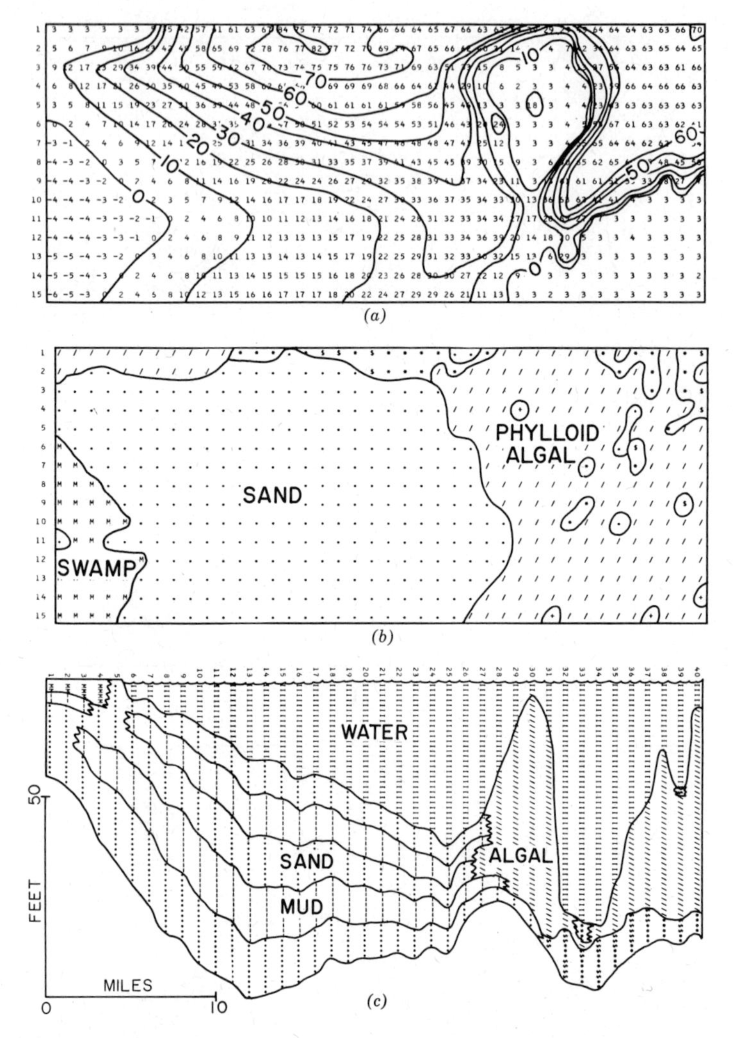

Figure 9-42 Depth map (*a*) facies map (*b*), and cross section (*c*), showing status of model at time increment 5 (see Figure 9-41 for explanation of symbols). From Harbaugh (1966).

sedimentary basin by terrestrially derived mud and sand, producing an extensive deltaic apron.

2. There was uneven tectonic downwarping, but the rate of sediment accumulation was much greater than the rate of downwarping, resulting in nearly complete filling of the marine basin.

3. There was progressive migration of organism communities,

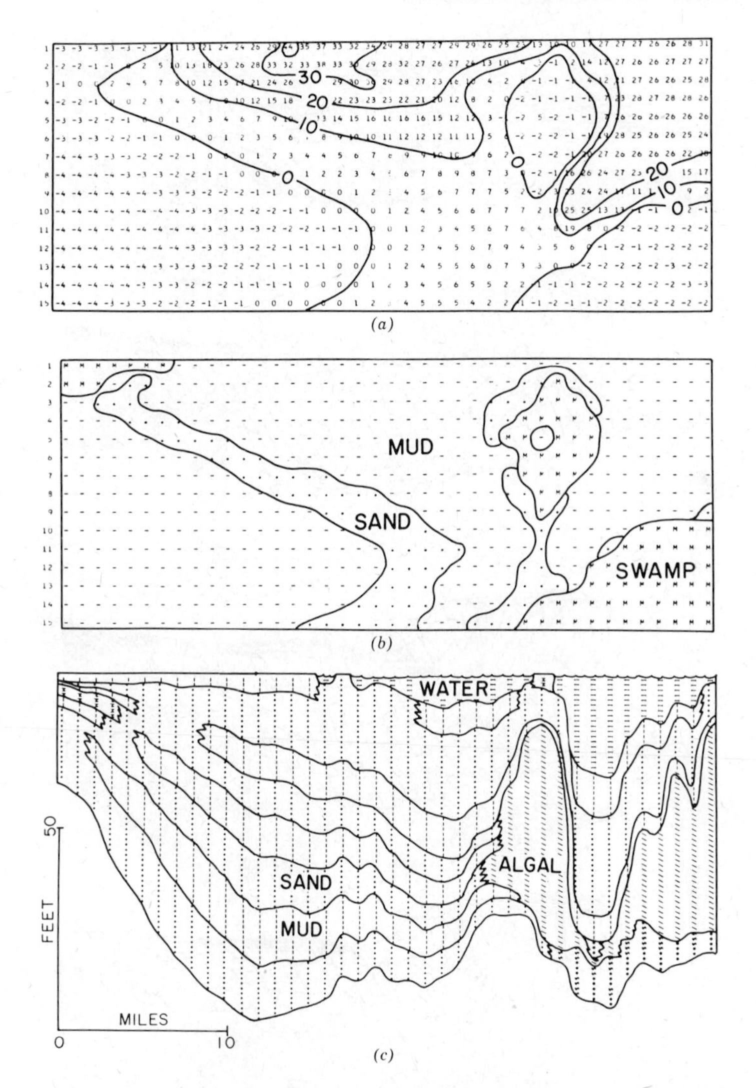

Figure 9-43 Depth map (*a*), facies map (*b*), and cross section (*c*), showing status of model at time increment 8 (see Figure 9-41 for explanation of symbols). From Harbaugh (1966).

from sponge and crinoid communities that populated much of the sea floor in the early part of the run (Figure 9-41), through widespread distribution of the phylloid-algal community in the middle part of the run (Figure 9-42), to ascendency of the swamp community late in the run (Figure 9-43).

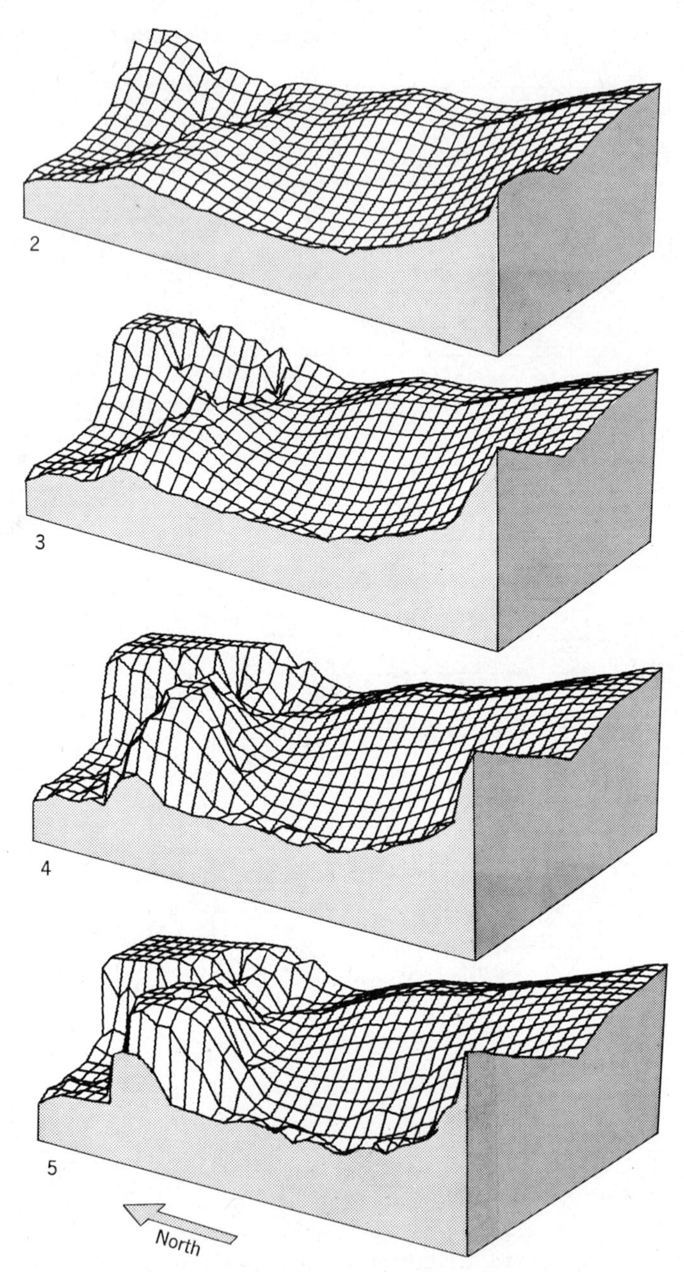

Figure 9-44 Series of computer-drawn perspective block diagrams showing topography of sea floor during simulation run represented in Figures 9-41 to 9-43. Surfaces for time increments 2, 3, 4, and 5 are shown. Block is 40 miles long and 15 miles wide. Orientation of block is reversed (north is toward left) as compared with maps and sections of Figures 9-41 to 9-43 and 9-45.

4. Prominent, steep-sided, flat-topped, submerged banks were formed by phylloid algae, which colonized initial shallow areas in the northern part of the area (Figure 9-44). One of the banks grew vigorously at the edge of the deltaic apron. Although bank growth was stimulated where shallow-water conditions prevailed initially, banks did not form in the southern part of the area because of the inhibiting influence of sand and mud, which poured into the basin from the south.

5. At the end of the run much of the basin had been filled, and water depths in most places lay close to sea level (Figure 9-43). In part of the area swamps spread over areas that lay slightly above sea level. The results of the simulation run shown in Figures 9-41 to 9-43 were judged to be moderately successful in some respects. They failed, however, to yield results that are considered to be sufficiently similar to observed geologic features, particularly with respect to the shape of the algal banks.

In a subsequent run the total amount of terrestrially derived sediment fed into the basin by the river was reduced and the amount of sand was also diminished. The result was contraction of the deltaic apron, accompanied by southward extension of one of the algal banks. The geometry of the algal banks (shown here in vertical sections only in Figure 9-45) accords better with the observed features of real deposits (Figure 9-39).

The usefulness of applying the simulation model to marine-bank origin was assessed as follows:

1. A series of numerical parameters have been derived to describe the environmental response of phylloid calcareous algae within the model. These parameters have little significance outside the model, but within it they seem to be useful indices of the behavior of algae.

2. An assumption that the phylloid algae are highly sensitive to both water depth and terrestrially derived sediment is reflected by the model. The numerical parameters assigned to the algae reflect these assumptions.

3. Further study is needed to determine whether the behavioral properties assigned to phylloid algae are useful in predicting the location and geometry of similar marine banks in Pennsylvanian strata in other areas, particularly where they form oil reservoirs in northwestern Oklahoma and in parts of the Permian basin in Texas and New Mexico.

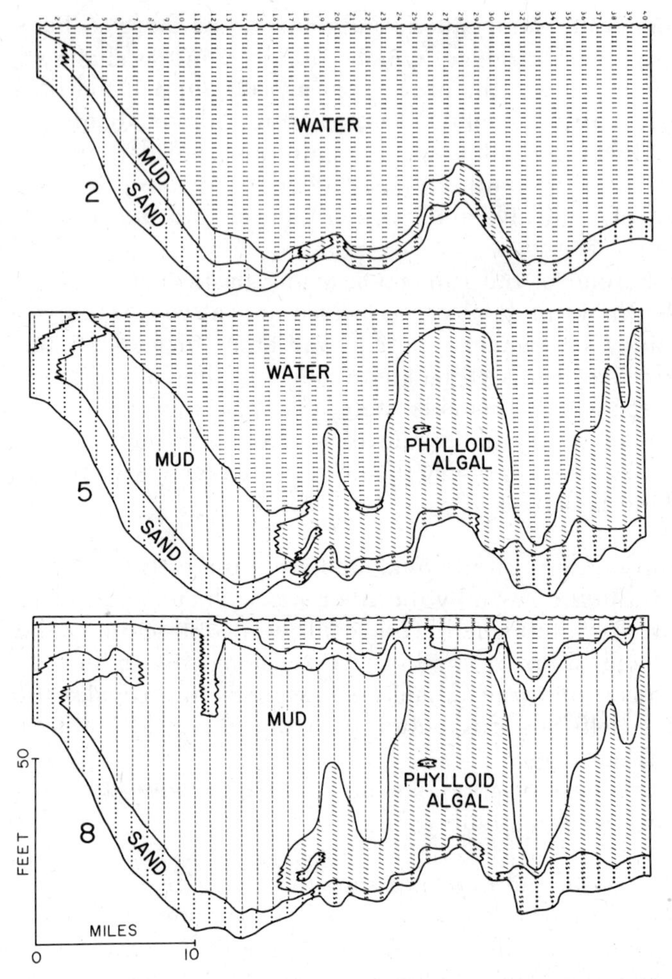

Figure 9-45. Series of three north–south sections that show status of simulation model at time increments 2, 5, and 8, respectively. Results may be compared with an interpretation of actual deposits (Figure 9-40). From Harbaugh (1966).

SUMMARY

The five sedimentation models described in this chapter cover a range of model types. The models contrast with each other in terms of programming complexity, in the methods of representing actual geologic processes, in the degree of success in achieving their objectives, and in the types of mathematical relationships that they employ.

Both the sedimentary-rock-mass model and the sedimentary-basin model are mathematically simple, employing relationships that are applied heuristically. By contrast, the evaporite-basin, delta, and carbonate-ecology models are more complex, both in the level of geological detail and in the mathematics and computer programming that they involve. Among these three models, however, we notice some striking differences. The evaporite model employs mainly fundamental theoretical relationships (fluid flow, evaporation, and precipitation of salt) and appears to be highly successful in simulating evaporite deposits that compare favorably with observed deposits. The delta model, on the other hand, includes mainly empirical relationships dealing with sediment transport. Although the output from the delta model does not compare wholly favorably with actual deltas, experimentation with the model answers a number of questions relating to the importance of variables and the possible origin of submerged levees and river-mouth bars. Furthermore, the fact that the model does not closely approximate any real delta suggests the importance of the variables excluded from the model, such as wave and current effects. Finally, although the carbonate-ecology model deals with problems that are highly relevant geologically, the relationships in the model are almost totally heuristic. Despite the apparently realistic behavior of the model, we should realize that many of the assumptions in the model are questionable. Nevertheless, if we are cautious in our appraisal of the results, the model is instructive, providing insight into relationships that are difficult to "simulate" mentally.

Annotated Bibliography

Agterberg, F. P., 1968, "The Use of Multivariate Markov Schemes in Petrology," *Journal of Geology*, v. 74, pp. 764–785.

Describes the use of Markov methods for treating chemical variation in sequences of strata.

Agterberg, F. P., 1966, *Markov Schemes for Multivariate Well Data*, Mineral Industries Experiment Station, Pennsylvania State University, Special Publication 2–65, pp. Y1–Y18.

Analytical application of Markov methods for treating variations in trace and minor elements in Devonian reef limestones encountered in wells penetrating the Swan Hills oil field in Alberta.

Albertson, M. L., Dai, Y. B., Jensen, R. A., and Rouse, Hunter, 1950, "Diffusion of Submerged Jets," *Transactions of the American Society of Civil Engineers*, v. 115, pp. 639–697.

Detailed development of mathematics pertaining to jet diffusion model.

Allegre, C., 1964, "Vers une logique mathematique des series sedimentaires," *Bulletin de Societé Geologique de France*, Series 7, Tome 6, pp. 214–218.

Employs Markov chains to characterize transitions in sedimentary sequences.

Bates, C. C., 1953, "Rational Theory of Delta Formation," *Bulletin of the American Association of Petroleum Geologists*, v. 37, pp. 2119–2162.

Proposes use of the jet model for deltaic sedimentation.

Bonham-Carter, G. F., and Harbaugh, J. W., 1970, "Stratigraphic Modeling by Computer Simulation," in *Data Processing for Geology and Biology*, J. L. Cutbill, ed., Systematics Association Symposium Volume, Academic Press (in press).

Bonham-Carter, G. F., and Sutherland, A. J., 1967, "Diffusion and Settling of Sediments at River Mouths: A Computer Simulation Model," *Transactions of the Gulf Coast Association of Geological Societies*, v. 17, pp. 326–338.

Bonham-Carter, G. F., and Sutherland, A. J., 1968, "Mathematical Model and FORTRAN IV Program for Computer Simulation of Deltaic Sedimentation," Computer Contribution 24, Kansas Geological Survey, 56 pp.

Bowen, A. J., and Inman, D. L., 1966, *Budget of Littoral Sands in the Vicinity of Point Arguello, Calif.*, U.S. Army Coastal Engineering Research Center, Technical Memorandum 19, 41 pp.

Tabulation of inputs, outputs, and net changes of littoral sand in cubic yards per year for segments of the California coast extending from Pismo Beach to Santa Barbara.

Briggs, L. I., 1958, "Evaporite Facies," *Journal of Sedimentary Petrology*, v. 28, pp. 46–56.

An earlier paper describing principles of the evaporite-basin model treated in this chapter.

Briggs, L. I., and Pollack, H. N., 1967, "Digital Model of Evaporite Sedimentation," *Science*, v. 155, pp. 453–456.

Pioneer paper dealing with the digital simulation of evaporite sedimentation.

Bruun, P., 1966, "Model Geology: Prototype and Laboratory Streams," *Bulletin of the Geological Society of America*, v. 77, pp. 959–974.

Provides resume of mathematical relationships in stream hydrology that pertain to actual streams, with a discussion of their application to small-scale models of streams.

Buttner, P. J. R., 1967, "Simulation Models in Stratigraphy," in *Data Processing and Computer Modeling for Geologists*, L. I. Briggs, ed., University of Michigan Engineering Summer Conference, May 8–19, 1967, 13 pp.

Outlines conceptual models for simulating nonmarine deposits.

Carr, D. C., Horowitz, A., Hrabar, S. V., Ridge, K. F., Rooney, R., Straw, W. T., Webb, W., and Potter, P. E., 1966, "Stratigraphic Sections, Bedding Sequences, and Random Processes," *Science*, v. 154, pp. 1162–1164.

Use of Markov chains to simulate stratigraphic sequences.

Conover, W. J., and Matalas, N. C., 1967, *A Statistical Model of Sediment Transport*, U.S. Geological Survey Professional Paper 575-B, pp. B60–B61.

Develops a statistical model describing the distribution of sediment particles in flowing water so as to obtain curve representing profile of suspended-sediment concentration with elevation above channel floor.

Daly, B. J., and Pracht, W. E., 1968, *A Numerical Study of Density Current Surges*, preprint from the Los Alamos Scientific Laboratory of the University of California, Los Alamos, N.M., U.S. Atomic Energy Commission Contract W-7505-ENG, 58 pp.

Uses marker-in-cell technique to simulate the flow behavior of a two-phase fluid in which the phases have different densities. The technique was used to simulate turbidity current flow, and theoretical results matched with results obtained from laboratory flume studies.

Dobson, R. S., 1967, *Some Applications of a Digital Computer to Hydraulic Engineering Problems*, Technical Report No. 80, Stanford University, Department of Civil Engineering, 172 pp.

Contains a computer method for wave refraction that could be employed in sedimentation applications.

Garrels, R. M., and Mackenzie, F. T., 1969, "Sedimentary Rock Types: Relative Proportions as a Function of Geological Time," *Science*, v. 163, pp. 570–571.

Pioneer paper dealing with mathematical models for rates of sediment recycling.

Gregor, C. B., 1968, "The Rate of Denudation in Post-Algonkian Time," *Nederlandsch Akademie von Wetenschappen*, Series B, Proc. 71, pp. 22–30.

Proposes model to explain distribution of sedimentary rock mass for the post-Algonkian part of the stratigraphic record. Suggests that denudation rates differ strongly before and after widespread occurrence of land plants.

Griffiths, J. C., 1967, *Scientific Method in Analysis of Sediments*, McGraw-Hill, New York, 508 pp.

Thorough, authoritative treatment of statistical tools in sampling and categorizing properties of clastic sediments.

Harbaugh, J. W., 1960, "Petrology of Marine Bank Limestones of Lansing Group (Pennsylvanian), Southeast Kansas," Kansas Geological Survey, Bulletin 142, Part 5, pp. 189–234.

Contains a description of marine-bank deposits that were subsequently represented in simulation models (Harbaugh 1966, 1967*a*, and Harbaugh and Wahlstedt, 1967).

Harbaugh, J. W., 1964, "Trend-Surface Mapping of Hydrodynamic Oil Traps with the IBM 7090/94 Computer," *Quarterly of the Colorado School of Mines*, v. 59, No. 4, pp. 557–578.

Describes ALGOL-58 program for simulating structural closure with respect to oil trapped under hydrodynamic conditions.

Harbaugh, J. W., 1966, "Mathematical Simulation of Marine Sedimentation with IBM 7090/7094 Computers," Computer Contribution 1, Kansas Geological Survey, 52 pp.

Contains computer program for simulating carbonate ecology and sedimentation.

Harbaugh, J. W., 1967*a*, "Computer Simulation as an Experimental Tool in Geology and Paleontology," in *Essays in Paleontology and Stratigraphy*, Raymond C. Moore Commemorative Volume, University of Kansas Department of Geology, Special Publication 2, pp. 368–389.

Provides a philosophical overview of digital simulation applied to sedimentation and to paleoecology.

Harbaugh, J. W., 1967*b*, "Computer Simulation of Petroleum Accumulation in Marine Sediments," in *Proceedings of the Seventh World Petroleum Congress*, Mexico City, pp. 625–632.

Describes method of mapping structural closure with respect to oil trapped under hydrodynamic conditions in dynamic marine-sedimentation model.

Harbaugh, J. W., and Bonham-Carter, G. F., 1968, *Computer Simulation of Shallow-Water Marine Sedimentation*, Summary Report, U.S. Office of Naval Research, Geography Branch, N00014-67-A-0112-0004, Task Number NR 388-081, 18 pp., 4 figures.

Illustrates output from sedimentary-basin model, with a brief description.

Harbaugh, J. W., and Merriam, D. F., 1968, *Computer Applications in Stratigraphic Analysis*, John Wiley and Sons, New York, 282 pp.

Chapter 8 describes several simulation applications, which include simple models for simulating the entrapment of oil under hydrodynamic conditions in sedimentary sequences.

Harbaugh, J. W., and Wahlstedt, W. J., 1967, "FORTRAN IV Program for Mathematical Simulation of Marine Sedimentation with IBM 7040 or 7094 Computers," Computer Contribution 9, Kansas Geological Survey, 40 pp.

Contains FORTAN program equivalent to ALGOL program described by Harbaugh (1966).

Harlow, F. H., 1964, "The Particle-in-Cell Computing Method for Fluid Dynamics," in *Computational Physics*, v. 3, B. Alder, ed., Academic Press, New York, pp. 319–343.

Well-written description of principles of finite-difference technique for solving Navier–Stokes equations describing viscous flow. Method could be used to represent fluid flow in sedimentation models, but consumes large quantities of computer time.

Jizba, Z. V., 1966, "Sand Evolution Simulation," *Journal of Geology*, v. 74, pp. 734–743.

Describes an iterative numerical method for simulating changes in sand-grain properties that may be expressed statistically.

King, C. A. M., 1968, "SPITSIM," in *The Use of Computers in Geomorphological Research*, British Geomorphology Research Group Symposium (Nottingham, England, November 1968), Occasional Paper No. 6, pp. 63–72.

Describes a stochastic model of coastal spit formation. The model represents the mechanisms involved in sand transport by wind-generated currents and storms in heuristic fashion, making extensive use of random numbers. Simulation of a shingle spit on the Hampshire coast of England is described.

Kinsman, D. J. J., 1969, "Modes of Formation, Sedimentary Associations, and Diagnostic Features of Shallow-Water and Supratidal Evaporites," *Bulletin of the American Association of Petroleum Geologists*, v. 53, pp. 830–840.

Authoritative discussion dealing with evaporite formation in the sabkha, or salt flat, environment.

Koch, G. S., and Link, R. F., 1967, "A Simulation of Ghost Stratigraphy," Computer Contribution 12, Kansas Geological Survey, pp. 22–25.

Uses a trend surface to analyze compositional variations in a granite body. Contrasting bands of positive and negative residuals are interpreted as the "ghosts" of former sedimentary strata that have been granitized.

Krumbein, W. C., 1964, *A Geological Process-Response Model for Analysis of Beach Phenomena*, Northwestern University Technical Report 8, ONR Contract 1228(26), Office of Naval Research, 15 pp.

Qualitative description of feedback controls in beach systems.

Krumbein, W. C., 1967, "FORTRAN IV Computer Programs for Markov Chain Experiments in Geology," Computer Contribution 13, Kansas Geological Survey, 38 pp.

Describes a number of short FORTRAN programs that may be used for generating stratigraphic sequences by using Markov models.

Krumbein, W. C., 1968*a*, "FORTRAN IV Computer Program for Simulation of Transgression and Regression with Continuous-Time Markov Models," Computer Contribution 26, Kansas Geological Survey, 38 pp. An important paper that describes continuous-time Markov models, which are described in Chapter 4 of this volume.

Krumbein, W. C., 1968*b*, "Statistical Models in Sedimentology," *Sedimentology*, v. 10, pp. 7–23.

Presents an overview of statistical models in sedimentation, many of which have relevance to simulation.

Krumbein, W. C., and Graybill, F. A., 1965, *An Introduction to Statistical Models in Geology*, McGraw-Hill, New York, 475 pp.

Detailed treatment of statistical applications in geology, with applications to sedimentation.

Miller, R. L., and Kahn, J. S., 1962, *Statistical Analysis in the Geological Sciences*, John Wiley and Sons, New York, 483 pp.

Pioneer synthesis of statistical methods applied in geology.

Oertel, G. and Walton, E. K., 1967, "Lessons from a Feasibility Study for Computer Models of Coal-Bearing Deltas," *Sedimentology*, v. 9, pp. 157–168.

The authors conclude that a realistic computer model is not feasible to develop with present restrictions imposed by computer memory size and computing speed. They do develop, however, a moderately detailed conceptual model, which suggests that the cyclic aspect of sediments in a typical large delta complex reflects the influence of negative-feedback loops that cause the system to oscillate in a sequence of superimposed frequencies.

Ojakangas, D. R., 1967, *Mathematical Simulation of Oil Trap Development*, Ph.D. Thesis, Branner Geological Library, Stanford University, 150 pp. Deals with a model for simulating the structural configuration of stratigraphic sequences in sedimentary basins through geologic time. The FORTRAN program representing the model provides for automated drafting of geologic sections. Output from the program is illustrated by Harbaugh and Merriam (1968, pp. 26–29).

Pitts, F. R., 1965, *Hager III and Hager IV: Two Monte Carlo Computer Programs for the Study of Spatial Diffusion Problems*, Northwestern University Geography Department Technical Report No. 4, Geography Branch, Office of Naval Research, Contract NONR 1228(33).

Contains computer program for probabilistic diffusion that could be used in ecological simulation.

Pollack, H. N., 1967, *Deterministic Modeling in Geology and Geophysics*, mimeographed notes for an engineering summer conference course given at the University of Michigan, Ann Arbor, May 1967, 21 pp.

Develops equations for fluid flow and transport of salt in the evaporite model described by Briggs and Pollack (1967).

Potter, P. E., and Blakely, R. F., 1967, "Generation of a Synthetic Vertical Profile of a Fluvial Sandstone Body," *Society of Petroleum Engineers Journal*, September 1967, pp. 243–251.

Uses Markov chains to generate synthetic sedimentary sequences.

Potter, P. E., and Blakely, R. F., 1968, "Random Processes and Lithologic Transitions," *Journal of Geology*, v. 76, pp. 154–170.

Describes the use of Markov transition matrices to characterize actual stratigraphic sequences and to synthesize artificial sequences for comparison purposes.

Ronov, A. B., 1959, "On the Post-Precambrian Geochemical History of the Atmosphere and Hydrosphere," *Geochemistry*, v. 5, pp. 493–506.

Source of data concerning the distribution of sedimentary rock masses with geologic age.

Schmalz, R. F., 1969, "Deep-Water Evaporite Deposition: A Genetic Model," *Bulletin of the American Association of Petroleum Geologists*, v. 53, pp. 798–823.

Discusses a deep-water evaporite model similar to the Briggs–Pollack model, but somewhat more complex in detail.

Schwarzacher, W., 1966, "Sedimentation in Subsiding Basins," *Nature*, v. 210, pp. 1349–1350.

Describes a simple control model that employs differential equations relating sedimentation rate to water depth and subsidence.

Schwarzacher, W., 1967, "Some Experiments To Simulate the Pennsylvanian Rock Sequence of Kansas," Computer Contribution 18, Kansas Geological Survey, pp. 5–14.

Uses both autocorrelation models and single- and double-dependence Markov chains to simulate stratigraphic sequences.

Schwarzacher, W., 1968, "Experiments with Variable Sedimentation Rates," Computer Contribution 22, Kansas Geological Survey, pp. 19–21.

Suggests ways of representing variable sedimentation rates with Markov transition matrices and explores outcomes of various models.

Sloss, L. L., 1962, "Stratigraphic Models in Exploration," *Journal of Sedimentary Petrology*, v. 32, pp. 415–422.

Presents details of simple conceptual models of sedimentation systems that relate crustal subsidence, sediment sources, and response of deposits formed in terms of facies geometry.

Sloss, L. L., 1969, "Evaporite Deposition from Layered Solutions," *Bulletin of the American Association of Petroleum Geologists*, v. 53, pp. 776–789.

Proposes model to explain synchronous deposition of shallow-water normal marine carbonate reefs with surrounding deep-water evaporite formation. Deposition occurs in density-layered waters with an upper layer of normal seawater salinity and lower layers saturated with respect to one or more evaporite salts. Carbonate deposition occurs in upper dilute waters, whereas salts deposit from deep-water brine, concentrated by evaporation on nearby shelf areas.

Vanoni, V. A., Brooks, N. H., and Kennedy, J. F., 1961, *Lecture Notes on Sediment Transportation and Channel Stability*, W. M. Keck Laboratory For Hydraulics and Water Research, California Institute of Technology, Technical Report No. KH-R-1.

Useful summary of a variety of mathematical models that pertain to sediment transport by rivers.

Van Straaten, L. M. J. U., 1961, "Some Recent Advances in the Study of Deltaic Sedimentation," *Journal of the Liverpool and Manchester Geological Society*, v. 2, pp. 411–442.

An excellent summary paper which gives a general overview on aspects of deltaic sedimentation and which gives various figures for foreset slope angles cited in the text.

Vistelius, A. B., 1967, *Studies in Mathematical Geology*, Consultants Bureau, New York, 294 pp.

A collection of 26 papers by Vistelius, most of which appeared originally in Russian. The book brings them together as a compact sequence of translations. Several of the papers are relevant to simulation, particularly those dealing with frequency distributions. Of particular relevance are papers describing the use of Markov transition matrices on pages 252–262.

Welch, J. E., Harlow, F. H., Shannon, J. P., and Daly, B. J., 1966, *The MAC Method: A Computing Technique for Solving Viscous, Incompressible Transient Fluid-Flow Problems, Involving Free Surfaces*, Los Alamos Scientific Laboratory Report LA-3425, 146 pp.

Detailed description of the particle-in-cell method for simulating viscous flow (employing finite-difference approximation to Navier–

Stokes equations) in which there are free surfaces, such as a wave on a sloping beach or a wave surging over a reef.

Wickman, F. E., 1961, "On the Distribution of Different Kinds of Geological Deposits in a Glaciated Region," *Arkiv für Mineralogi och Geologi*, v. 3, pp. 103–129.

Uses Markov transition matrices to characterize lateral transitions between different types of Quaternary deposits as recorded on maps.

CHAPTER 10
Other Applications

This chapter is intended to provide a survey of simulation applications in fields other than sedimentation. A substantial fraction of the chapter is devoted to a review of the literature. We should note that to date the applications of computer-simulation methods to geology are quite limited and that the literature is correspondingly small. On the other hand, there is a large body of literature that has relevance to simulation, although it does not directly involve computing. Accordingly the objective of this chapter is to touch on some of this literature to provide an overview of the geological simulation applications that have been undertaken to date and to suggest future applications.

The chapter is organized by classes of applications that are either within geology or touch on it. Instead of a consolidated bibliography at the end of the chapter, an annotated bibliography accompanies each class.

ECOLOGY AND PALEOECOLOGY

Ecology and paleoecology offer large opportunities for the application of simulation methods. We have touched on one form of ecological simulation in the carbonate-ecology model in Chapter 9. Other relevant aspects include (a) simulation of chemical, nutrient, and energy networks; (b) simulation of variations in population abundances as functions of physical-environment factors in which the populations do not interact with each other; and (c) simulation of populations that are interdependent, reacting with each other.

Chemical, Nutrient, and Energy Networks

Many ecological studies are concerned with chemical, nutrient, and energy networks. These can be summarized in diagrams (Figure 10-1) in which the width of the bands is proportional to the amount of energy or nutrients transferred per unit of time. Organisms in the ecological system participate at one of five relative positions in the flow circuit:

1. Plants carry out photosynthesis.
2. Animals eat plants (herbivores).
3. Carnivores eat herbivores.
4. Carnivores eat other carnivores.
5. Decomposers or microorganisms cause plant and animal material to be broken down.

Figure 10-1*a* represents an energy-balance relationship. An ecological system must adhere to the law of conservation of energy, because any losses or gains within the system must be represented by an aggregate increase or decrease of energy within the boundaries of the system. Much of the energy is supplied in the form of light and other radiant energy from the sun. In turn the energy is progressively degraded to heat within the system. Energy interrelationships in ecological systems can be represented quantitatively in simulation models, employing accounting methods similar to those used in materials-balance models. The transfer of nutrients and chemicals in an ecological system can be shown conveniently in network graph form, such as the "eating graph" shown in Figure 10-2. Alternatively these relationships can be represented by a Boolean matrix (Table 10-1), which specifies whether the animal listed in a particular row of the matrix does eat (1) or does not eat (0) the animal (or plants) listed in each column.

Simulation of Population Changes without Interaction between Organisms

Fox (1968) has developed a relatively simple dynamic simulation model to deal with variations in the abundance of fossils through time. He applied the model to variations in abundance of 10 brachiopod genera in a sequence of Upper Ordovician strata (Richmond

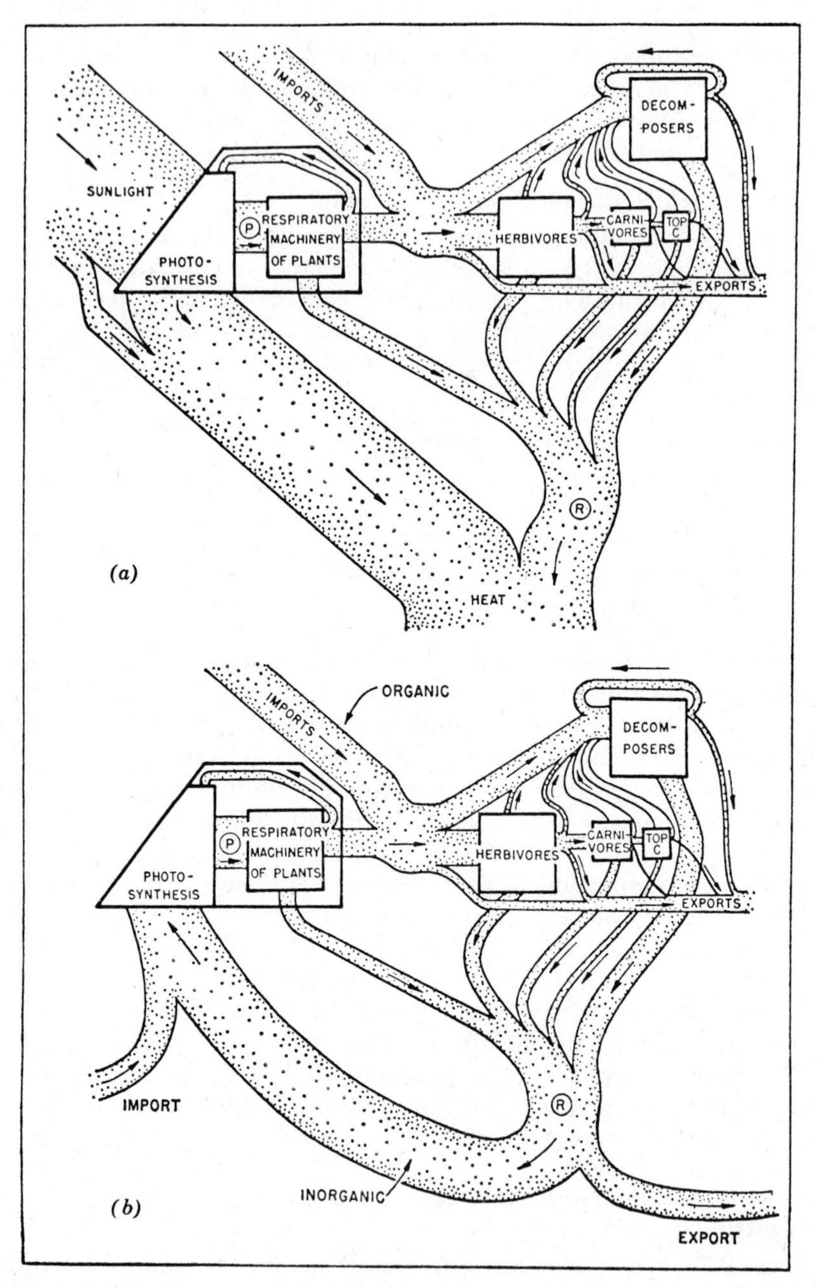

Figure 10-1 (*a*) Energy and (*b*) nutrient-flow diagrams that pertain to theoretical ecosystems. After Odum (1960), with permission from *American Scientist* and *Limnology and Oceanography.*

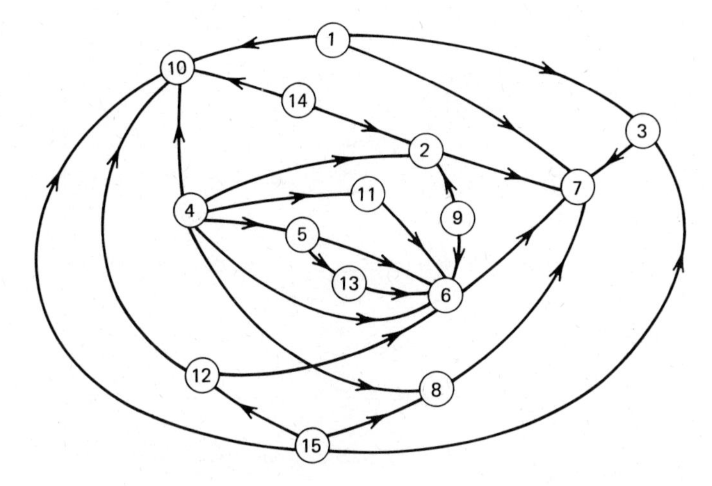

Figure 10-2 "Eating chart" consisting of network graph showing food chain involving plants and 15 classes of animals, as follows: (1) bear, (2) bird, (3) deer, (4) fox, (5) gartersnake, (6) insect, (7) plants, (8) rabbit, (9) raccoon, (10) rodent, (11) salamander, (12) skunk, (13) toad, (14) wildcat, (15) wolf. From Harary (1961), with permission from *General Systems*.

group) in southeastern Indiana. His model assumes that each of the brachiopod genera responds in a particular manner to changes in seawater temperature, salinity, and depth. By adjusting the assumed tolerances of these organisms to these environmental factors, their numerical distribution through the stratigraphic sequence (and thus through time) can be controlled in the model. The simulated responses are then compared with distributions of the brachiopods based on quantitative field sampling.

In spite of the tenuous assumptions involved in a simulation model of this type, there are a number of "anchor points" that do have a substantial degree of validity. For example, the relationship between light intensity in seawater and the photosynthetic activity of phytoplankton in modern seas is probably similar to that of plankton in ancient seas. Likewise the tolerances of modern marine animals to variations in salinity provide a rough guide to the salinity tolerances of comparable ancient forms.

Light intensity in seawater (ignoring seasonal and climatic fluctuations) is a function of the cloudiness, or turbidity, of the water. In turn the photosynthetic activity of phytoplankton is a function of light intensity. The two factors are interdependent: the presence of phytoplankton causes water to be clouded and brings about a decrease in light intensity at a given depth until an equilibrium relationship is

TABLE 10-1 "Eating" Matrix Corresponding to the Network Graph Shown in Figure 10-2[a]

Eating Animal	Eaten Animal 1	2	3	4	5	6	7	8	9	10	11	12	13	14	15	*Row Sum*
1	0	0	1	0	0	0	1	0	0	1	0	0	0	0	0	3
2	0	0	0	0	0	0	1	0	0	0	0	0	0	0	0	1
3	0	0	0	0	0	0	1	0	0	0	0	0	0	0	0	1
4	0	1	0	0	1	1	0	1	0	1	1	0	0	0	0	6
5	0	0	0	0	0	1	0	0	0	0	0	0	1	0	0	2
6	0	0	0	0	0	0	1	0	0	0	0	0	0	0	0	1
7	0	0	0	0	0	0	0	0	0	0	0	0	0	0	0	0
8	0	0	0	0	0	0	1	0	0	0	0	0	0	0	0	1
9	0	1	0	0	0	1	0	0	0	0	0	0	0	0	0	2
10	0	0	0	0	0	0	0	0	0	0	0	0	0	0	0	0
11	0	0	0	0	0	1	0	0	0	0	0	0	0	0	0	1
12	0	0	0	0	0	1	0	0	0	1	0	0	0	0	0	2
13	0	0	0	0	0	1	0	0	0	0	0	0	0	0	0	1
14	0	1	0	0	0	0	0	0	0	1	0	0	0	0	0	2
15	0	0	1	0	0	0	0	1	0	1	0	1	0	0	0	4
Column sum	0	3	2	0	1	6	5	2	0	5	1	1	1	0	0	27

[a] From Harary (1961), with permission from *General Systems*.

attained. Figure 10-3 is a curve relating photosynthetic activity to light intensity under equilibrium conditions. Alternatively photosynthetic activity and light intensity can be related to depth. The curves in Figure 10-4 relate photosynthetic activity to depth for various degrees of turbidity. Other environmental aspects, such as nutrient supply (nitrate and phosphate content of seawater) are avoided by expressing the photosynthetic activity in relative form.

The tolerance of a particular organism for an environmental factor can be expressed as a frequency-distribution curve. In effect the curves of photosynthetic activity (phytoplankton mass per unit of water) are graphs of frequency distributions. Similar curves can be prepared for other organisms, relating their tolerances to particular environmental factors, including salinity and water temperature (Figure 10-5). Figure 10-6 portrays a series of frequency-distribution curves that relate the tolerances of hypothetical organisms to combined variations in salinity, depth, and light intensity.

Fox's simulation experiments consist of making different assumptions as to the tolerances of brachiopods occurring in the Richmond group and progressively adjusting them until the emergence of a distribution pattern that accords reasonably well with the observed

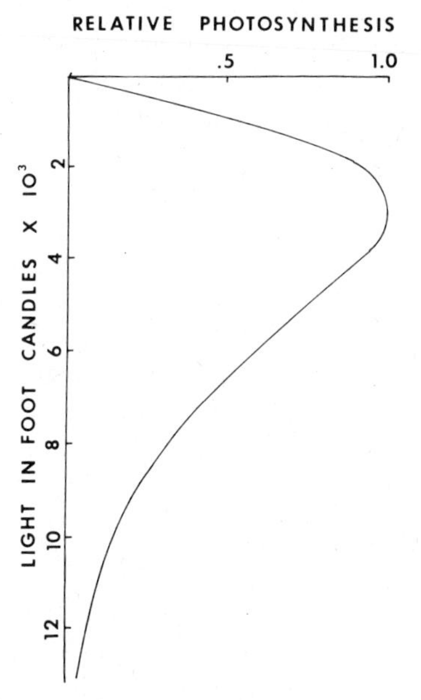

Figure 10-3 Curve relating relative photosynthetic activity of phytoplankton in seawater to light intensity. From Fox (1968).

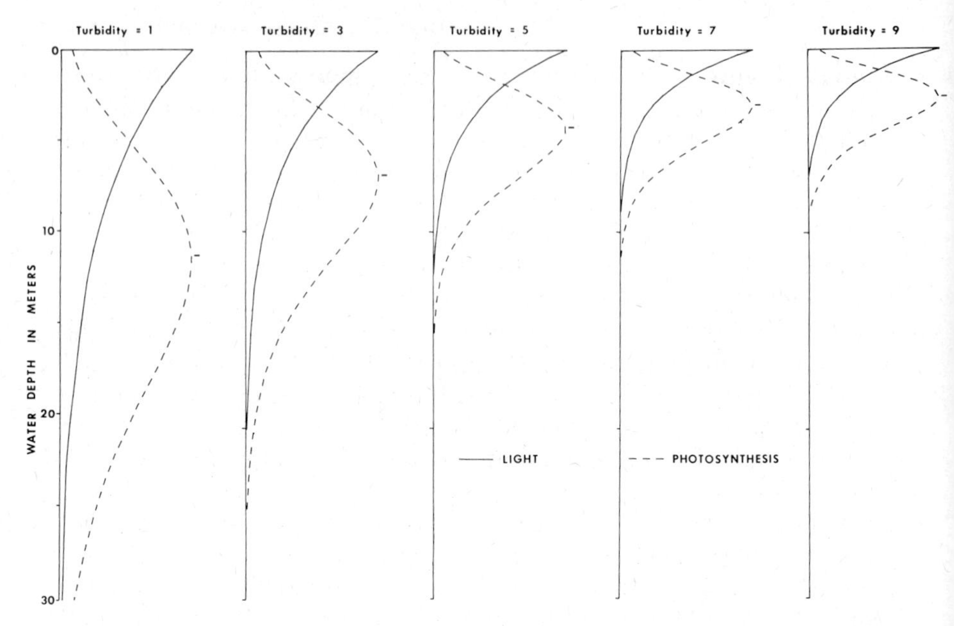

Figure 10-4 Series of curves that relate light intensity and relative photosynthetic activity to depth in water with varying degrees of turbidity. From Fox (1968).

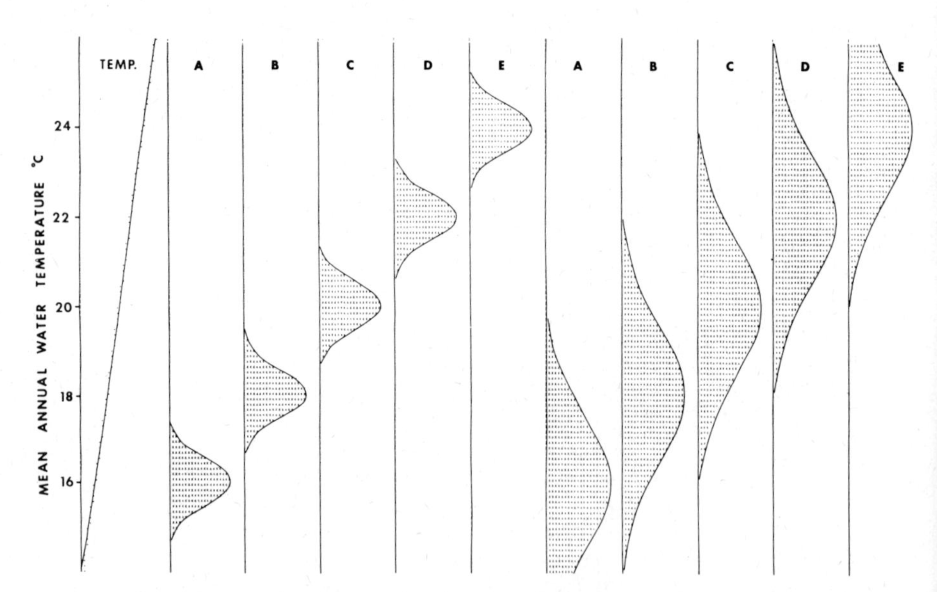

Figure 10-5 Frequency distribution (relative abundance) curves representing responses of hypothetical types of organisms (*A*, *B*, *C*, *D*, and *E*) to variations in water temperature. Curves in left half of diagram represent narrow tolerance ranges, whereas those on right represent broader ranges. From Fox (1968).

distribution. Some of the results of his experiments are shown in Figure 10-6. The observed stratigraphic distribution in the Richmond group is represented as a series of bar diagrams for 10 brachiopod genera. The three other diagrams represent simulated variations in the brachiopod genera within the stratigraphic sequence, adjusting the environmental factors and observing the response with assumed tolerances. As the curves reveal, it is possible to bring the simulated frequencies into rough accord with the observed ones. The importance of such an experiment is not that it provides a firm answer but rather that it brings a set of assumptions into accord with each other. It is possible that additional observations could be devised to test the validity of some of the assumptions.

Simulation of Population Changes with Interaction between Organisms

It is clear that in the real biological world populations are not only dependent on variations in physical aspects of the environment but also interdependent with respect to each other. Consequently simulation models that treat interdependent populations tend to be more realistic than those that do not.

Let us begin by considering the changes in population size in which change is a function of birthrate. A familiar example is provided by the growth of a bacterial population under ideal conditions. At the beginning there is a single bacterium, which divides to produce two bacteria, which in turn divide on a regular basis: 2, 4, 8, 16, · · ·. Thus the population N_T at the end of T generations is 2^T. If such a population is regarded as continuously varying in size, the rate of change in population size is

$$\frac{dN}{dt} = rN \qquad (10\text{-}1)$$

where N = number present
t = time
r = rate of increase

On integration this yields the equation pertaining to the law of growth and decay:

$$N_t = N_0 e^{rt} \qquad (10\text{-}2)$$

where N_0 = number at time zero
N_t = number at time t

When the logarithm of population size is graphed against time, a

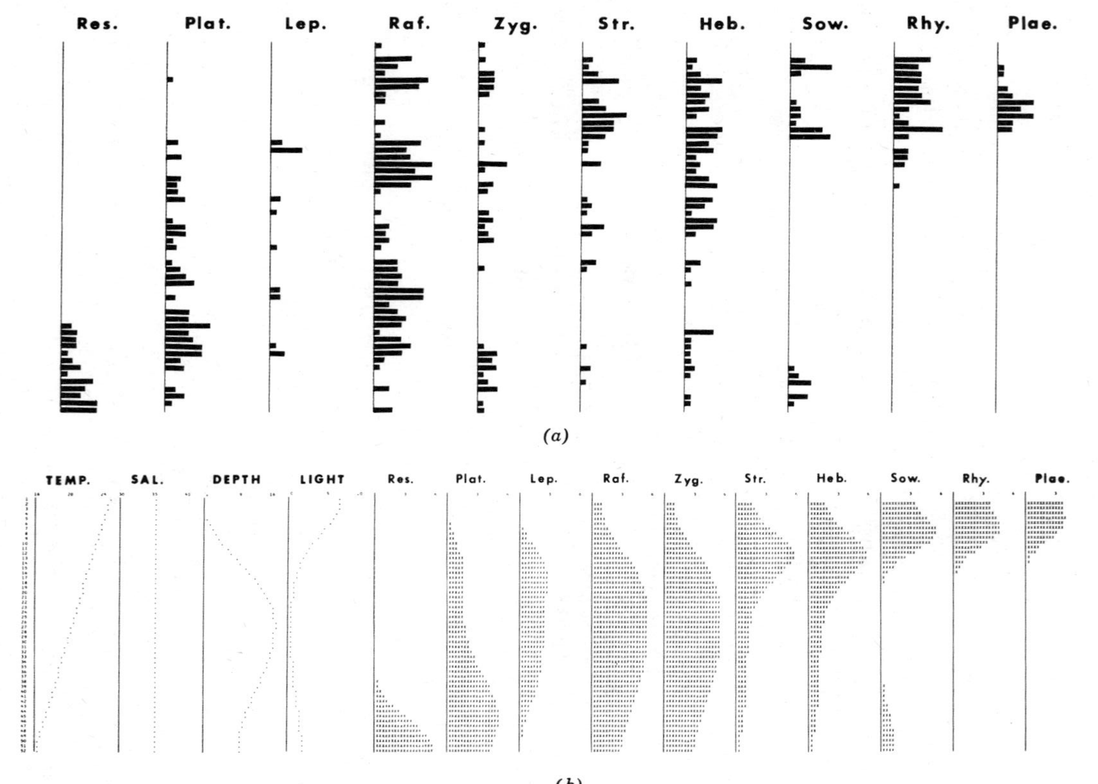

Figure 10-6 Series of diagrams pertaining to the paleoecological simulation experiments of Fox (1968). (*a*) Bar graphs for observed brachiopod distribution in Richmond group. Each bar represents average abundance for a 50-centimeter vertical interval in stratigraphic section. Generic names of brachiopods are abbreviated: Res. = *Resserella*; Plat. = *Platystrophia*; Lep. = *Leptaena*; Raf. = *Rafinesquina*; Zyg. = *Zygospira*; Str. = *Strophomena*; Heb. = *Hebertella*; Sow. = *Sowerbyella*; Rhy. = *Rhynchotrema*; Plae. = *Plaesiomys*. (*b*) Simulated response showing relative abundances in which the temperature

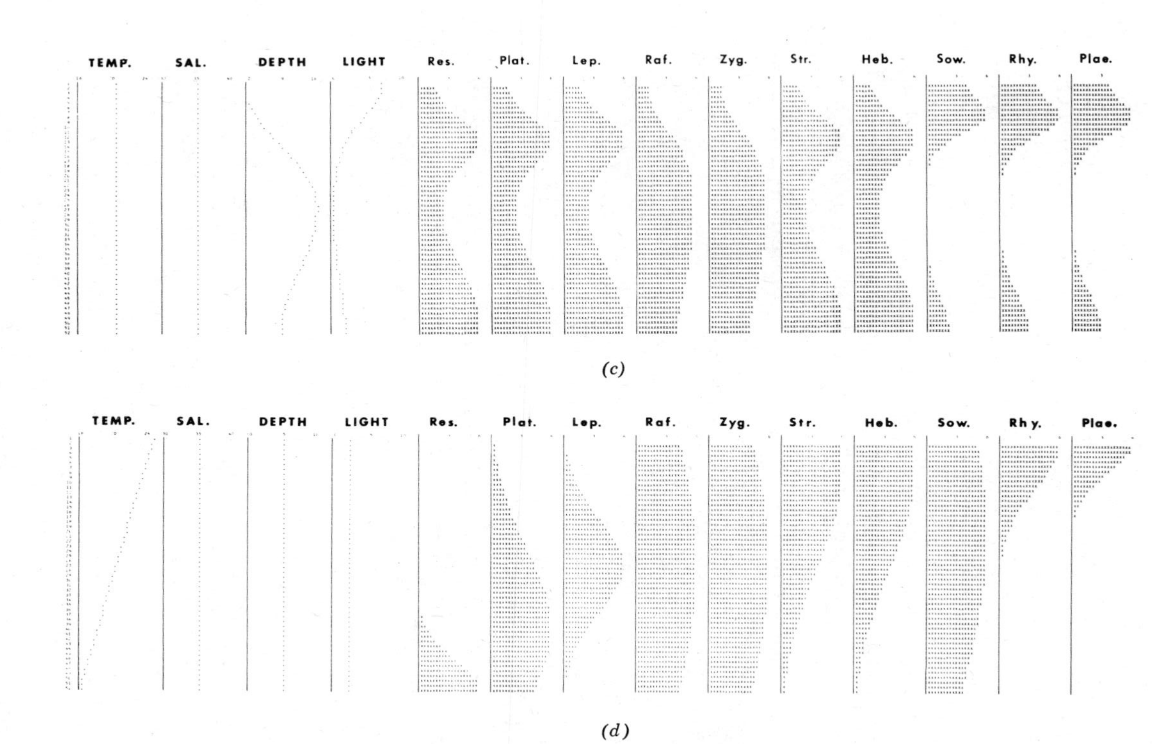

Figure 10-6 *Continued.* Series of diagrams pertaining to the paleoecological simulation experiments of Fox (1968). (*c*) Response when depth and light intensity are varied. (*d*) Response when temperature, depth, and light are varied.

straight line is obtained. In such a model changes in population size are proportional to the size of the population.

The simple birth model described above is unrealistic. No population can expand indefinitely under ideal conditions. Consider, for example, a prey–predator relationship. The population that serves as prey suffers losses as its individuals are killed by predators. On the other hand, the predator population is dependent on the presence of individuals in the prey population. If the prey population declines, the predator population will also decline.

Interactions between simple prey–predator populations can be represented mathematically. Kemeny and Snell (1962, p. 24) describe the mathematical theory pertaining to an idealized prey–predator relationship involving a single prey population. A rabbit population could serve as an example of the prey, and a fox population as the predator. The foxes can be assumed to eat nothing but rabbits. If there were no foxes and assuming an unlimited food supply for the rabbits (and ignoring other causes of death), the rate of increase in the rabbit population would be

$$\frac{dx}{dt} = ax \tag{10-3}$$

where x = number of individuals in the rabbit population
a = rabbit birthrate
t = time

Now let us assume that there are no rabbits, but there are foxes. Under these conditions the rate of decrease in the fox population can be expressed as follows:

$$\frac{dy}{dt} = -py \tag{10-4}$$

where y = number of individuals in the fox population
p = fox mortality rate

These equations are, of course, identical in form to Equation 10-1.

We can combine Equations 10-3 and 10-4 to represent the rate of change in the population sizes when rabbits and foxes interact. If we assume that the number of rabbits that are eaten by foxes is proportional to the product of the two population sizes, xy, we can modify Equations 10-3 and 10-4 to account for a decrease in rabbits due to foxes and an increase in foxes due to the availability of rabbits. The rate of change in the rabbit population (Lotka, 1924) can be written as

$$\frac{dx}{dt} = ax - bxy \tag{10-5}$$

where b is the rabbit death rate due to foxes, and x, y, t, and a have been already defined. The rate of change of the fox population is given by a similar equation:

$$\frac{dy}{dt} = cxy - py \tag{10-6}$$

where c is the per capita rate of increase in the prey population.

The behavior through time of prey and predator will be cyclic. Considering foxes and rabbits, we might begin at a particular point in time, (a) when there is an abundance of rabbits. This would bring about (b) an increase in the number of foxes which in turn brings about (c) a decrease in the number of rabbits. Then (d) the foxes decline due to the scarcity of rabbits, and (e) the rabbits subsequently increase because there are fewer foxes. Variations in the abundance of the two populations through time may be represented by curves (Figure 10-7).

We can consider these equations from another point of view. Consider the equilibrium point at which there is no change in either population with respect to time. This can be defined as

$$\frac{dx}{dt} = \frac{dy}{dt} = 0 \tag{10-7}$$

If the derivatives are equal to zero, we can rearrange Equations 10-5 and 10-6 to obtain

$$x = \frac{p}{c}; \qquad y = \frac{a}{b} \tag{10-8}$$

If we choose an equilibrium point as a starting position in time, the rates of change are zero, and so we remain at the starting point on a

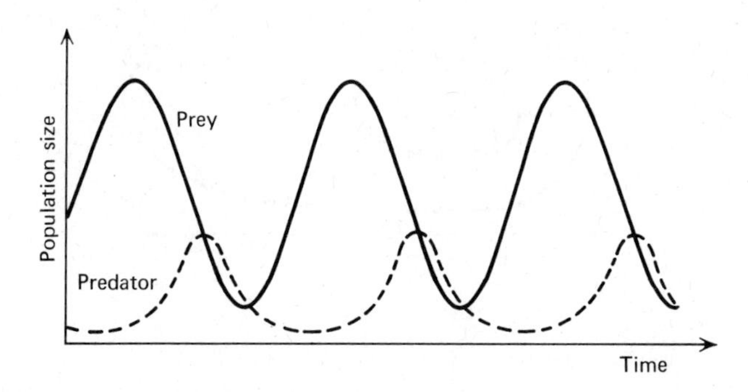

Figure 10-7 Graph of variations in population size of hypothetical prey and predator with respect to time.

graph of x with respect to y or vice versa. If, however, the starting point is not an equilibrium point, the locus of points forms a curve (Figure 10-8), or *trajectory*, which is a solution of Equations 10-5 and 10-6. This curve must be traversed in a fixed direction. To find the trajectories we can combine Equations 10-5 and 10-6 to eliminate t by dividing one by the other:

$$\frac{dy}{dx}=\frac{cxy-py}{ax-bxy} \tag{10-9}$$

Factoring, we obtain

$$\frac{dy}{dx}=\frac{y(cx-p)}{x(a-by)} \tag{10-10}$$

After rearranging, we have

$$\frac{a-by}{y}\cdot\frac{dy}{dx}+\frac{p-cx}{x}=0 \tag{10-11}$$

We shall not delve into the manner of obtaining solutions, but one form of a solution may be represented as

$$\frac{x^p}{e^{cx}}\cdot\frac{y^a}{e^{by}}=k \tag{10-12}$$

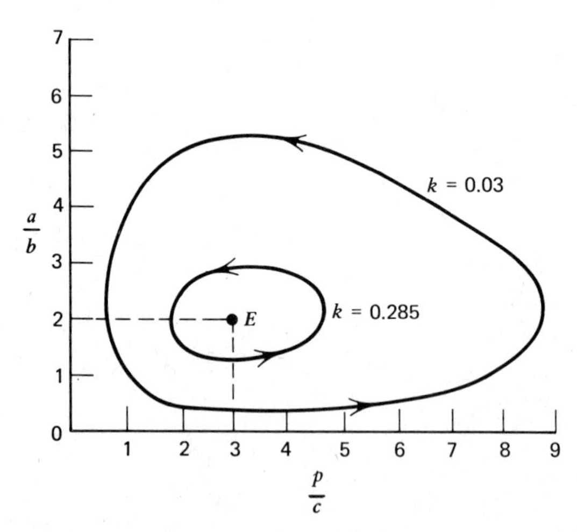

Figure 10-8 Graph of trajectories of prey–predator for different values of k. Coefficients of equations are $a=4$, $b=2$, $c=1$, and $p=3$; E denotes equilibrium point for these coefficients. Modified after Kemeny and Snell (1962). Copyright by Ginn and Company.

where k is a particular constant. For various values of k and prescribed values of the constants a, b, c, and p we can plot a series of closed curves that surround the equilibrium point (Figure 10-8). It is possible to obtain other solutions, which do not represent closed curves. Solutions of these equations may be manipulated numerically in simulation programs to mimic the effect of interdependent populations. Relatively complex interactions can be dealt with in this manner. Kemeny and Kurz (1967, p. 113) illustrate a simple program for obtaining numerical solutions to Equations 10-5 and 10-6. Reyment (1968) discusses the application of prey–predator models in paleoecology.

Fossil-Population Model

Next we describe the details of a generalized computer model dealing with variations in the populations of invertebrate marine organisms. The model deals with fluctuations in population size that stem from changes in "birth" and death rates in a single population. It also deals with variations in the physical size of organisms. Although the model deals with rate processes, these are not expressed as differential equations, as in the prey–predator examples described in the preceding section. Instead rate processes are embodied in an accounting system that statistically monitors groups of organisms that progress from birth to death through the system. Furthermore competition between different populations is not considered.

The model is a modification of a more complex computer model developed by Craig and Oertel (1966). Impetus to use such a model is provided by variations in fossil-population assemblages observed in sequences of beds. Interpretation of these variations is of course fraught with difficulty and uncertainty. It is possible, however, to employ a simulation model to synthesize variations in hypothetical populations, which in turn can be compared with the observed populations. Our version of the model is highly simplified and is included because it is useful as an exercise. It could, however, be readily expanded to include more realistic details.

Structure of Model

The basic structure of the model is illustrated by means of a flow chart (Figure 10-9). A population of living organisms (the nature of these organisms is unspecified, but they could be, for example, brachiopods) can be described by a size-frequency distribution, which records the frequency of individuals occurring in each of several classes according to the physical size of individuals (Figure 10-10). The distribution of individuals in the living population is

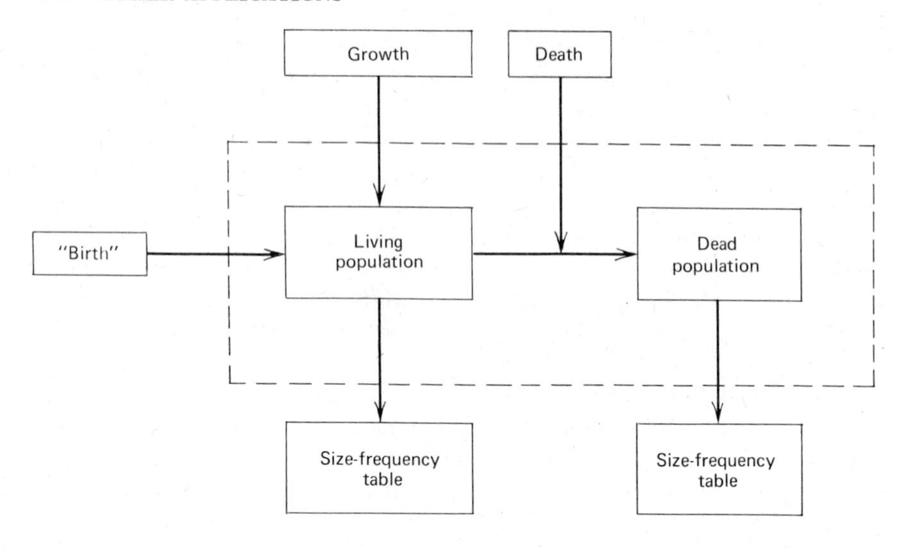

Figure 10-9 Flow chart of fossil-population model in which rates of birth, growth, and death (exogenous variables) control composition of living and dead populations. Output is in the form of size-frequency tables.

influenced primarily by three factors: "birth" of new individuals, growth, and death. The shape of the size-frequency histogram reflects these three factors. Birth introduces new individuals, which in time progress from small-size classes to large-size ones. Death removes individuals from all size classes in the living population and transfers them to the dead population. FORTRAN arrays are used to maintain the accounts of the frequency classes of the two populations.

The model has been designed to run for a period of 60 simulated months. Each month new individuals are born, some individuals die, and all surviving ones grow. The new "recruits" that are born each month form a *company* in the terminology of Craig and Oertel. The identity of each company is retained with the passage of each simulated month. Each company has a company number (the number of the month when it was created, ranging from 1 to 60) and an age in months. It also contains a number of individuals. Within each company it is assumed that all the individuals are the same size. The four characteristics (identification number, age, number of individuals, size of individuals) of each company are illustrated graphically in Figure 10-11.

Three exogenous variables—birth, growth, and death—are supplied as input to the model. The number of individuals born is specified for each month. In example runs illustrated here birth has been held

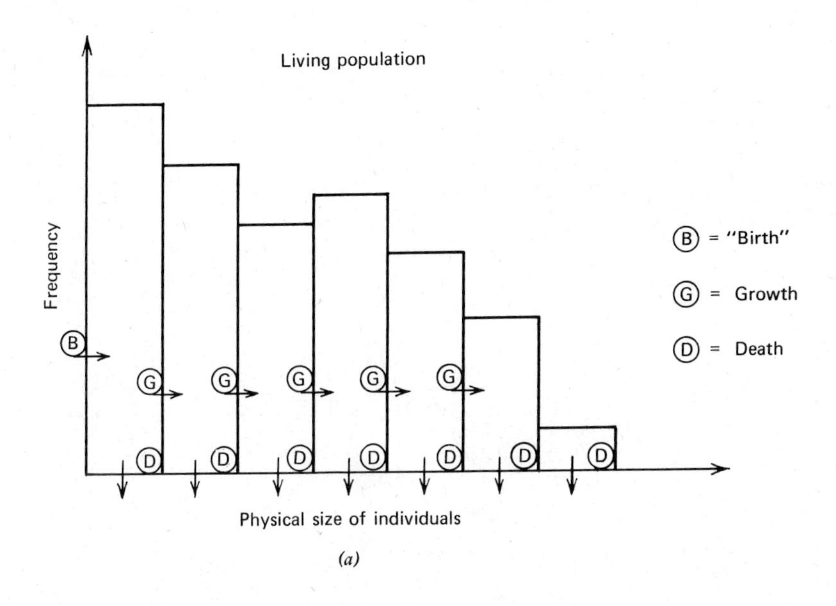

(a)

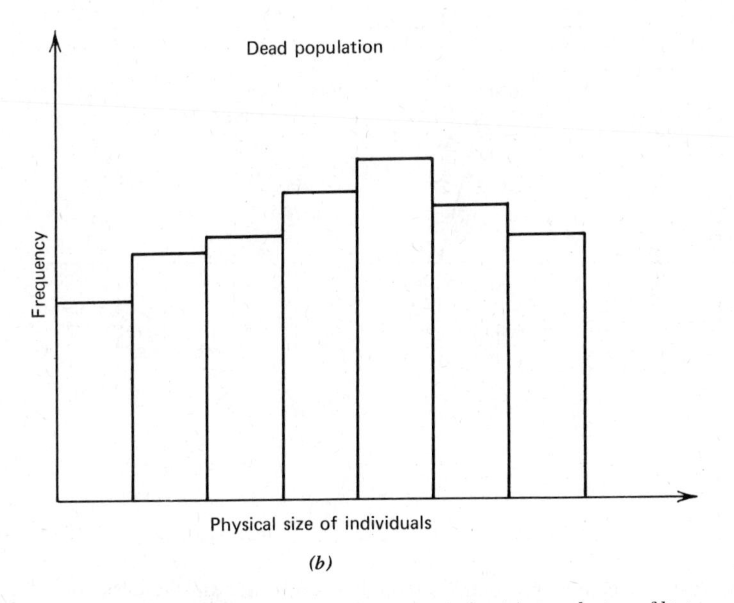

(b)

Figure 10-10 Diagram of size-frequency histograms for (*a*) population of living organisms and (*b*) population of dead organisms. Birth introduces individuals into smallest living size class, growth progressively moves individuals from smaller to larger size classes, and death removes individuals from all size classes in living population to equivalent size classes in dead population.

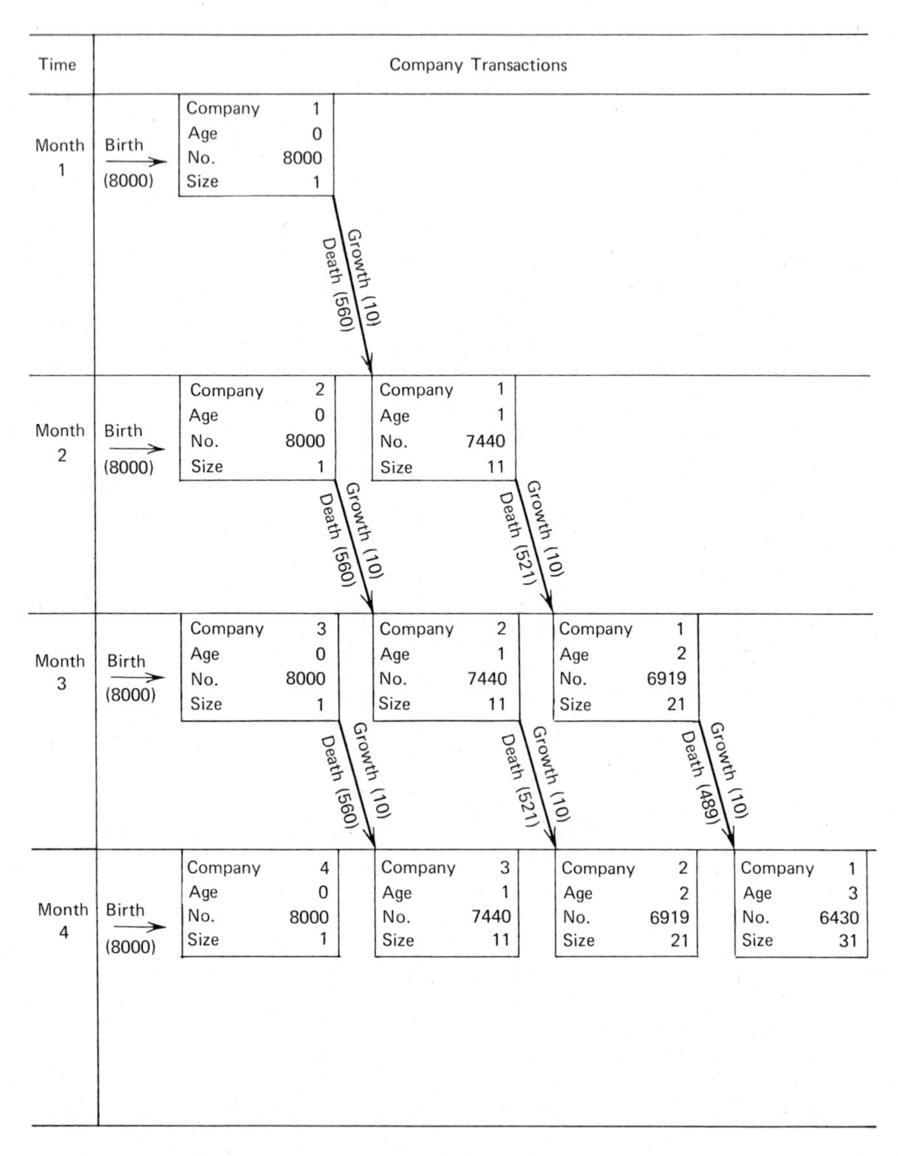

Figure 10-11 Diagram illustrating accounting steps in initial 4 months in operation of fossil-population model. Each month a new company is created containing that month's births. Individuals in each company grow by increments. Mortalities reduce number of live individuals in each company, causing dead individuals to be assigned to appropriate size classes of dead population. Size-frequency tables, which may be graphed as histograms, form output from model.

constant at 8000 individuals per month. This could be altered, for example, to represent seasonal variations in birthrate.

In its simplest form the physical growth of individuals is assumed to be a linear function of age (Figure 10-12). At birth individuals are assumed to be 1.0 unit in size (the nature of the unit is unspecified and arbitrary). Growth rate is described by the number of size increments added per company of individuals for a single month. Thus if during month N a company contains individuals that are 5.0 size units and the growth rate is 2.0 size units, their size during month $N+1$ is 7.0 units. Because the growth rate is linear, two constants, c_1 and c_2, describe a straight line representing the growth rate. Constant c_1 denotes the intercept, and c_2 denotes the slope of the line. If the growth rate does not vary with time, c_2 is zero and c_1 defines the growth rate. An unvarying growth rate is unrealistic, whereas a growth rate that decreases progressively with age would apply to many organisms.

Mortality rate is also assumed to be linearly related to age and is calculated as the number of deaths per hundred individuals (Figure 10-13). Thus, if a company contains 200 individuals and the death rate is 6, then 12 individuals in that company will die during that month. Constants c_3 and c_4 describe intercept and slope, respectively,

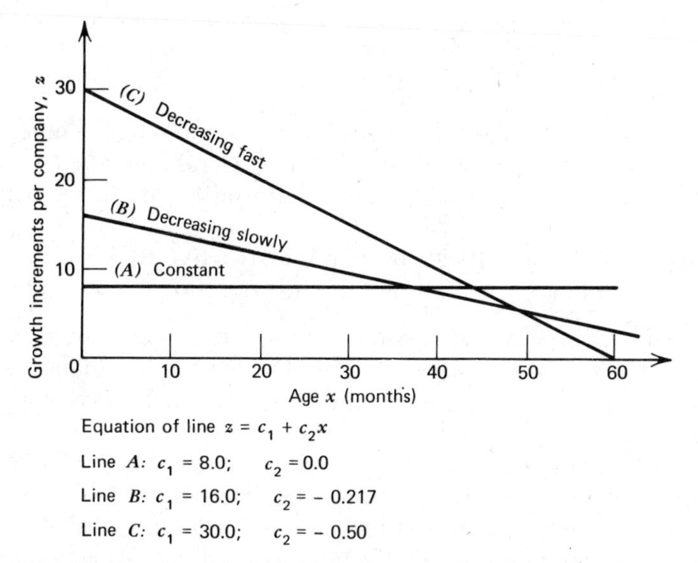

Figure 10-12 Various linear relationships for growth rate expressed as size increments per company according to age in months. Constants c_1 and c_2 define intercept and slope of each line and are input to simulation program.

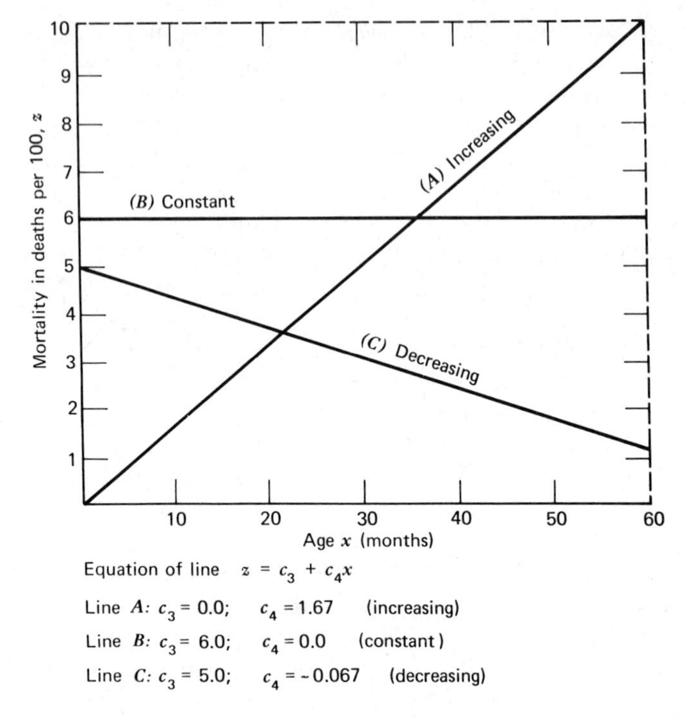

Figure 10-13 Various linear relationships for mortality rate expressed as deaths per hundred individuals according to age in months. Constants c_3 and c_4 define intercept and slope of each line and are input to program.

for any linear relationship between death rate and age. Figure 10-13 illustrates three such relationships: mortality rate constant with age, mortality rate decreasing with age, and mortality rate increasing with age.

Returning to Figure 10-11, it is now instructive to work through 4 months of simulated time to illustrate the operation of the model.

Month 1 Company 1 is created, age 0, with 8000 individuals and size 1.0 unit. The size-frequency table for dead individuals is empty, all 8000 individuals being in the smallest size class for living individuals.

Month 2 Company 2 is newly created. Surviving individuals in company 1 grow, and the remainder die. The amount of growth and the number of surviving individuals are recorded for company 2. The population account for both living and dead individuals is updated.

Month 3 Company 3 is created. Individuals in company 2 grow in physical size, and their numbers are reduced by death. Company 1, now aged 2 months, is likewise further modified by growth in physical size and by death.

Month 4 A new company is created, and the previous companies are modified.

At the end of 60 months the living and the dead individuals in each company are classified according to size and summed to give two size-frequency arrays. By this time the shape of the size-frequency curve is assumed to be more or less stable.

Stochastic Components

So far no mention has been made of stochastic elements in the model. Craig and Oertel's original model contains no random components, although variability in growth behavior is represented by means of a coefficient of variation. In the present model, however, the coefficients describing the linear relationship with age for both the growth and the mortality rates can be treated as normally distributed random variables.

At each simulated month the mean growth rate, labeled GROMN in the program, is calculated for each company by using the input values for intercept c_1 and slope c_2 of the linear relationship assumed between growth and age. The FORTRAN statement for this operation reads

```
GROMN=C1+C2*AGE
```

where AGE is the age of the company in months. The actual growth rate employed is then obtained by generating a normally distributed number, GRORAT, whose mean is GROMN as calculated above and whose standard deviation is GRODEV, assigned as input to the program. Similarly the mean death rate, DEDMN, is found by using the FORTRAN statement

```
DEDMN=C3+C4*AGE
```

where C3 and C4 are again the intercept and slope, respectively, for the linear death–age relationship. The actual death rate is then calculated by generating a normally distributed random number, DEDRAT, whose mean is DEDMN, calculated above, and standard deviation, DEDEV, assigned as input.

Computer Program

The FORTRAN program representing the fossil-population model is listed in Table 10-2, and a generalized flow chart describing it is illustrated in Figure 10-14. The program employs subroutine GAUSS

TABLE 10-2 FORTRAN Program for Fossil-Population Model

```
C.....PROGRAM TO GENERATE SIZE-FREQUENCY HISTOGRAMS FOR LIVE AND DEAD
C.....POPULATIONS OF ORGANISMS
C.....INSPIRED BY CRAIG AND OERTEL'S 1966 MODEL
C.....  CMNUM(N)      NO. OF INDIVIDUALS IN COMPANY RECRUITED IN N-TH MN
C.....  CMSIZ(N)      SIZE OF INDIVIDUALS IN SAME COMPANY
C.....  NMONTH       TOTAL OF MONTHS SIMULATED
C.....  DEDACC(K)     FREQ. OF DEAD INDIVIDUALS IN K-TH SIZE CLASS
C.....  LIVACC(K)     FREQ. OF LIVE INDIVIDUALS IN K-TH SIZE CLASS
C.....  NCLASS        NO. OF SIZE CLASSES IN SIZE FRQ. ACCOUNT
C.....  GROMAX(N)     MAX. GROWTH INCREASE (IN SIZE UNITS) FOR
C.....                  N-TH MONTH OLD INDIVIDUALS
C.....  C1, C2        CONSTANTS FOR LINEAR GROWTH VERSUS AGE RELATION
C.....  GRODEV        ST. DEV OF GROWTH RATE
C.....  C3,C4         CONSTANTS FOR LINEAR MORTALITY VS. AGE RELATION
C.....  DEDEV         ST. DEVIATION OF MORTALITY RATE
    1 FORMAT(20A4)
    2 FORMAT(3I10, 6F5.2)
    3 FORMAT(6F10.0)
    4 FORMAT(1H1, 'SIMULATION OF FOSSIL POPULATIONS'/1X,20A4)
    5 FORMAT(1H0,10X,2(14X, 'NUMBER')/ 8X,' SIZE',12X, 'LIVING',14X,
     1  'DEAD'/)
    6 FORMAT(1H , I10, 2I20)
      DIMENSION CMNUM(100),CMSIZ(100), DEDACC(20), TITLE(20), RECRUT(12)
      REAL LIVACC(20)
C.....READ INPUT PARAMETERS
  999 READ(5,1) TITLE
      WRITE(6,4) TITLE
      READ(5,2) NMONTH, NCLASS, IX, C1, C2, GRODEV, C3,C4,DEDEV
      READ(5,3) (RECRUT(M), M=1,12)
C.....SET ACCOUNTING ARRAYS TO ZERO
      DO 10 K=1,NCLASS
      LIVACC(K)=0.0
   10 DEDACC(K)=0.0
C.....BEGIN MONTHLY CYCLE
      DO 40 N=1,NMONTH
C.....ASSIGN  NO. OF INDIVIDUALS IN NEW COMPANY
C.....SIZE OF NEW INDIVIDUALS FIXED AT ONE UNIT
      M=MOD(N,12)
      IF (M.EQ.0) M=12
      CMNUM(N)=RECRUT(M)
      CMSIZ(N)=1.0
C.....ADD GROWTH INCREMENTS TO EACH COMPANY
      DO 30 NN=1,N
      AGE=N-NN
      GROMN=C1+C2*AGE
      CALL GAUSS(IX,GROMN,GRODEV,GRORAT)
      IF (GRORAT.LT.0.0) GRORAT=0.0
   30 CMSIZ(NN)=CMSIZ(NN)+GRORAT
C.....CALCULATE NO. OF DEATHS IN EACH COMPANY AND ASSIGN SKELETONS TO
C.....DEATH SIZE-FREQUENCY ACCOUNT
      DO 40 NN=1,N
      AGE=N-NN
      DEDMN=C3+C4*AGE
      CALL GAUSS(IX,DEDMN,DEDEV,DEDRAT)
      IF (DEDRAT.LT.0.0) DEDRAT=0.0
      DEDNUM=CMNUM(NN)*DEDRAT/100.0
      CMNUM(NN)=CMNUM(NN)-DEDNUM
C.....WHICH SIZE CLASS FOR SKELETONS?
      INDEX=CMSIZ(NN)/25.0+1.0
      IF (INDEX.GT.20) GO TO 40
      DEDACC(INDEX)=DEDACC(INDEX)+DEDNUM
   40 CONTINUE
C.....MAKE CENSUS ON LIVING INDIVIDUALS AND FILL LIVACC
      DO 50 N=1,NMONTH
      INDEX=CMSIZ(N)/25.0+1.0
      IF (INDEX.GT.20) GO TO 50
```

TABLE 10-2 *(contd.)*

```
      LIVACC(INDEX)=LIVACC(INDEX)+CMNUM(N)
   50 CONTINUE
C.....PRINT OUT SIZE FREQUENCY TABLES
      WRITE(6,5)
      DO 60 K=1,NCLASS
      IL=LIVACC(K)
      ID=DEDACC(K)
   60 WRITE(6,6) K, IL, ID
      WRITE(6,4) TITLE
      GO TO 999
      END
      SUBROUTINE GAUSS(IX,RMEAN,STDEV,RG)
C.....GENERATE RANDOM VARIABLE RG, MEAN=RMEAN, ST.DEV=STDEV
      SUM=0.0
      DO 10 I=1,12
      CALL RANDU(IX,IY,R)
      IX=IY
   10 SUM=SUM+R
      RG=(SUM-6.0)*STDEV+RMEAN
      RETURN
      END
```

to generate normally distributed random numbers and uses subroutine RANDU (Table 3-4) for generating uniformly distributed random numbers. Output from the program consists of tables showing the frequency of living and dead individuals in each size class. These tables can be effectively displayed as histograms.

Simulation Experiments with Population Model

Five computer experiments are illustrated, although the number of possible experiments, employing permutations of these exogenous variables, each capable of wide variation, is virtually infinite. The following results therefore provide only a small sample of those that are possible. Modifications of the model would provide still more flexibility.

Run 1 In this run birthrate is held constant at 8000 individuals per month, growth rate is zero, and mortality rate is zero. Figure 10-15*a* illustrates the result: all individuals accumulate in the lowest size class, none of them growing and none dying.

Run 2 Birth and growth rates are held constant, but zero mortality permits individuals to progress upward from the smallest size class. Since none dies, the curve is not curtailed to the right (Figure 10-15*b*).

Run 3 Constant birthrate, growth rate decreasing with age, and zero mortality produce the results shown in Figure 10-15*c*. Notice that no individual reaches a very large size and that the frequency of the older individuals is large.

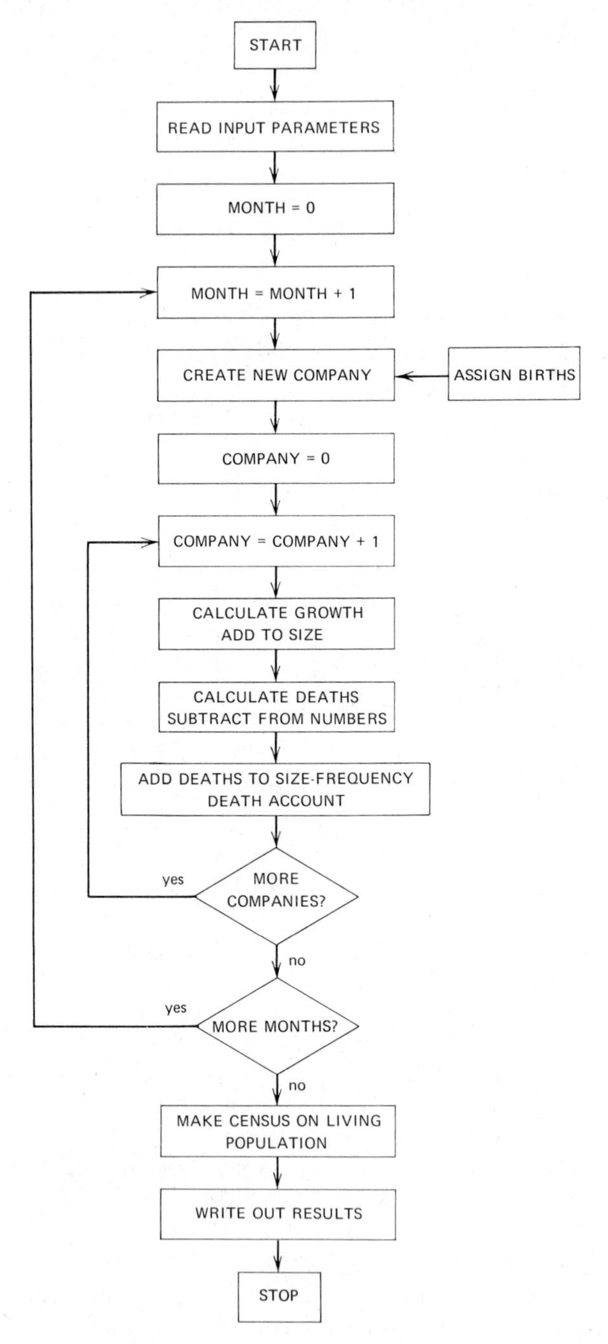

Figure 10-14 Flow chart of fossil-population simulation model.

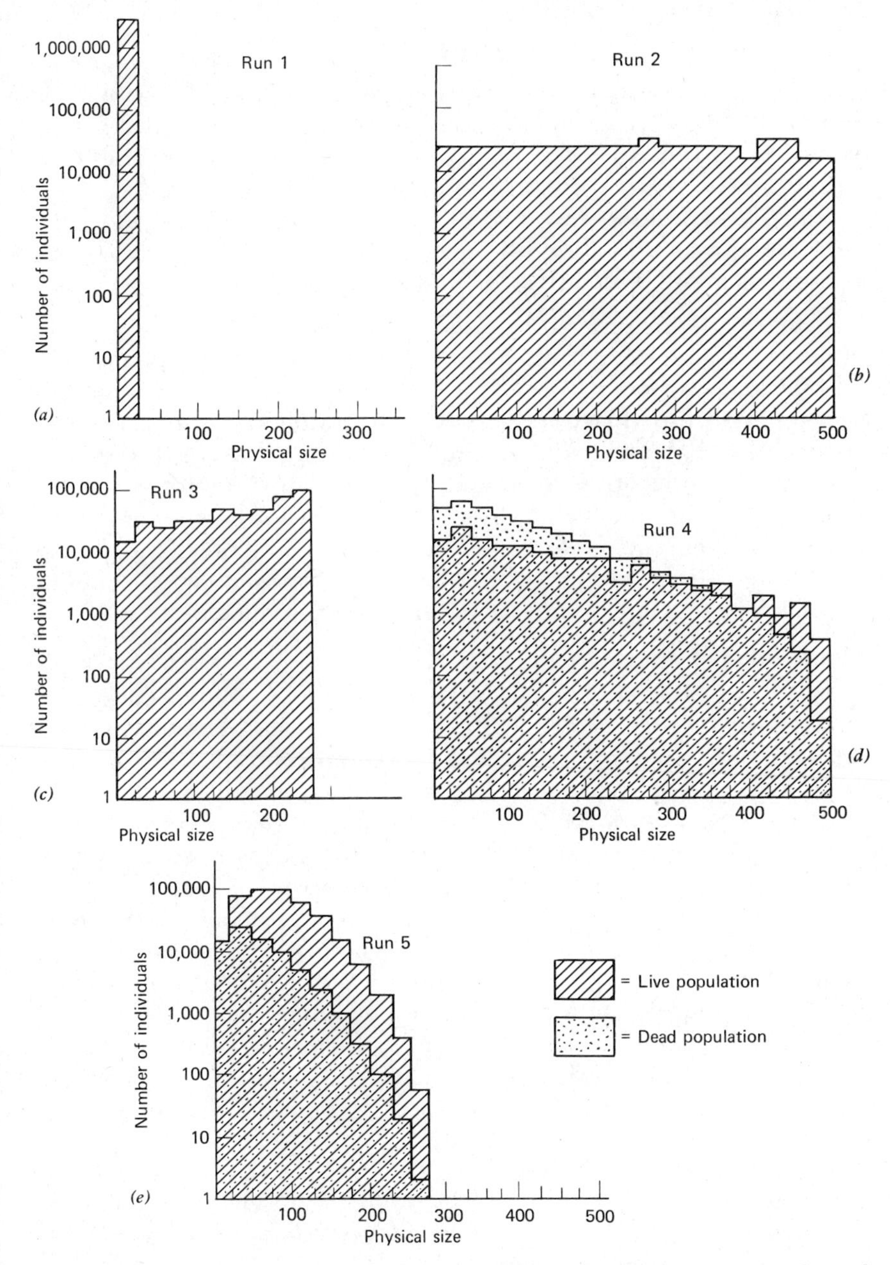

Figure 10-15 Histograms of frequency distributions of population versus physical size produced in experimental runs 1 to 5 (labeled *a* to *e*) of fossil-population model. Dead population is shaded, and live population is unshaded. Physical size is represented by arbitrary units. Note logarithmic scale for number of individuals in populations.

Run 4 Birthrate is constant with time, and growth and mortality rates are both constant with age (Figure 10-15*d*). The resulting life and death curves are similar in shape. With these particular parameter settings the curves cross over one another, the mortality rate permitting the survival of a large number of fully grown adults.

Run 5 Birth and growth rates are held constant, but mortality rate increases markedly with age (Figure 10-15*e*). As a result the right-hand tails of both distributions are truncated.

Readers are encouraged to make their own experiments with this program. Craig and Oertel (1966) illustrate the results of more than 40 experiments, and many more are possible. It would be interesting to take an observed size-frequency distribution and to optimize the fit between observed and simulated distributions, using the sequential-search optimization technique discussed in Chapter 8. Such a problem would involve at least three decision variables (birth, growth, and mortality rates), and the objective function could be the degree of fit. It would probably show that the same curves could be produced by making a variety of different assumptions.

Simulation of Fossil Forage Trails

Raup and Seilacher (1969) have developed a digital simulation model for studying the feeding patterns of ancient bottom-dwelling organisms. The model simulates the formation of trails by generating a sequence of points in an X–Y plane that are linked by means of a digital plotter (Figure 10-16).

Bottom-feeding organisms are faced with the optimization problem of obtaining maximum coverage of a given area of feeding ground without crossing existing tracks, in as compact a manner as possible. These requirements have been met in nature by a number of trail patterns, with meander systems being used most commonly. Three basic factors appear to apply to the development of meander trails: strophotaxis, which causes the animal to turn at 180 degrees at intervals; phobotaxis, which keeps it from crossing other tracks, including its own; and thigmotaxis, which causes it to keep close contact with former tracks.

The computer model is governed by a number of rules that control the movement of the "animal." The animal can move straight ahead, turn toward or away from a preexisting track, or make a full 180-degree turn. The choice of movement direction is controlled stochastically, employing a random-number generator. The principal parameters of the model are (a) turning radius for 180-degree turns, (b) mean distance between tracks, (c) relative intensities of thigmo-

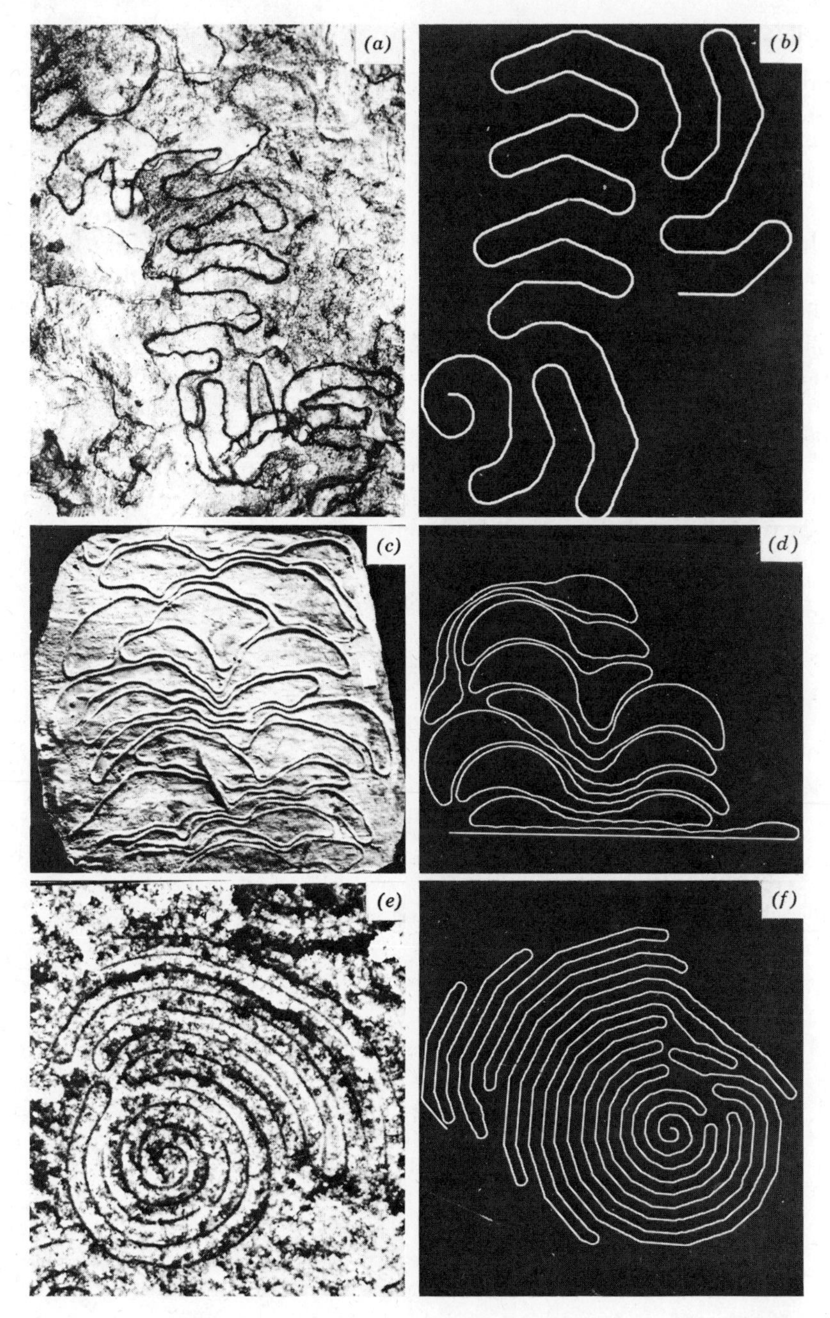

Figure 10-16 Fossil trails produced by ancient bottom-feeding organisms (left) are matched by simulated trails (right) drawn by digital plotter. From Raup and Seilacher (1969). Copyright by the American Association for the Advancement of Science.

taxis and phobotaxis, (d) mean length of meanders not terminated by obstructions, and (e) variability in the length of meanders.

Figure 10-16*b*, *d*, and *f* shows examples of output produced by the program, whereas the trails shown in Figure 10-16*a*, *c*, and *e* were produced by real animals. Examples shown in *a* and *b* illustrate the effect of rather weak thigmotaxis, giving rise to a sprawling feeding pattern. The cases shown in *c* and *d* illustrate the effect of somewhat stronger thigmotaxis, so that after making a 180-degree turn, the animal loses close contact with the earlier track for a period and then regains it again. The patterns shown in *e* and *f* exhibit strong thigmotaxis, giving rise to a compact and symmetrical spiral meander pattern.

The experiments show that many diverse foraging patterns can be attributable to one simple behavioral model. Besides clarifying the behavioral patterns of fossil animals, this model may be useful for aiding classification of fossil trails.

Annotated Bibliography

Bailey, N. T. J., 1964, *Elements of Stochastic Processes with Applications to the Natural Sciences*, John Wiley and Sons, New York, 249 pp.

Well-written, intermediate-level text dealing with stochastic processes. Applications include birth and death processes, epidemics, and diffusion processes.

Bartlett, M. S., 1957, "On Theoretical Models for Competitive and Predatory Biological Systems," *Biometrika*, v. 44, pp. 27–42.

A readable summary of simple mathematical models for interaction of biological populations.

Baum, R, F., 1967, "The Growth of a Population in a Restricted Environment," in *Some Mathematical Models in Biology*, H. M. Thrall, ed., University of Michigan, Report 4042-R-7, p. 4; "Interactions between Two Species," p. 7; "The Intensity of Reproduction as a Factor in the Growth of a Species," p. 5; "The Relationship between Host and Parasite," p. 6.

Growth of populations can be described by differential equations to simulate a variety of interactions, including (a) two species that compete for the same food, (b) prey–predator relationships, (c) host–parasite relationships, and (d) reproduction rate.

Craig, G. Y., and Oertel, G., 1966, "Deterministic Models of Living and Fossil Populations of Animals," *Quarterly Journal of the Geological Society of London*, v. 122, pp. 315–355.

Landmark paper dealing with the computer simulation of animal populations. Forty-two experimental simulation runs are illustrated and discussed.

Davidson, R. S., and Clymer, A. B., 1965, "The Desirability and Applicability of Simulating Ecosystems," in "Symposium on Advances in Biomedical Computer Applications," *Annals of the New York Academy of Sciences*, v. 128, pp. 790–794.

Simulation of phytoplankton population fluctuations.

Fox, W. T., 1968, "Simulation Models of Time-Trend Curves for Paleoecologic Interpretation," Colloquium on Time Series Analysis, Computer Contribution 18, Kansas Geological Survey, pp. 18–29.

Pioneer paper describing simulation of paleoecologic changes in a stratigraphic time sequence.

Garfinkel, D., MacArthur, R. H., and Sack, R., 1964, "Computer Simulation and Analysis of Simple Ecological Systems," in "Conference on Computers in Medicine and Biology," *Annals of the New York Academy of Sciences*, v. 115, pp. 943–951.

Simulation of interactive populations, including rabbit–grass model and fox–rabbit–grass model. Stochastic models, with Markov-chain representation, are touched on.

Garfinkel, D., and Sack, R., 1964, "Digital Computer Simulation of an Ecologic Interpretation," Colloquium on Time Series Analysis, Computer pp. 502–507.

Simulation model deals with six interacting species.

Gould, E. M., Jr., and O'Regan, W. G., 1965, "Simulation, a Step Toward Better Forest Planning," *Harvard Forest Papers*, No. 13, 36 pp.

Detailed description of computer programs for the simulation of economic factors in forest management.

Hagerstrand, T., 1965, "A Monte Carlo Approach to Diffusion," in *Spatial Analysis, a Reader in Statistical Geography*: B. J. Berry and D. F. Marble, eds., Prentice-Hall, Englewood Cliffs, N.J., pp. 368–384.

Describes stochastic simulation of diffusion processes. Although the applications described pertain to economic factors and practices, the numerical techniques can be applied in ecology and paleoecology.

Harary, F., 1961, "Who Eats Whom?," *General Systems*, v. 6, pp. 41–44.

Readable, brief paper describing an "eating matrix," which is a Boolean matrix describing interrelationships between animals that eat each other and animals that eat plants.

Harbaugh, J. W., 1966, "Mathematical Simulation of Marine Sedimentation with IBM 7090/7094 Computers," Computer Contribution 1, Kansas Geological Survey, 52 pp.

Description of computer programs for simulating some marine-sedimentation processes, with emphasis on interactions between carbonate-secreting marine-organism communities. Computer program is in ALGOL-58.

Harbaugh, J. W., and Wahlstedt, W. J., 1967, "FORTRAN IV Program for Mathematical Simulation of Marine Sedimentation with IBM 7040/7094 Computers," Computer Contribution 9, Kansas Geological Survey, 40 pp.

Describes FORTRAN IV program which parallels that described by Harbaugh (1966).

Holling, C. S., 1966, "The Strategy of Building Models of Complex Ecological Systems," pp. 195–214 in *Systems Analysis in Ecology*, K. E. F. Watt, ed., Academic Press, New York, 276, pp.

Readable discussion of philosophy and objectives in simulating population dynamics, illustrated with several prey–predator relationships.

Howard, R. A., Gould, E. M., and O'Regan, W. G., 1966, "Simulation for Forest Planning," *Simulation*, v. 7, pp. 44–52.

Readable, generalized description of the Harvard Forest simulation model for economic analysis of forest planning.

Kemeny, J. G., and Kurz, T. E., 1967, *Basic Programming*, John Wiley and Sons, New York, 122 pp.

Develops a simple program for solving prey–predator equations numerically in Chapter 15.

Kemeny, J. G., and Snell, J. L., 1962, *Mathematical Models in the Social Sciences*, Ginn and Company, New York, 144 pp.

Develops the differential equations of simple prey–predator model in Chapter 2.

Levins, R., 1966, "The Strategy of Model Building in Population Biology," *American Scientist*, v. 54, pp. 421–431.

Readable description of population simulation, including use of optimization methods.

Lotka, A. J., 1924, *Elements of Mathematical Biology*, reprinted in 1956 by Dover Publications, New York, 465 pp.

Classic, pioneer work dealing with the mathematics of population dynamics.

Matern, B., "Spatial Variation, Stochastic Models, and Their Application to Some Problems in Forest Surveys and Other Sampling Investigations," *Meddelanden Fran Statens Skogsorskningsinstitut*, v. 49, No. 5, 144 pp.

Thorough, advanced treatise on analytical stochastic models applied to problems in forest surveys.

Meier, R. L., Blakelock, E. H., and Hinomoto, H., 1964, "Computers in Behavioral Science: Simulation of Ecological Relationships," *Behavioral Science*, v. 9, pp. 67–76.

Representation of ecological interactions as a mathematical game.

Odum, E. P., 1960, "Ecological Potential and Analogue Circuits for the Ecosystem," *American Scientist*, v. 48, pp. 1–8.

Short paper describing ecological system in terms of electrical analogs.

Odum, E. P., and Odum, H. T., 1959, *Fundamentals of Ecology*, W. B. Saunders Company, Philadelphia, 546 pp.

Well-known textbook. Chapters 2 and 3 provide a readable introduction to ideas of ecological systems, emphasizing chemical and energy cycles.

Olson, J. S., and Uppuluri, V. R. R., 1967, "Ecosystem Maintenance and Transformation Models as Markov Processes with Absorbing Barriers," in *Some Mathematical Models in Biology*, H. M. Thrall *et al.*, eds., University of Michigan, Report 4042–R–7, 5 pp.

Describes the use of Markov transition-probability matrices to regulate transfers of energy or nutrients in an ecological system.

Patten, B. C., 1961, "Competitive Exclusion," *Science*, v. 134, pp. 1599–1601.

Application of Markov chains and elementary feedback-system concepts to interactions between populations of closely allied species.

Raup, D. M., and Seilacher, A., 1969, "Computer Simulation of Fossil Foraging Behavior," *Science*, v. 166, pp. 994–5.

Reyment, R. A., 1968, "Systems Analysis in Paleoecology," *Geologiska Foreningens i Stockholm Forhandlingar*, v. 89, pp. 440–447.

Deals with the representation of interactions between fossil populations, employing equations in finite-difference form.

Slater, H., 1967, "Competition among Species," in *Some Mathematical Models in Biology*, H. M. Thrall *et al.*, eds., University of Michigan, Report 4042–R–7, 4 pp.

Growth equations for populations of two competing species can be obtained from the solution of simultaneous differential equations.

Tasch, P., 1965, "Communications Theory and the Fossil Record of Invertebrates," *Transactions of the Kansas Academy of Science*, v. 68, pp. 322–329.

Application of elementary feedback-system concepts to fossil populations.

Teal, J. M., 1962, "Energy Flow in the Salt Marsh Ecosystem of Georgia," *Ecology*, v. 43, pp. 614–624.

Quantitative treatment of energy flow in ecological systems.

Valentine, J. W., 1967, "The Influence of Climatic Fluctuations on Species Diversity within the Tethyan Provincial System," *Aspects of Tethyan*

Biogeography, Systematics Association Publication No. 7, C. G. Adams and D. V. Ager, eds., pp. 153–166.

Develops theoretical models for explaining species diversity, employing different temperature gradients between equator and poles and assuming tolerance limits of theoretical organisms. Although this model has not been developed in mathematical form, it has the flavor of a simulation model and could certainly be elaborated considerably with a computer.

Watt, K. E. F., 1959, "A Mathematical Model for the Effect of Densities of Attacked and Attacking Species on the Number Attacked," *The Canadian Entomologist*, v. 91, pp. 129–144.

Employs sets of differential equations to represent interactions in host–parasite populations and prey–predator populations.

Watt, K. E. F., 1964*a*, "Computers and the Evaluation of Resource Management Strategies," *American Scientist*, v. 52, pp. 408–418.

Well-written discussion of philosophy and objectives of simulation of ecological systems with dynamic computer models.

Watt, K. E. F., 1964*b*, "The Use of Mathematics and Computers To Determine Optimal Strategy and Tactics for a Given Insect Pest Control Problem," *The Canadian Entomologist*, v. 96, pp. 202–220.

Description of principles and details of computer program for simulation of alternative forest-insect pest-control strategies.

Watt, K. E. F., 1966*a*, "The Nature of Systems Analysis," in *Systems Analysis in Ecology*, K. E. F. Watt, ed., Academic Press, New York, pp. 1–14.

General discussion of system analysis and computer simulation applied to ecology.

Watt, K. E. F., 1966*b*, "Ecology in the Future," in *Systems Analysis in Ecology*, K. E. F. Watt, ed., Academic Press, New York, pp. 253–267.

Describes simulation model for fishery industry.

Yuill, R. S. 1964, *A Simulation Study of Barrier Effects in Spatial Diffusion Problems*, Northwestern University Geography Department Technical Report No. 1, Geography Branch, Office of Naval Research, Contract NONR 1228(33), 47 pp.

Provides details of a probabilistic diffusion model that could be applied in ecological simulation.

EVOLUTION AND MORPHOLOGY

The simulation of organic evolution should be of strong interest to geologists, considering the highly interdependent relationships among evolution, paleontology, and stratigraphic geology. A number of geneticists have experimented with numerical simulation models involving interbreeding populations and gene pools. These models are of varying complexity and have been used to explore the results of varying natural selection pressure, population size, genetic drift, and effects of inbreeding. The role of these methods in dealing with problems in evolution from the paleontologist's point of view is an open question. Papers by Crosby (1963), Fraser (1962), Kojima (1965), Pimentel (1968), Wright (1964), and Young (1966) provide a sample of relevant papers.

Simulation of Shell Morphology

Simulation has been employed by Raup (1962, 1965, 1966) in exploring the morphological relationships in organisms of relative geometrical simplicity—namely, in the shapes of coiled shells. Raup's work is of special interest to paleontologists and geologists, considering the role that molluscs and other shelled organisms play as stratigraphic index fossils. Raup has pointed out that coiled shells that occur in a variety of taxonomic groups have important geometric characteristics in common, and this makes rigorous comparisons between them possible. In turn these geometrical relationships bear strongly on interpretation of their evolutionary lineage. There is general agreement that the bulk of coiled-shell forms in nature are variations of a rather simple model. Although there are some exceptions, particularly in the early ammonoids, a logarithmic spiral is generally regarded as a satisfactory basic model and is employed by Raup. Assuming a logarithmic form, four basic parameters describe the coiled shell:

1. Shape of the generating curve S
2. Whorl expansion rate W
3. Position of the generating curve in relation to the shell axis, D
4. The rate of whorl translation T

The coiled shell may be considered in a system of cylindrical coordinates (Figure 10-17). The y-axis coincides with the axis of coiling, θ pertains to revolution of the generating curve about the axis, and r is the distance of any point from the axis. If r_0 is the initial distance of a point from the axis, then r_θ, the distance after θ revolutions, may be expressed as

$$r_\theta = r_0 W^{\theta/2\pi} \qquad (10\text{-}13)$$

where W is the whorl expansion.

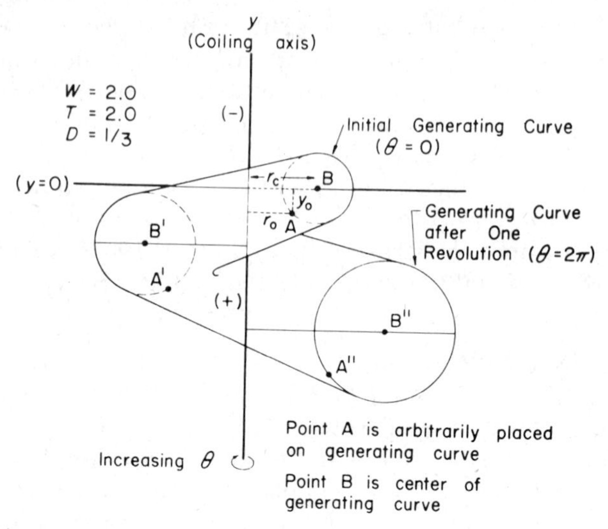

Point	Coordinate		
	θ	r	y
A	0	r_0	y_0
A'	π	$r_0 W^{1/2}$	$y_0 W^{1/2} + r_c T(W^{1/2}-1)$
A''	2π	$r_0 W$	$y_0 W + r_c T(W-1)$
B	0	r_c	0
B'	π	$r_c W^{1/2}$	$r_c T(W^{1/2}-1)$
B''	2π	$r_c W$	$r_c T(W-1)$

Figure 10-17 Diagram to illustrate method of generating hypothetical shell form by employing cylindrical coordinates. Reproduced from Raup (1966) with permission from the *Journal of Paleontology.*

In shells in which the coil lies in a plane (planispiral), or, in other words, in which the rate of whorl translation T is zero, the y value of a particular point is given by the expression

$$y_\theta = y_0 W^{\theta/2\pi} \tag{10-14}$$

Finally, in shells whose coils form a helix (helicoid) the generating curve moves along the y-axis, and the equation is

$$y_\theta = y_0 W^{\theta/2\pi} + r_c T(W^{\theta/2\pi} - 1) \tag{10-15}$$

where T = rate of whorl translation defined as dy/dr with respect to the center of the generating curve
r_c = r value of center of initial generating curve

It is to be emphasized that the four parameters S, W, D, and T do not completely describe the morphology of a shell. Such morphological factors as surface ornamentation, growth rings, internal structure, and shell thickness are not represented.

Given the four parameters, it is possible to visualize the total spectrum of geometrically possible shell forms. The spectrum can be regarded in the form of four-dimensional space, each of the parameters representing one dimension.

Some simulated examples are shown in Figure 10-18, which contains three series of hypothetical shell forms, each starting from the same form at the center of the figure. The initial form was chosen arbitrarily; although it does not represent any specific taxonomic group, it possesses features common to several groups. By adjusting

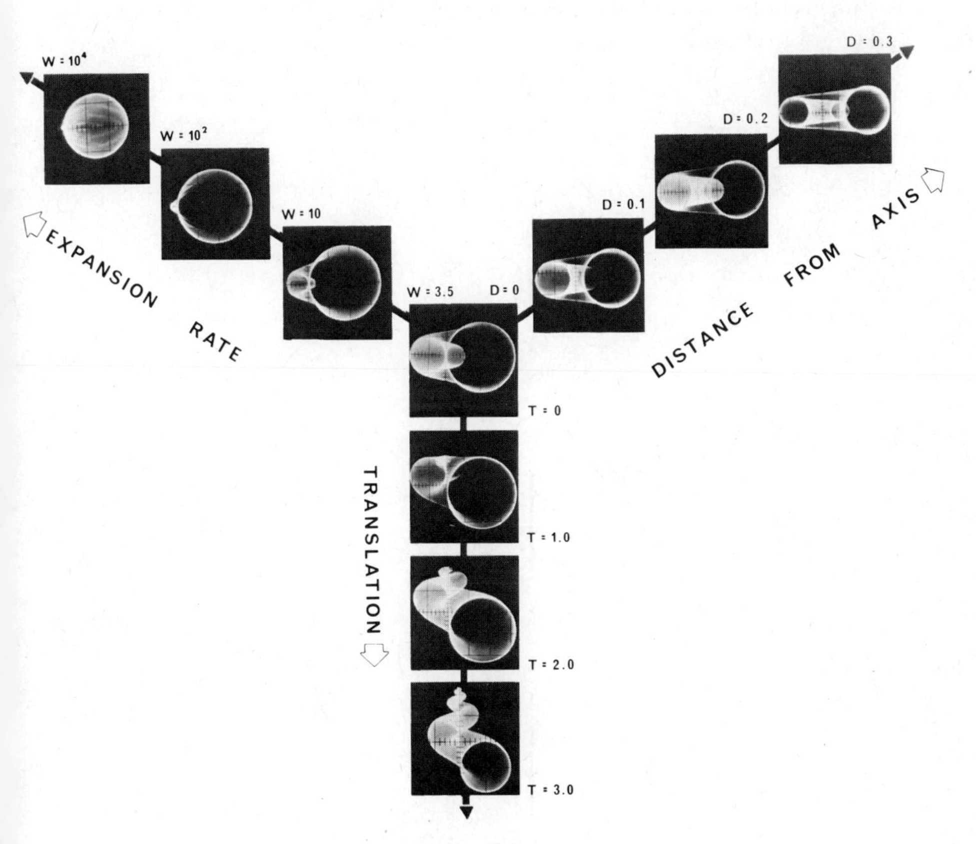

Figure 10-18 Three series of hypothetical shell forms produced by modifying arbitrary initial form in center of figure. Modified forms extending from this form have been produced by increasing the rate of translation T, increasing the expansion rate W, and increasing the distance from coiling axis, D. Reproduced from Raup (1966) with permission from the *Journal of Paleontology*.

the parameters and evaluating the generating equations, the other forms numerically "evolve" from this form. For example, an increase in the translation rate from $T = 0$ to $T = 3$ produces high-spired coiling types represented by some common gastropods. An increase in the expansion rate ($W = 3.5$ to 10^4 in the examples), on the other hand, changes the shell form from the univalve region of the spectrum to the bivalve region. Finally, increase in the distance from the coiling axis to the generating curve ($D = 0$ to $D = 0.3$) causes the shell to become more evolute and umbilicate. Effects of changing the fourth parameter, the shape of the generating curve S, are shown in Figure 10-19. Almost any configuration of the generating curve is possible. In the hypothetical shell shown in the middle of Figure 10-19 a circular generating curve has been employed, whereas the upper and

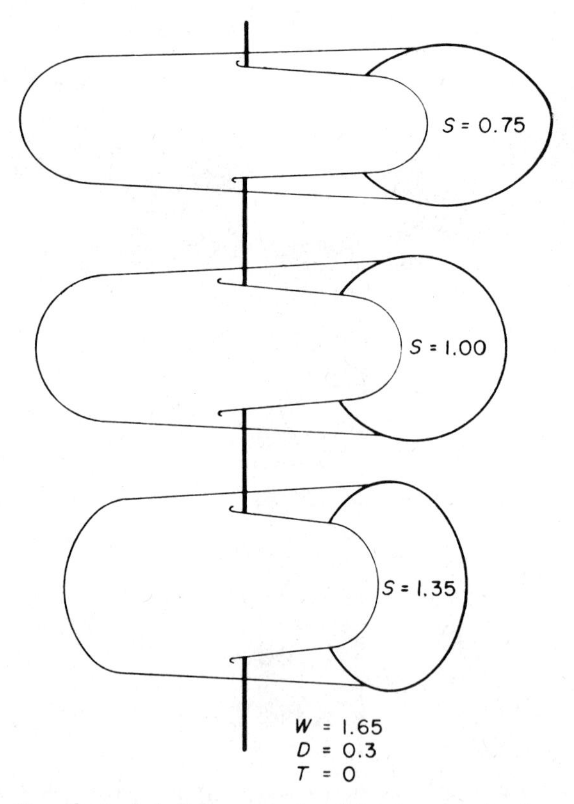

Figure 10-19 Hypothetical shell forms produced by changing the shape of the generating curve, S, while holding other parameters constant. In this example S is defined as the ratio between ellipse axes parallel to and perpendicular to coiling axis. Reproduced from Raup (1966), with permission from the *Journal of Paleontology*.

lower forms employ differently oriented ellipses. In the examples in this figure S is defined as the ratio between axes of the generating curve parallel and normal to the coiling axis.

Figure 10-20 portrays a three-dimensional continuum of theoretical forms in which parameters W, D, and T are free to vary, and the generating curve S is held constant as a simple circle. Comparable block diagrams could be constructed for other shapes of the generating curve. The parts of the three-dimensional continuum that are actually occupied are labeled. It should be noted that most of the species in the four taxonomic groups (gastropods, pelecypods, brachiopods, and ammonoids) are confined to nonoverlapping regions, which, if lumped together, occupy only a small part of the block's

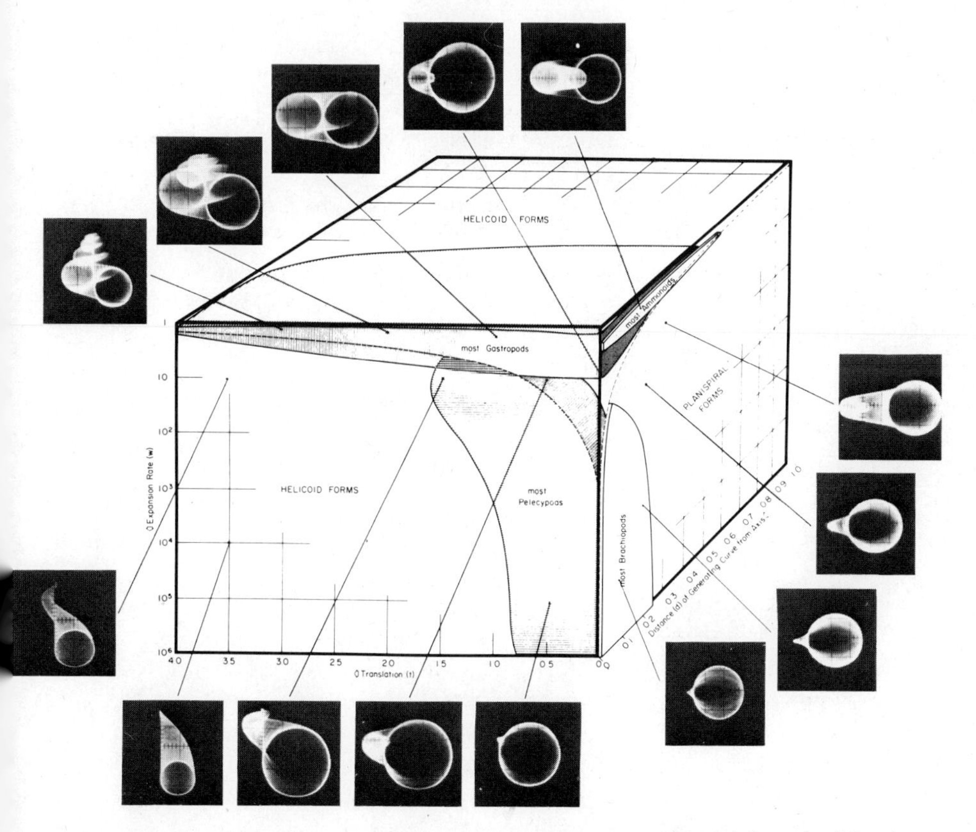

Figure 10-20 Three-dimensional continuum representing three (of total of four) shell parameters W, D, and T. Examples of shell that have been generated assume a circular generating curve. Reproduced from Raup (1966), with permission from the *Journal of Paleontology*.

total volume. The question arises as to whether the relatively unused regions represent physiologically impossible shell forms or whether evolution has simply failed to populate the entire block.

Annotated Bibliography

Crosby, J. L., 1963, "Evolution by Computer," *New Scientist*, No. 327, pp. 415–417.

A readable article, written for the layman, describing the fundamental details of a computer method for simulating evolution within an interbreeding population. A simple model involves selection of "parents" at random, picks a gene from a pair of genes (representing the particular gene under consideration) from each parent, and combines them in the offspring.

Fraser, A. S., 1962, "Simulation of Genetic Systems," *Journal of Theoretical Biology*, v. 2, pp. 329–346.

A detailed exposition of genetic simulation, touching on natural selection and genetic drift.

Kojima, K., 1965, "The Evolutionary Dynamics of Two-Gene Systems," in *Computers in Biomedical Research*, R. W. Stacy and B. D. Waxman, eds., Academic Press, New York, pp. 197–220.

A detailed, mathematical treatment of methods for simulating evolution.

Levins, R., 1961, "Mendelian Species as Adaptive Systems," *General Systems*, v. 6, pp. 33–39.

A discussion of natural selection with emphasis on the role of optimization.

Pimentel, D., 1968, "Population Regulation and Genetic Feedback," *Science*, v. 159, pp. 1432–1437.

Well-written exposition of role of genetic feedback in the control of numbers of a species.

Raup, D. M., 1962, "Computer as Aid in Describing Form in Gastropod Shells," *Science*, v. 138, pp. 150–152.

The first of a series of classic papers on geometrical analysis of coiled shells.

Raup, D. M., 1966, "Geometric Analysis of Shell Coiling: General Problems," *Journal of Pateontology*, v. 40, p. 1178–1190.

Extends algebraic and geometrical treatment of simulated shell forms and summarizes much of the author's earlier work in a lucid and detailed presentation.

Raup, D. M., and Michelson, A., 1965, "Theoretical Morphology of the Coiled Shell," *Science*, v. 147, pp. 1294–1295.

Describes the mathematical exploration of coiled-shell forms. Both digital and analog computers were employed. An analog computer coupled with an oscilloscope provides for graphic display of striking realism.

Shapiro, M. B., 1967, "An Algorithm for Reconstructing Protein and RNA Sequences," *Journal of the Association for Computing Machinery*, v. 14, pp. 720–731.

A well-written paper describing a computer program for simulating the construction of protein and ribonucleic acids. The paper is of interest to geologists because it provides an example of an approach that might be used in simulating molecular aggregates in the origin of life.

Wright, S., 1964, "Stochastic Processes in Evolution," in *Stochastic Models in Medicine and Biology*, J. Gurland, ed., University of Wisconsin Press, Madison, pp. 199–241.

A well-written exposition of the mathematics involved in the stochastic components of evolution.

Young, S. S. Y., 1966, "Computer Simulation of Directional Selection in Large Populations I. The Program, the Additive and the Dominance Models," *Genetics*, v. 53, pp. 189–205.

Presents results of evolution-simulation experiments.

GEOCHEMISTRY AND PETROLOGY

Application of computer-simulation models to geochemical processes is in relative infancy, although there are many potential opportunities. The geochemical components of many geologic processes—such as weathering, sedimentation, and ore deposition—can be represented as materials-balance models. Notable works in this connection are the iterative model of Horn and Adams (1966) dealing with geochemical balance and element abundances in the earth's crust, which is described in some detail below, and the gross geochemical crustal model of Shimazu and Urabe (1968). Another is the ore-mineral deposition model of Anttonen (1969), which is also discussed below.

Chemical and geochemical processes on a smaller scale also can be represented by simulation models—for example, various chemical equilibria (DeLand, 1967) and transformation of fatty acids in sediments to paraffin hydrocarbons (Kvenvolden and Weiser, 1967). Finally, the processes involved in geochemical sampling of rock bodies can be approached via simulation models (Eicher and Miesch, 1965; Miesch, Connor, and Eicher, 1964).

Adhering strictly to definition, computer-simulation methods appear not to have been applied to date in igneous petrology or igneous geology. There are, however, several papers that describe analytical applications that could be applied in a simulation sense. These include Wickman's (1966*a–e*) stochastic models of the frequency of volcanic eruptions, application of the Markov theory to the study of chemical variations in rock bodies by Agterberg (1966), application of Markov chains to the spatial distribution of mineral grains in igneous rocks (Vistelius, 1966), a method (McIntyre, 1963) of computer "synthesis" of the proportions of minerals in an igneous rock of specified chemical composition, and, finally, development (Whitten and Boyer, 1964) of conceptual models of igneous intrusive processes.

Geochemical Balances and Element Abundances

Horn and Adams (1966) describe a computer method for computing geochemical balances and element concentrations in different types of rock. The classical gross-crustal geochemical balance assumes that all sediments were ultimately derived from igneous rocks. Horn and Adams' method, incorporating a materials-balance accounting system, provides for input of (a) estimates of volume of sediments and (b) estimated minimum and maximum concentrations for each geochemical element in seven lithologies. In turn the method, through the use of an iterative-optimization technique, generates as output (a) the total mass of weathered igneous rock, (b) mass distribution of the elements within four sedimentary domains and the ocean, (c) a list of elements that cannot be balanced, and (d) new element-concentration tables derived from the optimized model.

An overall materials-balance reaction is assumed in which a chemical element leaves the primeval igneous source and subsequently resides either in sediments, in interstitial water in sediments, or in seawater. A series of materials-balance equations for individual elements may be written in general form for a particular time t as follows:

$$x_i X = y_i Y + z_i Z \tag{10-16}$$

where X = mass of igneous rock that has weathered in order to form masses Y and Z observed at time t

Y = total mass of sedimentary rock and sediments observed on and within the earth's crust at time t

Z = total mass of elements dissolved in the ocean at time t

x_i = the weight fraction or average concentration of element i in igneous rock at time t

y_i = the weight fraction or average concentration of element i in sediments or sedimentary rock at time t

z_i = the weight fraction of element i of the total dissolved solids in the ocean at time t

Recycling of sedimentary material, as discussed in the sedimentary-rock-mass models (Chapter 9) is not considered. Thus, setting $t =$ present, the value of X reflects the *minimum* mass of igneous rock that has weathered to form the present mass of sediments, and dissolved solids.

The term for sedimentary rocks, Y, may be expanded to include different lithologies and different sedimentary domains. The masses of sediment for four sedimentary domains are labeled C (continental domain), M (shelves and mobile-belt domain), H (oceanic-slope domain), and P (deep-ocean-basin domain). Thus we may write

$$y_i Y = c_i C + m_i M + h_i H + p_i P \tag{10-17}$$

where lower case letters refer to the respective element concentrations. Substituting Equation 10-17 into 10-16 and adding a new term γ, we obtain

$$x_i(X + \gamma) = c_i C + m_i M + h_i H + p_i P + z_i Z \tag{10-18}$$

where γ is the total surplus mass required to balance the equation, as discussed below; initially γ is set to zero.

Values for C, M, H, P, and Z are obtained from the estimated volume of each sedimentary domain, porosity of the rocks, and densities of the sedimentary grains and interstitial fluids. The total mass of weathered igneous rock, X, is assumed equal to the sum of the total mass of sedimentary rock plus the mass of dissolved solids in the ocean. Thus $X = C + M + H + P + Z$ and is assumed to be constant throughout the calculations.

The concentrations of 65 elements (denoted by lower case letters) were initially estimated by Horn and Adams from literature sources and are progressively revised during the calculation. Concentrations are expressed as minima and maxima, in parts per million. Thus 65 equations of the type shown in Equation 10-18 are obtained, one for each element considered.

The iterative-optimization procedure consists of two "nested" operations. First, Equation 10-18 for each element is balanced as nearly as possible by iteratively adjusting the element-concentration values, holding the values of γ, C, M, H, P, and Z (and thus X) constant. Some elements give a large surplus mass on the right side of the equation at this stage. In the second operation, the outer "loop" of the nested sequence, the value of γ is set to the sum of the outstandingly large surplus masses (for details see below) that remain unbalanced during the first operation, or inner "loop."

In the Horn and Adams model not only are the sedimentary terms on the right side of Equation 10-18 subdivided according to sedimentary domains but each sedimentary domain is further subdivided into six different lithologies. The original element concentrations are estimated for lithologies, and, by assuming the proportions of each lithology present in each sedimentary domain, average element concentrations for each domain are calculated. For example, it is assumed that continental-shield sediments are composed of 53 percent shale, 28 percent graywacke or sandstone, 16 percent carbonate, and 3 percent evaporite. Hence the average concentration c_i of element i in sediments of continental shields is

$$c_i = 0.53\text{sh}_i + 0.28\text{sg}_i + 0.16\text{ca}_i + 0.03\text{ev}_i \tag{10-19}$$

where sh_i = concentration of element i in continent-shield shales
sg_i = concentration of element i in continent-shield sandstones or graywackes
ca_i = concentration of element i in continent-shield carbonates
ev_i = concentration of element i in continent-shield evaporites

Similar equations, employing different coefficients, are written for sediments in the three other sedimentary domains. Sediments of oceanic slopes and of the deep-ocean floor are considered to consist entirely of oceanic clay and oceanic carbonate.

Steps in Computer Algorithm

The major steps in the algorithm are as follows:

1. The masses of the four sedimentary domains C, M, H, P, are calculated as well as the mass of total dissolved solids in the ocean, Z.
2. The value of X is set to the sum of C, M, H, P, and Z. Initially the total surplus mass γ is set to zero.
3. Average concentrations for each element are calculated for rocks in each sedimentary domain, given the maximum and minimum estimates of concentration in six lithologies and given the lithologic composition of each sedimentary domain.
4. For each element the right side of Equation 10-18 is calculated,

given the mass of each sedimentary domain and solids in the ocean (step 1), and the element concentrations (step 4). For element i the right side of the equation R_i is thus

$$R_i = c_iC + m_iM + h_iH + p_iP + z_iZ$$

5. For each element the left side of Equation 10-18 is evaluated. Thus for element i

$$L_i = x_i(X + \gamma) \tag{10-20}$$

6. Values obtained from each side of the equation are compared for each element. The difference between the two sides for element i is

$$D_i = R_i - L_i \tag{10-21}$$

7. The difference D_i is minimized for each element by adjusting the element-concentration data, iteratively repeating steps 3 to 6. Notice that at this stage the value of γ is not adjusted and that balancing is attempted by manipulation of the element-concentration values only.

8. After each element has been iterated 100 times, the remaining differences D_i are used to obtain the standard deviation S, where

$$S = \left[\frac{\sum_{i=1}^{n} (L_i - R_i)^2}{n-1} \right]^{1/2} \tag{10-22}$$

where n is the total number of elements considered.

9. The surplus mass γ is now defined as

$$\gamma = \sum_{i=1}^{n} D_i \qquad \text{for all } i \text{ with } D_i > S$$

In other words differences greater than one standard deviation are summed to form the surplus. The elements with large unbalanced differences are assumed to be partly derived from sources other than weathered igneous rock. Elements with small values of D_i are not included in this surplus, under the assumption that small differences are within the limits of error to be expected from the balance equations and estimates of element concentrations.

10. Steps 3 to 9 are repeated iteratively, progressively modifying the surplus mass γ and the element concentrations.

11. After about 12 iterations, the model becomes comparatively stable, and the iterations are stopped.

Using their method, Horn and Adams were able to bring 55 of the 65 elements into balance. Of the 10 remaining elements, seven (boron, sulfur, chlorine, antimony, selenium, bromine, and iodine) are probably not in balance because significant proportions of their total masses in sedimentary and oceanic domains were supplied as volcanic gaseous and liquid emanations. The failure of the three other elements—manganese, molybdenum, and lead—to balance is probably due to erroneous concentration estimates. Horn and Adams tabulate the revised element concentrations for each lithology and the total mass of each element in each sedimentary domain. Also, they find that the total mass of elements in sedimentary rocks and dissolved in seawater derived (a) from weathered igneous rock is 2040×10^{15} metric tons and (b) from sources other than weathered igneous rock (the final value of γ) is 61.2×10^{15} metric tons. These values are larger by about a factor of 2.5 than estimates by previous authors.

Synthesis of Minerals in Igneous Rocks

McIntyre (1963) describes a computer method for obtaining modal analyses of igneous rocks by recasting their chemical composition as the equivalent "normal" mineral composition, or *norm*. Input to the program consists of chemical analytical data, consisting of the major oxides as percentages (SiO_2, Al_2O_3, Fe_2O_3, FeO, MgO, CaO, Na_2O, K_2O, TiO_2, P_2O_5, and MnO) or expressed as percentages of the elements (Na, Mg, Al, Si, P, K, Ca, Ti, Mn, Fe^{2+}, and Fe^{3+}). These data may be obtained by various means, including X-ray fluorescence spectrography. Given the analytical data in either form, the program then proceeds to convert the weight proportions to the cation equivalents. Then the program "creates" various minerals in sequence by combining cations according to rules that relate the prescribed proportions of elements in each mineral.

The program contains a large number of mineral-forming steps. The actual steps employed depend in part on the cation proportions. The mineral-forming steps later in the program contain a number of branches so as to cause the cations to be distributed according to the material required to form each mineral that can be produced. Accordingly not all the mineral-forming steps are involved in treating a given set of analytical data.

The mineral-forming steps provided in the program are as follows:

1. Apatite and ilmenite are created.
2. If there is excess potassium, K_2SiO_3 is formed.
3. If there is excess sodium, acmite is created.

4. If there is excess sodium, $NaSiO_3$ is created.
5. Magnetite and hematite are formed.
6. Anorthite is formed; if excess aluminum remains, corundum is formed.
7. If excess titanium is present, either titanite or rutile is formed.
8. Diopside is formed; if excess calcium remains, this is converted to wollastonite.
9. Olivine and nepheline are formed provisionally.
10. Orthoclase is formed; if excess potassium remains, leucite is produced.
11. If there is insufficient silicon to permit wollastonite to be retained, wollastonite is reduced to Ca_2SiO_4.
12. If there still is insufficient silicon, diopside is reduced to Ca_2SiO_4 and olivine, as necessary.
13. If excess potassium remains at this point, leucite is reduced to kaliophilite.
14. Albite is then created; if excess sodium remains, nepheline is formed.
15. Hypersthene is then formed; if excess ferrous iron and magnesium remain, olivine is formed. Excess silica then forms quartz.
16. Finally the silicated forms of leucite, kaliophilite, and nepheline, are produced.

Although the program was written to treat data in an analytical sense, it is of interest from a simulation point of view because it simulates the mineral-forming processes in the formation of igneous rocks in accord with materials-balance requirements.

Stochastic Simulation of Igneous Textures

Vistelius (1966) describes the application of Markov chains to the simulation of the composition of granodiorites as observed in thin section. His model is based on the assumption that during crystallization minerals will precipitate from a magma until the residual melt no longer has a eutectic composition. As the temperature drops further, the whole of the residual melt crystallizes to form a mixture of quartz and potassium feldspar crystals. The individual grains and aggregates of grains of the earlier formed minerals (formed before crystallization of the eutectic) occur between the later formed quartz and potassium feldspar grains.

Vistelius proposes that the likelihood of transition from one mineral grain to the next in a spatial sense can be represented by a transition-probability matrix. A matrix of theoretical transition probabilities can be obtained by making these assumptions. A second matrix of empirical transition probabilities is obtained by counting actual transitions along traverses of a thin section of granodiorite. The two matrices are then compared. If the granodiorite is igneous in origin, the empirical matrix should closely resemble the observed matrix. Conversely, if there is poor agreement between the two transition matrices, the hypothesis that the rock was formed by crystallization from a magma is rejected, and a metasedimentary origin is preferred.

Hydrothermal Ore-Deposition Model

Anttonen (1969) has developed a digital dynamic model that simulates hydrothermal ore deposition and other related depositional phenomena in rocks. The model provides for deposition by both replacement and by filling of pore space. It employs the basic diffusion (time-dependent potential flow) equations for two dimensions, as described in Chapter 6. The objective is to permit the user to explore the effects of a variety of geological variables as they pertain to the deposition of a single mineral, which can be regarded as an ore mineral, or alternatively, as a pore-filling and rock-replacing material. The program is an initial effort that, with extensions, could be used to simulate the deposition of multiple minerals in three dimensions.

Although the model employs a diffusion model to represent the transportation of mineral material, it should be understood that actual ore-deposition processes involve several transport mechanisms. Fluid flow is undoubtedly much more important on a gross scale, whereas diffusion processes are probably effective only over relatively small distances, such as a few inches. Anttonen's model, however, employs a diffusion model to represent all transport processes for simplicity.

The geological variables incorporated into the model include temperature, pressure, porosity, permeability, rock type, and rock reactivity. As the computer program that embodies the model is presently organized, some of these variables (such as rock reactivity) are represented heuristically. Other variables, such as the role of temperature in the calculation of thermodynamic equations for equilibrium constants at elevated temperatures, are more formally treated. Fortunately the program is organized so that specific sections dealing with certain variables can be removed, rewritten, and reinstalled without affecting the rest of the program.

The program makes use of a series of two-dimensional arrays to represent variations in physical conditions within the rock body, which is represented by a grid of square cells. In the experimental run represented in Figures 10-21 to 10-24 a grid containing 21 rows and 26 columns was used. Each element in the array stores information about physical conditions and corresponds to a particular cell in the grid. One of the arrays contains information on rock types. The structural configuration—including stratified rocks, faults, folds, and intrusive bodies—can be represented by symbols representing different rock types, (e.g., Figure 10-21*a*).

Information in other arrays defines the remaining physical variables. The temperature array defines the temperature configuration across the rock body, and the pressure array defines the pressure configuration. Porosity is assigned to another array according to rock type and pressure conditions. Information on permeability is included in four arrays that represent permeability for the cation in the horizontal direction and the vertical direction, and permeability for the anion in the horizontal and vertical direction. Two arrays contain information about movement of the ore-forming fluid through the rock—one pertaining to anion concentration, and the other to cation concentration.

The ore-forming fluid moves through the rock by a diffusion process. The controls for the diffusion of the cation are independent of those for the anion, so that they can be allowed to diffuse at the same or at different rates, as the user chooses. Deposition of ore occurs when the cation and anion concentrations in a cell are great enough to exceed the value of the solubility product. Initially a solubility product is read into the program for standard conditions, (i.e. 1 atmosphere pressure and a temperature of 25°C). The initial solubility-product value is then used to calculate the solubility product K according to the equation

$$\log_e K = \frac{-\Delta H}{RT} + C \qquad (10\text{-}23)$$

where ΔH = heat of formation in kilocalories and may be obtained from handbooks (along with K at standard conditions)
R = constant of proportionality (equivalent to the gas constant)
T = temperature in degrees Kelvin
C = constant of integration

If the value of K, calculated with Equation 10-23 for the specific conditions in a cell is less than the product of the cation and anion concentrations, deposition occurs. On the other hand, if K exceeds the product of the two concentrations, deposition does not occur. Deposition is represented at a particular cell by replacing the rock-type

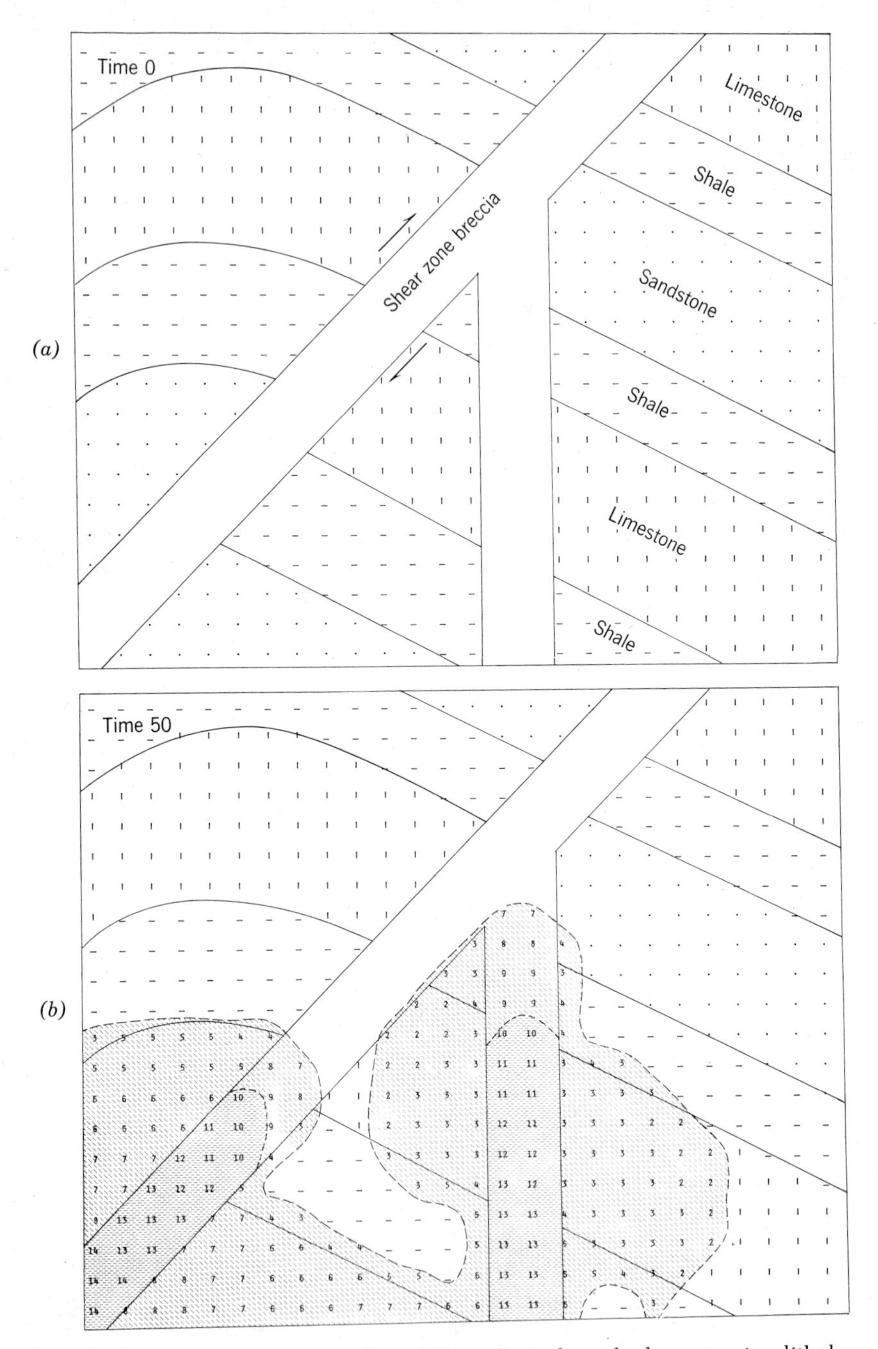

Figure 10-21 Vertical sections through hypothetical ore body portraying lithology and ore concentration at (*a*) time increment 0 and (*b*) time increment 50 during experimental simulation run. No scale is implied inasmuch as model is hypothetical. Sandstone is denoted by dots, limestone by vertical dashes, shale by horizontal dashes, and material in shear zone by blanks. Ore concentrations, in percent, are denoted by numbers and have been contoured. From Anttonen (1969).

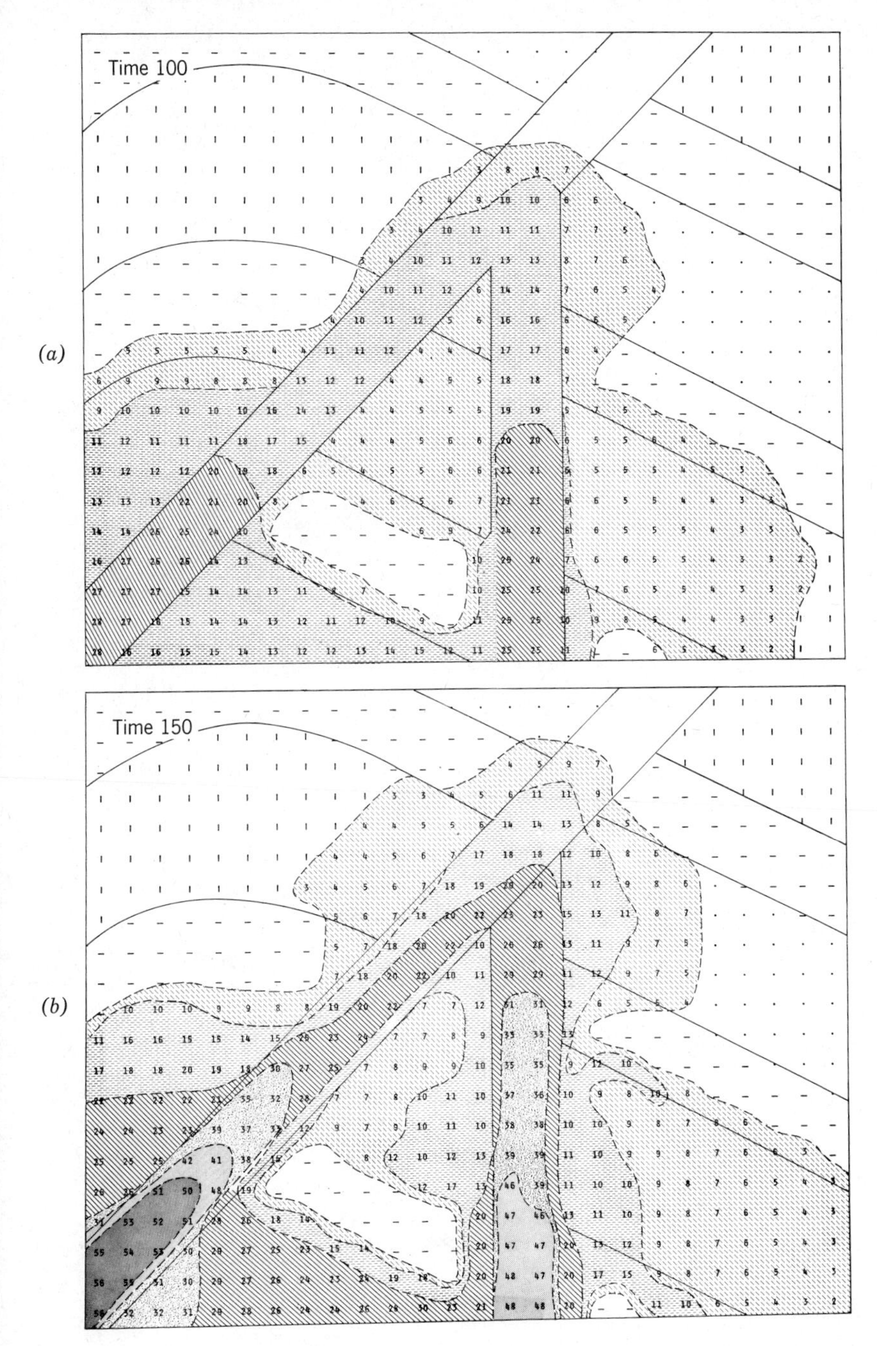

Figure 10-22 Vertical sections for (*a*) time increment 100 and (*b*) time increment 150, corresponding to those in Figure 10-21.

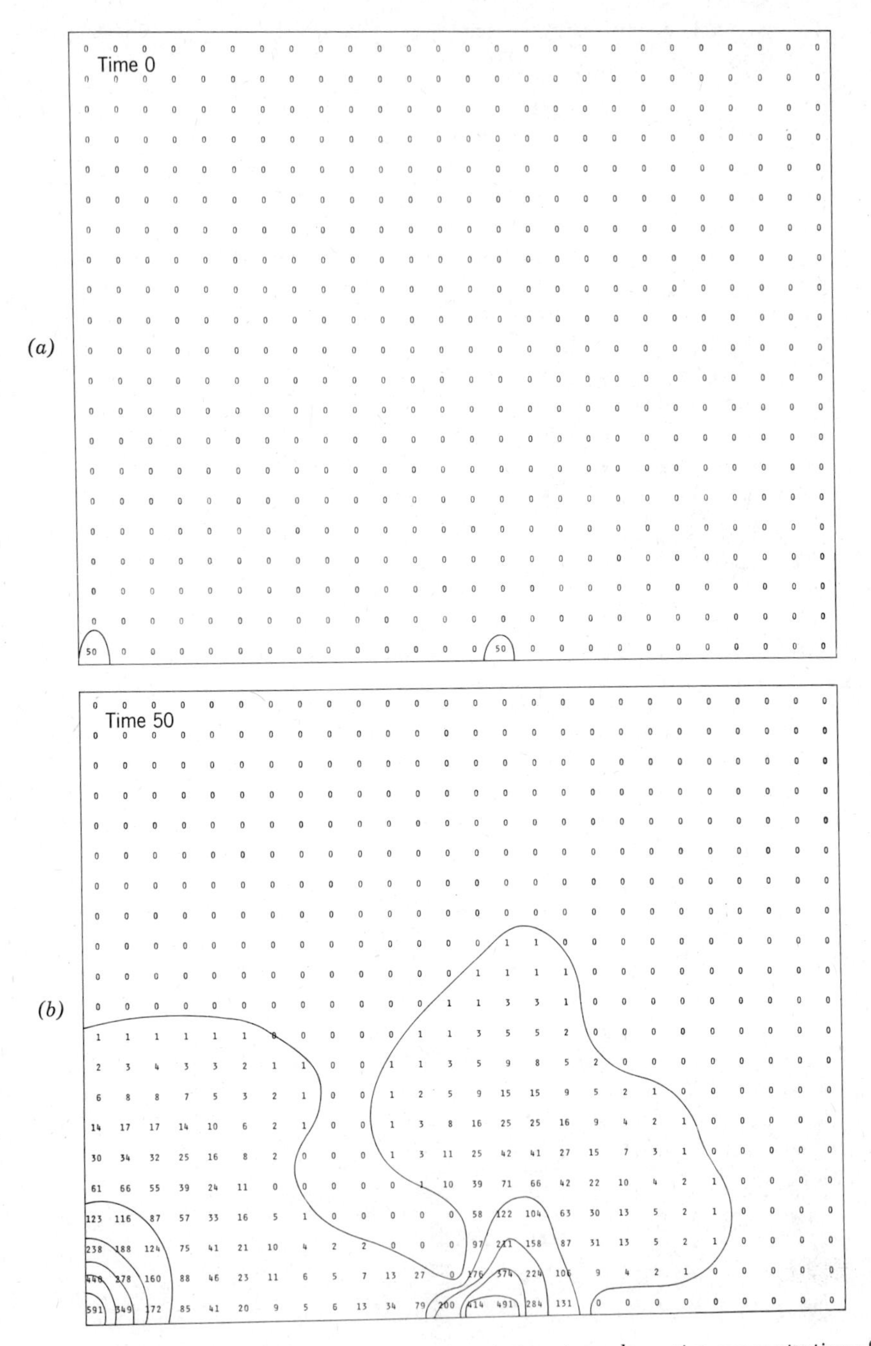

Figure 10-23 Vertical sections portraying variations in molar cation concentration of ore-forming fluid at (*a*) time increment 0 and (*b*) time increment 50. Numbers shown have been obtained by multiplying the calculated molar concentration by a constant and are shown in this form for display purposes. From Anttonen (1969).

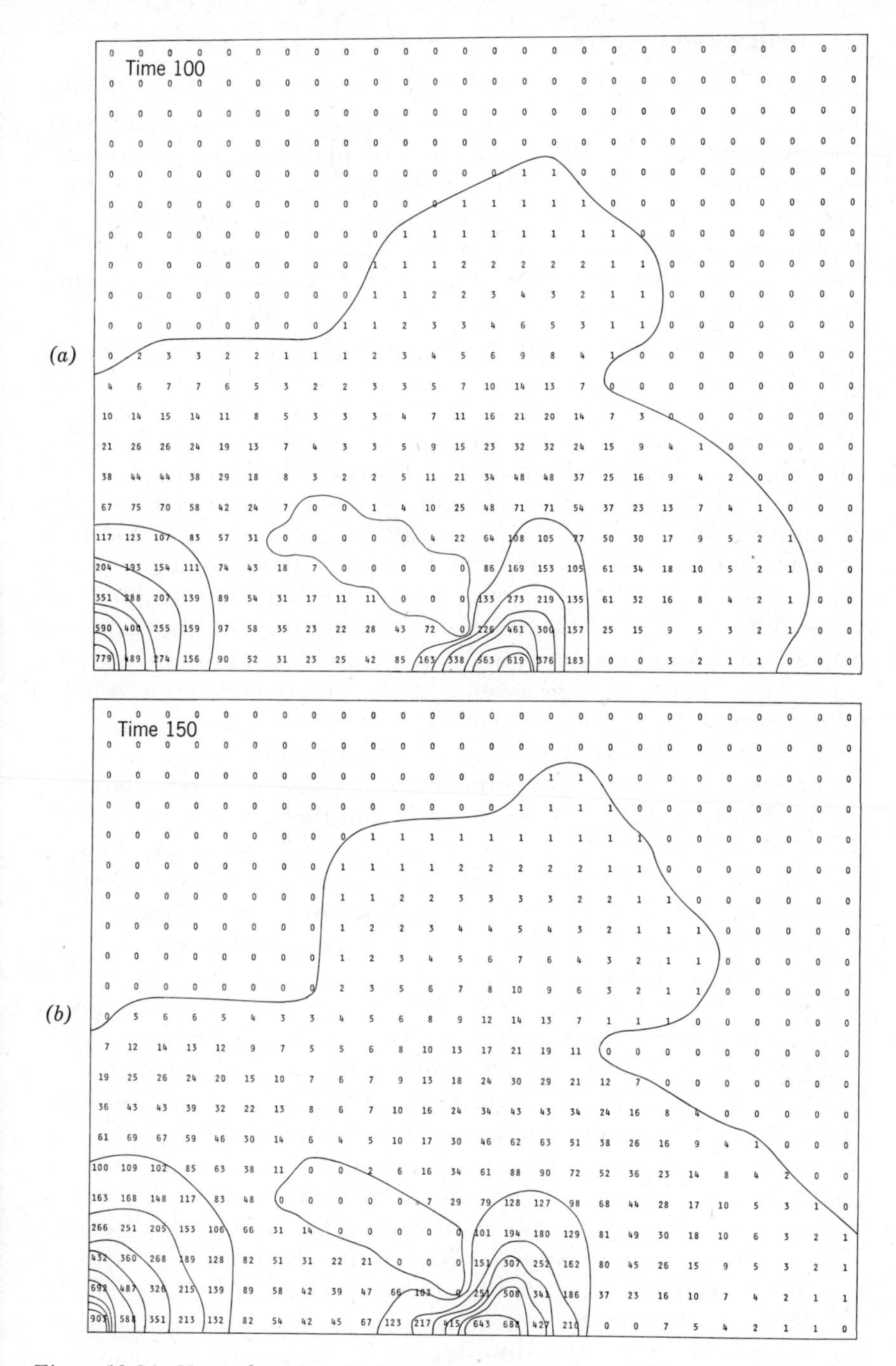

Figure 10-24 Vertical sections for (*a*) time increment 100 and (*b*) time increment 150, corresponding to those shown in Figure 10-23.

symbol (Figures 10-21 and 10-22) by a value between 0 and 100 that denotes the percentage of total volume of the cell occupied by ore. The principal factors that govern the calculation at a particular cell are available pore space, rock reactivity, and temperature. Rock reactivity is modified when deposition occurs, thus influencing deposition in subsequent time steps. Likewise porosity and diffusibility are continuously adjusted, causing the model to respond dynamically with the advance of each time increment.

The results of an experimental run with Anttonen's model are shown as a series of vertical sections in Figures 10-21 to 10-24. Figures 10-21 and 10-22 pertain to lithology at time increments 0, 50, 100, and 150, whereas Figures 10-23 and 10-24 pertain to the cation concentration of the ore-forming fluid at these time increments.

The structural configuration (Figure 10-21*a*) consists of a sequence of gently folded beds of sandstone, shale, and limestone that have been intersected by a reverse fault (accompanied by a prominent shear zone) dipping at 45 degrees toward the left. A vertical fracture in the lower center of the section intersects the fault but does not displace it. The fracture is also accompanied by a prominent shear zone. The shear zones are assumed to be highly permeable relative to the strata. Thus the fault and the fracture serve as veins that strongly influence the movement of the ore-forming fluids.

The ore-forming fluid is assumed to enter the rock body shown in the section from below. Both the fault and the fracture provide the inlets. The fluid progressively moves through the rock mass by diffusion, reacting with the different rock types according to reactivity coefficients that were supplied in advance of the run for each rock type. In the experiment limestone was assumed to be the most reactive, and shale the least reactive.

Flow of the ore-forming fluid is regulated by the permeability values supplied for the various rock types. Shale is defined as relatively impermeable, limestone is endowed with intermediate permeability, and sandstone is moderately to highly permeable. The permeabilities for shale, limestone, and sandstone are set so as to be the same in both horizontal and vertical directions. Permeability in the shear zones, accompanying the fault and the fracture, however, is greater in the vertical direction than in the horizontal direction. This has the effect of causing the ore-forming fluid to diffuse faster in the vertical direction. It is to be emphasized that the model employs the diffusion equation for transport, regardless of whether the actual transport is by molecular diffusion or by fluid flow. In reality, of course, fluid flow is strongly influenced by permeability, whereas molecular diffusion is not. The assumption that diffusion rate is influenced by permeability is strictly an artificial convenience within the model.

The results show that configuration of the hypothetical ore body is strongly influenced by the geological factors. The highest concentrations of ore (Figures 10-21*b* and 10-22) occur in the shear zones. Deposition in the host rock, beyond the shear zones, is important, however. The greater reactivity of limestone and sandstone, as contrasted with shale, strongly affects the shape of the ore body, which tends to expand outward in sandstone and limestone, whereas it is constricted opposite shale beds.

Although the present model is relatively crude in many respects, it suggests that more complex and much more realistic models could be developed. Three-dimensional models might prove to be useful guides in underground exploration, where various geological factors that include geologic structure, host rock type, ore composition and grade, wall-rock mineralogy, and permeability all need to be treated and interrelated. Furthermore, more advanced models could provide for the distinction between fluid flow and molecular diffusion in transportation of the ore-forming cations and anions.

Annotated Bibliography

Agterberg, F. P., 1966, "The Use of Multivariate Markov Schemes in Petrology," *Journal of Geology*, v. 74, pp. 764–785.

Describes the use of Markov methods to relate spatial and stratigraphic variations of major oxides in sequences of basaltic rocks.

Anttonen, G. J., 1969, *Ore Deposition Simulation Model*, unpublished master's degree report, Branner Geological Library, Stanford University.

Charyula, V., and Horn, M. K., 1968, "Geochemical and Logging Applications of a Computer Simulated Neutron Activation System," Computer Contribution 22, Kansas Geological Survey, pp. 35–42.

Description of a computer program for simulating neutron-activation experiments. The spectral responses of chemical elements occurring in different lithologies are simulated and in turn may be compared with observed spectral response. Input to the program includes the *irradiation* factors, such as neutron flux and irradiation time; the *sample* factors, which include sample volume, density, its composition in terms of chemical elements, and the isotopic composition of individual elements; and finally the *detection* factors.

DeLand, E. C., 1967, *Chemist–The Rand Chemical Equilibrium Program*, the Rand Corporation, Santa Monica, Calif., RM-5404-PR and AD-664045, 132 pp.

Explanation of principles and detailed operating instructions for computer programs to simulate chemical equilibrium constituents. For a given chemical milieu the program computes the final, steady equilibrium state by using the equilibrium constants. The method is based on the Gibbs theorem that the equilibrium composition is such that the total thermodynamic free energy of the system is minimized (or the entropy maximized) under conditions of the experiment. Using iterative-optimization methods, the program determines the composition that minimizes the system's free total energy, subject to constraints on the system.

Eicher, R. N., and Miesch, A. T., 1965, *Computer Simulation Program for Investigation of Geochemical Sampling Problems*, unpublished U.S. Geological Survey Report, 65 pp.

Describes a computer program for simulating sampling plans applied to igneous rock bodies. The assumption is made that spatial variations in the composition of rock bodies can be described by low-degree polynomial trend surfaces, plus a randomly distributed residual component with a prescribed frequency distribution. The coordinate values, designating the locations where the model is to be sampled, can be read in as data. In turn the sequence of simulated values corresponding to the prescribed sampling is produced as output.

Graf, D. L., Blyth, C. R. and Stemmler, R. S., 1967, *One-Dimensional Disorder in Carbonates*, Circular 408, Illinois State Geological Survey, 61 pp.

The diffraction effects expectable from calcium–magnesium carbonate minerals can be treated as Markov processes.

Helgeson, H. C., 1969, "A Chemical and Thermodynamic Model of Sulphide Deposition in Hydrothermal Systems," Geological Society of America Annual Meeting, Atlantic City, Abstracts with Programs for 1969, part 7, p. 98.

A model consisting of reversible and irreversible reactions describing the depositional process can be represented by a matrix of linear differential equations. Computer evaluation of the matrix permits quantitative and simultaneous prediction of the mass transfer attending silicate reactions and sulphide deposition in ore deposits containing large numbers of components and phases. The calculations define the solubilities of the minerals, as well as the amounts precipitated or destroyed, and the sequence in which they occur.

Horn, M. K., and Adams, J. A. S., 1966, "Computer-Derived Geochemical Balances and Element Abundances," *Geochimica et Cosmochimica Acta*, v. 30, pp. 279–297.

Hurley, P. M., Hughes, H., Faure, G., Fairbairn, H. W., and Pinson, W. H., 1962, "Radiogenic Strontium-87 Model of Continent Formation," *Journal of Geophysical Research*, v. 67, pp. 5315–5333.

Progressive change in the ratio of strontium isotopes, Sr 87/Sr 86, coupled with a materials-balance model, provides a measure of rate of formation of continental crustal material.

Kvenvolden, K. A., and Weiser, D., 1967, "A Mathematical Model of a Geochemical Process: Normal Paraffin Formation from Normal Fatty Acids," *Geochimica et Cosmochimica Acta*, v. 31, pp. 1281–1309.

Detailed description of a mathematical model of chemical transformations; makes extensive use of numerical solutions of differential equations.

McIntyre, D. B., 1963, *FORTRAN II Program for Norm and Von Wolff Computations*, Technical Report No. 14, Department of Geology, Pomona College, Claremont, Calif., 14 pp.

Miesch, A. T., Connor, J. J., and Eicher, R. N., 1964, "Investigation of Geochemical Sampling Problems by Computer Simulation," *Quarterly of the Colorado School of Mines*, v. 59, pp. 131–148.

Companion paper to the paper by Eicher and Miesch.

Shimazu, Y., and Urabe, T., 1968, "Dynamic Simulation of Planetary Evolution," *Journal of Physics of the Earth*, v. 16, pp. 129–136.

This paper is a brief description of a computer model that treats the gross geochemical cycle at the earth's surface as a complex circuit involving heat and energy transfer, much as in a chemical plant. The ocean is regarded as a combined tank reactor, condenser, and distillation vessel. The mantle yields silicate melts and volatiles that are input to the system and progressively recycled subsequently. These materials are allocated to the atmosphere, hydrosphere, and lithosphere according to stability conditions. The computer model is compartmented in a number of "unit process simulators," each of which has a prescribed, specialized function, as follows:

1. Unit process simulator EQCOMV gives the equilibrium composition of a multicomponent polyphase system at specified temperatures and pressures. Given the respective chemical elements, these are allocated to various chemical compounds in the gas, liquid, and solid phases, so as to minimize the total Gibbs free energy of the system. A nonlinear optimization technique is employed to find the minimum.

2. Unit process simulator MAGMA gives the rate of ascending magma, treating the progressive fractionation and enrichment of elements into melts as sequences involving partial melting, equilibrium freezing, fractionation freezing, and zone melting.

3. EQCOMS gives the equilibrium composition of seawater as a function of temperature, chloride-ion composition, and pH. The major negative ions (chloride, sulfate, carbonate, and bicarbonate) are supplied mainly from outgassing volatiles, whereas major positive ions are provided by eroding silicate rock.

4. EQDSTB gives the equilibrium surface temperature and the partition of mass of water and carbon dioxide among the atmosphere, hydrosphere, and lithosphere.

5. CARBON pertains to the geochemical cycle of carbon, treating removal of carbon from the atmosphere as carbon dioxide by photosynthesis and fixation as organic carbon in the lithosphere.

6. SEDSIM pertains to the transport of sediment from the land and deposition in the sea.

The program adheres to materials-balance requirements throughout.

Vistelius, A. B., 1966, "Genesis of the Mt. Belaya Granodiorite, Kamchatka (An Experiment in Stochastic Modeling)," *Doklady Akademii Nauk SSSR*, v, 167, pp. 48–50.

Whitten, E. H. T., and Boyer, R. E., 1964, "Process-Response Models Based on Heavy-Mineral Content of the San Isabel Granite, Colorado," *Bulletin of the Geological Society of America*, v. 75, pp. 841–862.

Develops a qualitative process-response model for explaining distribution of heavy minerals in a granite body.

Wickman, F. E., 1966*a–e*, "Repose Period Patterns of Volcanoes: I. Volcanic Eruptions Regarded as Random Phenomena; II. Eruption Histories of Some East Indian Volcanoes; III. Eruption Histories of Some Japanese Volcanoes; IV. Eruption Histories of Some Selected Volcanoes; V. General Discussion and a Tentative Stochastic Model," *Arkiv for Mineralogi och Geologi*, v. 4, No. 7, pp. 291–301, No. 8, pp. 303–317, No. 9, pp. 319–335, No. 10, pp. 337–350, No. 11, pp. 351–367.

A sequence of five papers that provide monographic treatment of the frequency of volcanic eruptions and the periods of repose between eruptions. Eruptive histories of a number of volcanoes are presented, including selected volcanoes in the East Indies, Japan, Hawaii, Iceland, Mexico, Italy, and the Philippines. The repose patterns for many volcanoes exhibit no memory effect and may be regarded as behaving according to a simple Poisson process. The eruptive activities of other volcanoes, however, do exhibit a memory effect and can be regarded as Markov processes.

STRUCTURAL GEOLOGY AND GEOPHYSICS

Structural geology and geophysics offer many opportunities for computer simulation. However, most papers that describe mathematical models in these fields deal with applications that lead to direct analytical solutions rather than solutions obtained by simulation. To attempt to survey the literature is out of the question here. It is feasible, however, to cite a few examples of papers that seem particularly relevant.

Many of the simulation applications in geophysics are concerned with the gravitative or magnetic response of a buried body, such as an ore body, an igneous intrusive, or a salt dome. For example, the interpretation of a gravity anomaly due to a buried body generally involves a series of alternative assumptions as to the volume, shape, density contrast, and depth of the buried body. We can consider these alternatives as a series of static simulation models, because the gravitative response of each assumed model is compared with the actual anomaly. The model by Isaacs (1967) is representative of simulation studies of this type. Other simulation models in geophysics include Farquhar's (1967) paper on seismic refraction, and that of Cherry and Hurdlow (1966) on the propagation of seismic disturbances.

There are a large number of papers dealing with scale-model structural simulation studies. Again, we shall not survey the literature, instead citing only examples of relatively recent papers, such as those of Cloos (1968), Hobbs (1967), and Ramberg (1967).

Finite-Element Analysis

One of the mathematical and geometrical tools that has been recently applied in structural geology is *finite-element* analysis. Finite-element methods have been widely applied by structural engineers in the numerical simulation of deformations in building structures (Zienkiewicz and Cheung, 1967). Dieterich (1969), and Dieterich and Carter (1969) have applied these methods in the numerical simulation of the folding of rocks.

The method they employ involves subdivision of the geologic body undergoing deformation into a series of finite elements or cells. Two-dimensional bodies are subdivided into triangular elements (Figure 10-25), and three-dimensional bodies into trapezoidal elements. The elements can be varied in size and arrangement, and are customarily made smaller in areas of interest.

At each node, or apex point of intersection between a triangle, the forces acting on the body are evaluated. These forces are then used to calculate the displacement of each node, using an equation of the type

$$\mathbf{F} = \mathbf{KD} \tag{10-24}$$

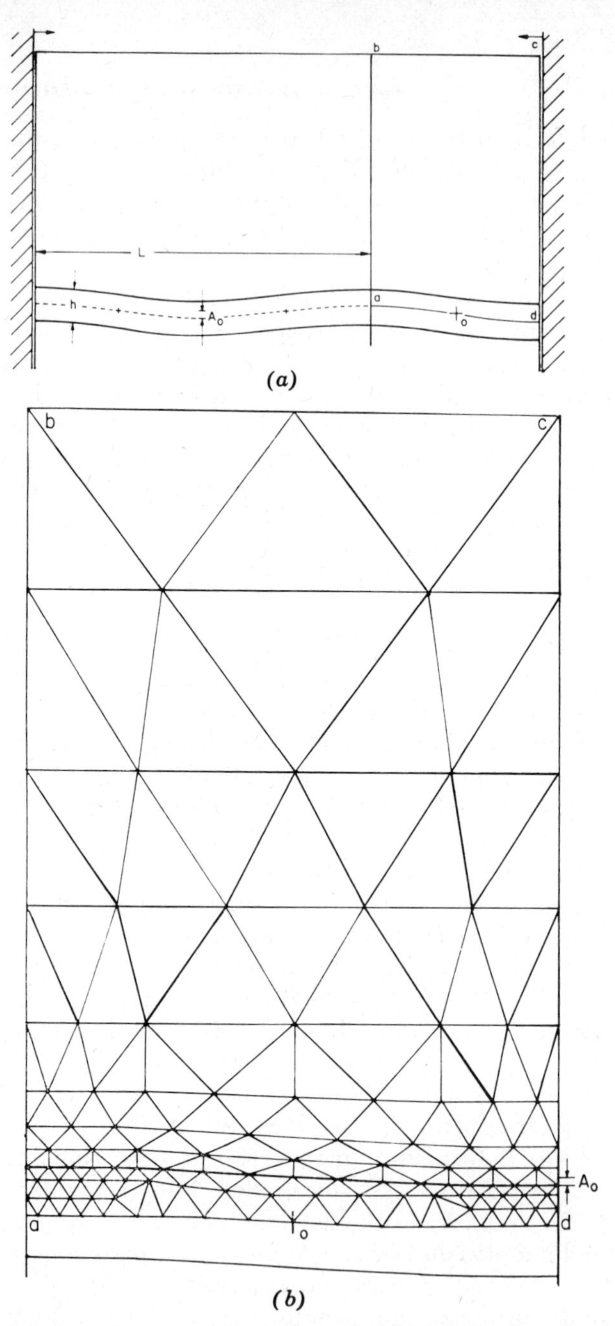

Figure 10-25 Finite-element analysis of simple fold. (*a*) Initial configuration of fold, showing boundaries, wavelength *L*, thickness *h*, and exaggerated amplitude A_0. Because of symmetry, only area enclosed by *abcd* need be studied. (*b*) Area *abcd* of (*a*) divided into 206 triangles. Triangles are smaller in areas of interest. From Dieterich and Carter (1969), with permission from the *American Journal of Science*.

where **F** is a matrix containing the forces acting at each node, **D** is a matrix containing the displacements at each node, and **K** is a matrix of constants (also known as a "stiffness" matrix).

For some structural problems the forces matrix **F** and stiffness matrix **K** are known, and the displacements **D** are calculated. By dividing time into discrete intervals, the progressive displacement of the nodes, and hence the whole body, can be simulated in a series of jerky steps.

In some cases, the displacements **D** are known, and the forces that produced the deformation can be calculated, given **K**. Besides being useful for modeling deformation of solid materials, the finite-element method also can be employed for modeling viscous-flow phenomena. In this case the **D**-matrix can represent velocities.

The finite-element method bears many similarities to the finite-difference method used for modeling fluid flow in Chapter 6. In both methods the region of interest is subdivided into discrete elements or cells, converting the problem from continuous space to discrete space. In finite-difference methods the individual elements are usually all the same shape and size, and are regular in arrangement. The finite-element method is more flexible in this respect, permitting variable element size and therefore variable resolution from place to place.

Both methods are approximation methods in that space as a continuous variable is represented discretely. In finite-difference methods differential equations that provide an exact, theoretical representation are approximated over finite intervals of space. In the finite-element method, however, the continuous body is first represented by an approximating body of finite elements. The equations describing the displacements within the approximating body are then evaluated exactly. Thus in the former case exact equations are solved approximately, whereas in the latter approximating equations are solved exactly. Readers interested in pursuing this topic are referred to the paper by Dieterich and Carter (1969), which contains a worked example to illustrate the method, and to the textbook by Zienkiewicz and Cheung (1967).

Salt-Dome Simulation

Howard (1968) has developed a model for the digital simulation of intrusive salt domes and their accompanying gravitative responses. Howard's model is two-dimensional and involves both deterministic and probabilistic relationships. Output from the program representing the model consists of a series of computer printouts (Figure 10-26) that depict vertical sections through the dome. A sequence of sections

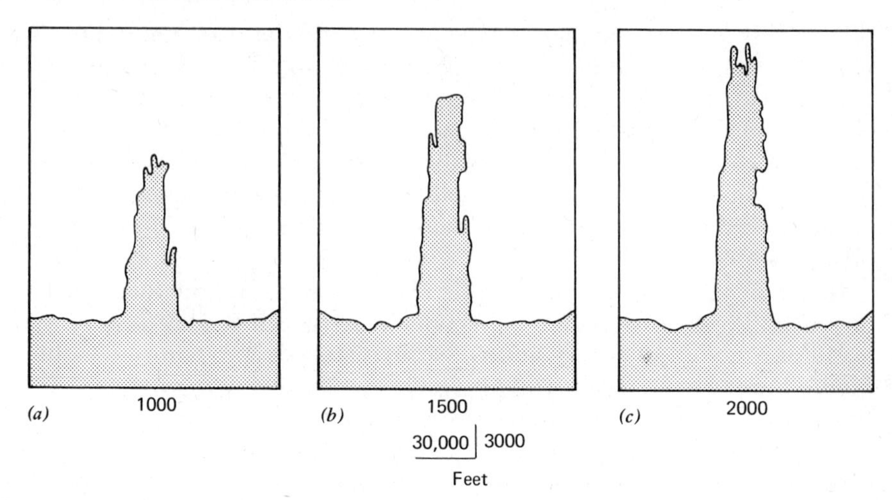

Figure 10-26. Simulated piercement dome at (*a*) 1000, (*b*) 1500, and (*c*) 2000 iterations. Salt is shaded and overburden is clear. From Howard (1968).

at different time increments is capable of showing the progressive development of the dome. The simulation experiments involve a thick layer of salt that is overlain by sediment. Because the density of the salt is less than that of the overlying sediment and because the salt can flow plastically, the salt migrates upward, piercing and displacing the overlying sediment. Movement of salt and sediment is on a cell-by-cell basis, each cell being represented in the computer-printed cross sections by a printed symbol (Figure 10-28).

The gross shape of the intrusive dome in its early stages can be represented by a continuous curve, approximated by straight-line segments (Figure 10-27). The curve, of course, represents the overall flow relationships in a dome: upward in the center part and downward and inward from the source or salt sink around the periphery of the dome. The various radii of the curve can be estimated on the basis of geophysical data, from drilling, and through scale-model studies. Transport of material in the model is governed by a random-walk technique, material in the high portion of the curve generally moving upward, with respect to that in the sink regions, where material migrates downward and thence toward the center of the dome. Volume of salt withdrawn from the sink is equal to that added to the dome, thus adhering to the law of conservation of mass (or volume in this case). Volumes are computed on a three-dimensional basis. Formulas for the computation of volume are given in Figure 10-27.

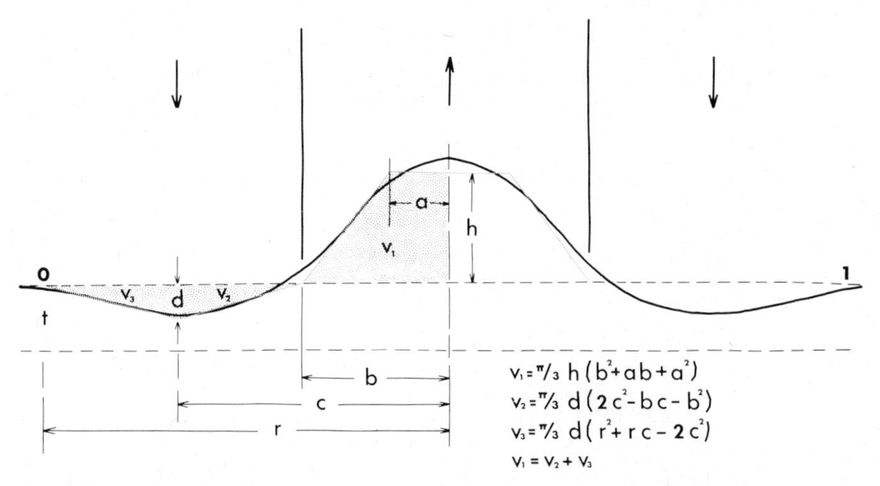

Figure 10-27 Generalized curve, approximated by straight-line segments, depicting configuration of a salt dome and its peripheral sink. Curve is employed as probability-density function, partly controlling movement of salt–sediment interface. From Howard (1968).

At each iteration of the program the position of the salt–sediment interface is moved upward or downward by using a two-step Monte Carlo technique. At the first step a location is picked at random along the salt–sediment interface, employing the salt–sediment interface curve in Figure 10-27 as a probability-density function. At the second step the actual displacement of the interface is calculated, employing a composite probability distribution that describes the summed effects of density, buoyancy, temperature, and the strength of overlying sediments. The actual displacement of the interface is determined by overlaying a nine-cell orthogonal grid centered about the point of displacement, with values in each cell giving the probability of movement of the interface in that particular direction. The sum of the probabilities equals unity. A random number in the range 0.0 to 1.0 is generated and used to select a particular cell, thus defining the direction of movement of the interface. This sampling procedure, similar in principle to the techniques outlined in Chapter 3, is repeated many times, causing progressive deformation of the salt–sediment interface.

The program is sufficiently versatile for a number of experiments to be performed. For example, the piercement effect of the salt dome may be assumed to keep pace with sedimentation (Figure 10-28) or alternatively piercement may be assumed to begin after a substantial

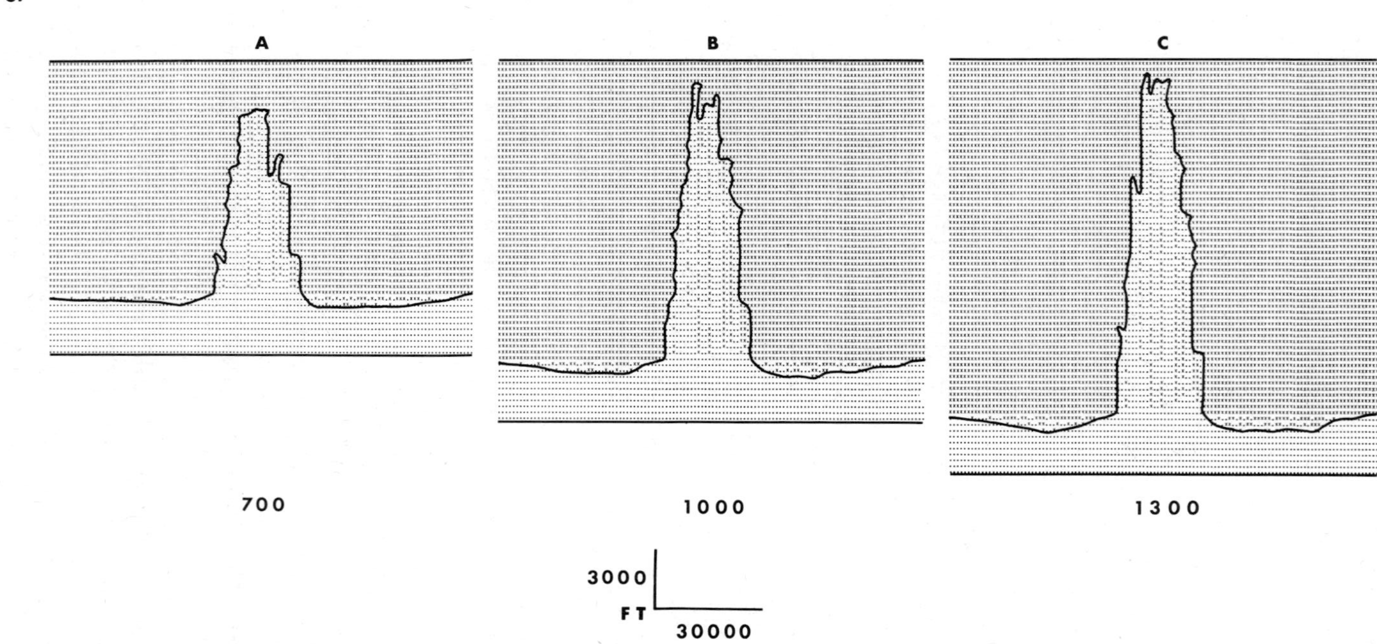

Figure 10-28 Sequence of simulated domes in which piercement has more or less kept pace with deposition and subsidence. Numbers pertain to numbers of iterations. From Howard (1968).

thickness of sediment has been deposited above the salt (Figure 10-26).

Such a model should be useful in petroleum exploration because it provides a means of representing the development of a complex dome. One of the features of Howard's program is that it provides a gravity curve (Figure 10-29) that is appropriate for the simulated

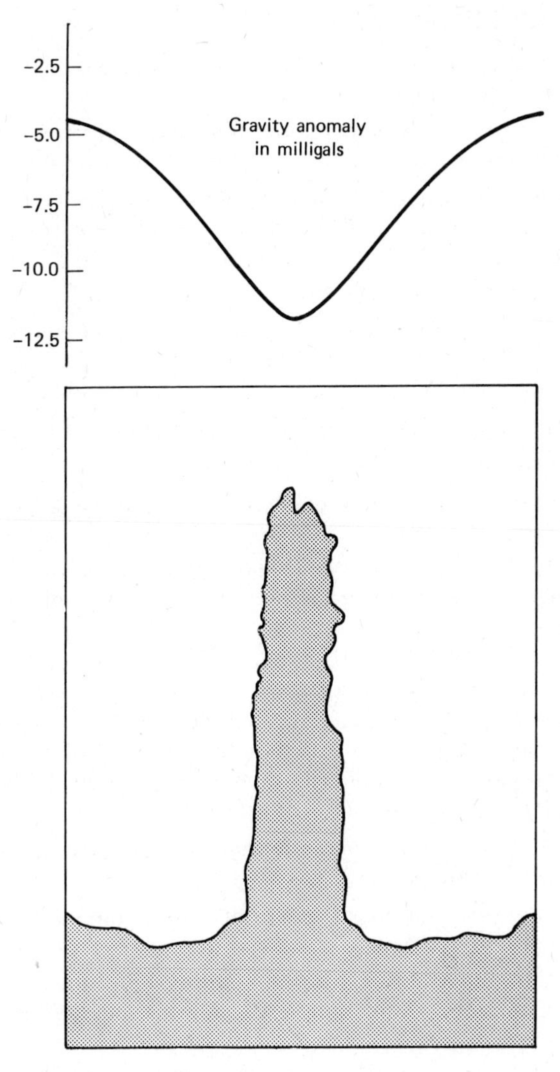

Figure 10-29 Calculated gravity profile over simulated salt dome. From Howard (1968).

dome. Since gravity profiles and contour maps provide important forms of data describing real salt domes, the user of the program can carry out successive experiments that bring simulated domes into accord with gravity data, as well as in accord with the geometry of the dome interpreted from drilling data.

Annotated Bibliography

Cherry, J. T., and Hurdlow, W. R., 1966, "Numerical Simulation of Seismic Disturbances," *Geophysics*, v. 31, pp. 33–49.

A technical paper dealing with finite-difference representation of deformation by means of a Lagrangian or distortable grid system.

Cloos, E., 1968, "Experimental Analysis of Gulf Coast Fracture Patterns," *Bulletin of the American Association of Petroleum Geologists*, v. 52, pp. 420–444.

Convincing application of physical scale models.

Dieterich, J. H., 1969, "Origin of Cleavage in Folded Rocks," *American Journal of Science*, v. 267, pp. 155–165.

Application of finite-element analysis in structural geology.

Dieterich, J. H., and Carter, N. L., 1969, "Stress-History of Folding," *American Journal of Science*, v. 267, pp. 129–154.

Pioneer paper applying finite-element analysis to the behavior of folded rocks.

Farquhar, R. P., 1967, *A Computer Display of a Seismic Refraction Model*, unpublished master's report, Branner Geological Library, Stanford University, 10 pp.

Deals with display of ray paths using a computer-linked cathode-ray tube. The tube displays the primary wave front of a simulated surface seismic shock as it propagates through stratified rocks. The user may adjust the parameter so as to simulate the structure of rocks of different velocities to yield a given distribution of arrival times of the wave front at different points along the earth's surface.

Hobbs, D. W., 1967, "Behavior and Simulation of Sedimentary Rocks," *Journal of Strain Analysis*, v. 2, pp. 307–316.

Describes physical compression experiments on the elastic and strength properties of sedimentary rocks.

Howard, J. C., 1968, "Monte Carlo Simulation Model for Piercement Salt Domes," Computer Contribution 22, Kansas Geological Survey, pp. 22–34.

Describes pioneer experiments in digital simulation of salt-dome structural development.

Isaacs, K. N., 1967, "The Simulation of Magnetic and Gravity Profiles by Digital Computer," *Geophysics*, v. 31, pp. 773–778.

Deals with the static simulation of the geologic structure in two-dimensional sections. A magnetic or gravity profile is produced for comparison with observed profiles.

Lee, W. H. K., 1968, "Effects of Selective Fusion on the Thermal History of the Earth's Mantle," *Earth and Planetary Science Letters*, v. 4, pp. 270–276.

Mathematical models describing the thermal history of the earth must yield heat-flow values for the surface of the earth that accord with observations. Assuming that the main source of energy is heat produced by radioactive decay, the distribution of heat sources in the earth will depend on the distribution of radioactive materials. Lee's models assume that radioactive elements are (a) uniformly distributed and (b) progressively concentrated at the earth's surface by convection currents. A diffusion-type equation describing heat flow from moving sources is solved by finite-difference methods, giving temperature–depth curves for successive periods of time after the earth was formed. The first model (uniform distribution, with no fractionation) is rejected because computed results show that the mantle should be largely molten and the surface heat-flow should be half of that actually observed. The second model provides a more acceptable temperature–depth distribution, implying that fractionation of radioactive elements has been active throughout geologic time.

Ramberg, H., 1967, *Gravity, Deformation, and the Earth's Crust*: Academic Press, New York, 214 pp.

Deals with physical scale models, with extensive theory.

Zienkiewicz, O. C., and Cheung, Y. K., 1967, *The Finite-Element Method in Structural and Continuum Mechanics*: McGraw-Hill, New York, 274 pp.

Authoritative textbook treating numerical techniques that can be adapted to the simulation of geologic structures.

FLUVIAL GEOMORPHOLOGY

Simulation methods are relatively advanced in fluvial geomorphology as compared with other branches of geomorphology. Applications to (a) longitudinal stream-profile development, (b) to slope and valley cross-section development, and (c) to stream-drainage-network development are discussed below.

Simulation of Stream Profiles

A random-walk simulation model can be used to simulate stream profiles (Figure 10-30). The elevation H of a point on the profile is

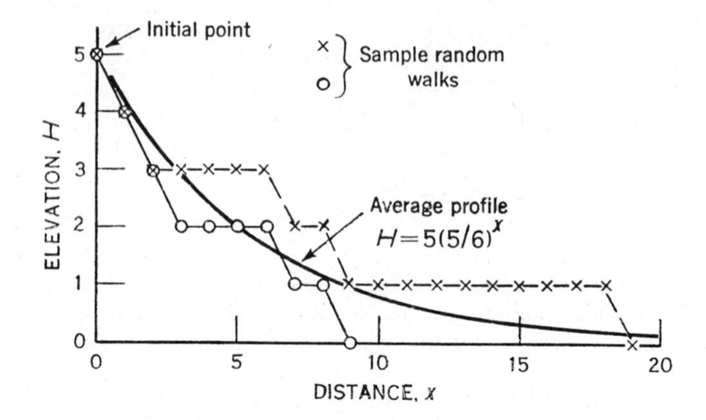

Figure 10-30 Ideal exponential stream profile (heavy line) and two random-walk simulated profiles. From Leopold and Langbein (1962).

measured with respect to the stream base level. The point is considered to move in unit steps, either to the right or downward, there being two choices at each step. There is a probability p that it will move downward and a probability q that it will continue at the same level. Since there are only two alternatives, $p+q=1$. Now the condition that the rate of expenditure of energy by the stream system be proportional to height above base level (Leopold and Langbein, 1962) can be adhered to by setting the probability p of a downward step proportional to the height above zero, decreasing to zero when the random walk reaches the base level. The progressive changes in the two probabilities p and q are represented in the model by two sample random-walk profiles in the simple model shown in Figure 10-30. The model is restricted to six discrete levels in equal increments (0, 1, 2, 3, 4, 5) above base level. The probability of a downward step at each level is $H/6$. Thus the probability of a downward step at level 5 is $\frac{5}{6}$ and zero at base level.

The assumption of no constraint on stream length, embodied in the simple model above, is not wholly realistic. It is possible to construct simple random-walk models that incorporate constraints on length. For example, the model can be organized so that the random walk reaches base level at or before reaching a given limiting distance. Figure 10-31 illustrates two ideal random-walk stream profiles (continuous lines) in which the random walk is constrained to reach base level at 6 and 10 stream-length units, respectively.

Constraints other than stream length can be incorporated within random-walk stream models. A temporary base level—such as that

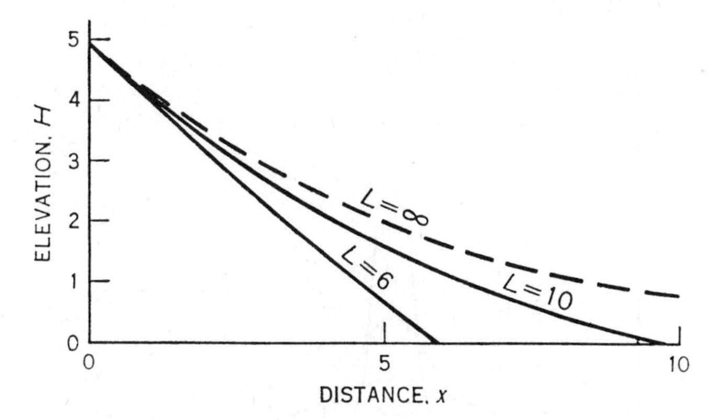

Figure 10-31 Ideal random-walk profiles (continuous lines) subject to constraints of stream length of 6 and 10 unit distances, respectively. Dashed line is ideal curve for profile without length constraint. From Leopold and Langbein (1962).

formed by the elevation of a trunk stream where it is joined by a tributary stream—could be represented by imposing an absorbing medium in the random-walk steps. Random walks terminate at a totally absorbing barrier, but the effect of a temporary base level can be modeled by a partial absorbing barrier that absorbs a given proportion of the random-walk steps, permitting the remainder to pass through unaffected.

Simulation of Slope Forms

The evolution of landscape can be regarded as a transport process in which material is moved from higher to lower places. The transport processes can be treated as either deterministic or stochastic processes, or as a combination of both. For example, the "decay" of a mountain range can be regarded as a diffusion process and can be represented by elementary diffusion equations, yielding a succession of topographic profiles whose shape is similar to Gaussian frequency-distribution curves (Figure 10-32). The diffusion equations can be applied at a smaller scale as well—for example, in models of soil creep and the development of local hillside slopes, as described by Culling (1963).

Pollack (1969) has made effective use of a modified form of the diffusion equation in modeling the development of deep valleys, simulating a succession of profiles through time. Initially he proposes three simple models, each of which involves the rate of lowering of the land surface through a generalized process of fluvial erosion and valley widening (Figure 10-33). In the first model (Figure 10-33*a*) the

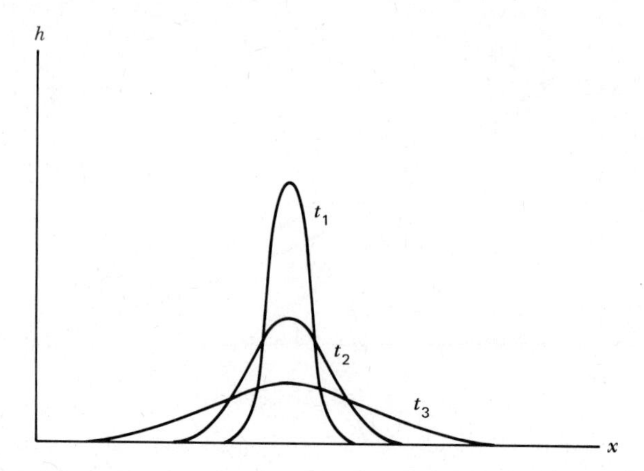

Figure 10-32 Series of three topographic profiles, shaped as Gaussian frequency-distribution curves, which could represent "decay" of mountain range by progressive diffusion of material from higher to lower elevation from time t_1 to t_3. From Scheidegger and Langbein (1966).

rate of erosion is uniform over the entire land surface, being set equal to a constant. Alternatively the rate of erosion can be considered to be proportional to elevation, as shown in the second model (Figure 10-33*b*), where high regions are more quickly eroded than lower regions. The constant *b* in the second model can be interpreted as resistance to erosion as related to, for example, lithology. The third

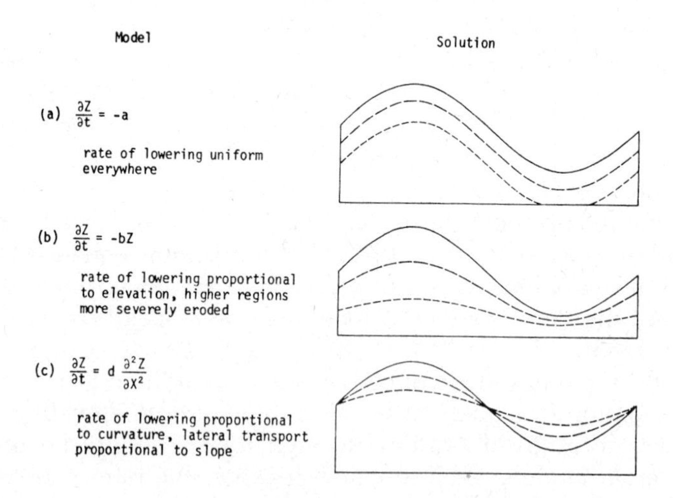

Figure 10-33 Three simple models, represented by differential equations (*a*) (*b*), and (*c*), for lowering of landscape. From Pollack (1969).

model (Figure 10-33*c*) involves the diffusion equation, in which the rate of erosion is proportional to curvature of the land surface. Pollack has combined some of the aspects represented by each of these three models into a composite diffusion model, represented by the following equation:

$$\frac{\partial Z}{\partial t} = \frac{\partial}{\partial X}\left[K(Z)\frac{\partial Z}{\partial X}\right] + A(X, Z) \tag{10-25}$$

where Z, X, and t are variables for elevation, distance, and time, respectively, and K and A are coefficients describing the erosional characteristics of the strata. It should be noted that Equation 10-25 is essentially a diffusion equation with an elevation-dependent diffusion coefficient $K(Z)$ and incorporates an extra term $A(X, Z)$, which is the rate of downcutting by a stream, with spatial dependence restricting action to selected stream sites. At these sites the rate is dependent on the resistance of the stratum in which the stream is flowing. The coefficient $K(Z)$ is a property, also proportional to the resistance of the strata, that plays a role in shaping the slopes exhibited in the valley walls and in determining the rate at which the valley widens.

To solve the equation for a particular erosional system it is necessary first to represent the characteristics of the stratigraphic section in numerical form, usually in a table of stratigraphic units with their respective thicknesses and properties. These properties of erosional resistance are shown graphically in Figure 10-34.

Finite-difference methods are used to obtain numerical solutions for Equation 10-25 under a variety of experimental conditions. Output from the model is given by the topographic elevation, or landscape profile, for a particular time. Evaluation of successive profiles provides an effective method of simulating a valley's development.

Pollack performed a series of experiments with the model to simulate the development of a hypothetical cross section through the Grand Canyon of Arizona (Figure 10-35). The experiment shown in Figure 10-35*a* involved several pulses, with rapid deepening occurring when a nonresistant bed is encountered. After each pulse of valley deepening, there is an interval of valley widening in which the profile "diffuses" outward from the stream site. The results are such that steep slopes develop on the resistant units, and gentle slopes on relatively nonresistant units. Sometimes, however, steep slopes occur where nonresistant strata crop out. As the succession of profiles reveals, the cliff and bench features come and go as erosion progresses.

The results of another experiment are shown in Figure 10-35*b*. This

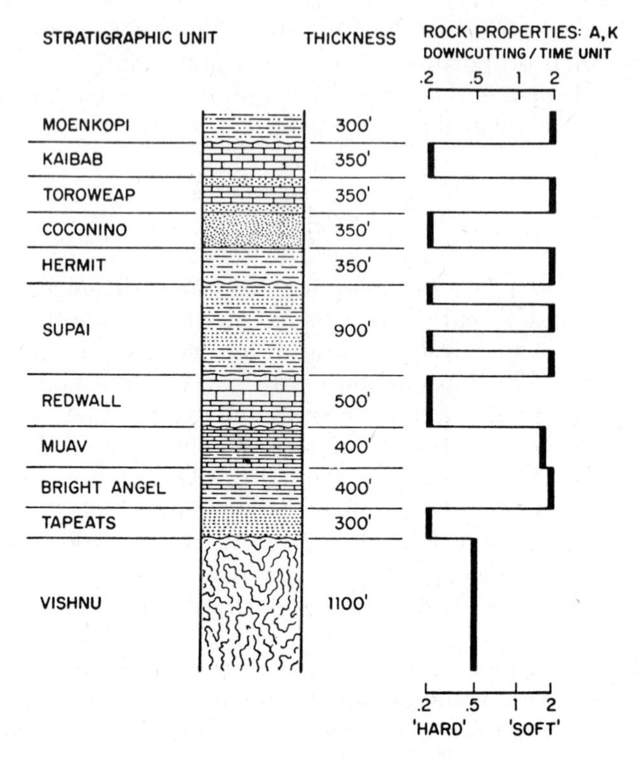

Figure 10-34 Stratigraphic sequence in Grand Canyon and graph of erosional resistance parameters employed in Equation 10-25. From Pollack (1969).

experiment involves a second small tributary stream that has half the downcutting ability of the trunk stream. As the two valleys widen, the adjacent sides of the two streams eventually meet, combining to carry away the interstream material. At various stages mesas appear and disappear. Eventually the minor stream is captured by the major stream, which continues to erode an asymetric canyon.

Although Pollack's model by no means perfectly approximates the Grand Canyon's profile, its performance is surprisingly realistic considering its stark mathematical simplicity. It suggests the feasibility of developing stream-erosion models in three dimensions.

Simulation of Drainage Networks

As has been shown, stream profiles may be approximated by random-walk models in which the steps in the random walk are constrained to be either downward or outward in a horizontal direction. Random-walk methods also can be used to simulate the evolution of

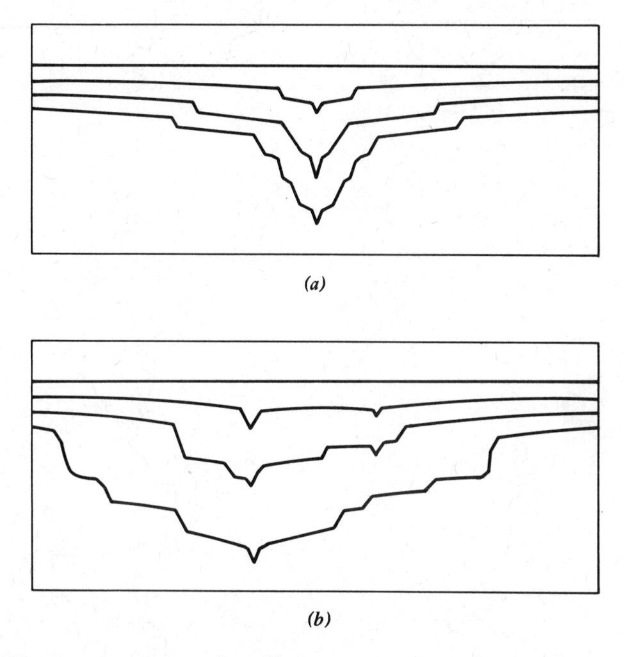

Figure 10-35 Two sequences of profiles representing evolution of a canyon through several time steps, employing Equation 10-25: (*a*) profiles produced by single stream; (*b*) profiles produced by trunk stream and smaller tributary stream. From Pollack (1969).

drainage network in map view. Here the constraints are somewhat different, but of course "uphill" steps in the random walk must be avoided. Random-walk drainage-network models have been developed by a number of hydrologists and geomorphologists. Leopold and Langbein (1962) describe a model that assumes the development of incipient rills that are uniformly spaced at the upper edge of a uniformly sloping plain (Figure 10-36). At the start of the simulation run the rills are assumed to progress in a general downhill direction as the model is advanced through uniform time increments. In each unit of time the random-walk process representing an individual rill moves away from the previous point by a unit distance. The movement at each step is constrained as either forward or to the right or left, but not backward (uphill). Movement in the various permissible directions can be regarded in probability terms. When two rills join, the resulting single rill can be treated with the same rules, and so on.

The result of a particular random-walk drainage network, the rules differing somewhat from those described above, is illustrated in

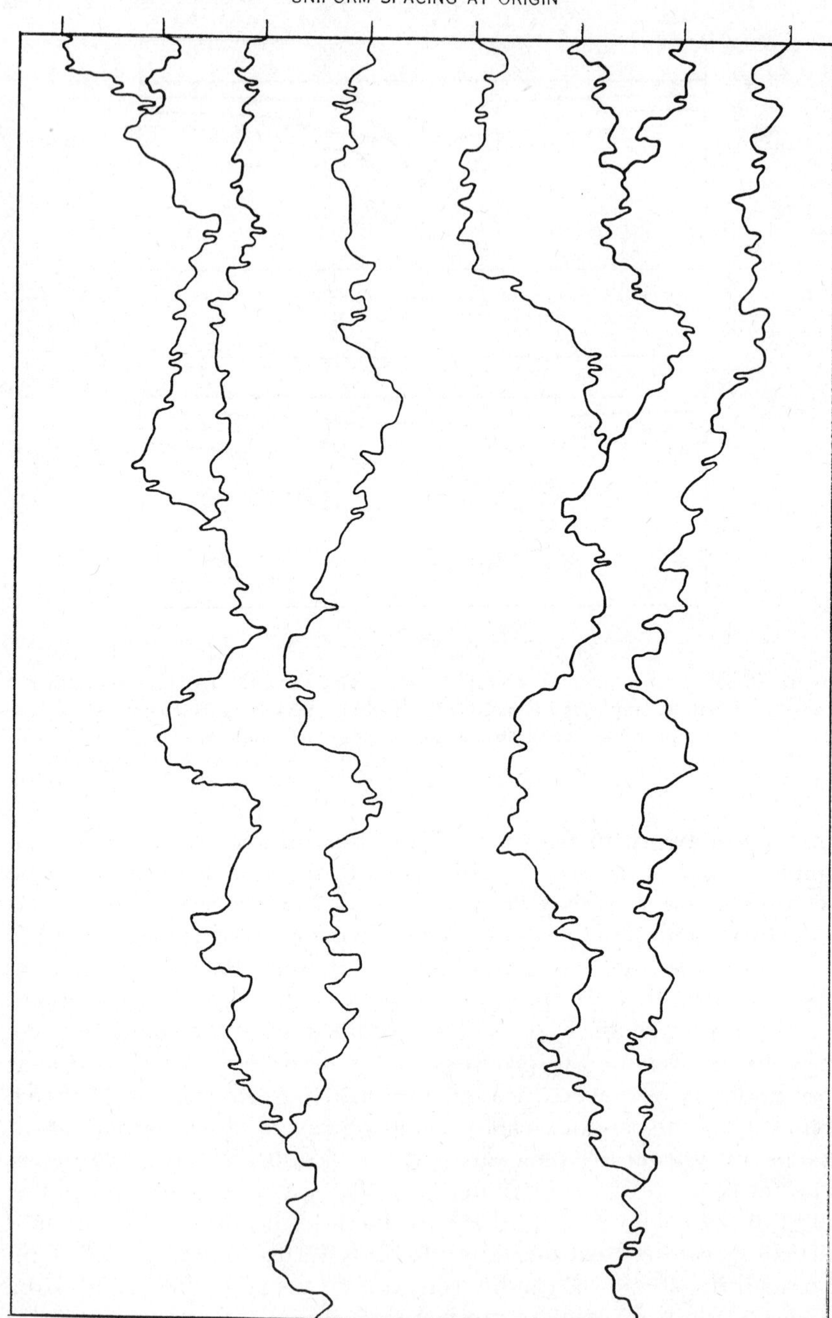

Figure 10-36 Random-walk stream network. Plain along which streams flow slopes gently toward bottom of figure. From Leopold and Langbein (1962).

Figure 10-37. The network is developed by using a meshwork of square cells. Each cell is to be drained, but the drainage channel from each square has an equal chance of leading off in any of the four cardinal directions, subject to the condition that, having made a choice of flow direction, flow in the reverse direction is impossible. Under these conditions it is possible for one or more streams to flow into a cell, but only one can flow out. By random selection of step directions

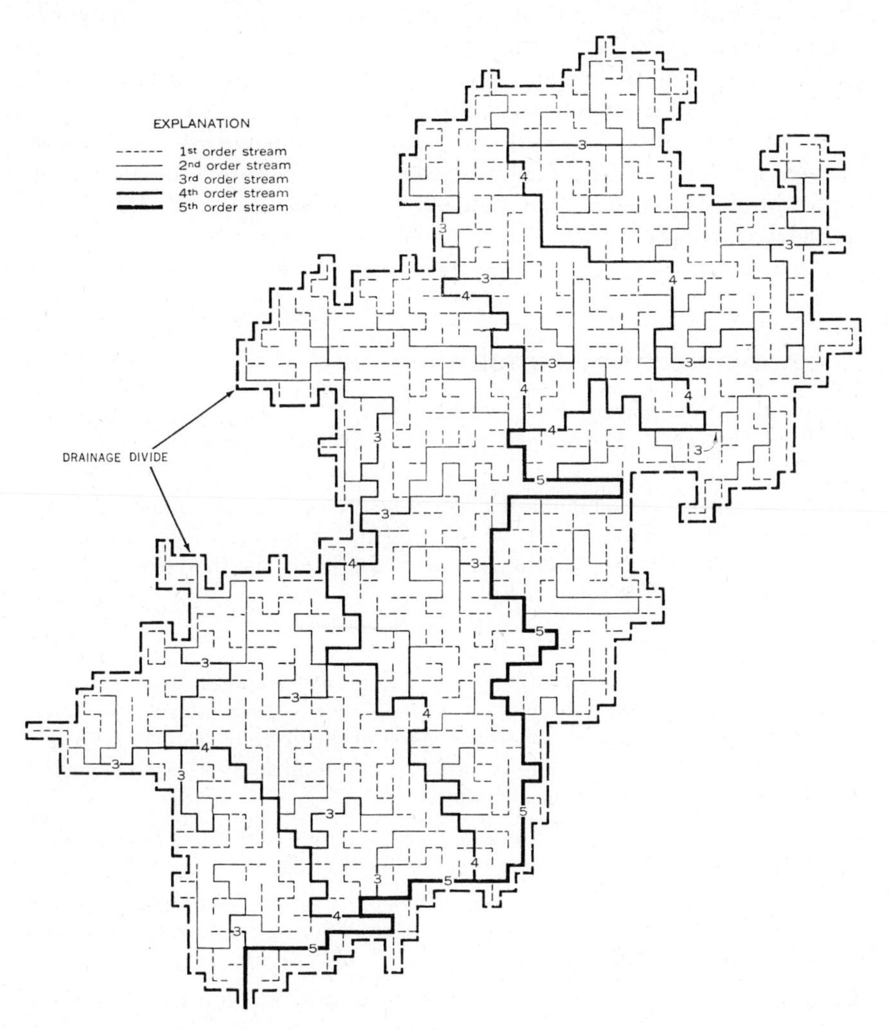

Figure 10-37 Drainage network simulated by employing random-walk technique. Streams are classified by order according to hierarchy of junctions of lesser streams to form larger streams. From Leopold and Langbein (1962).

under the specified conditions, a stream network is generated that is similar to certain actual drainage networks. Watershed divides are created, and streams are progressively joined to create streams of increasing size. The network created by simulation shown in Figure 10-37 represents a network that appears typical of a stream network in a region that is "homogeneous" in lithology and geologic structure. By suitably biasing the probabilities of steps in particular directions, the effect of structural "grain" on development of drainage networks can be imitated.

Computational techniques by which random-walk drainage networks can be produced by computer are described by Schenck (1963) and by Smart and others (1967). Furthermore the computer program described by Smart and others provides for the calculation and tabulation of statistical data after the simulation of a drainage network is completed.

Annotated Bibliography

Anderson, K. W., Handy, R. L., and Lohnes, R. A., 1968, *Final Report on Evolution of Drainage Patterns*, Engineering Research Institute, Iowa State University, 40 pp.

A review of development of drainage patterns from an analytical viewpoint. Many of the analytical relationships could be incorporated into simulation models. An important relationship is that of allometric growth:

$$y = ax^m$$

where a and m are constants. The graph of this equation forms a straight line when plotted on double logarithmic paper. The x-intercept is represented by the constant a, and the slope of the straight line is m. Length of drainage-basin perimeter plotted against total stream length reveals a relationship that is essentially allometric.

Chorley, R. J., 1962, *Geomorphology and General Systems Theory*, U.S. Geological Survey Professional Paper 500-B, 10 pp.

A generalized theoretical and philosophical discussion of applying concepts of dynamic systems to stream systems, emphasizing the self-adjusting, or equilibrium-seeking, aspects of systems.

Culling, W. E. H., 1960, "Analytical Theory of Erosion," *Journal of Geology*, v. 68, pp. 336–344.

A paper on the analytical theory of valley slopes, stream profiles, and alluvial fans.

Culling, W. E. H., 1963, "Soil Creep and the Development of Hillside Slopes," *Journal of Geology*, v. 71, pp. 127–161.

Soil creep can be treated as a stochastic process and can be represented by a one-dimensional random walk, as well as by deterministic diffusion methods.

Dacey, M. F., 1968, "Stream Length and Elevation for the Model of Leopold and Langbein," *Water Resources Research*, v. 4, pp. 163–166.

Employs a random-walk method as a model of a stream profile. The basic assumption is that in a unit of horizontal distance the elevation of the stream either remains constant or drops by a unit of elevation. The probability of either event depends on elevation.

Devdariani, A. S., 1966*a*, "A Plane Mathematical Model of the Growth and Erosion of an Uplift," *Soviet Geography*, v. 8, pp. 183–198.

A mathematical treatment of land forms produced by erosion. The stages in development of landscape are treated with a series of differential equations. For example, tectonic uplifts can be represented by a sine function with amplitude decreasing exponentially. The downcutting effects of streams flowing across an uplifted mass are treated as functions of elevation. Although the paper does not extend beyond a theoretical discussion, many of the concepts could be applied in dynamic simulation models.

Devdariani, A. S., 1966*b*, "The Profile of Equilibrium and a Regular Regime," *Soviet Geography*, v. 8, pp. 168–183.

A mathematical treatment of stream profiles, employing Fourier series as applied to heat conduction to represent stream gradients.

Leopold, L. B., and Langbein, W. B., 1962, *The Concept of Entropy in Landscape Evolution*, U.S. Geological Survey Professional Paper 500-A, 20 pp.

A theoretical and philosophical treatment of stream profiles, emphasizing the role of entropy in both a thermodynamic and probabilistic sense. The most probable condition of a river profile exists when energy in a river system is as uniformly distributed as may be permitted by physical constraints. The most probable state for certain profiles can be synthesized by random-walk models.

Leopold, L. B., Wolman, M. G., and Miller, J. P., 1964, "Drainage Pattern Evolution," in *Fluvial Processes in Geomorphology*, W. H. Freeman, San Francisco, 522 pp.

Chapter 10 deals with drainage-pattern evolution, including random-walk simulation of drainage networks.

Lohnes, R. A., and Handy, R. L., 1967, *The Drainage Density Envelope*, Engineering Research Institute, Iowa State University, Office of Naval Research Technical Report No. 4, 30 pp.

An analytical treatment of drainage networks.

Melton, M. A., 1958, "Correlation Structure of Morphometric Properties of Drainage Systems and Their Controlling Agents," *Journal of Geology*, v. 66, pp. 442–460.

A theoretical treatment of feedback control in fluvial drainage systems.

Pollack, H. N., 1968, "On the Interpretation of State Vectors and Local Transformation Operators," Computer Contribution 22, Kansas Geological Survey, pp. 47–51.

A short treatment of equations that can be applied to the deterministic representation of a sequence of events, illustrated with theoretical applications to changes in elevation of geomorphic profiles.

Pollack, H. N., 1969, "A Numerical Model of the Grand Canyon," in *Geology and Natural History of the Grand Canyon Region*; Four Corners Geological Society Guidebook to Fifth Field Conference, D. C. Baars, ed., pp. 61–62.

Scheidegger, A. E., 1968, "Horton's Laws of Stream Lengths and Drainage Areas," *Water Resources Research*, v. 4, pp. 1015–1021.

Includes a description of simulation of branching stream networks by random graph patterns.

Scheidegger, A. E., and Langbein, W. B., 1966, *Probability Concepts in Geomorphology*, U.S. Geological Survey Professional Paper 500-C, 14 pp.

An introduction to probabilistic analysis and simulation in fluvial systems, including random-walk and Markov-chain representation of slope development, and application of diffusion equations to the erosional reduction of upland areas. Optimization concepts are involved in river-channel geometry, where the principle of least work may be applied.

Schenck, H., Jr., 1963, "Simulation of the Evolution of Drainage-Basin Networks with a Digital Computer," *Journal of Geophysical Research*, v. 68, pp. 5739–5745.

A well-written paper that describes the computational details of random-walk simulation of drainage-basin networks.

Shreve, R. L., 1966, "Statistical Law of Stream Numbers," *Journal of Geology*, v. 74, pp. 17–37.

The broad generality of Horton's law of stream numbers seems to reflect the statistics of a large number of randomly merging stream channels.

Smart, J. S., 1968, "Statistical Properties of Stream Lengths," *Water Resources Research*, v. 4, pp. 1001–1014.

Discusses interrelationships between laws of stream numbers, stream lengths, and basin areas.

Smart, J. S., and Surkan, A. J., 1967, "The Relation Between Mainstream Length and Area in Drainage Basins," *Water Resources Research*, v. 3, pp. 963–974.

Geometrical and statistical analysis of drainage networks.

Smart, J. S., Surkan, A. J., and Considine, J. P., 1967, "Digital Simulation of Channel Networks," in *Symposium on River Morphology*, General Assembly of Bern, pp. 87–98.

About 600 stream-channel networks produced by simulation, using a random-walk technique, yield dimensionless parameters that are very similar to those of real stream systems. Rules governing the random-walk techniques are described.

Wickman, F. E., 1961, "The Distribution Pattern of Land and Water," *Arkiv für Mineralogi och Geologi*, v. 3, pp. 69–98.

A well-written, analytical treatment of the statistical relationships of land and water areas. The assumption is made that lengths of uninterrupted land or water as observed on straight traverses form an exponential frequency distribution.

Woldenberg, M. J., 1968, "Spatial Order in Fluvial Systems: Horton's Laws Derived from Mixed Hexagonal Hierarchies of Drainage Basin Areas," *Harvard Papers in Theoretical Geography*, No. 13, 36 pp.

The number of drainage basins developed on a homogeneous surface can be predicted by assuming that the basins form a series of nested hierarchies of hexagonal areas. A theoretical justification for the assumption of hexagonally shaped areas lies in the fact that a hexagon is the most compact shape possible that fills a two-dimensional space, or area. In turn this appears to accord with the concept of least work by the stream system within a drainage basin.

HYDROLOGY

Hydrology, which is concerned with much of the hydrologic cycle, provides outstanding opportunities for simulation. The processes involving the movement of water within a watershed are complex and highly interdependent. In spite of their complexity, they can be represented effectively by simulation models, employing familiar concepts. Materials-balance methods are very widely applied in watershed simulation, inasmuch as the models are generally concerned with inputs and outputs of volumes of water from place to place. Markov chains have been employed to represent the sequences of events in some models. Most models to date are not concerned with the representation of flow with formal fluid mechanics. Finally, optimization methods appear strongly applicable in some models, since the economic objectives in watershed-resource management concern rather complex tradeoffs involving the allocation of water for irrigation, recreation, power-generation, and flood-control purposes. We have illustrated watershed-hydrology simulation with two models—the Stanford watershed model and the Harvard Water Resources Group model.

Stanford Watershed Model

The Stanford watershed model provides an excellent example of the application of a simulation method to objectives that are both economic and scientific: it is a general-purpose simulation model dealing with hydrologic regimes of streams and rivers, which are in turn components of the gross hydrologic cycle. The hydrologic cycle is easy to describe in qualitative terms. Its components are readily identified, and the interactions between major components are well known. However, extension of qualitative relationships in the hydrologic cycle to yield quantitative results is much more difficult. For example, hydrology has not yet been advanced to the point where the outflow hydrograph response of a watershed to storm rainfall can be accurately predicted.

The motivation to develop watershed-simulation models is to better predict the responses of a watershed to variations in input. For example, given the hydrograph of a particular storm that occurred in a watershed, how would the hydrograph differ if the storm had occurred on a different date with much different initial specified conditions? Such a question is very difficult to answer. The processes of the hydrologic cycle that occur at any point in time are complexly interrelated. For example, infiltration rate depends on soil permeability and, in turn, on the distribution of moisture in the soil profile. Furthermore hydrologic processes tend to be time dependent. Runoff after a storm can be estimated for a point in time, and these estimates can be projected forward in time.

Structure of the Model

The Stanford watershed model is structured about the components of the hydrologic cycle (Figure 10-38). These components are represented in the simulation program, whose general organization is portrayed in a flow chart (Figure 10-39). Precipitation and potential evapotranspiration are the major inputs to the model. Actual evapotranspiration, streamflow, and soil-moisture levels are output from the program. Calculations may be carried out for any number of input stations. As in most digital dynamic simulation models, time is

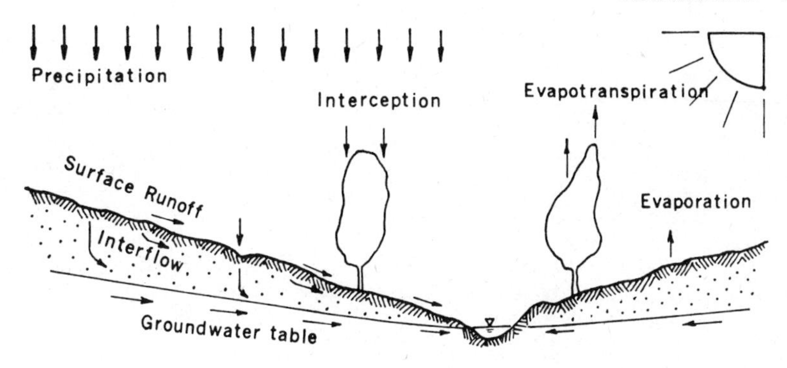

Figure 10-38 Schematic diagram of hydrologic cycle. From Crawford and Linsley (1966).

represented by finite intervals within the model. The roles of major components in the model are outlined below.

Processes at the Land Surface

The reaction of the land surface to rainfall has a major influence on streamflow. Infiltration of water into the soil, storage of water in surface depressions, overland flow, and interflow all interact at the land surface. Indeed, the interdependent effects of these processes are responsible for the complexity of hydrologic behavior. Precipitation over impervious areas, as well as that falling on the surface of lakes and streams, contributes directly to surface runoff. The model provides for establishment of an "impervious fraction" of total watershed area.

Infiltration is a key process at the land surface. Water may infiltrate immediately from rainfall into the soil profile (direct infiltration) or it may flow into temporary storage and infiltrate later (delayed infiltration). The interaction between the direct and delayed processes of infiltration is of major importance. As rainfall begins, flow enters soil fissures, loosely packed soil, and surface depressions. Flow into these features can be at rates that are relatively high and essentially independent of rainfall rates. However, if heavy rainfall continues, temporary storages fill, and overland flow, coupled with direct infiltration, begins to occur. When direct infiltration throughout the watershed governs runoff volume, variations in infiltration capacity strongly influence the behavior of the watershed.

Variations in infiltration capacity on an area basis can be expressed as a cumulative frequency-distribution curve (Figure 10-40). Such a curve represents, in effect, a large number of measurements of infiltra-

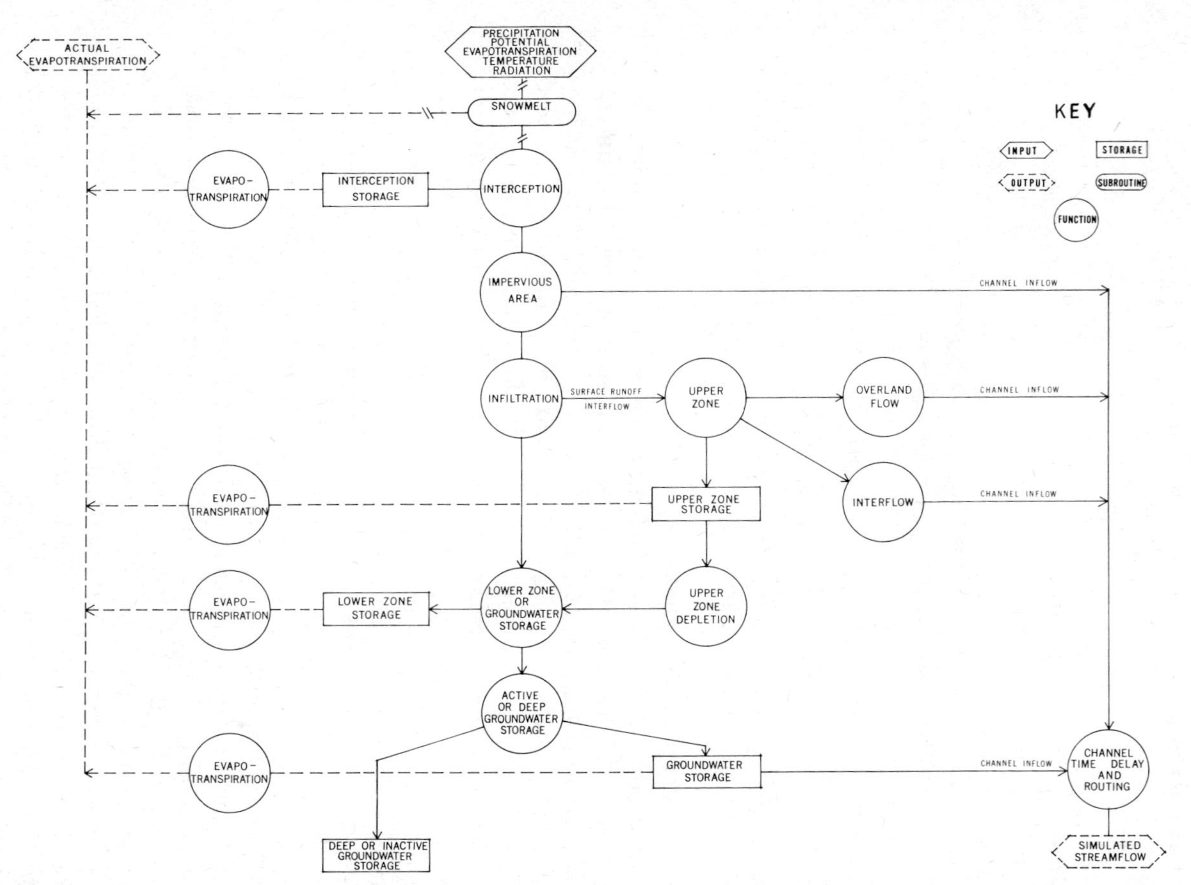

Figure 10-39 Generalized flow chart depicting major components of Stanford watershed model. From Crawford and Linsley (1966).

tion capacity and shows the proportions of watershed area with an infiltration capacity equal to or less than the measured values.

Disposition of rainfall at any point in the watershed depends on the local infiltration rates. If the mean moisture supply (Figure 10-40) is $\bar{x}$ inches at a particular time interval, the total volume of infiltration will be equal to or less than that indicated by the cumulative frequency-distribution curve. If the moisture supply increases, the infiltration volume will increase as long as some unfilled infiltration capacity remains. Moisture received above the infiltration capacity (represented by area *B* of Figure 10-40) is the volume of water that is free to move toward stream channels as overland flow. Water moving overland may or may not reach stream channels, some being diverted to temporary storage and in turn remaining subject to infiltration.

The movement of water in surface or overland flow is an important land-surface process in hydrology. Interactions between overland flow and infiltration need to be considered since both processes occur simultaneously. During overland flow water held in detention remains available for infiltration. Infiltration processes are approximated in the simulation model by two interdependent calculations that continuously determine the direct infiltration into the soil profile and the increases of water in temporary storage that result in delayed infiltration. Infiltration capacity is represented by a linear cumulative frequency distribution in the model, but the shapes of

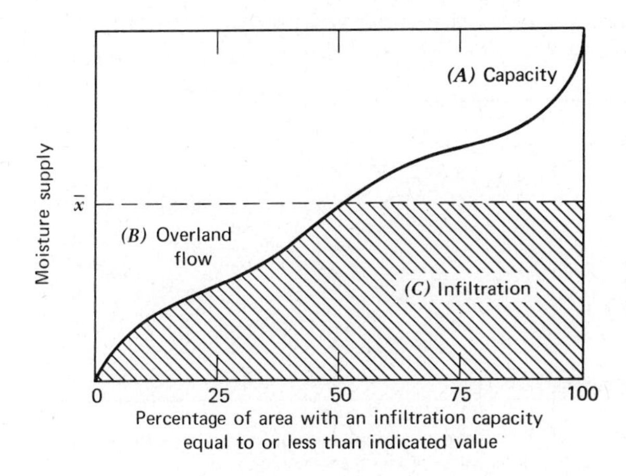

Figure 10-40 Diagram showing relationships between (*a*) cumulative-frequency-distribution curve of infiltration capacity over a hypothetical area, (*b*) volume of water, as a percentage, available for overland flow, and (*c*) volume of water that undergoes infiltration, as a percentage. From Crawford and Linsley (1966).

actual infiltration-capacity frequency distributions are poorly understood. In addition to a linear assumption, infiltration capacity is divided into two regions (Figure 10-41), a lower region representing water that moves into soil moisture storage and an upper region in which infiltration contributes to interflow. Thus the tendency for infiltrating water to become interflow is assumed to be proportional to the local infiltration capacity.

The response of the model to a given moisture supply $\bar{x}$ is graphically illustrated in Figure 10-42. Disposition of moisture is proportional to the three areas beneath the horizontal moisture-supply line. The response of the three components of land-surface response to variations in moisture supply are shown in Figure 10-43. As the curves in Figure 10-43 reveal, the proportion of net infiltration levels off as the moisture increases, the proportion detained in interflow increases and finally levels off, and the proportion entering into overland flow continues to increase.

Groundwater Movement and Evapotranspiration

The model provides for inflow to and outflow from groundwater storage. Inflow to groundwater is from a portion of the net infiltration and a portion of the delayed infiltration. Outflow is proportional to the product of the cross-sectional area of the groundwater volume and the gradient of flow.

The volume of water that leaves a watershed as evaporation and transpiration exceeds the total volume of streamflow in most hydrologic regimes. The model provides for evapotranspiration, and con-

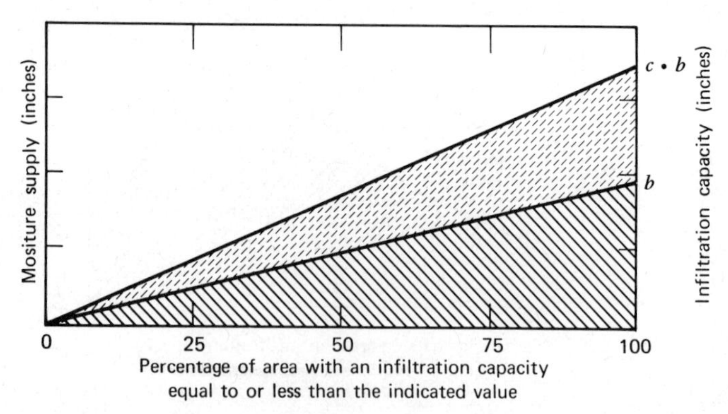

Figure 10-41 Linear cumulative frequency distribution for infiltration assumed in simulation model. Upper region represents infiltration contributed to interflow. Lower region represents infiltration that contributes to soil moisture storage. From Crawford and Linsley (1966).

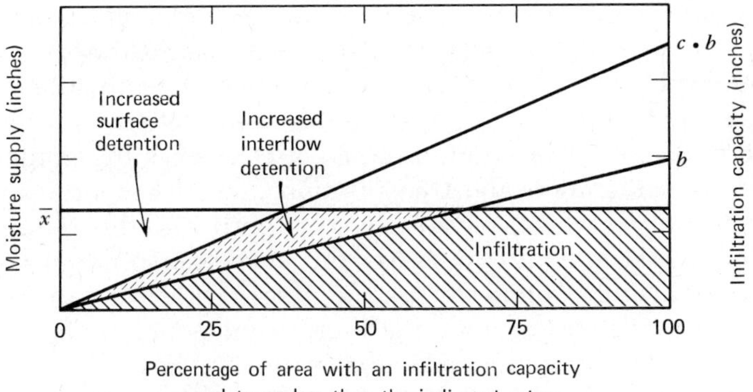

Figure 10-42 Graph illustrating response of simulation model to given average moisture supply $\overline{x}$. The three areas beneath horizontal line $\overline{x}$ are proportional to volumes of water assigned to infiltration, interflow detention, and detained in surface as overland flow. From Crawford and Linsley (1966).

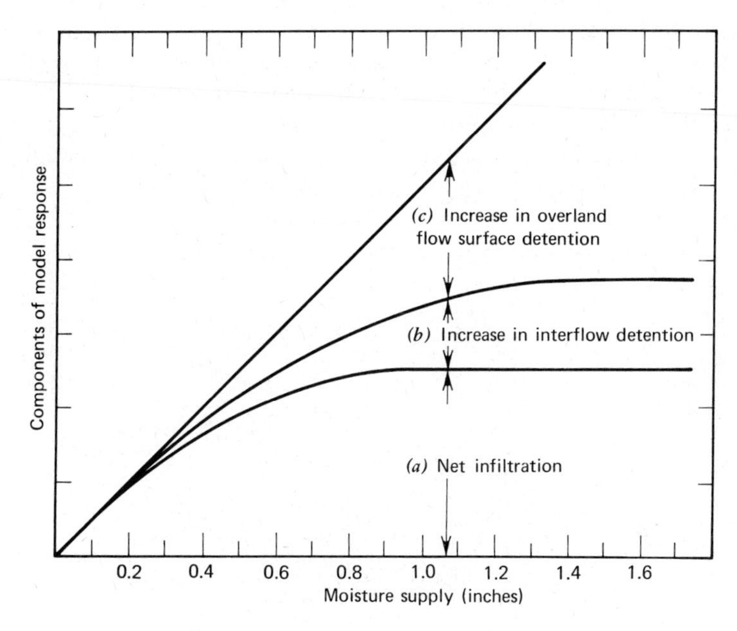

Figure 10-43 Response of simulation model as moisture supply varies. Intervals between curves represent proportion of (*a*) net infiltration, (*b*) water detained in interflow, and (*c*) water in overland flow. From Crawford and Linsley (1966).

siders water losses from interception storage and a minor proportion of loss due to evaporation from stream and lake surfaces and from evapotranspiration from groundwater storage. A distinction is made between potential evapotranspiration and actual transpiration. Potential evapotranspiration pertains, of course, to the potential for combined evaporation and transpiration. Actual evapotranspiration can occur only if water is available. To accord with this requirement the model first attempts to satisfy the evapotranspiration potential from water in interception storage and in turn from water in the upper zone. Since evapotranspiration at a given moment may be expected to vary over the area of a particular watershed, a linear cumulative distribution (Figure 10-44), similar to that representing infiltration capacity, is employed.

Simulation of Specific Watersheds

The outflow hydrograph as measured in a stream-channel system reflects the importance of the hydrological characteristics of the land surface relative to the time delay. A hydrograph of a watershed is shown in Figure 10-45, which compares observed and simulated outflow hydrographs and also the hydrograph of overland flow. In medium and larger watersheds, such as that shown in Figure 10-45, storage and flow in the channel system become large relative to those in overland flow. In other words, the channel system becomes the dominant factor in the outflow hydrograph, and separation of the overland flow and channel flow seems to identify the relative importance of different characteristics of the watershed.

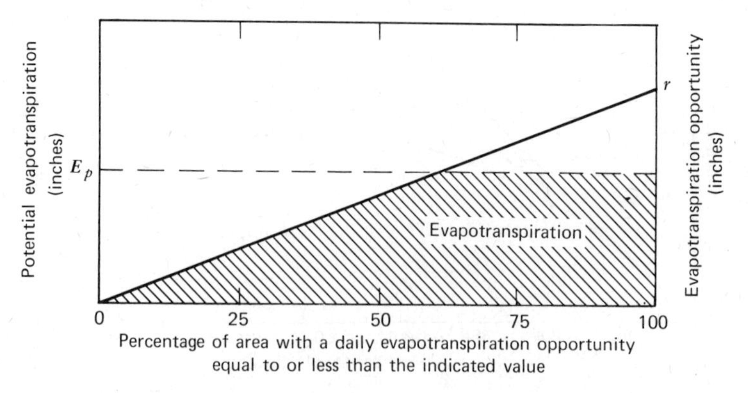

Figure 10-44 Graph representing hypothetical linear cumulative frequency distribution, relating evapotranspiration as percent of land area. From Crawford and Linsley (1966).

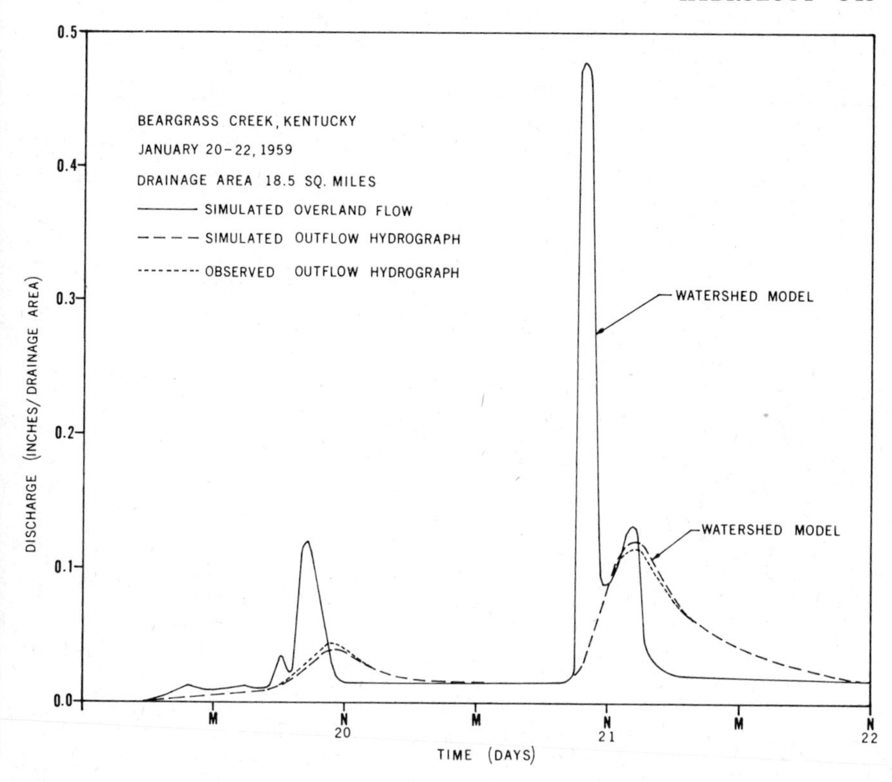

Figure 10-45 Hydrograph of Beargrass Creek, Kentucky, January 20–22, 1959, illustrating simulated overland flow and contrasting simulated stream outflow and observed stream outflow. From Crawford and Linsley (1966).

The model has been used to simulate stream-outflow volume over extended periods. Figure 10-46 is a graph of mean daily flow over a 7-month period of the Russian River at Hopland, California. The observed daily flow is compared with the simulated flow. The similarity of the simulated and observed outflow is a testimony of the ability of the simulation model to mimic the behavior of the Russian River watershed.

Harvard Water Resources Model

Another notable simulation application in hydrology is the Harvard Water Resources model (described by Fiering, 1965, and by Thomas and Burden, 1965) used to explore alternative strategies in dealing with "waterlogging" of irrigated areas in the Indus River Basin of West Pakistan, where leakage from irrigation canals has raised the

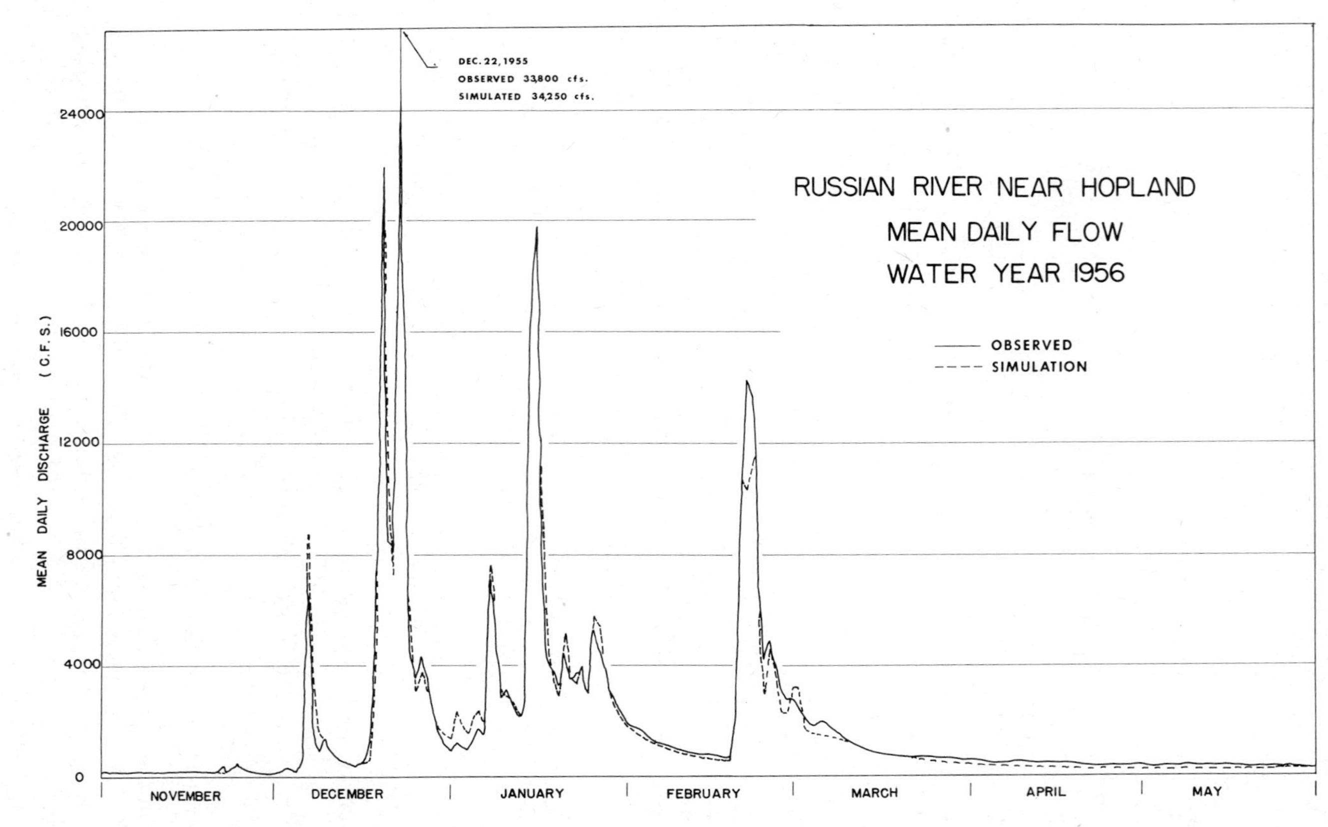

Figure 10-46 Hydrograph of Russian River near Hopland, California, over 7-month period. Curves contrast observed and simulated stream outflow. From Crawford and Linsley (1966).

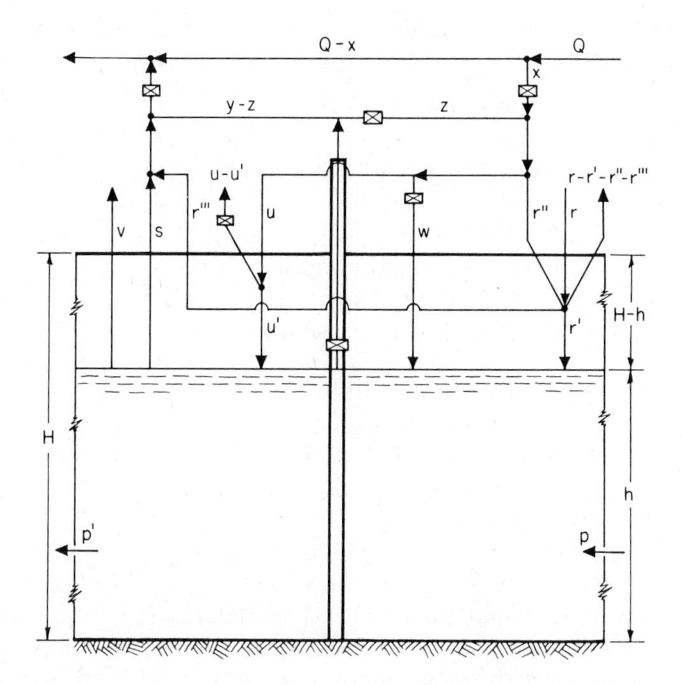

Figure 10-47 Schematic representation of water movement in Harvard Water Resources Group simulation model. From Thomas and Burden (1966).

water table so high that it lies at the land surface in places. The result is that the soil is saturated with water, and salts tend to accumulate at the surface due to evaporation, so that much of the land can no longer be cultivated. The problem is serious because more than 100,000 acres are being lost to cultivation each year, and the loss of agricultural products poses a potential food-shortage problem, particularly in view of the rising population.

The Harvard model was employed to ascertain the physical and economic effects of installing a system of tubed wells to pump groundwater, reducing the water table sufficiently far beneath the surface so that the soil will not be waterlogged and allowing the accumulated salt to be leached away. The remedy that emerges from the study is to install about 32,000 tube wells, each spaced about 6000 feet apart. The simulation model is an ingenious one, taking all strongly relevant factors into consideration. Figure 10-47 provides a schematic representation of the flow aspects of the model. As the diagram portrays, the overall water economy is complex, involving water moving through canals, water entering the ground as rainfall and via seepage from canals, and water exiting from the ground via pumping and evaporation. Symbols in the diagram are defined as follows:

Q = canal flow
r = rainfall
r' = rain-to-ground water flow
r'' = overland runoff-to-drain flow
w = canal leakage
u = irrigation water applied
u' = irrigation-water throughput
s = groundwater seepage to drain
v = evaporation from the groundwater
p = groundwater inflow
p' = groundwater outflow
H = height above datum (bottom of aquifer)
x = watercourse flow
y = well pumpage
z = tube well-to-irrigation flow
$y-z$ = tube well-to-drain flow

The six "valves" (small boxes with diagonals) shown in Figure 10-47 are economic decision points. For example, the "valve" of the leakage vector w represents the cost function of applying sealants to the walls

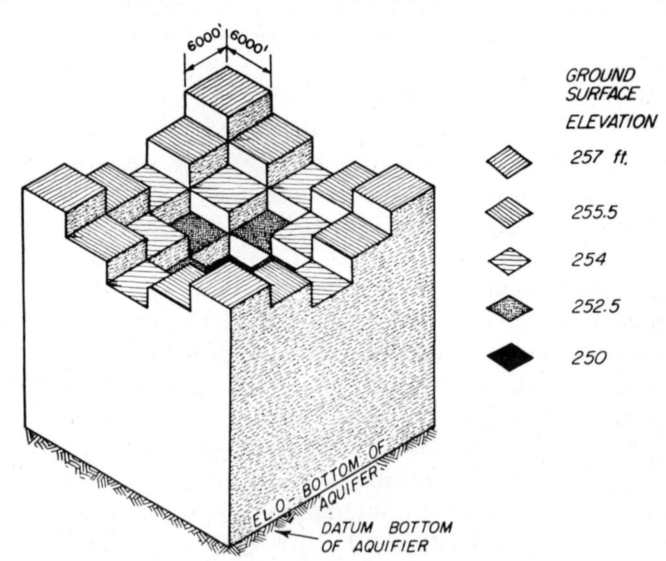

Figure 10-48 Detail of construction of multiwell model in which heights of columns with square cross sections represent thickness of aquifer. Different elevations of land surface are identified by graphic symbols as shown. From Thomas and Burden (1966).

and bottoms of canals to achieve various levels of efficiency in reducing leakage. The program makes use of a regular grid system to represent space. The two geographic dimensions are represented by square cells that are 6000 feet in width. The vertical dimension is represented by the height of each cell, which can be regarded as a prismatic column of variable length (Figure 10-48). The elevation of the land surface is extremely critical and is represented as an average for each cell.

The basic time unit is the 3-month season. The parameter that controls influx of water, evaporation, and other factors are specific for each season. Inputs representing fluctuations in rainfall and stream flow may include stochastic components that involve either normally or lognormally distributed random variables—or, as an option, these inputs can be purely deterministic.

Annotated Bibliography

Amorocho, J., and Hart, W. E., 1964, "A Critique of Current Methods in Hydrologic Systems Investigation," *Transactions of the American Geophysical Union*, v. 45, pp. 307–321.

A broad, generalized review of methods that may be applied to watershed hydrology, with emphasis on the role of stochastic processes. The merits of a linear versus nonlinear methods of analysis of hydrologic systems are discussed at length.

Carter, D. B., 1968, *Basic Data and Water Budget Computation for Selected Cities in North America*, Earth Science Curriculum Project Reference Series 8, Prentice-Hall, Englewood Cliffs, N.J., 34 pp.

A useful introduction to methods of computing the water budget. Data are given for 473 stations in the United States.

Crawford, N. H., and Linsley, R. K., 1966, *Digital Simulation in Hydrology: Stanford Watershed Model IV*, Department of Civil Engineering, Stanford University, Technical Report, No. 39, 210 pp.

One of the most thoroughly documented simulation models touching on geology that has been published to date.

Dawdy, D. R. and Lichty, R. W., 1968, "Methodology of Hydrologic Model Building," in *The Use of Analog and Digital Computers in Hydrology*, v. 2, Publication 81, International Association of Scientific Hydrology, UNESCO, pp. 347–355.

Deals with problems of comparing two alternative representations of rainfall-runoff simulation models and discusses criteria for comparing them with observed watershed data.

Fiering, M. B., 1965, "Revitalizing a Fertile Plain, a Case Study in Simulation and Systems Analysis of Saline and Waterlogged Areas," *Water Resources Research*, v. 1, pp. 41–61.

Well-written article describing the principles of the Harvard Water Resources Group simulation model.

Finnemore, E. J., and Perry, B., 1968, "Seepage through an Earth Dam Computed by the Relaxation Technique," *Water Resources Research*, v. 4, pp. 1059–1067.

Good discussion of finite-difference representation of fluid flow through porous media.

Hornberger, G. M., Fungaroli, A. A., and Remson, I., 1966, *A Computer Program for the Numerical Analysis of Soil-Moisture Systems*, Drexel Institute of Technology, Department of Civil Engineering, Philadelphia, 14 pp.

Provides listing of program.

Hufschmidt, M. M., 1963, "Simulating the Behavior of a Multi-Unit, Multi-Purpose Water-Resource System," in *Symposium on Simulation Models*, A. C. Haggett and F. E. Balderston, eds., Southwestern Publishing Co., Cincinnatti, pp. 202–219.

Concerned principally with the application of optimization methods in watershed-resource simulation. The objective function is defined as the maximization of the present value of benefits derived from irrigation, energy, and flood control over a 50-year interval. Present value of future benefits is discounted at a selected annual interest rate. The net-benefit response surfaces calculated were found to be surprisingly complex.

Hufschmidt, M. M., and Fiering, M. B., 1966, *Simulation Techniques for Design of Water-Resource Systems*: Harvard University Press, Cambridge, Mass., pp. 212.

A modern, detailed, and highly readable exposition of simulation techniques for water-resource systems, in which the prime objective is economic analysis of the system.

Lloyd, E. H., 1967, "Stochastic Reservoir Theory," in *Advances in Hydroscience*, Ven Te Chow, ed., v. 4, Academic Press, New York, pp. 281–339.

Provides thorough treatment of Markov chains applied to watershed hydrology, with emphasis on variations in reservoir inflow.

Loucks, D. P., and Lynn, W. R., 1966, "Probabilistic Models for Predicting Stream Quality," *Water Resources Research*, v. 2, pp. 593–605.

Use of Markov chains.

Manabe, S., Smagorinsky, J., and Strickler, R. F., 1965, "Simulated Climatology of a General Circulation Model with a Hydrologic Cycle," *Monthly Weather Review*, v. 93, pp. 769–798.

Principally concerned with a large-scale complex simulation model of the atmosphere, which includes a simple hydrologic cycle. Processes represented in the model include evaporation, precipitation, moisture balance, energy budget, and heat balance of the earth's surface. The paper is directed toward meteorologists.

Manzer, D. F., and Barnett, M. P., 1962, "Analysis by Simulation: Programming Techniques for a High-Speed Digital Computer," in *Design of Water Resource Systems, New Techniques for Relating Economic Objectives, Engineering Analysis, and Government Planning*, A. Maass, ed., Harvard University Press, Cambridge, Mass., pp. 324–381.

A detailed but readable description of a simulation model whose objectives are river management. The model can accommodate various combinations of reservoirs, hydroelectric power plants, and irrigated cropland areas. Input to the model includes desired volumes of water for irrigation purposes and power generation; allocation of reservoir capacity according to active, dead, and flood storage; and finally a series of monthly runoff values. An aspect of the model concerns the economic consequences of specific simulation results.

McGilchrist, C. A., Woodyer, K. D., and Chapman, T. G., 1968, "Recurrence Intervals between Exceedances of Selected River Levels. I. Introduction and a Markov Model," *Water Resources Research*, v. 4, pp. 183–189.

Uses Markov chains.

McGuinness, C. L., 1967, "Groundwater Research in the U.S.A.," *Earth Science Review*, v. 3, pp. 181–202.

Philosophical discussion of the rationale of treating groundwater problems as a dynamic system with both analog and digital computer.

Meier, W. L., Jr., and Beightler, C. S., 1967, "An Optimization Method for Branching Multistage Water Resource Systems," *Water Resources Research*, v. 3, pp. 645–652.

Application of dynamic programming techniques in optimizing complex water-resource systems.

Miller, D. H., 1968, *A Survey Course: The Energy and Mass Budget at the Surface of the Earth*, Association of American Geographers, No. 7, 142 pp.

An abbreviated, broad-scale survey and tabulation of concepts pertaining to the exchange of water, matter, and energy at the earth's surface. The section dealing with the exchanges of water at the earth's surface provides an extensive annotated bibliography.

Onstad, C. A., and Brakensiek, D. L., 1968, "Watershed Simulation by Stream Path Analogy," *Water Resources Research*, v. 4, pp. 965–971.

A short, readable description of a simulation application involving stream-outflow hydrographs that are synthesized by routing water input through a simulated drainage system. The simulated flow system is developed from a flow net.

Pattison, A., 1965, "Synthesis of Hourly Rainfall Data," *Water Resources Research*, v. 1, pp. 489–498.

A clear exposition of the application of Markov chains in simulating rainfall variations, which in turn may be used as input to watershed-simulation models. Hourly rainfall variations are simulated with a sixth-order Markov chain. Transition probabilities are estimated from actual data. The performance of the Stanford watershed model with actual rainfall data used as input, compared with synthetic data as input, is sufficiently similar to indicate that the variations in simulated rainfall closely approximate actual variations.

Pinder, G. F., and Bredehoeft, J. D., 1968, "Application of the Digital Computer for Aquifer Evaluation," *Water Resources Research*, v. 4, pp. 1069–1093.

Discusses finite-difference representation of groundwater flow. Provides review of literature pertaining to numerical and computational applications.

Remson, I., Fungaroli, A. A., and Hornberger, G. M., 1967, "Numerical Analysis of Soil-Moisture Systems," *Journal of the Irrigation and Drainage Division American Society of Civil Engineers*, v. 93, No. IR3, Proceedings Paper 5429, pp. 153–166.

Describe the mathematics of finite-difference representation of soil-moisture flow.

Shamir, U. Y., and Harleman, D. R. F., 1967, "Numerical Solutions for Dispersion in Porous Mediums," *Water Resources Research*, v. 3, pp. 557–581.

Develops mathematics for finite-difference representation of dispersion in three-dimensional flow fields in porous media. The technique may be applied in groundwater pollution studies.

Solomon, S. I., Denouvilliez, J. P., Chart, E. J., Woolley, J. A., and Cadou, C., 1968, "The Use of a Square-Grid System for Computer Estimation of Precipitation, Temperature, and Runoff," *Water Resources Research*, v. 4, pp. 919–929.

Discusses philosophy and techniques of applying a square-grid system for bookkeeping in watershed hydrology.

Thomas, H. A., Jr., and Burden, R. P., 1965, *Indus River Basin Studies*: Final Report to the Science Advisor to the Secretary of the Interior Harvard Water Resources Group, No. 14–08–0001–8305, 300 pp.

Detailed report dealing with the principles and details of the Harvard Water Resources Group simulation model applied to ground- and surface-water relationships in the Indus River Basin of West Pakistan.

Winslow, J. D., and Nuzman, C. E., 1966, *Electronic Simulation of Ground-Water Hydrology in the Kansas River Valley Near Topeka, Kansas*, Special Distribution Publication 29, Kansas Geological Survey, 24 pp.

Well-written description of analog-computer simulation of ground-water flow.

APPENDIX

Solution of Simultaneous Linear Equations

This appendix provides a brief introduction to some methods of solving simultaneous linear equations. Simultaneous linear equations appear in many applications, including regression methods, linear programming, and in methods for solving differential equations in finite-difference form. Let us begin by defining simultaneous linear equations.

A set of n algebraic equations with n unknowns is linear if each term in each equation contains no more than one unknown and if each unknown is confined to the first power. For example, if there are two simultaneous equations and two unknowns, then the graph of each equation is a straight line. The point at which the two lines intersect is a unique or simultaneous solution. The equations

$$x + y = 1 \tag{A-1}$$

$$2x - y = 4 \tag{A-2}$$

have a unique solution at $x = \frac{5}{3}$ and $y = -\frac{2}{3}$, as is revealed by their graphs in Figure A-1*a*. By contrast, the equations

$$4x + 2y = 1 \tag{A-3}$$

$$8x + 4y = -3 \tag{A-4}$$

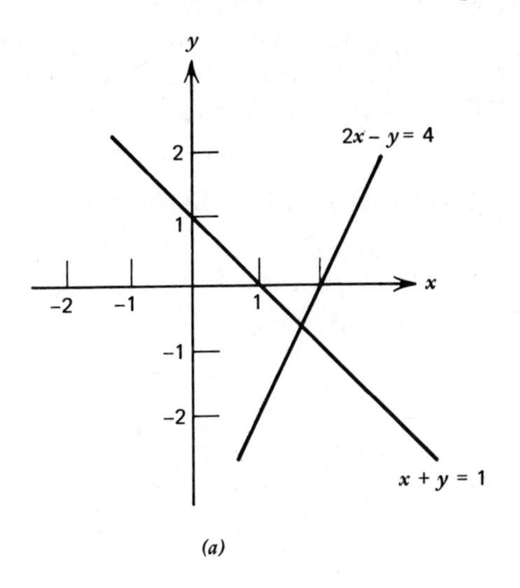

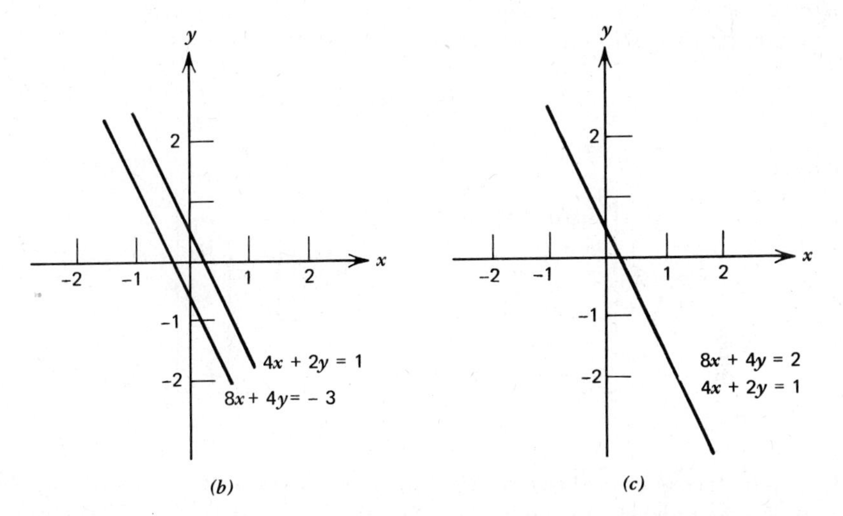

Figure A-1 Graphic solutions for simultaneous equations with two unknowns: (*a*) unique solution at intersection of lines; (*b*) no solution, since lines are parallel; (*c*) no unique solution, since both equations are defined by the same line.

have no unique solution, because their graphs are parallel lines and therefore do not intersect (Figure A-1*b*). The equations

$$4x + 2y = 1 \tag{A-5}$$

$$8x + 4y = 2 \tag{A-6}$$

also lack a unique solution (Figure A-1c), because Equation A-6 is simply Equation A-5 multiplied by 2. A system of equations for which no unique solution exists is said to be *singular*.

LINEAR EQUATIONS IN MATRIX FORM

A system of n linear equations with n unknowns may be written as

$$\begin{aligned} a_{11}x_1 + a_{12}x_2 + \cdots + a_{1n}x_n &= b_1 \\ a_{21}x_1 + a_{22}x_2 + \cdots + a_{2n}x_n &= b_2 \\ \cdots\cdots\cdots\cdots\cdots\cdots\cdots\cdots\cdots \\ a_{n1}x_1 + a_{n2}x_2 + \cdots + a_{nn}x_n &= b_n \end{aligned} \tag{A-7}$$

where a_{ij} and b_i are the known constants and x_i the unknowns. We may write this system in matrix notation as follows:

$$\mathbf{AX} = \mathbf{B} \tag{A-8}$$

where

$$\mathbf{A} = \begin{bmatrix} a_{11}a_{12} \cdots a_{1n} \\ a_{21}a_{22} \cdots a_{2n} \\ \cdot \\ \cdot \\ \cdot \\ a_{n1}a_{n2} \cdots a_{nn} \end{bmatrix} \quad \mathbf{X} = \begin{bmatrix} x_1 \\ x_2 \\ \cdot \\ \cdot \\ \cdot \\ x_n \end{bmatrix} \quad \mathbf{B} = \begin{bmatrix} b_1 \\ b_2 \\ \cdot \\ \cdot \\ \cdot \\ b_n \end{bmatrix}$$

The **A**-matrix is called the *coefficient matrix*. If the **B**-vector is joined to the **A**-matrix on the right to make the $(n+1)$th column of **A**, the resulting matrix is called the *augmented matrix* **A**|**B**.

$$\mathbf{A}|\mathbf{B} = \begin{bmatrix} a_{11}a_{12} \cdots a_{1n}b_1 \\ a_{21}a_{22} \cdots a_{2n}b_2 \\ \cdot \quad \cdot \quad \quad \cdot \quad \cdot \\ \cdot \quad \cdot \quad \quad \cdot \quad \cdot \\ \cdot \quad \cdot \quad \quad \cdot \quad \cdot \\ a_{n1}a_{n2} \cdots a_{nn}b_n \end{bmatrix} \tag{A-9}$$

For example, the system of equations

$$x_1 + x_2 = 1 \tag{A-10}$$

$$2x_1 - x_2 = 4 \tag{A-11}$$

can be described in matrix notation as

$$\begin{bmatrix} 1 & 1 \\ 2 & -1 \end{bmatrix} \cdot \begin{bmatrix} x_1 \\ x_2 \end{bmatrix} = \begin{bmatrix} 1 \\ 4 \end{bmatrix} \tag{A-12}$$

where

$$\mathbf{A} = \begin{bmatrix} 1 & 1 \\ 2 & -1 \end{bmatrix}; \quad \mathbf{X} = \begin{bmatrix} x_1 \\ x_2 \end{bmatrix}; \quad \mathbf{B} = \begin{bmatrix} 1 \\ 4 \end{bmatrix}$$

The augmented matrix is

$$\mathbf{A}|\mathbf{B} = \begin{bmatrix} 1 & 1 & 1 \\ 2 & -1 & 4 \end{bmatrix} \tag{A-13}$$

METHODS OF SOLUTION

There are two principal ways of solving sets of simultaneous linear equations—namely, by elimination and by iteration. Elimination is the method first encountered in solving small sets of simultaneous equations by hand. Matrix methods that employ the same principles are readily solved by computer and permit large sets of equations to be treated in this way. However, certain types of matrices, particularly those with a large number of zero elements, do not give satisfactory solutions by elimination methods but can sometimes be solved by iterative methods, such as the Gauss–Seidel method discussed in Chapter 5.

Elimination Methods

The equations

$$2x_1 + 5x_2 = 8 \tag{A-14}$$

$$6x_1 - 3x_2 = 4 \tag{A-15}$$

may be solved by elimination as follows: first we notice that the augmented matrix is

$$\mathbf{A}|\mathbf{B} = \begin{bmatrix} 2 & 5 & 8 \\ 6 & -3 & 4 \end{bmatrix} \tag{A-16}$$

and we wish to reduce this to

$$\begin{bmatrix} 1 & 0 & \frac{11}{9} \\ 0 & 1 & \frac{10}{9} \end{bmatrix}$$

or

$$\begin{bmatrix} 1 & 0 \\ 0 & 1 \end{bmatrix} \cdot \begin{bmatrix} x_1 \\ x_2 \end{bmatrix} = \begin{bmatrix} \frac{11}{9} \\ \frac{10}{9} \end{bmatrix} \tag{A-17}$$

where

$$\begin{bmatrix} 1 & 0 \\ 0 & 1 \end{bmatrix}$$

is called an *identity* matrix and consists of 1's in the principal diagonal and 0's elsewhere. Another matrix multiplied by an identity matrix is unchanged, so the values of x_1 and x_2 can be read directly from Equation A-17 as $x_1 = \frac{11}{9}$, $x_2 = \frac{10}{9}$.

Given this solution as our goal, let us return to the original equations and see how these solutions can be obtained by elimination. Our first step is to make the coefficient of x_1 equal to 1 in Equation A-14 by dividing through by 2 to give

$$1x_1 + \tfrac{5}{2}x_2 = 4 \tag{A-18}$$

If we now subtract six times the modified first equation from the second equation, we can make the coefficient of x_1 in the second equation equal to zero:

$$0x_1 - 18x_2 = -20 \tag{A-19}$$

We then divide by -18 to obtain

$$0x_1 + 1x_2 = \tfrac{10}{9} \tag{A-20}$$

Finally we subtract $\frac{5}{2}$ times Equation A-20 from Equation A-18 to obtain

$$1x_1 + 0x_2 = \tfrac{11}{9} \tag{A-21}$$

The equations are now

$$1x_1 + 0x_2 = \tfrac{11}{9} \tag{A-22}$$

$$0x_1 + 1x_2 = \tfrac{10}{9} \tag{A-23}$$

or, in other words,

$$\begin{bmatrix} 1 & 0 \\ 0 & 1 \end{bmatrix} \cdot \begin{bmatrix} x_1 \\ x_2 \end{bmatrix} = \begin{bmatrix} \frac{11}{9} \\ \frac{10}{9} \end{bmatrix} \tag{A-24}$$

and $x_1 = \frac{11}{9}$ and $x_2 = \frac{10}{9}$.

If, instead of performing these operations on the original equations, we perform them on the augmented matrix, we achieve exactly the same result:

$$\begin{bmatrix} 2 & 5 & 8 \\ 6 & -3 & 4 \end{bmatrix} \tag{A-25}$$

Divide the upper row by 2 to give

$$\begin{bmatrix} 1 & \frac{5}{2} & 4 \\ 6 & -3 & 4 \end{bmatrix} \tag{A-26}$$

Subtract six times upper row from lower row:

$$\begin{bmatrix} 1 & \frac{5}{2} & 4 \\ 0 & -18 & -20 \end{bmatrix} \tag{A-27}$$

Divide lower row by -18:

$$\begin{bmatrix} 1 & \frac{5}{2} & 4 \\ 0 & 1 & \frac{10}{9} \end{bmatrix} \tag{A-28}$$

Subtract $\frac{5}{2}$ times row 2 from row 1

$$\begin{bmatrix} 1 & 0 & \frac{11}{9} \\ 0 & 1 & \frac{10}{9} \end{bmatrix} \tag{A-29}$$

The same principles are involved in solving sets of equations containing more than two unknowns. The procedure can be generalized as the Gauss–Jordan elimination method and coded as a computer program. A number of alternative elimination methods are available for solving simultaneous linear equations, such as those described in McCalla (1967). Most computation centers maintain efficient library programs for the solution of simultaneous equations.

Iterative Solutions

For matrices with a large number of zero elements elimination procedures are prone to significant rounding errors, whereas iterative methods have certain advantages. Let us illustrate an iterative solution by using the following equations (which could also be solved by elimination):

$$3x_1 + 2x_2 = 2 \tag{A-30}$$

$$2x_1 + 3x_2 = 4 \tag{A-31}$$

First we rearrange the equations as follows:

$$x_1 = \frac{(2 - 2x_2)}{3} \tag{A-32}$$

$$x_2 = \frac{(4 - 2x_1)}{3} \tag{A-33}$$

If we now "guess" the initial value of x_2, say 3, and solve Equation A-32 for x_1, we obtain

$$x_1 = \frac{[2 - (2)(3)]}{3} = -1.33$$

Then, inserting this value of x_1 into Equation A-33, we solve for x_2:

$$x_2 = \frac{[4 - (2)(-1.33)]}{3} = 2.22$$

We now use the new value of x_2 to recalculate x_1:

$$x_1 = \frac{[2 - (2)(2.22)]}{3} = -0.81$$

which is in turn used to calculate x_2:

$$x_2 = \frac{[2 - (2)(-0.81)]}{3} = 1.87$$

This procedure is repeated iteratively until the values of x_1 and x_2 converge to values that are sufficiently accurate. The successive values for seven iterations are tabulated below and are displayed

Iteration	1	2	3	4	5	6	7
x_1	−1.33	−0.81	−0.58	−0.51	−0.44	−0.41	−0.40
x_2	2.22	1.87	1.72	1.67	1.62	1.61	1.60

graphically in Figure A-2. This method is known as the Gauss–Seidel method and is employed in Chapters 5 and 6.

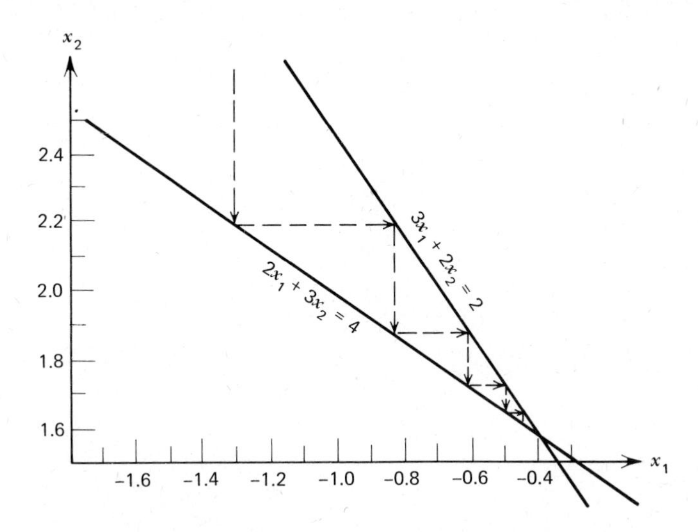

Figure A-2 Graph showing solution of two simultaneous linear equations. Successive iterations converge to true solution at $x_1 = -0.4$, $x_2 = 1.6$. From Franks (1967).

Annotated Bibliography

Carlile, R. E., and Gillett, B. E., 1968, "Gaussian Elimination Method for Solving a System of Linear Equations: Computer Programming and Mathematical Techniques for Engineers," *Oil and Gas Journal*, v. 66, No. 35, pp. 82–84.

Part of a series of articles by the same authors on numerical methods of solving linear equations. An excellent and well-written introduction. The other articles in this series are also recommended.

Forsythe, G., and Moler, C. B., 1967, *Computer Solutions of Linear Algebraic Systems*, Prentice-Hall, Englewood Cliffs, N.J., 148 pp.

A detailed but clearly written discussion of computational techniques for solving linear simultaneous equations by matrix methods.

Franks, R. G. E., 1967, *Mathematical Modeling in Chemical Engineering*, John Wiley and Sons, New York, 285 pp.

Chapters 1 and 2 contain a useful introduction to the various types of equations used in modeling, as well as a general overview of computer methods of solution.

McCalla, T. R., 1967, *Introduction to Numerical Methods and FORTRAN Programming*, John Wiley and Sons, New York, 359 pp.

A medium-level text for numerical analysis; contains numerous FORTRAN algorithms.

Index